STUDIES ON GRAPHS AND
DISCRETE PROGRAMMING

annals of discrete mathematics

General Editor

Peter L. HAMMER, University of Waterloo, Ont., Canada

Advisory Editors

C. BERGE, Université de Paris, France
M.A. HARRISON, University of California, Berkeley, CA, U.S.A.
V. KLEE, University of Washington, Seattle, WA, U.S.A.
J.H. VAN LINT, California Institute of Technology, Pasadena, CA, U.S.A.
G.-C. ROTA, Massachusetts Institute of Technology, Cambridge, MA, U.S.A.

50
1931 N·H 1981
P C

NORTH-HOLLAND PUBLISHING COMPANY – AMSTERDAM • NEW YORK • OXFORD

NORTH-HOLLAND
MATHEMATICS STUDIES **59**

Annals of Discrete Mathematics (11)

General Editor: Peter L. Hammer

University of Waterloo, Ont., Canada

Studies on Graphs and Discrete Programming

Editor:

P. HANSEN

Façulté Universitaire
Catholique de Mons
Mons, Belgium

1931 N·H 1981

NORTH-HOLLAND PUBLISHING COMPANY – AMSTERDAM • NEW YORK • OXFORD

ISBN: 0 444 86216 1

Publishers

NORTH-HOLLAND PUBLISHING COMPANY
AMSTERDAM • NEW YORK • OXFORD

Sole distributors for the U.S.A. and Canada

ELSEVIER NORTH-HOLLAND, INC.
52 VANDERBILT AVENUE,
NEW YORK, N.Y. 10017

Library of Congress Cataloging in Publication Data

Main entry under title:

Studies on graphs and discrete programming.

(Annals of discrete mathematics ; 11) (North-Holland mathematics studies ; 59)
Includes proceedings of the Workshop on Applications of Graph Theory to Management, held at the European Institute for Advanced Studies in Management, Brussels, Mar. 20-21, 1979, and also papers from the 10th Mathematical Programming Symposium, Montreal, Aug. 27-31, 1979.
Bibliography: p.
1. Networks analysis (Planning)--Congresses. 2. Graph theory--Congresses. 3. Integer programming--Congresses. I. Hansen, P. (Pierre) II. Workshop on Applications of Graph Theory to Management (1979 : European Institute for Advanced Studies in Management) III. International Mathematical Programming Symposium (10th : 1979 : Montreal, Québec) IV. Series. V. Series: North-Holland mathematics studies ; 59.
T57.85.S78 519.7'7 81-14027
ISBN 0-444-86216-1 AACR2

PRINTED IN THE NETHERLANDS

FOREWORD

This volume contains the proceedings of the Workshop on Applications of Graph Theory to Management, held at the European Institute for Advanced Studies in Management, Brussels, March 20–21, 1979. In addition, contributed papers are included, many of which were presented at the Xth Mathematical Programming Symposium, Montreal, August 27–31, 1979.

Research on Graph Theory and on Discrete Programming has been very active during the last decade; the intersection of both areas constitutes a particularly fruitful research domain. A variety of studies in the present volume illustrates the point, both as far as theory and applications are concerned. Some progress is made towards proving the deep Chvátal–Hammer conjecture on the aggregation of inequalities in integer programming. The intriguing relationship between the optimal values of variables in linear 0–1 programs and their continuous relaxations is further analysed, for matching and for packing problems. Polynomial algorithms are proposed for new or little studied problems such as coloring claw-free perfect graphs or optimizing totally dual integral systems. Improved resolution methods are suggested for classical problems, e.g. assignment and quadratic assignment. Several new theorems provide original perspectives on important applications to school timetabling and scheduling the games of a sports league. Other applications concern basis restructuring and reinversion in linear programming—for which a surprising connection with a machine maintenance problem is exhibited—network synthesis with non-simultaneous multicommodity flows and testing the reliability of a tree system. Studies on recent developments of discrete programming without direct ties with graphs are also presented: solving non-separable non-convex programs, a noteworthy yet neglected subject is taken up; parametric multicriteria integer programming is explored. Complexity theory is used to analyse the administration of standard length telephone cable reels and to provide polynomial algorithms for some continuous locational problems.

To all those who have generously given of their time during this volume's preparation, allowing all papers to be refereed, sincere thanks are due.

P. HANSEN
1981

CONTENTS

P. Hansen, ed., Studies on Graphs and Discrete Programming

INTEGER AND FRACTIONAL MATCHINGS*

Egon BALAS

Graduate School of Industrial Administration, Carnegie-Mellon University, Pittsburgh, PA 15213, USA

We examine the connections between maximum cardinality edge matchings in a graph and optimal solutions to the associated linear program, which we call maximum f-matchings (fractional matchings). We say that a maximum matching M separates an odd cycle with vertex set S, if M has no edge with exactly one end in S. An odd cycle is separable if it is separated by at least one maximum matching. We show that (1) a graph G has a maximum f-matching that is integer, if and only if it has no separable odd cycles; (2) the minimum number q of vertex-disjoint odd cycles for which a maximum f-matching has fractional components, equals the maximum number s of vertex-disjoint odd cycles, separated by a maximum matching; (3) the difference between the cardinality of a maximum f-matching and that of a maximum matching in G is one half times s; (4) any maximum f-matching with fractional components for a minimum number s of vertex-disjoint odd cycles defines a maximum matching obtainable from it in s steps; and (5) if a maximum f-matching has fractional components for a set Q of odd cycles that it not minimum, there exists another maximum f-matching with fractional components for a minimum-cardinality set S of odd cycles, such that $S \subset Q$, $|Q \setminus S|$ is even, and the cycles in $Q \setminus S$ are pairwise connected by alternating paths. Statement (4) duplicates an earlier result of J.P. Uhry [7].

1. Introduction

Let $G = (V, E)$ be an undirected graph with $|V| = n$ and $|E| = m$, let A be the vertex-edge incidence matrix of G, and e_p the p-vector of 1's. Consider the problem of finding a maximum-cardinality edge-matching in G

$$\text{EM}(G) \qquad \max\{e_m x \mid Ax \leq e_n,\ x \in \{0, 1\}^m\}$$

and the associated linear program

$$\text{LEM}(G) \qquad \max\{e_m x \mid Ax \leq e_n,\ x \geq 0\}.$$

We will call EM(G) the *matching problem*, and LEM(G) the fractional matching, or shortly the *f-matching* problem. A feasible solution to EM(G) will be called a *matching*, a feasible solution to LEM(G), an *f-matching*. Clearly, every matching is an f-matching.

* Research supported by the National Science Foundation through grant MCS76-12026 A02, and by the U.S. Office of Naval Research through contract N0014-C-0621 NR 047-048.

We wish to investigate the relationship between maximum matchings and maximum f-matchings in a graph G, i.e., between optimal solutions to EM(G) and LEM(G). For a problem P, we denote by $v(P)$ the value of (an optimal solution to) P. For vertex sets $S, T \subseteq V$, we denote by (S, T) the set of edges of G with one end in S and the other in T, and by $\langle S \rangle$ the subgraph of G induced by S. Also, $\bar{S} = V \setminus S$.

An alternating path relative to a matching M is a path whose edges alternate between M and $\bar{M}$. An M-augmenting path is an alternating path whose end vertices are distinct and are not incident with M. A matching M in G is of maximum cardinality if and only if G contains no M-augmenting path (Berge [2]).

If G has no odd cycles (i.e., is bipartite), A is totally unimodular and every basic solution to LEM(G) is integer, i.e.,

$$v(\text{LEM}(G)) = v(\text{EM}(G)). \tag{1}$$

If G is not bipartite, LEM(G) has fractional basic solutions, and the convex hull of feasible integer solutions, i.e., the matching polytope, is defined (together with the constraints of LEM(G)) by the inequalities

$$\sum_{e \in (S, S)} x_e \leq \tfrac{1}{2}(|S| - 1), \qquad S \subseteq V, \quad |S| \geq 3 \text{ and odd.} \tag{2}$$

(Edmonds [3]). Any basic feasible solution to LEM(G) has components equal to 0, 1 or $\frac{1}{2}$ (Balinski [1]). Furthermore, if $x_e = \frac{1}{2}$, then e belongs to some odd cycle C, and $x_f = \frac{1}{2}$, $\forall f \in C$.

Let M be a maximum matching in G. If G is bipartite, M is a maximum f-matching; but the converse if false, as illustrated by the graph in Fig. 1, where the maximum matching shown in heavy lines is also a maximum f-matching, in spite of the presence of a 5-cycle.

It would seem that whether a maximum matching M is also a maximum f-matching, depends on whether M covers all vertices of every odd cycle; but this is not so. The matching in the graph of Fig. 2 leaves a vertex of an odd cycle exposed, yet it is a maximum f-matching. On the other hand, the maximum matching M in the graph of Fig. 3(a) covers all vertices of the only odd cycle of the graph, yet it is not a maximum f-matching, as shown by the f-matching M' of Fig. 3(b): $|M| = 3$, while the value of M' is $3\frac{1}{2}$.

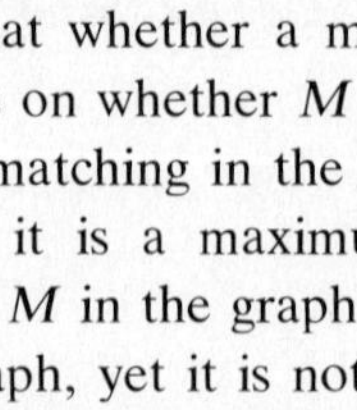

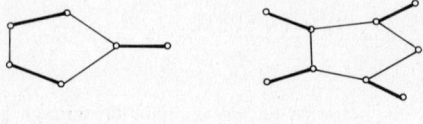

Fig. 1.

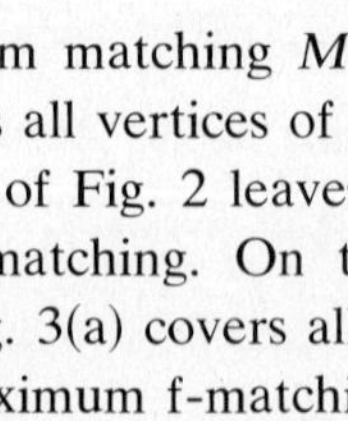

Fig. 2.

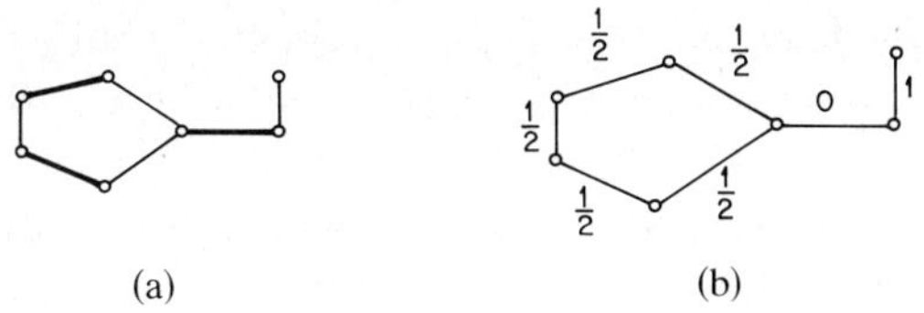

Fig. 3.

2. Separable odd cycles

The key concept for understanding the connection between maximum matchings and maximum f-matchings is the following.

Let M be a maximum matching and C an odd cycle in G, and let S be the vertex set of C. We will say that M *separates* C if

$$M \cap (S, \bar{S}) = \emptyset. \tag{3}$$

An odd cycle which is separated by at least one maximum matching in G will be called *separable*. If C is a separable odd cycle in G, with vertex set S, then G has a maximum matching of the form $M = M_1 \cup M_2$, where M_1 and M_2 are maximum matchings in $\langle S \rangle$ and in $\langle \bar{S} \rangle$ respectively. If C is a nonseparable odd cycle with vertex set S of cardinality $2k+1$, then every maximum matching contains at least $k+1$ edges incident with S.

Theorem 1. *G has a maximum f-matching that is integer if and only if G has no separable odd cycle.*

Proof. *Necessity.* Let M be a maximum matching in G which separates an odd cycle C with vertex set S, and $|S| = 2k+1$. Since M is maximal and $M \cap (S, \bar{S}) = \emptyset$, M contains k edges of (S, S). Let $\bar{x}$ be the incidence vector of M, and define $\hat{x}$ by

$$\hat{x}_e = \begin{cases} \frac{1}{2} & e \in C, \\ 0 & e \in (S, S) \setminus C, \\ \bar{x}_e & e \in E \setminus (S, S). \end{cases}$$

Clearly, $\hat{x}$ is a feasible solution to LEM(G), and

$$v(\text{LEM}(G)) \geq e_m \hat{x} = |M| - k + \tfrac{1}{2}(k+1) > v(\text{EM}(G)),$$

hence G has no maximum f-matching that is integer.

Sufficiency. Let $v(\text{LEM}(G)) > v(\text{EM}(G))$, and let $\hat{x}$ be a basic maximum f-matching in G having fractional components for a minimum number of odd cycles, say $C_1, \ldots, C_q$. For $i = 1, \ldots, q$, let S_i be the vertex set of C_i, with

$|S_i|=2k_i+1$. Clearly, $S_i \cap S_j=\emptyset$, $\forall i, j \in \{1, \ldots, q\}$. Further, let $S=\bigcup_{i=1}^{q} S_i$. The components $\hat{x}_e$ of $\hat{x}$ such that $\hat{x}_e=1$ define a matching M' in $\langle \bar{S} \rangle$ which is clearly maximum, or else $\hat{x}$ itself could not be a maximum f-matching.

For $i=1, \ldots, q$, let M_i be a set of k_i mutually nonadjacent edges of C_i, i.e., a maximum matching in $\langle S_i \rangle$. Then the edge set

$$M=M' \cup M_1 \cup \cdots \cup M_q \tag{4}$$

is a matching in G such that $M \cap (S_i, \bar{S}_i)=\emptyset$, $i=1, \ldots, q$.

Next we prove by contradiction that M is of maximum cardinality, which implies that the odd cycles $C_1, \ldots, C_q$ are all separable and thus completes the proof the theorem.

Suppose M is not maximum. Then there exists an M-augmenting path P in G. Moreover, we claim that the maximum matchings M_i in $\langle S_i \rangle$, $i=1, \ldots, q$, which are not unique, can always be chosen such that P has no edge in common with any of the cycles C_i, $i=1, \ldots, q$. For suppose $P \cap C_i \neq \emptyset$ for some $i \in \{1, \ldots, q\}$. Since C_i has only one vertex exposed with respect to M, at least one of the end vertices of the path P, say v, is not contained in C_i. Let w be the first vertex of C_i encountered when traversing P starting from v. Then the maximum matching M_i in $\langle S_i \rangle$ can be replaced by $\tilde{M}_i$, containing those k_i mutually nonadjacent edges of C_i that leave exposed w. This replaces the matching M in G by

$$\tilde{M}=(M \setminus M_i) \cup \tilde{M}_i,$$

such that $|\tilde{M}|=|M|$, and the portion of P between w and v is then an $\tilde{M}$-augmenting path $\tilde{P}$, with $\tilde{P} \cap C_i=\emptyset$. If $\tilde{P}$ still has some edge in common with some C_j, $j \in \{1, \ldots, q\}-\{i\}$, the above procedure can be repeated. After at most q applications of this procedure, we obtain a matching $\hat{M}$ in G such that $|\hat{M}|=|M|$,

$$\hat{M}=M' \cup \tilde{M}_1 \cup \cdots \cup \tilde{M}_q$$

(where each $\tilde{M}_i$ is a maximum matching in $\langle S_i \rangle$), and an $\hat{M}$-augmenting path $\hat{P}$, not containing any edge of any C_i, $i=1, \ldots, q$. This proves that w.l.o.g., the path P can be assumed to have no edge in common with any C_i, $i=1, \ldots, q$.

If we now reverse the assignment of edges in P with respect to M, we obtain a matching M^* in G, such that $|M^*|=|M|+1$. But from M^* one can construct an f-matching $\bar{x}$ in G such that $e\bar{x} \geqslant e\hat{x}$ and $\bar{x}$ has fractional components for fewer odd cycles than $\hat{x}$, contrary to our assumptions on $\hat{x}$.

To construct $\bar{x}$, first note that since P contains no edge of any C_i and is an alternating path, P contains no edge of $\langle S \rangle$, and the only vertices of P contained in S are one, or possibly both, end vertices of P. The fact that P has at least one vertex in S follows from the maximality of the matching $M'=M \cap (\bar{S}, \bar{S})$ in $\langle \bar{S} \rangle$.

Let v_1 and v_2 be the two end vertices of P. We consider two cases.

Case 1. $v_1 \in S_i$ for some $i \in \{1, \ldots, q\}$, $v_2 \notin S$. Let $e_1, \ldots, e_{2k+1}$ be the edges of C_i, and let e_1 and e_{2k+1} be incident with v_1. Define $\bar{x}$ by

$$\bar{x}_e = \begin{cases} 1-\hat{x}_e & e \in P, \\ 1 & e \in \{e_2, e_4, \ldots, e_{2k}\}, \\ 0 & e \in \{e_1, e_3, \ldots, e_{2k+1}\}, \\ \hat{x}_e & \text{otherwise.} \end{cases}$$

Case 2. $v_1 \in S_i$ and $v_2 \in S_j$ for some $i, j \in \{1, \ldots, q\}$. Let e_r, $r = 1, \ldots, 2k_i + 1$, and d_s, $s = 1, \ldots, 2k_j + 1$, be the edges of C_i and C_j respectively, where e_1 and e_{2k_i+1} are incident with v_1, and d_1 and d_{2k_j+1} are incident with v_2. Then define $\bar{x}$ by

$$\bar{x}_e = \begin{cases} 1-\hat{x}_e & e \in P, \\ 1 & e \in \{e_2, e_4, \ldots, e_{2k_i}\} \cup \{d_2, d_4, \ldots, d_{2k_j}\}, \\ 0 & e \in \{e_1, e_3, \ldots, e_{2k_i+1}\} \cup \{d_1, d_3, \ldots, d_{2k_j+1}\}, \\ \hat{x}_e & \text{otherwise.} \end{cases}$$

In Case 1, $\bar{x}$ has fractional components for $q-1$ odd cycles, and $e_m\bar{x} = e_m\hat{x} + \frac{1}{2}$, since the reversal of edge assignments in P adds 1 to, while the change of values for $e \in C_i$ subtracts $\frac{1}{2}$ from, the value of $e_n\hat{x}$. In Case 2, $\bar{x}$ has fractional components for $q-2$ odd cycles, and $e_m\bar{x} = e_m\hat{x}$. In both cases the existence of $\bar{x}$ contradicts our assumption that $\hat{x}$ is a maximum f-matching which has fractional components for a minimum number (q) of odd cycles.

Thus M is a maximum cardinality matching in G, which separates the odd cycles $C_1, \ldots, C_q$. □

The graphs in Figs. 1 and 2 both have maximum f-matchings that are integer, since in each case the unique maximum matching shown in heavy lines does not separate the unique odd cycle of the graph, hence the latter has no separable odd cycle. The graph of Fig. 3, on the other hand, has a maximum matching (other than the one shown in Fig. 3(a)) which separates the odd cycle: it contains two edges of the cycle, and a third edge not incident with any vertex of the cycle. Therefore there exists no maximum f-matching that is integer.

It is interesting to note that if LEM(G) has an integer optimum, it does not follow that the linear program dual to LEM(G), namely the fractional vertex covering problem LVC(G), also has an integer optimum. In other words, the absence of a separable odd cycle in G does not guarantee that the cardinality of a maximum edge matching is equal to that of a minimum vertex cover. For example the graph of Fig. 4 has no separable odd cycle and therefore it has a

(a) (b)

Fig. 4.

maximum f-matching of value 3 that is integer. On the other hand, the cardinality of a minimum vertex cover is 4, and the fractional vertex cover that assigns value $\frac{1}{2}$ to every vertex v is the only one with value 3.

In proving the 'if' part of Theorem 1, we have rediscovered what turns out to be a recent result of J.P. Uhry:

Corollary 1.1 (J.P. Uhry [7]). *Let $\hat{x}$ be a maximum f-matching in G having fractional components for a minimum number of odd cycles C_i, $i = 1, \ldots, q$. Let $M' = \{e \in E \mid \hat{x}_e = 1\}$, and for $i = 1, \ldots, q$, let M_i be a maximum matching in C_i. Then*

$$M = M' \cup M_1 \cup \cdots \cup M_2$$

is a maximum matching in G.

This result shows how to obtain a maximum matching in an arbitrary graph G from a maximum f-matching in G that satisfies an additional requirement (having fractional components for a minimum number of odd cycles). It turns out that this procedure can be reversed, to obtain a maximum f-matching from a maximum matching that, again, satisfies an additional requirement: to separate a maximum number of disjoint odd cycles. (Two odd cycles are disjoint if they have no common vertex.) The next Corollary and its proof state this relationship more precisely.

For a maximum matching M, let $\sigma(G, M)$ be the maximum number of disjoint odd cycles separated by M. For a maximum f-matching x, let $\gamma(G, x)$ be the number of odd cycles for which x has fractional components. Finally, let $\mathcal{M}$ be the set of maximum matchings, and χ the set of maximum f-matchings in G.

Corollary 1.2.

$$\max_{M \in \mathcal{M}} \sigma(G, M) = \min_{x \in \chi} \gamma(G, x).$$

Proof. In the sufficiency part of the proof of Theorem 1, a maximum f-matching $\hat{x}$ such that

$$\gamma(G, \hat{x}) = q = \min_{x \in \chi} \gamma(G, x)$$

was used to construct the matching defined by (4), which was shown to be maximum in G, and which separates q cycles of G. We claim that $q = \max_{M \in \mathcal{M}} \sigma(G, M)$. For suppose not, then there exists a maximum matching M^* in G that separates $p \geqslant q+1$ odd cycles $C_1, \ldots, C_p$. For $i = 1, \ldots, p$, let S_i be the vertex set of C_i, $|S_i| = 2k_i + 1$, let $S = \bigcup_{i=1}^{p} S_i$, and let $\bar{x}$ be the incidence vector of M^*. Then setting $\tilde{x}_e = \frac{1}{2}$ for $e \in C_i$, $i = 1, \ldots, p$, $\tilde{x}_e = \bar{x}_e$ otherwise, defines an f-matching $\tilde{x}$ such that $e_m \tilde{x} = |M^*| + \frac{1}{2}p$. On the other hand, $e_m \tilde{x} = |M| + \frac{1}{2}q$, where M is the maximum matching defined by (4). Since $|M| = |M^*|$ and $p \geqslant q+1$, this contradicts the assumption that $\hat{x}$ is a maximum f-matching in G. $\square$

Next we turn to the relationship between the value of a maximum f-matching and that of a maximum matching in an arbitrary graph G. Let $\Delta(G)$ denote the difference of these two values, i.e.,

$$\Delta(G) = v(\text{LEM}(G)) - v(\text{EM}(G)). \tag{5}$$

It is not hard to derive bounds on $\Delta(G)$. If C is an odd cycle with vertex set S, clearly $\Delta(\langle S \rangle) = \frac{1}{2}$. Since in the absence of odd cycles in G, $\Delta(G) = 0$, if $\gamma(G)$ is the maximum number of disjoint odd cycles in G, it is easy to see that

$$\Delta(G) \leqslant \tfrac{1}{2}\gamma(G). \tag{6}$$

This also yields a bound on $\Delta(G)$ independent of the number of odd cycles. Indeed, since a cycle has at least 3 edges, $\gamma(G) \leqslant \frac{1}{3}|E|$, i.e.,

$$\Delta(G) \leqslant \tfrac{1}{6}|E|, \tag{7}$$

a bound which is attained for any graph G which is the union of triangles.

Further, since $\gamma(G) \leqslant v(\text{EM}(G))$ obviously holds for any graph G, from (6) one also has

$$v(\text{LEM}(G))/v(\text{EM}(G)) \leqslant \tfrac{3}{2}. \tag{8}$$

The bounds (6), (7) and (8) are more or less trivial. The proof of Theorem 1 yields a considerably stronger bound, namely the actual value of $\Delta(G)$.

We define the *separation number* $\sigma(G)$ of a graph G as the maximum number of disjoint odd cycles separated by any maximum matching in G, i.e.,

$$\sigma(G) = \max_{M \in \mathcal{M}} \sigma(G, M).$$

Corollary 1.3. $\Delta(G) = \frac{1}{2}\sigma(G)$.

Proof. Let $\hat{x}$ be a maximum f-matching in G such that $\gamma(G, \hat{x}) = \sigma(G)$. Then the matching M defined by (4) with respect to $\hat{x}$ is maximum, as shown in the proof of Theorem 1. Clearly, $|M| = e_m \hat{x} - \frac{1}{2}\sigma(G)$. $\square$

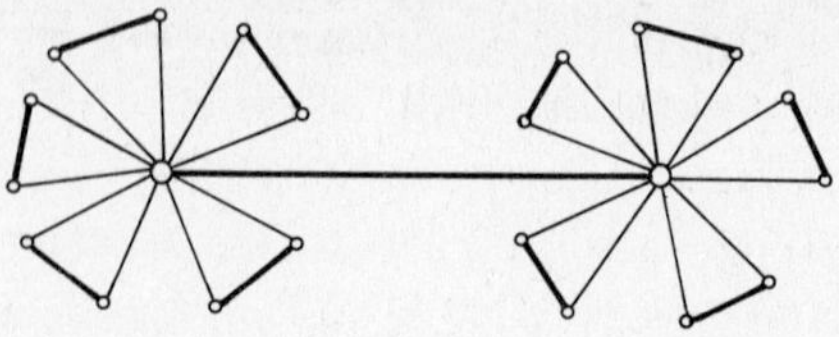

Fig. 5.

The difference between the value of a maximum f-matching and that of a maximum matching can be considerably smaller than $\frac{1}{2}\gamma(G)$, the bound given by (6), since the separation number $\sigma(G)$ of a graph can be much smaller than the number $\gamma(G)$ of its odd cycles. Fig. 5 shows a graph G with $\sigma(G)=0$ and $\gamma(G)=2p$, where $p=5$. Similar graphs can be constructed for arbitrary large p. The unique maximum matching, which is also a maximum f-matching in G, is shown in heavy lines.

For $S\subset V$, we denote by $N(S)$ the set of vertices in $\bar{S}$ adjacent to some vertex in S. For $v\in V$, we denote by $I(v)$ the set of edges incident with v in G.

Corollary 1.4. *Every graph G has a maximum f-matching $\hat{x}$ such that*

$$\hat{x}_e=\begin{cases}\frac{1}{2} & \forall e\in C_i, \quad i=1,\ldots,\sigma(G),\\ 0 & \forall e\in (S_i,V)\setminus C_i, \quad i=1,\ldots,\sigma(G),\\ 0 \text{ or } 1 & \forall e\in(\bar{S},\bar{S}),\end{cases} \tag{9}$$

$$M'=\{e\in E\mid \hat{x}_e=1\} \text{ is a maximum matching in } \langle\bar{S}\rangle, \text{ and} \tag{10}$$

$$\sum_{e\in I(v)}\hat{x}_e=1 \quad \forall v\in N(S), \tag{11}$$

where $C_1,\ldots,C_{\sigma(G)}$ is a maximum-cardinality set of disjoint separable odd cycles in G, S_i is the vertex set of C_i, $i=1,\ldots,\sigma(G)$, and $S=S_1\cup\cdots\cup S_{\sigma(G)}$.

Conversely, every feasible solution $\hat{x}$ of LEM(G) *that satisfies* (9) *and* (10) *is a maximum f-matching in G, and also satisfies* (11).

Proof. From Corollary 1.2, G has a maximum f-matching $\hat{x}$ such that $\gamma(G,\hat{x})=\sigma(G)$. Such an $\hat{x}$ clearly satisfies (9) and (10). Conversely, if $\hat{x}$ satisfies (9) and (10), then the matching M defined by (4) relative to $\hat{x}$ is maximum in G, hence $e_m\hat{x}=|M|+\frac{1}{2}\sigma(G)=v(\text{EM}(G))+\frac{1}{2}\sigma(G)$ and thus $\hat{x}$ is maximal.

To show by contradiction that every maximum f-matching that satisfies (9), (10) also satisfies (11), suppose (11) is violated for some $v\in N(S)$. Let $u\in S_i$ be a vertex of S adjacent to v, and let $e_1,\ldots,e_{2k_i+1}$ be the edges of C_i, with e_1 and e_{2k_i+1} incident with u.

Since (11) is violated for v, and $\hat{x}_e = 0$ or 1 for $e \in I(v)$, it follows that $\hat{x}_e = 0$, $\forall\, e \in I(v)$. But then the vector $\tilde{x}$ defined by

$$\tilde{x}_e = \begin{cases} 1 & e = (u, v), \\ 1 & e \in \{e_2, e_4, \ldots, e_{2k_i}\}, \\ 0 & e \in \{e_1, e_3, \ldots, e_{2k_i+1}\}, \\ \hat{x}_e & \text{otherwise} \end{cases}$$

is an f-matching with $e_m\tilde{x} = e_m\hat{x} + \frac{1}{2}$, contrary to the assumption that $\hat{x}$ is a maximum matching. □

While every graph has a maximum f-matching $\hat{x}$ satisfying (9), (10) (and hence (11)), it is not true that *every* (basic) maximum f-matching satisfies (9). In case of alternative optima there may be only one that satisfies the condition of Corollary 1.4. The graph G of Fig. 4, for instance, has the maximum f-matching shown in (a), which satisfies (9), since G has no separable odd cycle; but it also has the one shown in (b), which does not satisfy the condition of Corollary 1.4.

3. Generating maximum matchings from maximum f-matchings

Let $\hat{x}$ be a basic maximum f-matching, let

$$\mathscr{C}(\hat{x}) = \{C_1, \ldots, C_q\}$$

be the set of odd cycles C_i such that $\hat{x}_e = \frac{1}{2}$, $\forall e \in C_i$, and for $i = 1, \ldots, q$, let S_i be the vertex set of C_i.

We will denote by $\mathscr{F}(\hat{x})$ the family of matchings associated with $\hat{x}$, introduced in the proof of Theorem 1, i.e., the set of all matchings of the form

$$M = M' \cup M_1 \cup \cdots \cup M_q,$$

where $M' = \{e \in E \mid \hat{x}_e = 1\}$, and for $i = 1, \ldots, q$, M_i is a maximum matching in $\langle S_i \rangle$.

We define two operations on the components of $\hat{x}$ associated with certain edge sets of G.

By *complementing* on $E' \subseteq E$, we mean replacing $\hat{x}_e$ by $\hat{x}'_e = 1 - \hat{x}_e$, $\forall e \in E'$.

By *alternate rounding* on $C_i \in \mathscr{C}(\hat{x})$, $C_i = \{e_1, \ldots, e_{2k_i+1}\}$, we mean replacing $\hat{x}_e$, $e \in C_i$, by $\hat{x}'_e = 0$, $e \in \{e_1, e_3, \ldots, e_{2k_i+1}\}$, and $\hat{x}'_e = 1$, $e \in \{e_2, \ldots, e_{2k_i}\}$.

From the results of the previous section it follows that one way of finding a maximum matching in a graph is to find a maximum f-matching $\hat{x}$ with a minimum-cardinality $\mathscr{C}(\hat{x})$ and then use alternate rounding on the odd cycles in $\mathscr{C}(\hat{x})$ to generate a matching $M \in \mathscr{F}(\hat{x})$.

Remark 1. The matchings in $\mathscr{F}(\hat{x})$ are of maximum cardinality if and only if $\mathscr{C}(\hat{x})$ is of minimum cardinality.

Proof. From Corollary 1.3. □

Next we address the problem of finding a maximum f-matching $\hat{x}$ with a minimum-cardinality $\mathscr{C}(\hat{x})$.

Lemma 1. *If $\hat{x}$ is a basic maximum f-matching and $q = |\mathscr{C}(\hat{x})| > \sigma(G)$, then*

$$q \equiv \sigma(G) \pmod 2.$$

Proof. Let $M \in \mathscr{F}(\hat{x})$. Then $e_m\hat{x} = |M| + \frac{1}{2}q$, and from Corollary 1.3, $e_m\hat{x} = v(\mathrm{EM}(G)) + \frac{1}{2}\sigma(G)$. Since both $|M|$ and $v(\mathrm{EM}(G))$ are integer, subtracting the second equation from the first one produces

$$\tfrac{1}{2}(q - \sigma(G)) \equiv 0 \pmod 1)$$

and multiplication by 2 yields the congruence in the Lemma. □

We define an *alternating path* relative to an f-matching $\hat{x}$ as a path whose edges e alternate between $\hat{x}_e = 0$ and $\hat{x}_e = 1$. We say that a path P connects two odd cycles C_1, C_2, if P connects a vertex of C_1 to a vertex of C_2. A vertex v of G is *exposed* relative to $\hat{x}$ if $\hat{x}_e = 0$ for all edges e incident with v.

Theorem 2. *Let $\hat{x}$ be a basic maximum f-matching such that $q = |\mathscr{C}(\hat{x})| > \sigma(G)$. Then G contains $\alpha = \frac{1}{2}(q - \sigma(G))$ vertex-disjoint alternating paths P_k relative to $\hat{x}$, each of which connects two odd cycles $C_{i_k}, C_{j_k} \in \mathscr{C}(\hat{x})$, and the cycles C_{i_k}, C_{j_k}, $k = 1, \ldots, \alpha$, are all distinct. Furthermore, alternate rounding on the odd cycles C_{i_k}, C_{j_k}, and complementing on the paths P_k, $k = 1, \ldots, \alpha$, produces a maximum f-matching $\bar{x}$ with $|\mathscr{C}(\bar{x})| = \sigma(G)$.*

Proof. Let M^* be a maximum matching in G and $M \in \mathscr{F}(\hat{x})$. Since $|M| = e_m\hat{x} - \frac{1}{2}q$ and $|M^*| = e_m\hat{x} - \frac{1}{2}\sigma(G)$ (Corollary 1.3), $|M^*| - |M| = \frac{1}{2}(q - \sigma(G)) = \alpha$. It is therefore known (see Theorem 1 of [5]) that G contains α vertex-disjoint M-augmenting paths, $\pi_1, \ldots, \pi_\alpha$.

We show by contradiction that each π_k connects two cycles, $C_{i_k}, C_{j_k} \in \mathscr{C}(\hat{x})$, $i_k \neq j_k$. Suppose this is false; then at least one end-vertex of π_k, say u, is in $\bar{S}$, where $S = \bigcup_{i=1}^{q} S_i$. Since $\langle \bar{S} \rangle$ contains no M'-augmenting path (or else $\hat{x}$ would not be a maximum f-matching), π_k is incident with some vertex of S. Let v be the first vertex of S encountered when π_k is traversed starting from u, and let π'_k be the subpath of π_k connecting u to v. Clearly, π'_k is an alternating path relative to $\hat{x}$. Then using alternate rounding on C_i, the cycle whose vertex set S_i contains v, and complementing on π'_k, produces an f-matching $\bar{x}$ such that $e_m\bar{x} = e_m\hat{x} + \frac{1}{2}$, contrary to the assumption that $\hat{x}$ is maximum.

This proves that both end-vertices of π_k belong to S. Further, since each S_i contains only one vertex exposed relative to M, the sets S_{i_k}, S_{j_k} containing the two end-vertices of π_k are distinct—hence so are the two cycles C_{i_k}, C_{j_k}; and for the same reason (that each S_i contains only one exposed vertex) the α pairs of odd cycles C_{i_k}, C_{j_k}, connected by the α paths π_k are all distinct.

Now using alternate rounding on the 2α odd cycles connected by the paths π_k reduces the value of the f-matching for every such cycle by $\frac{1}{2}$, i.e., for every pair of cycles C_{i_k}, C_{j_k} connected by a path π_k, by 1. On the other hand, using complementing on the paths π_k increases the value of the f-matching by 1 for every path π_k. Thus the alternate rounding and complementing produces an f-matching $\bar{x}$ of the same value as $\hat{x}$, hence maximum; and it reduces the number of odd cycles for which the f-matching has fractional components, by $2\alpha = q - \frac{1}{2}\sigma(G)$. □

Two cycles C_i, C_j with vertex sets S_i, S_j will be called *adjacent*, if there exists $u \in S_i$ and $v \in S_j$ such that $(u, v) \in E$. Since the edge (u, v) is an alternating path relative to $\hat{x}$, we have

Corollary 2.1. *If two odd cycles $C_i, C_j \in \mathscr{C}(\hat{x})$ are adjacent, with $u \in C_i$, $v \in C_j$ and $(u, v) \in E$, then alternate rounding on C_i and C_j and complementing on (u, v), produces a maximum f-matching $\bar{x}$ such that $|\mathscr{C}(\bar{x})| = |\mathscr{C}(\hat{x})| - 2$.*

Also, from Theorem 2 we have

Corollary 2.2. *If $|\mathscr{C}(\hat{x})| > \sigma(G)$, there exists a maximum f-matching $\bar{x}$ such that $\mathscr{C}(\bar{x}) \subset \mathscr{C}(\hat{x})$ and $|\mathscr{C}(\bar{x})| = \sigma(G)$.*

Theorem 2 can be used to obtain from an arbitrary basic maximum f-matching, one with fractional components for a minimum number of odd cycles. The alternating paths relative to $\hat{x}$ can be found by a scanning and labeling procedure of the type used for finding augmenting paths relative to a matching. The complexity of such a scanning and labeling routine is $O(n^2)$ for each path found (where $n = |V|$), and since there are $\alpha = \frac{1}{2}(q - \sigma(G))$ paths to be found, obtaining from $\hat{x}$ a maximum f-matching $\bar{x}$ such that $|\mathscr{C}(\bar{x})| = \sigma(G)$ requires at most $0(n^2 \cdot \alpha)$ steps.

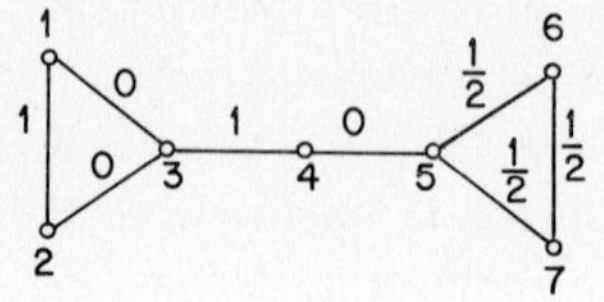

(a)

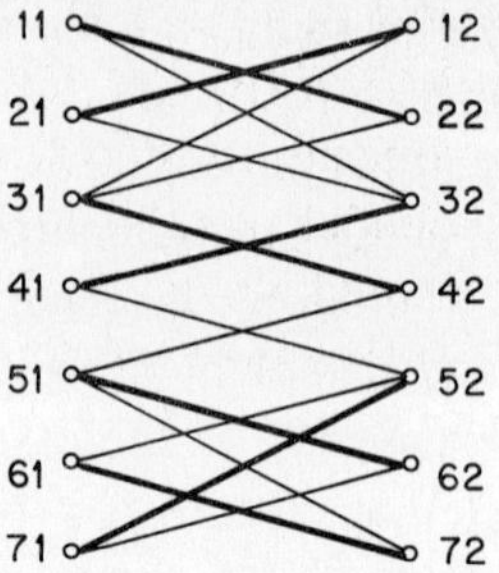

(b)

Fig. 6.

The problem of finding a maximum f-matching in an arbitrary graph G is equivalent to the problem of finding a maximum matching in a bipartite graph $\hat{G}=(V_1 \cup V_2, E_1 \cup E_2)$, defined as follows. For every vertex $v_i \in V$, $\hat{G}$ has a pair of vertices $v_{i1} \in V_1$ and $v_{i2} \in V_2$. For every edge $(v_i, v_j) \in E$, G has a pair of edges $(v_{i1}, v_{j2}) \in E_1$ and $(v_{i2}, v_{j1}) \in E_2$. We say that the vertices v_{i1} and v_{i2}, as well as the edges (v_{i1}, v_{j2}) and (v_{i2}, v_{j1}), are copies of each other. If we associate the 0–1 vectors y^1 and y^2 with the edge sets E_1 and E_2 respectively, then a matching $(\hat{y}^1, \hat{y}^2)$ in $\hat{G}$ defines an f-matching $\hat{x}$ in G via the rule

$$\hat{x}_i = \begin{cases} 1 & \text{if } \hat{y}_i^1 = \hat{y}_i^2 = 1, \\ 0 & \text{if } \hat{y}_i^1 = \hat{y}_i^2 = 0, \\ \frac{1}{2} & \text{if } \hat{y}_i^1 = 0,\ \hat{y}_i^2 = 1, \text{ or } \hat{y}_i^1 = 1,\ \hat{y}_i^2 = 0. \end{cases}$$

Obviously, a maximum matching M in $\hat{G}$ then defines a maximum f-matching of cardinality $\frac{1}{2}|M|$ in G. Fig. 6 shows a graph G with a maximum f-matching, and the associated graph $\hat{G}$ with the corresponding maximum matching.

A bipartite graph of the above type was introduced by Edmonds and Pulleyblank (see Nemhauser and Trotter [6]) as an equivalent for the fractional vertex packing problem, but the equivalence obviously holds for the fractional matching problem LEM(G) as well.

Thus finding a maximum f-matching in an arbitrary graph G with n vertices is reduced to finding a maximum matching in a bipartite graph $\hat{G}$ with $2n$ vertices. Hopcroft and Karp [5] give an easy to implement $O(n^{\frac{5}{2}})$ algorithm for finding a maximum matching in a bipartite graph. An algorithm of the same complexity ($O(n^{\frac{5}{2}})$), but much harder to implement, for maximum matching in an arbitrary graph, has been proposed by Even and Kariv [4]. Thus the worst case behavior of algorithms for maximum matching in a bipartite graph seems to be no better than that of their counterparts for arbitrary graphs. Nevertheless, the worst case is not the only, and perhaps not the most relevant one. The need for handling odd cycles in a matching algorithm for arbitrary graphs either by shrinking or by special labeling rules, makes these algorithms considerably more difficult to implement than matching algorithms for bipartite graphs. Thus the idea of using the above described equivalence has some merit, even though it could not improve the worst-case bound. However, working out the details of an algorithm based on this approach is beyond the scope of our paper.

Acknowledgements

I wish to thank Gerard Cornuéjols, Alan Hoffman and Robert Tarjan for helpful comments on an earlier version of this paper.

References

[1] M.L. Balinski: On maximum matching, minimum covering and their connections, in H.W. Kuhn, ed., Proc. Princeton Symposium on Mathematical Programming (Princeton Univ. Press, 1970) 303–312.
[2] C. Berge, Two theorems in graph theory, Proc. Nat. Acad. of Sciences (U.S.A.) 43 (1975) 842.
[3] J. Edmonds, Paths, trees and flowers, Canad. J. Math. 17 (1965) 449–467.
[4] S. Even and O. Kariv, An $O(n^{2.5})$ algorithm for maximum matching in general graphs, Proc. 16th Annual Symposium on Foundations of Computer Science (IEEE, New York, 1975) 100–112.
[5] J.E. Hopcroft and R.M. Karp, An $n^{\frac{5}{2}}$ algorithm for maximum matchings in bipartite graphs, SIAM J. Comput. 2 (1973) 225–231.
[6] G.L. Nemhauser and L.E. Trotter, Jr. Vertex packings: structural properties and algorithms, Math. Programming 8 (1975) 232–248.
[7] J.P. Uhry, Sur le problème du couplage maximal, R.A.I.R.O. 9 (1975) V-3, 13–20.

P. Hansen, ed., Studies on Graphs and Discrete Programming

PACKING PROBLEMS

Claude BERGE
University of Paris VI, Paris, France

1. Introduction

Most of the optimization problems in integers can reduce to a large class of combinatorial problems, often called 'packing problems', which are defined as follows:

Let $H=(E_1, E_2, \ldots, E_m)$ be a finite family of sets; let $H'=(E_{i_1}, E_{i_2}, \ldots, E_{i_k})$ be a subfamily of H whose members are pairwise disjoint ('matching'): find a matching of maximum cardinality. We shall denote by $\nu(H)$ the maximum number of members of H which are pairwise disjoint.

A subset T of $\bigcup E_i$ is called a *transversal set* of H if T meets all the E_i's: find a transversal set of minimum cardinality. We shall denote by $\tau(H)$ the *transversal number* of H, i.e., the smallest size of a transversal set of H.

Clearly, for every matching H' and for every transversal set T, we have $|T| \geq |H'|$. So, if we have obtained a matching H' with k elements and a transversal set T with cardinality k, we know that this matching H' is a *maximum matching* (and that T' is a *minimum transversal*).

However, this criterium is not valid for every set family H. In fact, a family H for which $\nu(H)=\tau(H)$ is said to have the 'König property', and a large part of hypergraph theory is devoted to the structural properties which yield the König property.

In this paper, we shall survey a refinement of the classical Hypergraph Theory which can be used to show that a matching of H is maximum even if H does not satisfy $\nu=\tau$, i.e., if H has not the König property. The numerous applications to number theory (Turán numbers, the Erdös problem, etc. . . .) will not be considered here.

Example 1. *The personnel assignment.*

In a certain firm, p workmen $x_1, x_2, \ldots, x_p$ are available to fill q positions $y_1, y_2, \ldots, y_q$, each workman being qualified for one or more of these jobs. Can each man be assigned to a position for which he is qualified?

This problem can also be formulated as follows: given a bipartite graph G defined by two disjoint sets of vertices X and Y and by a set of edges having one end-point in X and the other in Y, determine a *maximum matching* of G, that is a maximum set of independent edges. A set $T \subseteq X \cup Y$ is a *transversal* of G if every edge has at least one end-point in T.

For a bipartite graph G, it is known that the maximum size $\nu(G)$ of a matching is equal to the minimum size $\tau(G)$ of a transversal set: this is a famous theorem of König, which can be used to check that a given matching is maximum.

Unfortunately, it is not true that every graph G satisfies $\nu(G) = \tau(G)$; the graph G consisting of an elementary cycle of length 5 has a matching number $\nu(G) = 2$, but its transversal number is $\tau(G) = 3$. For those graphs, we need more elaborated theorems to study the maximum matchings.

Example 2. *The truncated chessboard.*

Consider first an 8×8 chessboard, whose upper left and lower right corner have been removed, can we cover the 62 remaining squares of the chessboard with 31 dominoes, each domino covering exactly two adjacent squares of the chessboard?

Here, it is easy to see that no solution can exist; because if we color the squares of the chessboard black and white alternately, each arrangement of the dominoes covers the same number of black and white squares, *but the truncated chessboard has not the same number of black and white squares*!

However, for other types of truncated chessboards, we need more. The maximum number of dominoes is the matching number $\nu(H)$ of a family $H = (E_1, E_2, \ldots, E_m)$ of 2-element sets; if we notice that H is a bipartite graph, it follows from the König theorem that $\nu(H) = \tau(H)$. This equality can be used to show when an arrangement of dominoes is optimal.

Now, consider the case of *tetraminoes*, which are pieces of wood, like the dominoes, but a tetramino covers exactly four adjacent squares of the chessboard *in line*. What is the maximum number of tetraminoes that we can place in a truncated chessboard? This number is in fact the matching number $\nu(H)$ for a family $H = (E_1, \ldots, E_m)$, where each E_i consists of 4 consecutive adjacent squares of the chessboard. If we have found an arrangement with k tetraminoes, we know that this number k is maximum if we see a transversal set with cardinality k. However, for some truncated chessboard, the corresponding set-family has not the König property.

A companion problem is: given a rectangular box of size $a \times b \times c$, what is the maximum number of $\alpha \times \beta \times \gamma$ bricks that can be packed into the box? The answer to this problem is known only if the bricks are 'harmonic', i.e. if α is a divisor of β, and β a divisor of γ [5]. In this case, the König property holds (Brualdi–Foregger). The wastage has been determined for $1 \times b$ bricks [2] and for $2 \times b$ bricks [6], but the general problem is still unsolved.

2. Fractional transversal number of a simple graph

In this section, we consider a simple graph G, with vertex-set $X=\{x_1, x_2, \ldots, x_n\}$ and edge-set $E=\{e_1, e_2, \ldots, e_m\}$. Let $\nu(G)$ denote the size of a maximum matching, i.e., the largest number of edges of G having no common end-vertex. Let $\tau(G)$ denote the size of a minimum transversal T, i.e., so that every edge has at least one end-vertex in T. Clearly, $\nu(G) \leqslant \tau(G)$.

A *fractional transversal* is a weight $t(x) \geqslant 0$, associated with each vertex x, such that

$$[x, y] \in E \Rightarrow t(x)+t(y) \geqslant 1.$$

The *value* of this fractional transversal is

$$t(X)=\sum_{x \in X} t(x).$$

Denote by $\tau^*(G)$ the minimum possible value for a fractional transversal of G. Clearly, the weight $t(x)=\frac{1}{2}$ is fractional transversal, but is not always optimal.

A 2-*matching* of G is a partial hypergraph G' of the multigraph $2G$ (obtained from G by replacing each edge by two parallel edges) whose maximum degree is $\leqslant 2$. Let $\nu_2(G)$ denote the maximum number of edges for a 2-matching of G.

Also, a 2-*transversal* of G is a weight $t(x) \in \{0, 1, 2\}$ such that the total weight of an edge is always $\geqslant 2$; $\tau_2(G)$ denotes the minimum value of a 2-transversal.

Theorem 1 (Tutte, Lovász, Berge etc...). *For every graph G,*

$$\nu(G) \leqslant \tfrac{1}{2}\nu_2(G)=\tau^*(G)=\tfrac{1}{2}\tau_2(G) \leqslant \tau(G).$$

Theorem 2. *The weight $t(x)=\frac{1}{2}$ is an optimal fractional transversal if and only if G has a 2-matching which is 2-regular.*

Theorem 3. *The weight $t(x)=\frac{1}{2}$ is the only optimal fractional transversal if and only if G is regularisable with no bipartite connected components.*

A graph G is *regularisable* if a regular multigraph can be obtained from G by replacing each edge by a non-empty set of parallel edges. The graph $G=\{ab, bc, cd, de, ae, be\}$ is not regular, but it is regularisable (duplicate the edges ab, ae, and multiply by 3 the edge cd). So $t(x)=\frac{1}{2}$ is the only optimal fractional transversal.

The graph $G=\{ab, bc, cf, fd, de, ae, be\}$ has a regular 2-matching but is not regularisable; so we have an other optimal fractional transversal t' defined by:

$$t'(a)=t'(c)=t'(d)=\tfrac{1}{3}; \qquad t'(b)=t'(e)=t'(f)=\tfrac{2}{3}.$$

Theorem 4 (Lovász). *For a graph G,*

$$\tau^*(G) \leqslant \tfrac{1}{2}[\nu(G)+\tau(G)].$$

Corollary. *For a graph G, the following properties are equivalent:*

(i) $\tau^*(G)=\tau(G)$.

(ii) $\nu(G)=\tau(G)$.

Proof. By Theorem 1, (ii) implies (i). Furthermore, (i) implies (ii), because if $\nu(G)<\tau(G)$, we obtain a contradiction by writing:

$$\tau(G)=\tau^*(G)=\tfrac{1}{2}\tau_2(G)\leqslant\tfrac{1}{2}[\nu(G)+\tau(G)]<\tau(G).$$

Consider a maximum matching M of $G=(X,E)$; so two edges in M are disjoint. An *alternating chain* is an elementary chain whose edges are alternating in M and $E-M$. A *blossom* is an odd cycle of length $2k+1$ containing k edges of M, such that the vertex which is not incident with one of those k edges is linked to an unsaturated vertex by an alternating chain. A *posy* consists of two (not necessarily disjoint) blossoms joined by an odd alternating chain (with first and last edge belonging to M). We say that the maximum matching M *separates* an odd cycle with vertex-set S if M has no edge between S and $X-S$.

Theorem 5 [11]. *A graph G, with maximum matching M, satisfies $\nu(G)=\tau(G)$ if and only if there is no flower and no posy.*

Theorem 6 [1]. *A graph G, with maximum matching M, satisfies $\nu(G)=\tau^*(G)$ if and only if there is no odd cycle separated by M.*

Theorem 7 [1]. *If s is the maximum number of vertex-disjoint odd cycles separated by M, then*

$$\tau^*(G)-\nu(G)=\tfrac{1}{2}s.$$

3. Fractional transversal number of a hypergraph

In order to extend the above results to a general family of sets, we shall always keep in mind the classical theorems of graph theory and we shall use the same terminology.

A family $H=(E_1,E_2,\ldots,E_m)$ of subsets of a finite set X is called a *hypergraph* on X if

(i) $E_i\neq\emptyset$, for $i=1,2,\ldots,m$;

(ii) $\bigcup_{i=1}^m E_i=X$.

The elements of X are called the *vertices* and the sets E_i are called the *edges* of H. The *order* of the hypergraph H is $|X|$ and is denoted by $n(H)$. The number of edges of H is denoted by $m(H)$. The *rank* of H is $r(H)=\max_i |E_i|$. Also, we put

$$s(H)=\min_i |E_i|.$$

The *degree* $d_H(x)$ of a vertex x is

$$d_H(x)=|\{i\colon x\in E_i\}|.$$

The maximum degree of H is denoted by $\Delta(H)=\max_{x\in X} d_H(x)$. If each $x\in X$ has the same degree, H is *regular.*

A *partial hypergraph of* H is a subfamily $H'\subseteq H$. For $A\subseteq X$, the *subhypergraph* of H *induced by* A is

$$H_A=(E_i\cap A\colon i\leqslant m, E_i\cap A\neq\emptyset).$$

A *matching* is a partial hypergraph H' of H with $\Delta(H')\leqslant 1$, and $\nu(H)$ denotes the maximum size of a matching. For any positive integer k, a k*-matching* is a partial hypergraph H' of kH with $\Delta(H')\leqslant k$, and $\nu_k(H)$ denotes the maximum size of k-matching.

A *transversal* of H is a set $T\subseteq X=\{x_1, x_2, \ldots, x_n\}$ with $T\cap E_i\neq\emptyset$ for all i; equivalently, it is a (0, 1)-vector $t=(t_1, t_2, \ldots, t_n)$, where t_i is the 'weight' of the vertex x_i, such that

$$\sum_{x_j\in E_i} t_j\geqslant 1 \quad \text{for all } i.$$

A *fractional transversal* is a vector t, whose coordinates are non-negative real numbers, which satisfies the same condition. A k*-transversal* is a vector $t\in\{0, 1, 2, \ldots, k\}^n$ such that

$$\sum_{x_j\in E_i} t_j\geqslant k \quad \text{for all } i.$$

We denote by $\tau_k(H)$ the minimum value of a k-transversal, and, for $k=1$, by $\tau(H)$ the minimum value of a transversal. Also, we denote by $\tau^*(H)$ the minimum value of a fractional transversal.

Clearly, $\nu(H)\leqslant\tau^*(H)\leqslant\tau(H)$ (by the duality principle in linear programming). But we have more:

Theorem 8. *For a hypergraph H,*

$$\begin{aligned}\nu(H)&\leqslant\min_{k\geqslant 1}\frac{1}{k}\nu_k(H)\leqslant\max_{H'\subseteq H}\frac{m(H')}{\Delta(H')}\\ &\leqslant\max_{k\geqslant 1}\frac{1}{k}\nu_k(H)=\tau^*(H)=\min_{k\geqslant 1}\frac{1}{k}\tau_k(H)\\ &\leqslant\min_{A\subseteq X}\frac{|A|}{s(H_A)}\leqslant\max_{k\geqslant 1}\frac{1}{s}\tau_k(H)=\tau(H).\end{aligned}$$

These inequalities, which constitute the main tool for discrete optimization problems, are called the 'fundamental inequalities'. It is easy to deduce directly that for a hypergraph H satisfying the König property (like the interval hypergraphs), *a necessary and sufficient condition for the existence of a matching with k edges is that, for every* $A \subseteq X$,

$$ks(H_A) \leq |A|.$$

For many hypergraphs which arise in combinatorics or number theory, some of the above inequalities hold with equality.

Theorem 8 can be completed by an upper bound for $\tau(H)$:

Theorem 9 (Lovasz, Stein). *For a hypergraph H,*

$$\tau(H) \leq [1 + \operatorname{Log} \Delta(H)]\tau^*(H).$$

Other upper bounds exist in the literature (Henderson–Dean Menger, etc. ...), and for specific types of hypergraphs, these results can sometimes be improved a great deal.

Consider now an r-uniform hypergraph $H = (E_1, E_2, \ldots, E_m)$; so

$$|E_i| = r \quad \text{for all } i.$$

The vector $t = (\frac{1}{r}, \frac{1}{r}, \ldots, \frac{1}{r})$ is a fractional transversal for H, but not necessarily with minimum value. We can show:

Theorem 10. *Let H be a r-uniform hypergraph on X; the following conditions are equivalent:*

(i) $t = (\frac{1}{r}, \frac{1}{r}, \ldots, \frac{1}{r})$ *is an optimal fractional transversal;*

(ii) *for some integer* $k \geq 1$, *H has a regular k-matching;*

(iii) *no real-valued function* $p(x)$ *on X satisfies:*

$$\sum_{x \in X} p(x) = 0, \qquad \sum_{x \in E_i} p(x) > 0 \quad \text{for all } i.$$

The result gives the value of $\tau^*(H)$ for those hypergraphs which are regular; for instance, the finite projective plane with seven points is a 3-uniform regular hypergraph, and satisfies

$$\nu(H) = 1, \qquad \tau^*(H) = \tfrac{7}{3}, \qquad \tau(H) = 3.$$

The *complete r-uniform hypergraph* K_n^r, which consists of all the r-subsets of a set X with cardinality n, satisfies

$$\nu(H) = \frac{n}{r}, \qquad \tau^*(H) = \frac{n}{r}, \qquad \tau = n - (r-1).$$

Note that the minimum value of k such that $(1/k)\tau_k(H)=\tau^*(H)$, which is known for graph, (by Theorem 1), is unknown, even for triple systems.

The above theorem shows also the $\tau^*(H)=n/r$ for *regularisable hypergraphs; H* is said to be *regularisable* if a regular hypergraph can be obtained from H by multiplying each edge E_i by a positive integer $z_i \geq 1$.

Theorem 11. *H is regularisable if and only if no real valued function $p(x)$ on X satisfies:*

$$\sum_{x\in X} p(x)=0,$$

$$\sum_{x\in E_i} p(x)\geq 0 \quad \textit{for all } i,$$

$$\sum_{x\in E_i} p(x)>0 \quad \textit{for some } i.$$

In general, the optimal fractional transversal of a regularisable hypergraph is not unique.

References

[1] E. Balas, Integer and fractional matchings, Management Sciences Report No. 426, Carnegie-Mellon University, PA 15213 (February 1979).
[2] S. Barnett and S. Kynch, Brick packings, Oper. Res. 15 (1967) 1051–1056.
[3] C. Berge, Regularisable graphs I, II, Discrete Math. 23 (1978) 85–89, 91–95.
[4] C. Berge, Packing problems and hypergraph theory: a survey, Ann. Discrete Math. 4 (1979) 3–37.
[5] N.J. de Bruijn, Bricks packings, Amer. Math. Monthly 76 (1969) 37–40.
[6] D.A. Hilton and J.A. Richard, Brick packings, Combinatorial Math. Lecture Notes in Math. 686 (Springer-Verlag) 174–183.
[7] L. Lovász, 2-matchings and 2-covers of hypergraphs, Acta Math. Acad. Sci. Hungar. 26 (1975) 433–444.
[8] L. Lovász, On the ratio of optimal integral and fractional covers, Discrete Math. 13 (1975) 383–390.
[9] G.L. Nemhauser and L.E. Trotter, Vertex packings, Math. Programming 8 (1975) 232–248.
[10] W.R. Pulleyblank, Dual integrality in p-matching problems, Discussion paper 7717 CORE, Lovain (1977).
[11] F. Sterboul, A characterization of the graphs in which the transversal number equals the matching number, to appear in J. Combinatorial Theory (B) (1979).

P. Hansen, ed., Studies on Graphs and Discrete Programming

ADMISSIBLE TRANSFORMATIONS AND THEIR APPLICATION TO MATCHING PROBLEMS

Rainer E. BURKARD

Mathematisches Institut, Universität zu Köln, Köln, Fed. Rep. Germany

Ulrich DERIGS

Industrieseminar, Universität zu Köln, Köln, Fed. Rep. Germany

For special combinatorial optimization problems different kind of objective functions as for instance the classical sum objective or the bottleneck objective are of practical interest. Mostly these objective functions are treated separately and independently. The introduction of an 'algebraic objective function' the cost coefficients of which are elements of an ordered commutative semigroup allows a unified treatment. For standard problems like shortest path, matching and matroid intersection problems this approach has led to efficient algorithms for a broad class of objective functions.

One specific type of algorithm is based on transformations of the cost coefficients and on a purely combinatorial motivated optimality criterion. We will introduce this principle of 'admissible transformations' for the general combinatorial optimization problem with algebraic objective. Then we demonstrate how this general principle combined with some problem specific combinatorial observations leads to efficient and transparent algorithms for the problem of finding an optimal perfect matching.

1. Introduction: Objective functions in combinatorial programming

The commonly used objective functions in combinatorial programming are sum objectives. For example, in the travelling salesman problem the sum of distances between the towns of a tour is minimized, in scheduling theory often the sum of completion times is minimized, in assignment problems the sum of costs for single assignments is minimized and so on.

Bottleneck objectives found increasing interest in the last years. A natural bottleneck problem arises in location theory where one wants to place a fire station such that the maximal distance from the fire station to the buildings which are to be protected becomes minimal. In scheduling theory frequently the maximal completion time is minimized. Assignment problems with bottleneck objectives occur also in production line balancing or in a more complicated form in hospital planning under emergency aspects.

A similar objective arises in time–cost transportation problems. A certain amount of (perishable) goods is to be transported from the suppliers to the consumers in the shortest possible time and subject to this condition in a way such that the amount of goods which are shipped along the longest way is minimal. This model is not only important for the shipping of flowers or fruits, but plays also an important role at military operations.

The time–cost objective is a special multicriteria objective. Other multicriteria objectives arise for example in personnel assignment where persons are assigned to free positions according to their qualification and experiences. Here n criteria $c_1, \ldots, c_n$ are considered such that criterion c_i dominates all criteria c_j, $j > i$. We can model this by lexicographically ordered vectors.

Multicriteria objectives do not only arise in assignment problems, on the contrary, most of real life problems lead to multicriteria objectives and often the criteria can be ordered lexicographically. (We do not consider here problems with partially ordered criteria, since these lead to an other solution concept: to the concept of efficient or undominated solutions, which we exclude from the present discussion).

Now the question arises if there is a mathematical model which covers all the examples discussed above. And we will restrict the possibilities by the requirement: if a problem with sum objective is efficiently solvable, then it should also be efficiently solvable for the general objective.

In the next section we introduce the general combinatorial optimization problem and discuss an algebraic way to define objective functions. It will turn out that all the problems mentioned in the introduction will be special cases of such combinatorial optimization problems with algebraic objectives. In Section 3 we discuss admissible transformations as a solution method for solving combinatorial optimization problems with algebraic objectives and apply this method to solve perfect matching problems. For this purpose some combinatorial properties of matching problems are reviewed in Section 4. In Section 5 and 6 we describe the Hungarian method and the shortest augmenting path method for solving perfect matching problems with algebraic objectives before we close with concluding remarks.

2. Combinatorial optimization problems with algebraic objectives

Let $N = \{1, 2, \ldots, n\}$ be a finite set and let $\mathcal{S}$ be a class of subsets S of N. We call S a feasible solution. Let H be a totally ordered set and $z : \mathcal{S} \to H$ a mapping of $\mathcal{S}$ in H. $z(S)$ are the cost of the feasible solution S. The general combinatorial programming problem can now be written in the form

$$\min_{S \in \mathcal{S}} z(S).$$

For example in an assignment problem $\mathcal{S}$ is the set of all permutations of the set $\{1, \ldots, n\}$ and $z(S)$ is the objective value of a special permutation, which may be expressed in the linear case by

$$z(S) = \sum_{\substack{i=1 \\ (i,j)\in S}}^{n} c_{ij}.$$

To meet the requirement of the existence of good algorithms in the case of efficiently solvable sum problems we assign to every element in N a cost value in H and we require some structure for the set H. It should be possible to compose two elements in H and this composition should be associative, commutative and compatible with the order relation on H, i.e. an (internal) composition $*$ is defined on H which fulfills

$$a*(b*c) = (a*b)*c,$$

$$a*b = b*a,$$

$$a \leq b \Rightarrow a*c \leq b*c \quad \text{for all } c \in H.$$

A set H with composition $*$ and order relation $\leq$ fulfilling these axioms is called a commutative ordered semigroup $(H, *, \leq)$.

Now we are able to describe a combinatorial programming problem with an algebraic objective function (ACP) in the following way. Let $(H, *, \leq)$ be an ordered commutative semigroup and let $c: N \to H$ be a mapping which assigns to every element $i \in N$ some cost $c_i \in H$. For $S \subseteq E$ we define

$$c(S) = \mathop{\ast}_{i\in S} c_i$$

where $\ast_{i\in S} c_i$ denotes $c_1 * c_2 * \cdots * c_k$ for $S = \{1, 2, \ldots, k\}$.

The combinatorial optimization problem with algebraic objective function becomes now

$$\min_{S\in\mathscr{S}} \mathop{\ast}_{i\in S} c_i.$$

As was pointed out earlier, we are not only interested in a mathematical formulation of the problems but even more in their solution. It turns out that most algorithms use a reduction of some cost $b \in H$ by cost $a \in H$ with $a < b$. Therefore we require the following divisor rule:

$$a \leq b \Rightarrow \exists c \in H\colon a*c = b. \tag{2.1}$$

The more complicated the combinatorial problem or the algorithm, the more structure is needed for H. For many algorithms we need therefore the following weak cancellation rule in H:

$$a*b = a*c \ \Rightarrow\ b = c \text{ or } a*b = a \tag{2.2}$$

Now let us discuss the semigroups $(H, *, \leqslant)$ underlying the examples mentioned in the introduction.

(1) $H = \mathbb{R}$, $*$ is the addition of real numbers, $\leqslant$ is the natural order of real numbers. This system leads to sum objectives.

(2) $H = \mathbb{R} \cup \{-\infty\}$, $a * b := \max(a, b)$ and $\leqslant$ is the natural order of real numbers. This system leads to bottleneck objectives. If $a \leqslant b$, then $\max(a, b) = b$, therefore the divisor axiom holds with $c = b$. Further the weak cancellation rule is true, since either $a < b$ which implies $a < c$ and $b = c$ or $a \geqslant b$, then $\max(a, b) = a$.

(3) $H := (\mathbb{R} \cup \{-\infty\}) \times \mathbb{R}_+$,

$$a = \begin{pmatrix} a_1 \\ a_2 \end{pmatrix}, \qquad b = \begin{pmatrix} b_1 \\ b_2 \end{pmatrix}.$$

$$a * b := \begin{cases} \begin{pmatrix} a_1 \\ a_2 \end{pmatrix} & \text{if } a_1 > b_1, \\ \begin{pmatrix} b_1 \\ b_2 \end{pmatrix} & \text{if } b_1 > a_1, \\ \begin{pmatrix} a_1 \\ a_2 + b_2 \end{pmatrix} & \text{if } a_1 = b_1. \end{cases}$$

Further we define $a \leqslant b$ if $a_1 < b_1$ or $a_1 = b_1$ and $a_2 \leqslant b_2$. This system leads to time–cost objectives. The first component of a represents a time and the second component of a represents some cost. Note that again the divisor axiom and the weak cancellation rule hold.

(4) $H = \mathbb{R}^n$, $*$ is the vector addition and $\leqslant$ the lexicographical order of vectors with n components. This system leads to multicriteria objectives.

All systems mentioned so far are special cases of commutative ordered semigroups with divisor rule and weak cancellation law.

The following example shows an ordered commutative semigroup with divisor rule, in which the weak cancellation law does not hold.

Example. $H = [0, 1] \cup \{2\}$,

$$a * b := \begin{cases} a + b, & \text{if } a + b \leqslant 1, \\ 2, & \text{if } a + b > 1, \end{cases}$$

and $\leqslant$ is the order in H induced by the natural order of $\mathbb{R}$. This is an example for an ordered commutative semigroup in which the divisor rule holds, but not the weak cancellation law, since

$$\tfrac{3}{4} * \tfrac{1}{2} = \tfrac{3}{4} * \tfrac{3}{4}, \quad \text{but } \tfrac{1}{2} \neq \tfrac{3}{4} \quad \text{and} \quad \tfrac{3}{4} * \tfrac{1}{2} \neq \tfrac{3}{4}.$$

For a detailed discussion of the underlying algebraic structures we refer to the classical book of Fuchs [15] on ordered algebraic structures and the recent monograph on these problems by Zimmermann [19].

The examples above show that the algebraic treatment of objective functions leads to a unifying framework for different models. Apart from shortest route problems in networks, which have been treated since about ten years in an axiomatic way (see e.g. the books of Carré [6], Gondran–Minoux [17] and Zimmermann [19] on this subject) linear assignment problems with algebraic objectives were the first, which have been solved. Since 1974 most of the efficiently solvable problems (in the sense of Edmonds) have been investigated in this direction and solution procedures were derived for matroid intersection problems, matching problems, transportation problems and network flow problems. By using duality arguments recently LPs with algebraic objectives were treated. We refer to [5] as a recent survey on the state of the art and to [19] as a monograph of this field.

3. Admissible transformations

For solving combinatorial optimization problems with algebraic objectives a thorough knowledge of the underlying combinatorial and semigroup structures is necessary. Whereas the combinatorial structure varies from problem to problem (see e.g. Section 4), we need the following result from semigroup theory: Let $(H, *, \leq)$ be a weakly cancellative, ordered commutative semigroup fulfilling the divisor rule (2.1). Then there exists a unique partition of H into a maximal family $(H_\lambda)_{\lambda \in \Lambda}$, where Λ is a nonempty linearly ordered set and for all $a \in H_\lambda$, $b \in H_\mu$ with $\lambda, \mu \in \Lambda$, $\lambda < \mu$ holds $a < b$ and $a * b = b * a = b$.

In particular Λ has a minimum λ_0 and the neutral element σ of H is an element of H_{λ_0}. Further the strong cancellation rule holds in H_λ, λ fixed, i.e.

$$a * b = a * c \Rightarrow b = c \quad \text{for } a, b, c \in H_\lambda.$$

A proof of this *ordinal decomposition* of H is e.g. given in [19].

Now we will outline the concept of admissible transformations which has turned out to be a central method for tackling ACP. We assume that $(H, *, \leq)$ is a commutative ordered semigroup. Further let $c : N \to H$ be a cost function. A transformation T of c in $c' = T(c)$ is called *admissible*, if there exists a constant $z(T)$ such that for all $S \in \mathcal{S}$ the following equation holds

$$\mathop{*}_{i \in S} c_i = z(T) * \mathop{*}_{i \in S} c'_i.$$

Now the following optimality criterion can be shown

Theorem 3.1. *Let T be an admissible transformation of c in $c' = T(c)$, such that*

$c_i' \geq \sigma$ *for all* $i \in N$. *If there exists a feasible solution* $S_0 \in \mathcal{S}$ *with*

$$\mathop{*}_{i \in S_0} c_i' * z(T) = z(T),$$

then S_0 *is an optimal solution of*

$$\min_{S \in \mathcal{S}} \mathop{*}_{i \in S} c_i$$

with objective value $z(T)$.

Proof. Let $S \in \mathcal{S}$ arbitrary

$$\mathop{*}_{i \in S} c_i = z(T) * \mathop{*}_{i \in S} c_i' \geq z(T)$$

$$= z(T) * \mathop{*}_{i \in S_0} c_i' = \mathop{*}_{i \in S_0} c_i.$$

Theorem 3.1 holds for any ordered commutative semigroup, but to find admissible transformations it will be necessary to make use of the divisor rule. Admissible transformations for assignment problems, which lead to an $O(n^4)$ algorithm for algebraic assignment problems were given in [4]. For more complicated combinatorial programs such as matching problems, or for algorithms with better complexity, e.g. an algorithm of $O(n^3)$ steps for assignment problems, the notion of admissible transformations has to be generalized. For this we assume that H is an ordered commutative semigroup with divisor rule (2.1) and weak cancellation law (2.2).

We consider

$$\min_{S \in \mathcal{S}} \mathop{*}_{i \in S} c_i$$

and let z_{opt} be the optimal value of this problem. A transformation T of c in $c' = T(c)$ is called a *general admissible transformation*, if there are elements $\alpha(S)$, $\beta(S) \in H$ such that for all $S \in \mathcal{S}$:

$$\mathop{*}_{i \in S} c_i * \alpha(S) = \beta(S) * \mathop{*}_{i \in S} c_i', \qquad \lambda(\alpha(S)) \leq \lambda(\beta(S)) \leq \lambda(z_{\text{opt}}).$$

For general admissible transformations the following optimality criterion can be shown (cf. [7]).

Theorem 3.2. *Let H be an ordered commutative semigroup with divisor rule and weak cancellation law. If a general admissible transformation T and a feasible solution* $S_0 \in \mathcal{S}$ *have the properties*

$$c_i' \geq \sigma \quad \textit{for all } i \in N, \tag{3.3}$$

$$\beta(S_0) * \mathop{\ast}_{i \in S_0} c_i' = \beta(S_0), \tag{3.4}$$

$$\alpha(S_0) \geqslant \alpha(S) \quad \textit{for all } S \in \mathcal{S}, \tag{3.5}$$

$$\beta(S_0) \leqslant \beta(S) \quad \textit{for all } S \in \mathcal{S}, \tag{3.6}$$

then S_0 *is an optimal solution.*

It is easy to derive from the assumptions of the theorem the inequality

$$\mathop{\ast}_{i \in S_0} c_i * \alpha(S_0) \leqslant \mathop{\ast}_{i \in S} c_i * \alpha(S) \quad \text{for all } S \in \mathcal{S},$$

but to show that now

$$\mathop{\ast}_{i \in S_0} c_i \leqslant \mathop{\ast}_{i \in S} c_i$$

needs a detailed discussion of the underlying algebraic structure.

These general admissible transformations with constant values α, β were used for solving algebraic assignment and transportation problems in an efficient way. For solving perfect matching problems we shall use general admissible transformations with constant β but variable $\alpha(S)$, where the value of $\alpha(S)$ depends on the blossoms of the perfect matching S.

It is easy to see that in this case condition (3.4) is equivalent to

$$\beta * c_i = \beta \quad \text{for all } i \in S_0. \tag{3.4'}$$

Therefore we can define the set $N' := \{i \in N \mid \beta * c_i = \beta\}$ of *admissible elements.* Then (3.4) is equivalent to

$$S_0 \subseteq N'. \tag{3.4''}$$

Example. Let N be the edge set of a graph and let $\mathcal{S}$ be the set of all perfect matchings. Then N' defines a partial graph G' and (3.4) specializes to the condition that S_0 has to be a perfect matching in G'.

For the algorithmic treatment of special types of problems the concept of admissible transformations has to be combined with some problem-specific combinatorial arguments to obtain combinatorially natured and motivated optimality criteria. The conditions (3.5) and (3.6) are then replaced by specific combinatorial conditions (3.5′) and (3.6′) such that (3.5′) implies (3.5) and (3.6′) implies (3.6). Furtheron the validity of these combinatorial conditions should be easy to check.

We will now demonstrate the applicability of this concept on a specific problem—the perfect matching problem, and show how the combination of different admissible transformations and combinatorial arguments lead to different types of algorithms for the same problem.

In the next chapter we will introduce the basic definitions and theorems for matchings in graphs.

4. Combinatorial properties

In this section we consider some combinatorial properties of matchings in graphs which are used in the next sections.

Let $G=(V, E, \psi)$ be a loopless graph with vertex set V and edge set E. For every $e \in E$, $\psi(e) \subseteq V$ is the pair of vertices which meet e. Whenever we can do so without loss of clarity we will omit the incidence function ψ and write $G=(V, E)$ resp. $e=\{i, j\}$ with $i, j \in V$. Now a *matching* M is a subset of E with the property that the edges in M are pairwise disjoint. M is called *perfect* if for $V(M):=\bigcup_{e \in M} e=V$ holds.

For any $V_1 \subseteq V$ we define the coboundary of V_1

$$\delta(V_1):=\{e \in E \mid |\psi(e) \cap V_1|=1\}$$

and the set of edges having both ends in V_1

$$\gamma(V_1):=\{e \in E \mid \psi(e) \subseteq V_1\}.$$

The graph $G[V_1]:=(V_1, \gamma(V_1))$ is the subgraph *induced* by V_1. A subgraph H of G having the same nodeset as G is said to *span* G.

An *alternating path* with respect to a matching M in G is a path the edges of which are alternately elements of M and not. *Alternating trees* and *circles* are defined analogously. An *augmenting path* is an alternating path between two unmatched nodes. With an augmenting path P we define

$$M \oplus P:=(M \setminus P) \cup (P \setminus M).$$

$M \oplus P$ is again a matching in G. For alternating circles K we define $M \oplus K$ analogously.

The following theorem is fundamental for every algorithmic treatment of matchings in graphs.

Theorem 4.1 (Berge [1]). *A matching M contains a maximum number of edges iff it admits no augmenting path.*

Such matchings are called *maximum cardinality matchings* (m.c.-matching).

For $B \subseteq V$ we define the graph $G \times B=(V_B, E_B, \psi_B)$ obtained from G by *shrinking* B by

$$V_B:=(V-B) \dot{\cup} \{v_B\},$$

$$E_B:=E-\gamma(B),$$

$$\psi_B(e):=\begin{cases}\psi(e), & \text{if } e \in E_B \setminus \delta(B),\\ (\psi(e)-B) \dot{\cup} \{v_B\} & \text{if } e \in \delta(B).\end{cases}$$

The node v_B is called pseudonode of B. For $M \subset E$ we define $M_B := M \cap E_B$. Let $\mathcal{A}$ be a set of subsets of V. We say $\mathcal{A}$ is *nested* if $|A| \geq 3$ for all $A \in \mathcal{A}$ and $A, B \in \mathcal{A}$, $A \cap B \neq \emptyset \Rightarrow A \subset B \vee B \subset A$.

For any $A \in \mathcal{A}$ we define $\mathcal{A}[A] := \{B \in \mathcal{A} \mid B \subsetneq A\}$.

If $\{A_1, \ldots, A_n\}$ is the set of maximal elements of $\mathcal{A}$ we define

$$G \times \mathcal{A} := (\cdots (G \times A_1) \times A_2) \times \cdots \times A_n).$$

The order of the sets $A_1, A_2, \ldots, A_n$ has no effect on $G \times \mathcal{A}$. With $E_{\mathcal{A}}$ the edge set of $G \times \mathcal{A}$ we denote $M_{\mathcal{A}} := M \cap E_{\mathcal{A}}$ for $M \subseteq E$.

We are interested in nested families $\mathcal{A}$ having the additional property

$$G[A] \times \mathcal{A}[A] \text{ is spanned by an odd circle for each } A \in \mathcal{A}. \qquad (4.2)$$

Those nested families are called *shrinking families.* A (maximal) set A from a shrinking family with $|M \cap \gamma(A)| = \frac{1}{2}(|A| - 1)$ is called (outermost) *blossom* with respect to M. The rôle of blossoms and shrinking families in connection with matching problems is studied in [14].

The following property is fundamental for an algorithmic treatment of matchings in nonbipartite graphs.

Theorem 4.3 (Edmonds [11]). *Let M be a matching in G and $\mathcal{A}$ a shrinking family such that every $A \in \mathcal{A}$ is a blossom with respect to M. Then every augmenting path $P_{\mathcal{A}}$ in $G \times \mathcal{A}$ with respect to $M_{\mathcal{A}}$ induces an augmenting path P with respect to M.*

Corollary 4.4. *Let $\mathcal{A}$ be a shrinking family of G and $M_{\mathcal{A}}$ a (perfect) matching in $G \times \mathcal{A}$. Then $M_{\mathcal{A}}$ induces (uniquely) a (perfect) matching M in G such that every $A \in \mathcal{A}$ is a blossom with respect to M.*

To determine an augmenting path with startnode s we construct an alternating tree with root s. This tree leads to a bicoloring of its nodes. Therefore we can attach an 'S'-label to nodes which bear the same color as the root and a 'T'-label to the other nodes of the tree.

Now a blossom is detected whenever there is an edge $e \in M$ between two 'S'-labeled nodes. This blossom is then shrunken to a pseudonode. The pseudonode receives an 'S'-label and we try to enlarge the tree by adding appropriate nodes and edges. If an 'S'-labeled node is joint with an unmatched node an augmenting path has been detected and the matching can be augmented by changing the rôle of matching and nonmatching edges on this path. For this purpose pseudonodes have to be expanded and the path through these blossoms has to be restored.

Lawler [18] and Gabow [16] describe appropriate labeling techniques to provide backtracing through nested blossoms.

If all 'S'-labeled nodes are only connected with 'T'-labeled nodes the tree is said to be hungarian and no augmenting path starting from node s exists.

To determine an m.c. matching alternating trees are built from each unmatched node s as root using the above mentioned labeling and shrinking steps. An augmenting path is detected if two 'S'-labeled nodes out of two different trees are connected by an edge.

If all trees become hungarian the actual matching is an m.c. matching in the shrunken graph. After expanding all pseudonodes this matching can be extended to an m.c. matching in the original graph due to Corollary 4.4.

At the end of the algorithm the nodeset V is partitioned into three classes $V = \mathcal{S} \dot{\cup} \mathcal{T} \dot{\cup} \mathcal{U}$ where $\mathcal{S}$ is the set of all 'S'-labeled nodes or nodes contained in 'S'-labeled pseudonodes. Further every unmatched node belongs to the class $\mathcal{S}$. $\mathcal{T}$ is the set of all 'T'-labeled nodes. The class $\mathcal{U}$ consists of $2k$ ($k \geq 0$) unlabeled nodes which are joined by k matching edges.

The following definition and theorem are fundamental for the algorithm in Section 5.

Let $G = (V, E)$ be a given graph and let $\mathcal{N} = \{N_1, \ldots, N_p\}$ be a set of subsets of nodes $N_i \subseteq V$ such that each N_i contains an odd number of elements ($i = 1, \ldots, p$).

If $|N_i| = 1$, then N_i is said to cover $\delta(N_i)$ the set of all edges incident with the node in N_i, and the capacity of N_i is 1.

If $|N_i| = 2k+1$, $k \in \mathbb{N}$, then N_i is said to cover $\gamma(N_i)$, all edges in the subgraph induced by N_i, and the capacity of N_i is k.

The family $\mathcal{N}$ is said to be an odd-set-cover of G if every edge is covered by at least one subset $N_i \in \mathcal{N}$. The capacity of $\mathcal{N}$, denoted by cap($\mathcal{N}$), is the sum of the capacities of the odd sets contained within it.

The following theorem is due to Edmonds [11].

Theorem 4.5. *For any graph G the maximum number of edges in a matching is equal to the minimum capacity of an odd-set-cover.*

Edmond's algorithm for determining an m.c. matching yields simultaneously a suitable odd-set-cover.

Consider the partition of V at the termination of the algorithm. Then the family $\mathcal{N}$ can be constructed in the following way. Every T-labeled node becomes a singleton set in $\mathcal{N}$. There are exactly as many such nodes as matching edges in the hungarian trees.

The nodes contained in a pseudonode become an odd set in $\mathcal{N}$ and its capacity is equal to the number of edges of the m.c. matching contained in the expanded blossom.

If $\mathcal{U} = \emptyset$, the cover is complete. If $|\mathcal{U}| = 2$ we choose arbitrarily one of the two nodes as singleton sets in $\mathcal{N}$.

If $|\mathscr{U}|=2k$, $k\geqslant 2$ we choose arbitrarily one of the nodes as singleton set in $\mathscr{N}$ and the remaining $2k-1$ nodes as odd subsets with capacity $k-1$. This completes the odd-set-cover.

5. The hungarian method (Blossom-algorithm)

The blossom algorithm for solving the matching problem with sum objective was developed by Edmonds [12] in 1965. That work initialized the research in 'polyhedral combinatorics'. This method has been modified for algebraic objective functions by Derigs [8] formulating the method in terms of admissible transformation and using only combinatorially natured optimality criteria. We will outline this approach in this section. Again we assume that $(H, *, \leqslant)$ is an ordered commutative semigroup with divisor rule and weak cancellation law. We consider a complete graph $G=(V, E)$ with an even number $|V|$ and edgeweights $c_{ij}\in H$ for $e_{ij}\in E$.

Defining $\mathscr{M}$ to be the set of all perfect matchings we obtain the *algebraic perfect matching problem*

$$\min_{M\in\mathscr{M}} \mathop{*}_{e_{ij}\in M} c_{ij}. \tag{5.1}$$

Now let $\mathscr{R}$ be the set of all subsets $R_k\subset V$ with $|R_k|\geqslant 3$, odd. With every $i\in V$ we associate two *node weights* $u_i, w_i\in H$ and with every $R_k\in\mathscr{R}$ a *blossom weight* $d_k\in H$. Then we define a transformation $T: c_{ij}\to c'_{ij}$ for $e_{ij}\in E$ by

$$c'_{ij}*u_i*u_j=c_{ij}*w_i*w_j*\mathop{*}_{e_{ij}\in\gamma(R_k)} d_k \tag{5.2}$$

with $c'_{ij}:=\sigma$ if

$$u_i*u_j=w_i*w_j*\mathop{*}_{e_{ij}\in\gamma(R_k)} d_k.$$

For every $M\in\mathscr{M}$ we obtain by cumulating the equation (5.2)

i.e.

$$c'(M)*\underbrace{\mathop{*}_{i\in V} u_i}=c(M)*\underbrace{\mathop{*}_{i\in V} w_i*\mathop{*}_{e\in M}\left(\mathop{*}_{e\in\gamma(R_k)} d_k\right)}$$

$$c'(M)*\beta(M)=c(M)*\qquad\alpha(M).$$

It is easy to see that such a transformation is admissible if

$$\lambda(w_j)\leqslant\lambda(u_j)\leqslant\lambda(z_{\text{opt}})\quad\text{for all } j\in V, \tag{5.4}$$

and

$$\lambda(d_k)\leqslant\lambda(z_{\text{opt}})\qquad\text{for all } R_k\in\mathscr{R}. \tag{5.5}$$

In particular we obtain $\beta(M)=\beta=\text{const.}$ for all $M\in\mathscr{M}$.

Let us define by $E':=\{e_{ij}\mid c'_{ij}=\sigma\}$ the set of admissible edges and $G'=(V, E')$ the *admissibility graph*.

Definition 5.6. An admissible transformation of type (5.2) is called *suitable* if there exists a shrinking family $\mathscr{A}$ in G' s.t.

$$d_k > \sigma \Rightarrow R_k \in \mathscr{A}. \tag{5.7}$$

The following theorem gives a sufficient optimality criterion for suitable transformations.

Theorem 5.8. *Let $M'_{\mathscr{A}}$ be a perfect matching in $G' \times \mathscr{A}$. Then the induced perfect matching M' in G' is an optimal solution for* (5.1).

Proof. (cf. Derigs [8]).

Starting with the dummy transformation with $u_i = w_i = d_k = \sigma$ and $\mathscr{A} = \emptyset$ alternately maximum cardinality matchings in $G' \times \mathscr{A}$ are determined resp. new transformations are performed. The first time a perfect matching can be found in $G' \times \mathscr{A}$ the optimal solution is obtained. For determining the maximum cardinality matching $M'_{\mathscr{A}}$ we use the labeling technique outlined in Section 4 which yields simultaneously a minimal odd-set-cover $\mathscr{N}'$. If the matching is not perfect the new transformation is defined by the odd-set-cover in such a way, that $\mathscr{N}'$ is no longer an odd-set-cover in the resulting admissibility graph. After at most $|V|^2$ transformations an optimal solution is obtained in this way. The complete labeling algorithm can be found in [8].

6. The shortest augmenting path algorithm

In this section we present another approach for solving the algebraic perfect matching problem (5.1). Again we assume that the underlying system is an ordered commutative semigroup with divisor rule and weak cancellation law. Now let M be any matching in G.

Definition 6.1. An alternating circle K with respect to M is called *negative* with respect to M if $c(M \oplus K) < c(M)$.

The following theorem gives a combinatorially natured optimality criterion.

Theorem 6.2. *$M \in \mathscr{M}$ solves* (5.1) *iff it admits no negative cycle.*

Proof. (cf. Derigs [10]).

We will now describe a method based on this optimality criterion. Let M be any matching in G and $s \in V$ unmatched with respect to M. Then we define $\mathbb{P}_s(M)$ to be the set of all augmenting paths with startnode s. For $P \in \mathbb{P}_s(M)$ we define

$$c(P) := \underset{e_{ij} \in M \oplus P}{*} c_{ij} \tag{6.3}$$

the *length* of P. $P_0 \in \mathbb{P}_s(M)$ is called *shortest augmenting path* with respect to M if $c(P_0) \leq c(P)$ for all $P \in \mathbb{P}_s(M)$.

Theorem 6.4. *Let M_k be a matching in G which does not allow a negative alternating circle and $P \in \mathbb{P}_s(M_k)$ the shortest augmenting path. Then $M_{k+1} := M_k \oplus P$ does not allow a negative circle.*

Proof. (cf. Derigs [10]).

This theorem motivates the following algorithm for solving the algebraic perfect matching problem (5.1).

Shortest Augmenting Path Algorithm 6.5 (SAP-algorithm)

(0) $M = \emptyset \to (1)$.

(1) Determine an unmatched vertex $s \in V$. If none exists $\to$ Stop: M is optimal; else $\to$ (2).

(2) Determine a shortest augmenting path $P \in \mathbb{P}_s(M)$.

(3) $M := M \oplus P \to (1)$.

The shortest path problem (2) can now be solved using the concept of admissible transformations.

Again we associate with every $i \in V$ two node weights $u_i, w_i \in H$ and with every $R_k \in \mathcal{R}$ a blossom weight $d_k \in H$. The transformation $T: c_{ij} \to c'_{ij}$ is now defined by

$$c'_{ij} * u_i * u_i * \underset{e_{ij} \in \delta(R_k)}{*} d_k = c_{ij} * w_i * w_j \tag{6.6}$$

with

$$c'_{ij} := \sigma \quad \text{if } u_i * u_j * \underset{e_{ij} \in \delta(R_k)}{*} d_k = c_{ij} * w_i * w_j.$$

For $P \in \mathbb{P}_s(M)$ with terminalnode t we obtain cumulating all equations (6.6) for $e_{ij} \in M \oplus P$

$$c'(P) * \beta(P) = c(P) * \alpha(P)$$

with

$$\alpha(P) = w_s * \underset{i \in V(M)}{*} w_i * w_t,$$

$$\beta(P) = u_s * \underset{i \in V(M)}{*} u_i * u_t * \underset{R_k \in \mathcal{R}}{*} d_k^{|(M \oplus P) \cap \delta(R_k)|}$$

where $d_k^t := d_k * \cdots * d_k$ (t times). Again we see that (6.6) is admissible if (5.4) and (5.5) hold.

Definition 6.7. An admissible transformation of type (6.6) is called *suitable* with respect to $\mathbb{P}_s(M)$ if there exists a shrinking family $\mathcal{A}$ s.t.

(1) $i \neq s$ unmatched $\Rightarrow u_i = w_i = \sigma$,

(2) $d_k > \sigma \Rightarrow R_k \in \mathcal{A}$ is a blossom with respect to M s.t.

$$s \notin R_k \;\Rightarrow\; |M \cap \delta(R_k)| = 1.$$

Now we obtain the following optimality criterion.

Theorem 6.8. *Let* $T: c_{ij} \to c'_{ij}$ *be a suitable transformation and* $\mathcal{A}$ *the associated shrinking family (i.e.* $\mathcal{A} = \{R_k \in \mathcal{R} \mid d_k > \sigma\}$*). If* $\bar{P} \in \mathbb{P}_s(M)$ *s.t.*

$$c'(\bar{P}) = \sigma, \tag{6.9}$$

$\bar{P}_{\mathcal{A}}$ *is an augmenting path with respect to* $M_{\mathcal{A}}$*, then* $\bar{P}$ *is a shortest augmenting path with respect to* $\mathbb{P}_s(M)$. (6.10)

Proof. For every $P \in \mathbb{P}_s(M)$ and $R_k \in \mathcal{A}$ holds

$$|(M \oplus P) \cap \delta(R_k)| \geq 1.$$

Since $\bar{P}_{\mathcal{A}}$ is augmenting with respect to $M_{\mathcal{A}}$

$$|(M \oplus \bar{P}) \cap \delta(R_k)| = 1$$

holds for all $R_k \in \mathcal{A}$. Since for all unmatched nodes $t \neq s$ the relation $u_t = w_t = \sigma$ holds we obtain

$$\beta(\bar{P}) = \beta_0 * \mathop{\ast}_{R_k \in \mathcal{A}} d_k \leq \beta_0 * \mathop{\ast}_{R_k \in \mathcal{A}} d_k^{|(M \oplus P) \cap \delta(R_k)|} = \beta(P)$$

with $\beta_0 = u_s * \ast_{i \in V(M)} u_i$, and

$$\alpha(P) = w_s * \mathop{\ast}_{i \in V(M)} w_i \quad \text{for all } P \in \mathbb{P}_s(M).$$

Together with (6.9) we obtain the optimality for $\bar{P}$ applying Theorem 3.2.

Based on this optimality criterion the SAP-algorithm 6.5 could be formulated as a labeling method for the sum and bottleneck objective (cf. [10, 9]). This labeling technique is a modification of an algorithm of Edmonds and Johnson [13] for solving the Chinese Postman Problem.

We have shown two different algorithms for solving the perfect matching problem with algebraic objective function. Both algorithms are motivated by combinatorially natured arguments and are rather transparent.

If we define a '$*$'-operation or the solution of an equation $a * c = b$ for $a < b$ to be of complexity O(1), then both algorithms can be performed with an amount of $O(|V|^3)$ steps. Yet computational tests have shown that the SAP-method is superior to the hungarian method for the sum objective as well as for the bottleneck objective.

FORTRAN-implementations of the SAP-method for both cases—sum objective and bottleneck objective—can be found in [3]. Our tests have shown that the empirical order of these SAP implementations is approximately $O(|V|^{2.8})$ in the sum-case and $O(|V|^2)$ in the bottleneck-case. For solving the perfect matching problem in a complete graph on 100 nodes these programs consume about 0.2 CPU-sec in the sum-case and 0.05 CPU-sec in the bottleneck case on a CDC-CYBER 76.

7. Conclusive remarks

In this paper we were concerned with a specific efficiently solvable problem. In general we can observe that at these kind of combinatorial optimization problems the complexity does not increase if we consider algebraic objective functions. The theory of algebraic objectives, however, is also useful at NP-hard combinatorial optimizations problems as for example quadratic assignment problems, three dimensional assignment problems or some scheduling problems. Since the algebraic nature of the objective function is only of interest for deriving bounds, the enumeration scheme remains unchanged for all problems of the same type, differing only in the form of their objectives. Algorithms for solving NP-hard problems require in most cases a long and involved computer program. And this can be written thus that at the start only the respective algebraic operation is specified and the program solves the problem with the corresponding objective function. This was used for example in a computer code for some scheduling problems [2], which equally solves problems with different kind of objectives such as the sum of completion times, the maximal completion time and others.

These examples show that the algebraic approach for objective functions leads to *schemes of algorithms* which then yield special algorithms if the algebraic operation is specified. This kind of treatment is a powerful tool for handling combinatorial optimization problems which often presents new insights and possibilities and leads to new efficient algorithms.

References

[1] C. Berge, Two theorems in graph theory, Proc. Nat. Acad. Sci. U.S.A. 43 (1957) 842–844.
[2] R.E. Burkard, Remarks on some scheduling problems with algebraic objective functions, Oper. Res. Verfahren 32 (1979) 63–77.

[3] R.E. Burkard, and U. Derigs, Assignment and matching problems: Solution methods and FORTRAN-programs, Lecture Notes in Economics and Mathematical Systems 184 (Springer-Verlag, Berlin, 1980).
[4] R.E. Burkard, W. Hahn and U. Zimmermann, An algebraic approach to assignment problems, Math. Programming 12 (1977) 318–327.
[5] R.E. Burkard and U. Zimmermann, Combinatorial optimization in linearly ordered semimodules: a survey, to appear in: B. Korte ed., Modern Applied Mathematics: Optimization and Operations Research, (North-Holland, Amsterdam, 1981).
[6] B.A. Carre, Graphs and Networks (Oxford Univ. Press, 1979).
[7] U. Derigs, Duality and admissible transformations in combinatorial optimization, Z. Oper. Res. 23 (1979) 251–267.
[8] U. Derigs, A generalized hungarian method for solving minimum weight perfect matching problems with algebraic objective, Discrete Appl. Math. 1 (1979) 167–180.
[9] U. Derigs, On two methods for solving the bottleneck matching problem, in: K. Iracki et al., ed., Optimization Techniques, Part 2, Lecture Notes in Control and Information Sciences 23 (Springer-Verlag, Berlin, 1980) 176–184.
[10] U. Derigs, A shortest augmenting path method for solving minimal perfect matching problems, Report 79–06, Mathematisches Institut der Universität zu Köln, Köln (1979), to appear in Networks.
[11] J. Edmonds, Paths, trees, and flowers, Canad. J. Math. 17 (1965) 449–467.
[12] J. Edmonds, Maximum matching and a polyhedron with 0,1 vertices, J. Res. Nat. Bur. Standards 69B (1965) 125–130.
[13] J. Edmonds, and E. L. Johnson, Matching, Euler tours and the chinese postman, Math. Programming 5 (1973) 88–124.
[14] J. Edmonds, and W. Pulleyblank, Facets of 1-matching polyhedra, in: Hypergraph Seminar, Lecture Notes in Mathematics No. 411 (Springer-Verlag, Berlin, 1974) 214–242.
[15] L. Fuchs, Partially ordered algebraic systems (Addison Wesley, Reading, MA, 1963).
[16] H. Gabow, An efficient implementation of Edmond's algorithm for maximum matching on graphs, J. ACM 23 (2) (1975) 221–234.
[17] M. Gondran, and M. Minoux, Graphes et Algorithmes (Eyrolles, Paris, 1979).
[18] E.L. Lawler, Combinatorial Optimization: Networks and Matroids (Holt, Rinehart and Winston, New York, 1976).
[19] U. Zimmermann, Linear and Combinatorial Optimization in Ordered Algebraic Structures, Ann. Discrete Math. 10 (North-Holland, Amsterdam, 1981).

P. Hansen, ed., Studies on Graphs and Discrete Programming

POLYNOMIAL ALGORITHMS FOR TOTALLY DUAL INTEGRAL SYSTEMS AND EXTENSIONS

R. CHANDRASEKARAN
University of Texas at Dallas, Richardson, TX 75080, USA

1. Introduction

In this paper we consider integer (linear) programs that have the property that: for every integral right-hand side vector for which the linear program (obtained by dropping the integrality requirement) has an optimal value for the objective function the optimal value of the integer program differs from that for the linear program by at most a fraction. This class properly includes totally dual integral (TDI) systems along with many other cases. We provide a unified basis for algorithms that involve the solving of one linear program for TDI systems. This work draws very much upon the work of D.R. Fulkerson especially on antiblocking systems.

More precisely, we consider two classes of integer programs that are well known and occur frequently. They are:

I: *Covering Problems:*

$$\min b^t y: \quad y \geq 0,\ A^t y \geq w,\ y \text{ int.}$$

II: *Packing Problems:*

$$\max b^t y: \quad y \geq 0,\ A^t y \leq w,\ y \text{ int.}$$

In both these problems, we assume A, b, and w are integral. If in addition A is 0/1 and $w \geq 0$ we have *set covering* and *set packing* problems respectively. The cases in which A is 0/1, $w \geq 0$, $b = e$ have received a great deal of attention, most notably in the work of Fulkerson et al. It is of considerable interest to know whether or not the integrality requirement can be dropped i.e. whether or not there exists an integral solution that is optimal to the linear program, whenever the linear program has an optimal solution. Even when this is the case, it is not easy to get an integer optimal solution; for example the simplex method might not produce it. Fulkerson gives an outline of an algorithm using the knowledge of extreme points of the dual polyhedron; this is done in the case of antiblocking (covering) systems. We extend this to the blocking (packing) systems, unify both and render the approach usable i.e. we do it without having to enumerate extreme points (this is quite possibly what Fulkerson intended in the first place). The hope is that this might also throw some light on the conditions necessary and sufficient for the existence of integral solutions that are optimal to the linear program for all values of the right-hand side for which the linear program has an optimal solution.

All undefined terms in this paper are exactly as in Fulkerson's papers [1–3]. We start by treating the case in which A is 0/1, $w \geqslant 0$, $b = e$. Then we generalize step-by-step until all cases are treated.

2. Set covering problems

Assumptions. (1) A is 0/1, w is non-negative integral, $b = e$.

(2) Problem Î has an integral optimal solution whenever it has an optimal solution.

$$\hat{\mathrm{I}}: \qquad \min b^t y: \quad y \geqslant 0, A^t y \geqslant w.$$

Definition. Let B be a matrix whose rows are the extreme points of $\mathscr{B} = \{x \mid Ax \leqslant e, x \geqslant 0\}$. B is called the *antiblocker of A* and conversely.

Results from Fulkerson [1–3]

Theorem 1. *Problem* I *has integer optimal solutions for all integral w for which is has optimal solutions iff, the antiblocker B of A is* 0/1.

Theorem 2. *If A and B are* 0/1 *matrices that form an antiblocking pair, then*

(i) $A_{i\cdot} \cdot B_{j\cdot} \leqslant 1 \; \forall i, j$.

(ii) $\{\min e^t y: A^t y \geqslant w, y \geqslant 0\} = \max_j (B_{j\cdot} \cdot w)$.

(iii) *If* $(\bar{A}, \bar{B})$ *are column submatrices of A and B generated by the same set of columns, then* $(\bar{A}, \bar{B})$ *is an antiblocking pair of matrices.*

Using these theorems *and the knowledge of B Fulkerson described the following algorithm for solving* Î *as well as* I.

Algorithm 1

Step 0: Let $k = 0$; $A^0 = A$; $B^0 = B$; $w^0 = w$; and to to Step 1.

Step 1: (a) Find $\Omega^k = \max_j (B^k_{j\cdot} \cdot w^k)$; if $\Omega^k = 0$ go to Step 2.

(b) Let $S^k = \{j: B^k_{j\cdot} \cdot w^k = \Omega^k\}$.

(c) Find

$$i^k \ni A^k_{i^k} \cdot B^k_{j\cdot} = 1, \quad j \in S^k,$$
$$\leqslant 1, \quad j \notin S^k.$$

Such an index exists, and corresponds to row i_*^k of the original matrix A.

(d) Let $\bar{w}^{k+1} = (w^k - A^k_{i^k \cdot})^+$; $[(x^+ = (x_1^+, \ldots, x_n^+)$ where $x_i^+ = \max(0, x_i)]$. If any component of $\bar{w}^{k+1} = 0$, delete that component of $\bar{w}^{k+1}$, the corresponding columns of A^k and B^k to get w^{k+1}, A^{k+1} and B^{k+1} respectively. Go to Step 1(a).

Step 2: Stop; an optimal solution to $\hat{\mathrm{I}}$ and I is $y^* = \sum_{j=1}^{k-1} e_{i_*^j}$.

Remarks. (1) It is assumed that to begin with $w^0 > 0$; for otherwise we simply drop redundant constraints and get a reduced system that does satisfy this condition.

(2) The set S^k of indices in Steps 1(b) and 1(c) denotes the set of extreme points that are optimal for the dual of $\hat{\mathrm{I}}$ at that stage. Hence we suggest the following modification of Algorithm 1 that circumvents the requirement that B be known explicitly in advance.

Algorithm 2

Step 0: Set $k = 0$, $A^0 = A$; $w^0 = w$ and go to Step 1.

Step 1: (a) Solve the linear program

$$\max w^k x^k: \quad x^k \geqslant 0, S^k \geqslant 0, A^k x^k + IS^k = e.$$

If this maximum value $= 0$ go to Step 2.

(b) Let the submatrix of (A^k, I) that corresponds to columns having a reduced cost factor of 0 be denoted by B^k.

(c) Find $i^k \ni A^k_{j^k \cdot} \cdot x = 1 \, \forall x \in \{x \mid x \geqslant 0, B^k x = e\}$. Such an index exists and corresponds to row i_*^k of the original matrix A.

(d) Let $\bar{w}^{k+1} = (w^k - A^k_{i^k \cdot})^+$; If any component of $\bar{w}^{k+1} = 0$ delete that component of $\bar{w}^{k+1}$, the corresponding column of A^k to get w^{k+1} and A^{k+1} and go to Step 1(a).

Step 2: Stop; an optimal solution to $\hat{\mathrm{I}}$ and I is $y^* = \sum_{j=1}^{k-1} e_{i_*^j}$.

Comparison of the two algorithms

It should be clear that Algorithm 2 is really a usable version of Algorithm 1. Steps 1(a) and (b) of Algorithm 2 give us precisely the set S^k of 1(b) and 1(c) of Algorithm 1; but without explicitly enumerating them. *Such an algorithm is in some sense a primal dual type algorithm except that it starts with an optimal solution to the dual instead of just a feasible solution in the primal-dual algorithm of linear programming.* As soon as a feasible solution to the primal is obtained, we stop.

Implementation of Steps 1(a), (b) *and* (c): While it may appear that Steps 1(a), (b) and (c) have to be repeated again and again, the process involves much less. First we wish to point out that an optimal solution at kth iteration of Step 1(a) remains optimal for $(k+1)$st step. Hence no further iterations are required in solving the linear programs of Step 1(a) except those involved in the first step. This is exactly what happens in Algorithm 1; the sets S^k keep enlarging (except for redundant ones that are thrown away). Step 1(b) is therefore quite easy and involves very little computation. Step 1(c) asserts that $\{x \geq 0, B^k x = e\} \Rightarrow A^k_{i^k} \cdot x = 1$. We already know from antiblocking theory that $A^k_{i^k} \cdot x \leq 1$ for $x \geq 0$, $B^k x = e$. So all we need to do is to make sure that $\{\min A^k_{i^k} \cdot x : x \geq 0, B^k x = e\} = 1$. This does not involve much work since we already have a feasible canonical form from Step 1(a) whose current basic solution is optimal if i^k is properly chosen.

3. Set packing problems

Assumptions. (1) Same as under set covering problems on A, w and b.

(2) Problem II has an integral optimal solution whenever it has an optimal solution.

$$\text{II:} \qquad \max b^t y: \quad y \geq 0, A^t y \leq w.$$

Definition. Let $\hat{B}$ be a matrix whose rows are the extreme points of $\{x \mid x \geq 0, Ax \geq e\} = \hat{B}$. Then $\hat{B}$ is called the *blocker* of A and conversely.

Results from Fulkerson [1–3]

Theorem 3. II *has the integer optimal solutions for all integral w for which it has optimal solutions only if* $\hat{B}$ *is* 0/1. *This condition, however, is not sufficient.*

Definition. (1) A *deletion* operation on j in a matrix A is that of dropping jth column and all the rows which have a non-negative entry in that column. The order in which deletion is carried out on a set of columns is irrelevant and we may talk of deleting a subset of columns.

(2) A *contraction* operation on a column j in matrix A is that of dropping jth column of A. Again we may talk of *contracting* a subset of columns of A.

(3) The order in which successive deletions and contractions are carried out is irrelevant as long as the set of columns contracted and the set deleted remains the same.

(4) By $A(I_1, I_2)$ we mean the matrix obtained by deleting I_1 and contracting I_2; $(I_1 \cap I_2 = \emptyset)$. Such a matrix is called a *minor* of A.

Theorem 4. *If* A *and* $\hat{B}$ *are blockers, then* $A(I_1, I_2)$ *and* $\hat{B}(I_2, I_1)$ *are blockers, blocker of a minor of* A *is a minor of the blocker of* A.

Theorem 5. *If* A *and* $\hat{B}$ *are* 0/1 *matrices that form a blocking pair, then*

(i) $A_{i\cdot} \cdot \hat{B}_{j\cdot} \geq 1 \ \forall i, j$.

(ii) $(\max e^t y\colon y \geq 0, A^t y \leq w) = \min_j \hat{B}_{j\cdot} \cdot w$.

Because of the negative result contained in Theorem 3 Fulkerson did not pursue this case at all. We present one more result and use these to get an algorithm similar to Algorithm 1 for covering problems. This uses the knowledge of $\hat{B}$.

Theorem 6. *If Problem* II *has integer optimal solutions for all integral* w *for which it has optimal solutions, then* II *with* A *replaced by* $A(I_1, I_2)$ *has the same property.*

Proof. Let the components of w corresponding to I_1 be set to 0 and those corresponding to I_2 be set ∞.

Algorithm 3

Step 0: Set $k=0$, $A^0 = A$, $\hat{B}^0 = \hat{B}$; $w^0 = w$ and to to Step 1.

Step 1: (a) Find $\Omega^k = \min_j \hat{B}^i_{j\cdot} \cdot w^k$; If $\Omega^k = 0$ go to Step 2.

(b) Let

$$\begin{aligned} \hat{B}^k_{j\cdot} \cdot w^i &= \Omega^k, & j \in S^k, \\ &= \Omega^k + \delta^k_j, & j \notin S^k \ \delta^k_j \geq 1. \end{aligned}$$

(c) Find

$$\begin{aligned} i^k \ni A^k_{i^k} \cdot \hat{B}^k_{j\cdot} &= 1, & j \in S^k, \\ &\leq \delta^k_j + 1, & j \notin S^k. \end{aligned}$$

Such an index exists and corresponds row i^k_* of the original matrix A.

(d) Let $\bar{w}^{k+1} = (w^k - A^k_{i^k\cdot})^+$. If any component of $\bar{w}^{k+1} = 0$, delete that component to get w^{k+1} and perform a deletion operation on those indices in A^k to get A^{k+1}; perform a contraction on those indices in $\hat{B}^k$ to get $\hat{B}^{k+1}$. Go to Step 1(a) with A^{k+1}, $\hat{B}^{k+1}$ and w^{k+1}.

Step 2: Stop; an optimal solution to II is $y^* = \sum_{i=1}^{k-1} e_{i^i_*}$.

Our attempts at developing an algorithm similar to Algorithm 2 for set packing have not been successful. However, in the process we were able to discover the following type of algorithm for both these problems and extend these to other problems as well. Indeed, the idea is very simple and would be the very first to occur to anyone dealing with this set of problems. It would also explain the success of enumeration schemes on set packing and set covering problems at least for this class.

Algorithm 4 (for Problem Î and I)

Step 0: Set $k=0$; $A^0=A$; $w^0=w$ and go to Step 1.

Step 1: (a) Solve the linear program

$$\max w^k \cdot x: x \geqslant 0; A^k x \leqslant e.$$

Let this maximum value equal Ω^k; If $\Omega^k=0$ go to Step 2.

(b) Solve the linear program (not from scratch)

$$\max(w^k - A^k_{1.}) \cdot x: \quad x \geqslant 0, A^k x \leqslant e.$$

Let this maximum value equal $\bar{\Omega}^k$.

(c) If $\bar{\Omega}^k \neq \Omega^k - 1$, then drop $A^k_{1.}$ of A^k to get A^{k+1} let $w^{k+1}=w^k$ and go to Step 1(a).

(d) If $\bar{\Omega}^k = \Omega^k - 1$, then let $\bar{w}^{k+1} = (w^k - A^k_{1.})^+$. If any components of $\bar{w}^{k+1} = 0$, delete these to get w^{k+1}; contract the corresponding columns of A^k to get A^{k+1}; go to Step 1(a).

Step 2: Stop; each time Step 1(d) was used, the row $A^k_{1.}$ corresponds to row i^k of original matrix A. An optimal solution to Î and I is given by $y^* = \sum_{j=1}^{k-1} e_{i^j}$.

Algorithm 5 (for Problem ÎI and II)

Step 0: Set $k=0$; $A^0=A$; $w^0=w$; and go to Step 1.

Step 1: (a) Solve the linear program:

$$\min w^k \cdot x: \quad x \geqslant 0, A^k x \geqslant e.$$

Let this minimum value equal Ω^k. If $\Omega^k=0$ to to Step 2.

(b) Solve the linear program (not from scratch)

$$\min(w^k - A^k_{1.} \cdot x: \quad x \geqslant 0, A^k x \geqslant e.$$

Let this minimum value be equal to $\bar{\Omega}^k$.

(c) If $\Omega^k \neq \bar{\Omega}^k - 1$, then drop $A^k_{1.}$ of A^k to get A^{k+1}; let $w^{k+1}=w^k$ and go to Step 1(a).

(d) If $\bar{\Omega}^k = \Omega^k - 1$, let $\bar{w}^{k+1} = (w^k - A^k_{1.})^+$. If any component of $\bar{w}^{k+1}=0$, delete it to get w^{k+1}; delete the corresponding colums of A^k to get A^{k+1}; go to Step 1(a).

Step 2: Stop; each time Step 1(d) was used, the row $A^k_{1\cdot}$ corresponds to row i^k of the original matrix A. An optimal solution to $\hat{\text{I}}$ and II is given by $y^* = \sum_{j=1}^{k-1} e_{i^j}$.

Remark. These algorithms at first sight may seem to involve the solution of many linear programs, and hence quite complicated. But really not much work is involved. The reason is outlined in the following results. They also show why these algorithms work.

Discussion on Algorithm 4

Lemma 1. *Let an optimal solution at Step* 1(a) *in the kth cycle be* x^k. *(We might as well suppose* x^k *is an extreme point of the appropriate polyhedron.) If* $\bar{\Omega}^k = \Omega^k - 1$, *then* x^k *is also optimal to the linear program in Step* 1(b).

Proof.

$$\begin{aligned}(w^k - A^k_{1\cdot})x^k &= w^k x^k - A^k_{1\cdot} \cdot x^k \\ &= \Omega^k - A^k_{1\cdot} \cdot x^k \\ &\geq \Omega^k - 1 \\ &= \bar{\Omega}^k \qquad \text{(under hypotheses of the lemma).}\end{aligned}$$

Also $A^k x^k \leq e$ since x^k is feasible to linear program in Step 1(a).

Corollary 1. *If* x^k *is an optimal solution to the linear program in Step* 1(a) *at kth cycle, and kth cycle does not go to Step* 1(d), *then* $x^{k+1} = x^k$.

Proof. Since kth cycle does not go to Step 1(d),

$$\bar{\Omega}^k \neq \Omega^k - 1.$$

It is clear that $\bar{\Omega}^k \geq \Omega^k - 1$ since

$$\begin{aligned}\bar{\Omega}^k &\geq (w^k - A^k_{1\cdot})x^k \\ &= \Omega^k - A^k_{1\cdot} x^k \geq \Omega^k - 1.\end{aligned}$$

Also $\bar{\Omega}^k \leq \Omega^k$ since

$$\begin{aligned}\bar{\Omega}^k &= (w^k - A^k_{1\cdot})x^* \\ &= w^k \cdot x^* - A^k_{1\cdot} \cdot x^* \\ &\leq w^k x^* \leq w^k x^k = \Omega^k,\end{aligned}$$

where x^* is optimal to the linear program in Step 1(b) at the kth cycle.

Combining these two results we have $\bar{\Omega}^k = \Omega^k$ and hence

$$w^k x^k = w^k x^* = w^k x^* - A^k_{1.} x^*.$$

Hence $A^k_{1.} x^* = 0$ and x^* is also optimal to linear program in Step 1(a) at the kth cycle. Since for x^* the constraint $A^k_{1.} x^* \leqslant 1$ is not tight, x^* is also optimal for the linear program at Step 1(a) in the $(k+1)$st cycle. This then implies x^k is also optimal to this linear program.

Corollary 2. *If x^k is an optimal solution to the linear program at Step* 1(a) *in the kth cycle and the kth cycle goes to Step* 1(d), *then again $x^{k+1} = \hat{x}^k$ ($\hat{x}^k$ is obtained by deleting components from x^k corresponding to deleted components of w^k).*

Proof. Since kth cycle did go to Step 1(d) we have $\bar{\Omega}^k = \Omega^k - 1$. Also

$$(w^k - A^k_{1.}) x^k = \Omega^k - A^k_{1.} x^k \geqslant \Omega^k - 1$$

Hence

$$(w^k - A^k_{1.}) x^k = \Omega^k - 1 \quad \text{and} \quad A^k_{1.} x^k = 1.$$

Hence

$$w^{k+1} \cdot \hat{x}^k = \bar{w}^{k+1} \cdot x^k = \Omega^k - 1 = \bar{\Omega}^k = w^{k+1} \cdot x_{k+1}.$$

Since the solution to the linear program in Step 1(a) with $\bar{w}^{k+1}$ (and the extended matrix) is the same as to that with w^{k+1}. (They differ only by addition of some zeroes). Hence $\hat{x}^k$ is an optimal solution to the linear program in Step 1(a) at $(k+1)$st cycle.

Remark. Thus, Steps 1(a), (b), (c) and (d) involve very little work after the first cycle of 1(a). We wish to point out that although these optimal solutions are easy to find, it may take some iterations of the simplex method to establish the optimality or non-optimality of these solutions. But all these will be iteration at optimality. One could also further speed this up by clever devices. (See Algorithm 6.)

Discussion on Algorithm 5

This is completely parallel to that on Algorithm 4. We present it for the sake of completeness and due to the fact that some of the proofs are based on different arguments.

Lemma 2. *Let an optimal solution at Step* 1(a) *in the kth cycle be x^k. If $\bar{\Omega}^k = \Omega^k - 1$, then x^k is also optimal to the linear program in Step* 1(b).

Proof

$$(w^k - A^k_{1\cdot})x^k = w^k x^k - A^k_{1\cdot} x^k$$
$$= \Omega^k - A^k_{1\cdot} x^k \leqslant \Omega^k - 1 = \bar{\Omega}^k.$$

Also $A^k x^k \geqslant e$, since x^k is feasible to the linear program in Step 1(a).

Corollary 3. *If x^k is an optimal solution to the linear program in Step 1(a) at the kth cycle and kth cycle does not go to Step 1(d), then $x^{k+1} = x^k$.*

Proof. Consider $(A^k)^t y \leqslant w^k$; $y \geqslant 0$; max $e^t y$. Since A^k is a minor of A, this problem has an integer optimal solution whenever it has an optimal solution, y. *Case* 1: $y_1 \geqslant 1$. In this case

$$\Omega^k = \max e^t y: \quad (A^k)^t y \leqslant w^k,\ y \geqslant 0,$$
$$\bar{\Omega}^k = \Omega^k - 1 = \max e^t y: \quad (A^k)^t y \leqslant (w^k - A^k_{1\cdot}),\ y \geqslant 0.$$

Hence this cannot occur. Therefore in every integral optimal solution (such exist by above) $y_1 = 0$. Hence

$$\Omega^k = \max e^t y: \quad (A^{k+1})^t y \leqslant w^{k+1} = w^k;\ y \geqslant 0$$
$$= \min w^{k+1} \cdot x: \quad x \geqslant 0;\ A^{k+1} \cdot x \geqslant e.$$

But x^k satisfies $A^{k+1} \cdot x^k \geqslant e$, $x^k \geqslant 0$ and $w^{k+1} \cdot x^k = \Omega^k$: Hence $x^{k+1} = x^k$.

Corollary 4. *If x^k is an optimal solution to the linear program at Step 1(a) in the kth cycle and kth cycle goes to Step 1(d), then $x^{k+1} = \hat{x}^k$ obtained by dropping from x^k the components corresponding to the dropped elements in w^{k+1} as compared to $\bar{w}^{k+1}$.*

Proof. Since $\bar{\Omega}^k = \Omega^k - 1$ and

$$\bar{\Omega}^k = \max e^t y: \quad (A^k)^t y \leqslant (w^k - A^k_{1\cdot});\ y \geqslant 0,$$

and by assumption there is an integer optimal solution to this linear program, there is an integer optimal solution to

$$\max e^t y: \quad (A^k)^t y \leqslant w^k;\ y \geqslant 0$$

with $y_1 \geqslant 1$. Hence $A^k_{1\cdot} x^k = 1$. Hence

$$(w^k - A^k_{1\cdot})x^k = w^k x^k - A^k_{1\cdot} x^k$$
$$= \Omega^k - 1 = \bar{\Omega}^k$$
$$= \bar{w}^{k+1} \cdot x^k = w^{k+1} \cdot \hat{x}^k.$$

Since

$$w^{k+1}\cdot\hat{x}=\bar{w}^{k+1}x \quad \text{for any } x:\quad x\geqslant 0,\ A^{k+1}x\geqslant e,$$
$$=w^{k}x-A^{k}_{1.}x\leqslant\bar{\Omega}^{k}$$

it follows that $x^{k+1}=\hat{x}^{k}$.

Remark. Thus, once again Steps 1(a), (b), (c) and (d) involve very little work after the first cycle of 1(a).

General Case: So far we have assumed that A is 0/1, w is non-negative integral and $b=e$. We now turn to the general case: w integral, b integral and A integral. We also relax the second assumption. We now assume that the optimal values of the objective functions of I and Î differ by no more than a fraction whenever they exist (similar statements on II and II). Our algorithms are based on the following simple lemma.

Lemma 3. *If there is an optimal solution to* Î *with* $y_1\geqslant y_1^0$, *then there exists an optimal solution to* I *with* $y_1\geqslant\lfloor y_1^0\rfloor$ *under the above assumptions.*

Proof. Let y^* be an optimal solution to Î satisfying the hypothesis of the lemma. Then $y^*-\lfloor y_1^0\rfloor e_1$ is an optimal solution to Î with the right hand side equal to $w-\lfloor y_1^0\rfloor A^{t}_{.1}$. Hence there exists an optimal solution $\hat{y}$ to I whose right hand side equals $w-\lfloor y_1^0\rfloor A^{t}_{.1}$ such that the objective value equals $\lceil b^{t}(y^*-\lfloor y_1^0\rfloor e_1)\rceil$. Hence $(y+\lfloor y_1^0\rfloor e_1)$ is the required integral solution to I whose right hand side is w.

Algorithm 6 (general algorithm for problem I)

Assumptions: (1) A, b, w integral.

(2) The difference between the objective values for a pair of optimal solutions to I and Î is strictly less than 1 for all w for which optimal solutions exist for Î.

Step 0: Solve Î; let the objective value equal Z^*.

Step 1: Solve the linear programs

$$\begin{aligned}&\max\ y_j,\\ &\text{s.t.}\ \ A^{t}y-IS=0,\\ &\qquad b^{t}y=0,\\ &\qquad y\geqslant 0,\ S\geqslant 0,\ y_j\leqslant 1\end{aligned}$$

for $j = 1, 2, \ldots, m$ and the linear programs

$$\max S_i,$$
$$\text{s.t.}\quad A^t y - IS = 0,$$
$$b^t y = 0,$$
$$y \geq 0,\ S \geq 0,\ S_i \geq 1$$

for $i = 1, 2, \ldots, n$. Let the corresponding optimal solutions be $(y^0, S^0)^j$ and $(\bar{y}, \bar{S})^i$ respectively. Let

$$(y, S) = \sum (y^0, S^0)^j + \sum (\bar{y}, \bar{S})^i + (0, \ldots, 0, y_{k+1}, \ldots, y_m; 0, \ldots, 0, S_{(l+1)}, \ldots, S_n)$$

(after rearranging the coordinates) where $y_{k+1}, \ldots, y_m$ and $S_{l+1}, \ldots, S_n$ are positive. Go to Step 2.

Step 2: Set $j = 1$, $w^1 = w$; and $z_1 = \lceil Z^* \rceil$ and go to Step 3.

Step 3: Solve the (bounded) linear program

$$A^t y - IS = w^j,$$
$$b^t y = Z_j,$$
$$y \geq 0,\ S \geq 0,$$
$$\max y_j.$$

Let the value of the objective be y_j^0. If $j = k$, let $\bar{w} = w^j - \lfloor y_j^0 \rfloor A_j$ and $\bar{Z} = Z_j - b_j \lfloor y_j^0 \rfloor$ and go to Step 4. If $j \leq k$, increase j by 1, define $w^{j+1} = w^j - \lfloor y_j^0 \rfloor A_j$. and $Z_{j+1} = Z_j - b_j [y_j^0]$ and go to Step 3.

Step 4: Set $l = 1$, $w^l = \bar{w}$, and go to Step 5.

Step 5: Solve the (bounded) linear program

$$A^t y - IS = w^l,$$
$$b^t y = \bar{Z},$$
$$y \geq 0,\ S \geq 0,$$
$$\max S_l.$$

Let the objective value be S_l^0. If $l = r$, let $\bar{\bar{w}} = w - \lfloor S_l^0 \rfloor e_l$ and go to Step 6. If $l < r$, increase l by 1, define $w^{l+1} = w^l - \lfloor S_l^0 \rfloor e_l$ and go to Step 5.

Step 6: Solve the system

$$A^t y - IS = \bar{\bar{w}},$$
$$b^t y = \bar{Z},$$
$$y, S \text{ integer},\ y_j = 0,\ 1 \leq j \leq k;\ S_i = 0,\ 1 \leq i \leq l$$

by the use of Smith normal form. Let the solution be $(\hat{y}, \hat{S})$. [If such a solution does not exist, terminate; Assumption 2 is not satisfied.] Go to Step 7.

Step 7: Let $\theta \geqslant 0 \ni \theta(y, S)$ be integral and $\theta(y, S) + (\hat{y}, \hat{S}) \geqslant 0$, where (y, S) is the solution from Step 1. [Such a θ does exist and is easy to find.] Define

$$(y^*, S^*) = \theta(y, S) + (\hat{y}, \hat{S}) + (\lfloor y_1^0 \rfloor, \ldots, \lfloor y_k^0 \rfloor, 0, \ldots, 0; \lfloor S_1^0 \rfloor, \ldots, \lfloor S_l^0 \rfloor, 0, \ldots, 0)$$

(y^*, S^*) is the required optimal solution to I. Stop.

Discussion of Algorithm 6

Lemma 3 offers a quick way of building up the optimal vector for problem I. Instead of increasing a variable's value one at a time, we can fix the optimal value by solving one linear program. This is possible only if there is a bound to the value of the variable among the optimal solutions to Î. Step 1 identifies the variables that do not have such a bound. These are the variables $y_{k+1}, \ldots, y_m$ and $S_{l+1}, \ldots, S_n$. Thus the value of the remaining variables can be obtained by using the logic in Lemma 3 and this is done in Steps 3 and 4 of the algorithm. There is an infinite set of values for the variables $y_1, \ldots, y_k$ and $S_1, \ldots, S_l$ (if one exists) and one such set is found by Steps 6 and 7. Since Lemma 3 justifies Steps 3 and 4, if there is no solution at Step 6, it is clear that no nonnegative solution exists with $y_j \geqslant \lfloor y_j^0 \rfloor, i = 1, \ldots, k$ and $S_i \geqslant \lfloor S_i^0 \rfloor, i = 1, \ldots, r$; but this contradicts the assumption and Lemma 3. On the other hand, the existence of a solution at Step 6, also gives us a nonnegative solution at Step 7. While it may appear that several linear programs have to be solved at Step 1, one could save some work by not solving the linear programs for those j for which the previous solution has a positive y_j, and a similar strategy is to be used for the S_i as well. A similar algorithm can be developed for problem II under similar assumptions but is not necessary since problem II under such assumptions can be converted to a case of problem I.

Conclusion

The resemblance to enumeration schemes such as branch and bound and implicit enumeration using LP is striking especially for the latter algorithms. Indeed they are proper implementations of these algorithms under the hypotheses we began with. Since Algorithm 6 is the general one, we restrict our remarks to its efficiency. Steps 0, 1, 3 and 5 involve at most $2(m+n)+1$ linear programs. (Note that this is a generous bound.) For those interested in polynomialness, this work is polynomial if one uses the Russian algorithms [5]. (Note that the data of these problems do not grow very much.) For those interested in practical efficiency the work involved is not much if we used the simplex method. (Note that the constraint matrix is the same everytime.) Step 6 involves finding the Smith normal form whose effort is polynomial in the input size (see for example [4]) and this algorithm is also practically efficient. Step 7 involves very little work. Thus, these algorithms is reasonably efficient. This may explain the success of enumeration algorithm for these problems. The closeness of the integer optimal value to the linear optimal value eliminates quite a bit of the search. This is not too surprising. These algorithms are of a primal dual variety: the only difference between these and primal dual algorithms for linear programs is that we use optimal dual solutions in building up optimal primal solutions. We hope that these procedures will lead to some characterizations of the conditions necessary to assure the hypotheses under which these algorithms work.

References

[1] D.R. Fulkerson, Blocking polyhedra in: B. Harris, ed., Graph Theory and its Applications (Academic Press, New York, (1970) 93–111.
[2] D.R. Fulkerson, Blocking and antiblocking pairs of polyhedra, Math. Programming 1 (1971) 168–194.
[3] D.R. Fulkerson, Antiblocking polyhedra, J. Combinatorial Theory 12 (1972) 50–71.
[4] R. Kannan and A. Bachem, Polynomial algorithms for computing the Smith and Hermite normal forms of an integer matrix, SIAM J. Comput. 8 (1979) 499–507.
[5] L.C. Khachian, A polynomial algorithm in linear programmings, Doklady Adademii Nauk SSSR 244 (1979) 1093–1096; English translation: Soviet Math. Dokl. 20 (1979) 191–194.

P. Hansen, ed., Studies on Graphs and Discrete Programming

MINIMAL COST-RELIABILITY RATIO SPANNING TREE*

R. CHANDRASEKARAN

School of Administration and Management, University of Texas at Dallas, Richardson, TX 75080, U.S.A.

Y.P. ANEJA and K.P.K. NAIR

Faculty of Adminstration, University of New Brunswick, Fredericton, N.B., E3B 5A3, Canada

The problem of finding a minimal cost-reliability ratio spanning tree in a network is considered. The optimal solution to this problem is shown to map into an efficient extreme point of the convex hull of spanning trees in a bicriteria problem of minimizing the negative of the logarithm of reliability and cost. Based on this result a polynomial algorithm consisting of an indirect search (without requiring to find logarithms) in the set of efficient extreme points is given for computing the solution to the cost-reliability ratio problem.

1. Introduction

Let $[N; E]$ be an undirected network in which for each edge $i \in E$ there are two numbers, namely a non-negative integral cost and a positive probability of functioning. We consider the problem of finding a minimal cost-reliability ratio spanning tree. Given a weight for each edge a minimal (maximal) spanning tree can be found efficiently [3, 5, 6]. Further, given a probability of functioning for each edge, a maximally reliable spanning tree also can be found by the same algorithm available for the maximal spanning tree [5]. If both the numerator and denominator in the ratio are linear, the minimal (maximal) ratio spanning tree problem can be solved by a polynomial algorithm as shown first by Chandrasekaran [1] and subsequently in a more general context by Megiddo [4]. However, these approaches are not applicable to the present problem because of the nonlinearity in the denominator. However, the problem is of practical interest in the design of communication networks [2], in addition to its theoretical value. Linearizing the denominator by a logarithmic operation is not helpful since the problem so obtained does not solve the original one. We study the problem with the help of a bicriteria spanning tree problem in which the objectives are minimization of cost and the negative of the logarithm of reliability. It is shown that an optimal solution to the original problem is an

* This work was partially supported by NSERC Grant A3368 and UNB Grant 12-1.

efficient extreme point of the convex hull of spanning trees in the bicriteria space. The algorithm involves an implicit search, without finding any logarithm, in the set of efficient extreme points of the convex hull. From a recent work of Chandrasekaran [1] it follows that the number of efficient extreme points of the convex hull is polynomially bounded, and therefore our algorithm for the minimal cost-reliability ratio spanning tree is polynomially bounded.

2. The problem and certain properties

In $[N; E]$ for each edge $i \in E$ let p_i $(0 < p_i \leqslant 1)$ and c_i, respectively, be the probability of functioning and the nonnegative integral cost. Let Ω be the set of all spanning trees in $[N; E]$ and for each $T \in \Omega$ define

$$C(T) = \sum_{i \in T} c_i, \qquad R(T) = \prod_{i \in T} p_i, \qquad S(T) = \sum_{i \in T} \sigma_i \tag{1}$$

where $\sigma_i = -\ln \cdot p_i$.

Now the problem that we consider is

$$\operatorname*{Min}_{T \in \Omega} m(T) = C(T)/R(T). \tag{2}$$

Associated with (2) we define two bicriteria spanning tree problems as follows:

$$\operatorname*{Min}_{T \in \Omega}(-R(T), C(T)) \tag{3}$$

and

$$\operatorname*{Min}_{T \in \Omega}(S(T), C(T)). \tag{4}$$

It should be noted that (3) is obtained from (2) by replacing the fractional objective by the two objectives given by the negative of the denominator and the numerator. Also (4) is derived from (3) by replacing $-R(T)$ by $S(T)$ defined in (1).

Appealing to the concept of efficient solutions [7] we observe that with respect to the objectives in (3) a tree $\hat{T} \in \Omega$ is efficient if and only if there does not exist a tree $\tilde{T} \in \Omega$ such that $-R(\tilde{T}) \leqslant -R(\hat{T})$ and $C(\tilde{T}) \leqslant C(\hat{T})$ with a strict inequality in at least one case. An identical definition holds with respect to the objectives in (4). Now from the logarithmic relation between $R(T)$ and $S(T)$ defined in (1) it follows that a tree $\hat{T} \in \Omega$ is efficient with respect to the objectives in (3) if and only if $\hat{T}$ is efficient with respect to the objectives in (4). Let H denote the convex hull of all spanning trees in the bicriteria space (SC-plane) of (4). The mapping of a tree T_k in the SC-plane will have its coordinates as $D(T_k)$ and $C(T_k)$ and for convenience of notation these, respectively, are denoted by S_k and C_k. In general the coordinates of a point

$h \in H$ will be represented by S_h and C_h. A point $\hat{h} \in H$ is efficient if and only if there does not exist a $\tilde{h} \in H$ such that $S_{\tilde{h}} \leqslant S_{\hat{h}}$ and $C_{\tilde{h}} \leqslant C_{\hat{h}}$ with a strict inequality in at least one case.

Theorem 1. *An optimal solution T^* in* (2) *with value $m(T^*)$ maps into an efficient extreme point of the convex hull H in the bicriteria space of* (4).

Proof. For any point $h \in H$ define $m_j = C_h/e^{-S_h}$. Clearly given h in the interior of H, then there exists a point $\hat{h}$ on the boundary of H such that $m_{\hat{h}} \leqslant m_h$. Now if $\bar{h}$ is a point on the boundary of H but not an extreme point, then there exist two extreme points k and l such that $S_{\bar{h}} = \alpha S_k + \overline{1-\alpha} S_l$ and $C_{\bar{h}} = \alpha C_k + \overline{1-\alpha} C_l$ where $0 < \alpha < 1$. Also assume

$$C_k/e^{-S_k} = m_k \leqslant C_l/e^{-S_l} = m_l \tag{5}$$

Now by convexity and (5)

$$e^{-(\alpha S_k + \overline{1-\alpha} S_l)} < \alpha e^{-S_k} + \overline{1-\alpha e^{-S_l}} < \alpha \cdot \frac{C_k}{m_k} + \overline{1-\alpha}\frac{C_l}{m_k};$$

equivalently

$$[\alpha C_k + \overline{1-\alpha} C_l]/e^{-[\alpha S_k + \overline{1-\alpha} S_l]} > m_k,$$

that is

$$m_{\bar{h}} = C_{\bar{h}}/e^{-S_h} > m_k.$$

Therefore, the minimum of m_h for $h \in H$ is attained at an extreme point of H. All extreme points of H are indeed mapped by trees. Moreover T^* is clearly an efficient tree in (3) and hence in (4) also. Therefore T^* maps into an efficient extreme point of H.

Similar to Theorem 1 it can be shown that there exists an optimal solution to (2) which maps into an efficient extreme point of the convex hull in the bicriteria space (RC-plane) of (3). However, it is not helpful in obtaining an algorithm since the mapping in the RC-plane involves a nonlinear function. Moreover, the result of Chandrasekaran [1] is not applicable directly to the convex hull in the RC-plane. Therefore, we will use only Theorem 1 as above.

Theorem 1 shows that T^* can be found by searching in the set of efficient extreme points of H in the SC-plane. A direct search in this set will need finding logarithms which is an infinite process. Therefore, we propose to develop an indirect search without the use of logarithms. Each efficient extreme point of H has an image in the RC-plane, the reliability-cost plane. The image of a tree T_k in the RC-plane will have it coordinate $R(T_k)$ and

$C(T_k)$, respectively, denoted by R_k and C_k. The search that we propose will be with the help of these images.

For all $i, j \in E$ set $a_{ij} = p_i^{c_j}$. It should be noted that since c_i for all $i \in E$ is a nonnegative integer, $p_i^{c_j}$ is obtained by a finite process. Using the notions of computer science this process is treated as a single operation while claiming polynomiality for the algorithm we propose. However, the algorithm will hold even without the integrality assumption.

Lemma 1. *Let T_k and T_l be two efficient trees that map into two efficient extreme points of H. Also assume $C_k > C_l$ and $S_k < S_l$. Then finding a minimal spanning tree with edge weights $d_i = (C_k - C_l)\sigma_i + (S_l - S_k)c_i$ is equivalent to finding a minimal spanning tree with edge weights*

$$w_i = \prod_{j \in T_l} (a_{ij}/a_{ji}) \cdot \prod_{j \in T_k} (a_{ji}/a_{ij}).$$

Proof. We need to show that $d_i \leq d_j$ if and only if $w_i \leq w_j$. Clearly,

$$(C_k - C_l)\sigma_i + (S_l - S_k)c_i \leq (C_k - C_l)\sigma_j + (S_l - S_k)c_j$$

iff

$$e^{(C_k - C_l)\sigma_i} \cdot e^{(S_l - S_k)c_i} \leq e^{(C_k - C_l)\sigma_j} \cdot e^{(S_l - S_k)c_j},$$

i.e.

$$p_i^{C_l - C_k}(R_k/R_l)^{c_i} \leq p_j^{C_l - C_k}(R_k/R_l)^{c_j},$$

i.e.

$$(p_i^{C_l}/R_l^{c_i})(R_k^{c_i}/p_i^{C_k}) \leq (p_j^{C_l}/R_l^{c_j})(R_k^{c_j}/p_j^{C_k}),$$

i.e.

$$w_i \leq w_j.$$

3. Algorithm and validation

In this section we provide a polynomial algorithm for computing T^* with value $m(T^*)$, and its validation. The algorithm consists of a search in the set of efficient extreme points of the convex hull H in (4). Obviously, all the efficient extreme points of H are on the efficient frontier of H. Now a systematic search is carried out along this efficient frontier. A set W whose elements are of the form (k, l) is introduced. Here $(k, l) \in W$ denotes two efficient extreme points h_k and h_l, respectively, mapped by trees T_k and T_l. The coordinates of h_k and h_l are respectively (S_k, C_k) and (S_l, C_l). An element $(k, l) \in W$ signifies that efficient extreme points if any, other than h_k and h_l, on the efficient frontier between h_k and h_l are yet to be examined. Set Q has its element $q_{(k,l)} = C_l/R_k$ and it allows to decide whether or not $(k, l) \in W$ can be discarded. Further,

$q_{(k,l)} \in F$ signifies that $(k, l) \in W$ is to be discarded. Initially both W and Q are empty.

Algorithm

Step 0: Set $r = 1$ and find

$$R_1 = \operatorname*{Max}_{T\in\Omega}\left[\prod_{i\in T} p_i\right] \quad \text{and} \quad C_1 = \operatorname*{Min}_{T\in\Omega}\left[\sum_{i\in T} c_i \,\middle|\, \prod_{i\in T} p_i = R_1\right]$$

attained at tree T_1. Find

$$\bar{C} = \operatorname*{Min}_{T\in\Omega}\left[\sum_{i\in T} c_i\right] \quad \text{and} \quad \bar{R} = \operatorname*{Max}_{T\in\Omega}\left[\prod_{i\in T} p_i \,\middle|\, \sum_{i\in T} c_i = \bar{C}\right]$$

attained at tree $\bar{T}$. If $\bar{C} = C_1$ and $\bar{R} = R_1$, terminate setting $T^* = T_1$ and $m(T^*) = C_1/R_1$. Otherwise set, $r = 2$, $T_2 = \bar{T}$, $C_2 = \bar{C}$, $R_2 = \bar{R}$, $m_1 = C_1/R_1$, $m_2 = C_2/R_2$, $q_{12} = C_2/R_1$ and $\hat{m} = \text{Min}(m_1, m_2)$ and the corresponding tree as $\hat{T}$.

Also, set $W = \{(1, 2)\}$ and $Q = \{q_{(1,2)}\}$. Go to Step 1.

Step 1: Choose any $(k, l) \in W$, and for all $i \in E$ find

$$w_i = \prod_{j\in T_l} (a_{ij}/a_{ji}) \cdot \prod_{j\in T_k} (a_{ji}/a_{ij}).$$

Find

$$V = \operatorname*{Min}_{T\in\Omega}\left[\sum_{i\in T} w_i\right].$$

Find

$$\bar{C} = \operatorname*{Min}_{T\in\Omega}\left[\sum_{i\in T} c_i \,\middle|\, \sum_{i\in T} w_i = V\right]$$

attained at $\bar{T}$ with $\bar{R} = \prod_{i\in\bar{T}} p_i$. If $\bar{C} = C_l$ and $\bar{R} = R_l$, set $W = W - \{(k, l)\}$ and $Q = Q - \{q_{(k,l)}\}$, and go to Step 5. Otherwise set $r = r + 1$, $T_r = \bar{T}$, $C_r = \bar{C}$, $R_r = \bar{R}$, $m_r = C_r/R_r$, $q_{(k,r)} = C_r/R_k$ and $q_{(r,l)} = C_l/R_r$. Also, set

$$W = [W - \{(k, l)\}] \cup \{(k, r), (r, l)\}$$

and

$$Q = [Q - \{q_{(k,l)}\}] \cup \{q_{(k,r)}, q_{(r,l)}\}.$$

If $m_r < \hat{m}$, set $\hat{m} = m_r$, $\hat{T} = T_r$ and go to Step 4; otherwise go to Step 2.

Step 2: If $q_{(k,r)} \geq \hat{m}$, set $W = W - \{(k, r)\}$ and $Q = Q - \{q_{(k,r)}\}$. Go to Step 3.

Step 3: If $q_{(r,l)} \geq \hat{m}$, set $W = W - \{(r, l)\}$ and $Q = Q - \{q_{(r,l)}\}$. Go to Step 5.

Step 4: Find $F=\{q_{(k,l)} \mid q_{(k,l)} \geqslant \hat{m}\}$. Set $W=W-\{(k,l) \mid q_{(k,l)} \in F\}$ and $Q=Q-F$. Go to Step 5.

Step 5: If $W \neq \emptyset$, go to Step 1. Otherwise terminate setting $T^*=\hat{T}$ and $m(T^*)=\hat{m}$.

A few remarks on certain steps in the algorithm are in order. Trees T_1 and $\bar{T}$ at Step 0 which appears only in the first iteration, and $\bar{T}$ at Step 1 of every iteration can be found by introducing a minor change in the algorithms [3, 5, 6] for computing minimal and maximal spanning trees in a network. Tree T_1 is found as follows. With edge weights as p_i for all $i \in E$ find a maximal spanning tree using Kruskal's algorithm [3]; but with a slight difference in the rules regarding the choice of an edge. While choosing an edge with weight p_i, examine all alternative edges with the same p-values and among these choose the one that has the least c-value. Analogous changes in the rules are adequate for finding the trees denoted by $\bar{T}$'s in Step 0 and Step 1 respectively. The validation of these remarks can be easily obtained by the method of contradiction.

Validation of the Algorithm: If T is a tree, let its mapping (S, C) in the SC-plane be denoted by β and the image of β in the RC-plane by γ. Thus $\gamma=(R, C)$, where the relation $e^{-S}=R$ holds.

If the algorithm terminates at Step 0, then there exists a tree which attains maximum reliability as well as minimal cost, and hence is the optimal solution to the problem. Equivalently, H has a unique efficient extreme point.

For the case when the algorithm does not terminate at Step 0, for each $(k, l) \in W$, define

$$\delta_{(k,l)}=\{(S, C) \mid S \geqslant S_k, C \geqslant C_l, (S, C) \leqslant \lambda(S_k, C_k)+(1-\lambda)(S_l, C_l); 0 \leqslant \lambda \leqslant 1\}$$

and

$$\Delta_{(k,l)} = \text{interior of } \delta_{(k,l)}.$$

Set

$$\Delta_W = \bigcup_{(k,l) \in W} \Delta_{(k,l)}.$$

Lemma 2. *The trees T_r $(r=1,2,\ldots)$ generated by the algorithm are efficient trees and their mappings in the SC-plane β_r $(r=1,2,\ldots)$ are distinct efficient extreme points of H. Moreover, for each $(k,l) \in W$, $S_k < S_l$ and $C_k > C_l$.*

Proof. Consider the start of Step 1 of the first iteration of the algorithm. At this point $W=\{(1,2)\}$. Clearly, the way T_1 and T_2 are defined, β_1 and β_2 are distinct efficient extreme points, for otherwise the algorithm would have terminated in Step 0. Also $S_1 < S_2$ and $C_1 > C_2$. Thus the lemma holds at the start of Step 1 of the first iteration. Now assume that the lemma holds at the start of Step 1 of some iteration. For any (k, l) chosen from W, by assumption, β_k and β_l are distinct efficient extreme points. Also, by convexity of H, for any two elements $(k_1, l_1) \in W$, and $(k_2, l_2) \in W$, $\Delta_{(k_1,l_1)} \cap \Delta_{(k_2,l_2)} = \emptyset$. Now by Lemma 1, finding a tree $\bar{T}$ which minimizes $\sum_{i \in T} w_i$ is equivalent to finding a tree which minimizes a positively weighted sum of the two objectives in (4). Thus $\bar{T}$ is an efficient spanning tree. Furthermore, the way T is defined at Step 1, $\bar{\beta}$ is an efficient extreme point of H. Also, if $\bar{\beta} \neq \beta_l$, then from noting that β_l and β_k are efficient extreme points and the convexity of H, it follows that $\beta_r (= \bar{\beta}) \in \Delta_{(k,l)}$ and $\beta_r \notin \Delta_{(p,q)}$ for $(p, q) \in W - \{(k, l)\}$. Thus β_r is distinct from the efficient points generated so far. Also $S_k < S_r < S_l$ and $C_k > C_r > C_l$. Thus the lemma holds at the start of Step 1 of the next iteration also. This completes the proof.

Lemma 3. *At any iteration, if there exists an efficient tree $\tilde{T}$, with mapping $\tilde{\beta}$ as an efficient extreme point of H, such that $m(\tilde{T}) < \hat{m}$, then $\tilde{\beta} \in \Delta_W$.*

Proof. Consider the first iteration. While entering Step 1, $W = \{1, 2\}$ and from the way T_1 and T_2 are defined, it follows that if $m(\tilde{T}) < \hat{m} = \min(m_1, m_2)$, then $\tilde{\beta} \in \Delta_W$. Thus the lemma holds at the start of the first iteration. Now assume that the lemma holds at the start of some iteration. Suppose (k, l) is chosen from W at Step 1. At this step, if $\bar{C} = C_l$ and $\bar{R} = R_l$, then for each T, $\beta \notin \Delta_{(k,l)}$, and the removal of (k, l) from W is justified and the lemma still holds. Otherwise $\bar{\beta} = \beta_r \in \Delta_{(k,l)}$. Efficient extreme points in $\Delta_{(k,l)}$, other than β_r, must be in $\Delta_{(k,r)} \cup \Delta_{(r,l)}$, for β_r is an efficient extreme point and H is convex. Therefore the removal of (k, l) and insertion (k, r) and (r, l) in W is justified. From Lemma 2, for $(k, l) \in W$, we have $S_k < S_l$ and $C_k > C_l$. Thus if a tree maps in $\Delta_{(k,l)}$, then $m(T) > C_l / e^{-S_k} = q_{(k,l)}$. So if $q_{(k,r)} > \hat{m}$, clearly $\Delta_{(k,r)}$ can not contain $\tilde{\beta}$ such that $m(\tilde{T}) < \hat{m}$. Thus the removal of (k, r) and or (r, l) from W as stated in Steps 2 and 3 does not affect the lemma. Removal of elements from set W at Step 4 is also justified for the same reasons. Thus the lemma holds at the next iteration of the algorithm also.

Lemmas 2 and 3 together establish that upon termination $\Delta_W = \emptyset$ since $W = \emptyset$, and therefore $T^* = \hat{T}$ and $m(T^*) = \hat{m}$.

Lemma 4. *The algorithm terminates in polynomial number of iterations and the order of the algorithm is* $O(|E|^3 \cdot \log |N|)$.

Proof. As revealed by Lemma 2, each efficient tree T_r generated by the algorithm maps into a distinct efficient extreme point of H. In an iteration of the algorithm if a distinct efficient extreme point of H is generated the cardinality of the set W may increase at most by one; otherwise it decreases by one. Thus the algorithm is finite. We show now that the efficient extreme points of H are polynomially bounded.

Chandrasekaran [1] has considered the problem of

$$\underset{T\in\Omega}{\text{Min}}\left[\sum_{i\in T} c_i - \lambda \sum_{i\in T} \sigma_i\right], \quad -\infty<\lambda<+\infty, \tag{6}$$

and has shown that an optimal spanning tree to this problem for a given λ remains optimal for all λ in a closed interval. Moreover the number of such closed intervals of λ is bounded by $O(|E|^2)$. Clearly, each efficient extreme point of the convex hull H in the bicriteria space in (4) is mapped by an optimal solution to (6) for some λ, $-\infty<\lambda<0$. This follows from the fact that solving (6) for any λ, $-\infty<\lambda<0$ is equivalent to finding a spanning tree that minimizes a positively weighted sum of the two objectives in (4). Thus the number of efficient extreme points of H is bounded by $O(|E^2|)$. Since the algorithm consists of a search in the set of efficient extreme points of H, clearly the number of iterations is bounded by $O(|E^2|)$. In solving a spanning tree problem as stated in Step 0 or Step 1, say using Kruskal's algorithm [3], we need to carry out some additional effort as remarked earlier. However this does not change the order of the spanning tree algorithm, that is, $O(|E|\log|N|)$. Thus our algorithm is bounded by $O(|E|^3\log|N|)$.

References

[1] R. Chandrasekaran, Minimal ratio spanning trees, Networks 7 (1977) 335–342.

[2] H. Frank and I.T. Frisch, Communication, Transmission, and Transportation Networks (Addison-Wesley, Reading, MA, 1971).

[3] J.B. Kruskal, On a shortest spanning subtree of a graph and the travelling salesman problem, Proc. Amer. Math. Soc. 7 (1956) 45–50.

[4] N. Megiddo, Combinatorial optimization with rational objective functions, Math. Operations Research (1979) 414–424.

[5] R.C. Prim, Shortest connection networks and some generalizations, Bell System Tech. J. 36 (1957) 1389–1401.

[6] A.C. Yao, An $O(|E|\log\log|V|)$ algorithm for finding minimum spanning tree, Information Processing Letters 4 (1975) 21–23.

[7] M. Zeleny, Linear Multiobjective Programming (Springer-Verlag, New York, 1974).

P. Hansen, ed., Studies on Graphs and Discrete Programming

A GRAPH THEORETIC ANALYSIS OF BOUNDS FOR THE QUADRATIC ASSIGNMENT PROBLEM

Nicos CHRISTOFIDES and M. GERRARD

Department of Management Science, Imperial College of Science and Technology, Exhibition Road, London SW7 2BX, England

We express the Quadratic Assignment Problem in terms of graph multiplication of a flow graph G^{f} with a distance graph G^{d}. By decomposing G^{f} into simpler graphs, we give a general unified procedure to calculate bounds. This procedure generates the two previously known bounds as special cases and also generates some new bounds. By enumerating all practical decompositions, we show that no better bounds can be computed in reasonable time—by solving linear assignment subproblems—other than the bounds pointed out in this paper.

1. Introduction

Consider s machines $1, \ldots, \alpha, \ldots, s$ with a known flow of material $f_{\alpha\beta}$ between every pair of machines (α, β). Let there be $t \geqslant s$ locations $1, \ldots, i, \ldots, t$ with known distances d_{ij} between every pair of locations (i, j). We will henceforth consider the flow and distance matrices to be symmetric.

An assignment of machines to locations is a one to one mapping ρ of the set of machines into the set of locations, so that $\rho(\alpha)$ is the location that machine α is assigned to.

The cost of a mapping ρ is defined as

$$z(\rho) = \sum_{\alpha<\beta} f_{\alpha\beta} d_{\rho(\alpha)\rho(\beta)}. \tag{1}$$

Given the two matrices $[f_{\alpha\beta}]$ and $[d_{ij}]$, the quadratic assignment problem (QAP) is that of finding a mapping ρ^* which minimizes $z(\rho)$ as given by (1).

The QAP appears in a number of spacial location problems such as the allocation of machines to locations—used above to introduce the QAP—the location of electronic components on circuit boards [13], the ordering of interrelated data on magnetic tape, etc. Other examples not involving spacial location, but which can be formulated as QAP's include the triangularization of economic input–output matrices [1], the minimization of average job completion time in machine scheduling [9] and extensions of the travelling salesman problem [8].

A survey of exact algorithms for the general QAP is given by Pierce and Crowston [12], and an improved algorithm is described in [3]. Exact algorithms, however, are unable to solve general QAP's of even moderate size [3].

All the exact algorithms described to date use branch and bound procedures, where at each node of the decision tree bounds are calculated by solving linear assignment problems. The basic reason why the exact algorithms perform badly, is that the quality of the bounds thus computed is, in general, poor and the computing time for the bound is high.

In this paper we concentrate on the question of calculating bounds. The QAP is recast in graph theoretic terms as the dot product of a distance graph with a flow graph. We then consider the decomposition of these graphs into elementary graphs for which the QAP can be solved exactly in a simple manner [4]. For each such decomposition a bound can be calculated. We examine, by means of equivalence classes, all the decompositions which can concievably be useful in computing bounds. This leads to a unifying procedure and the bounds proposed by Lawler [8], Gilmore [6], Land [7] and Gavett and Plyter [5] are generated in this way as special cases. Some new bounds are also produced. Although the quality of the new bounds is better than that of the existing ones, the main conclusion is that no better bounds—derived by solving linear assignment problems—exist for the QAP.

2. Graphical representation

A graph G is defined by the doublet (X, A) where X is a set of vertices and A a set of links. Unless otherwise specified we will use 'graph' to mean a 'non-directed graph without loops'. The terminology used is from [2].

Given a graph $G'=(X', A')$ with a link cost matrix $[c'_{ij}]$, $i, j \in X'$, an isomorphic graph $G''=(X'', A'')$ with a cost matrix $[c''_{ij}]$ and an adjacency preserving mapping ρ of X' onto X'', the *dot-product graph* is written—using the product operator $\dot{\rho}$—as:

$$G' \dot{\rho} G''$$

and is defined as the graph $G=(X, A)$ isomorphic to G' with costs given by

$$c_{ij} = c'_{ij} \cdot c''_{\rho(i)\rho(j)}.$$

The *value* of a graph $G=(X, A)$ is defined as:

$$V(G) = \sum_{(x_i, x_j) \in A} c_{ij}.$$

An *image* of a graph G' in a graph G'' is any partial subgraph of G'' which is

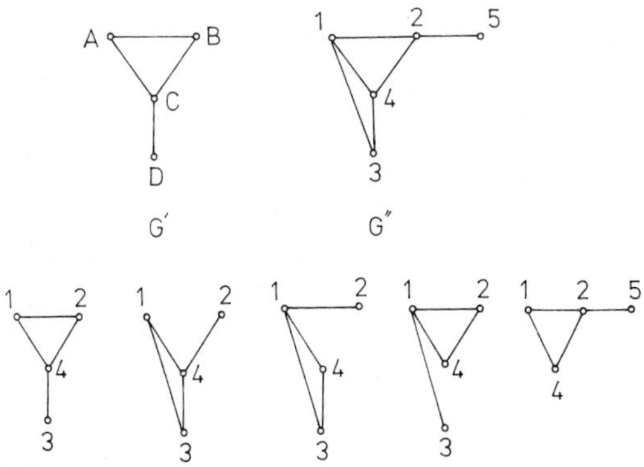

Fig. 1. Images of G' and G''.

isomorphic to G'. We will denote by $M(G', G'')$ the family of all such image graphs. The cardinality of the set $M(G', G'')$ is called the *image number* of G' in G''.

Fig. 1 shows all possible images of a graph G' in another graph G''.

The QAP can now be restated in the following way.

Let $G^f = (X^f, A^f)$ be a flow-graph, whose vertices X^f represent the set of machines and the link costs are the flows between the corresponding machines. Similarly, let $G^d = (X^d, A^d)$ be a distance-graph, whose vertices represent the set of locations and the link costs are the distances between the corresponding locations. We will assume (without loss of generality) that $|X^f| \leq |X^d|$.

The QAP is then the problem of finding a graph G and a mapping ρ of G^f on G which minimizes the expression:

$$\min_{G \in M(G^f, G^d)} \left[\min_{\rho} V(G^f \dot{\rho}\, G) \right]. \tag{2}$$

The number of different mappings ρ of G^f onto an isomorphic graph is the *isomorphic number* $s(G^f)$ of G^f. Thus, the inner minimization of (2) is over a set of cardinality $s(G^f)$ and the outer minimization is over a set of cardinality $|M(G^f, G^d)|$.

3. Bounds from graph decomposition

Let the flow graph G^f be decomposed into p distinct isomorphic partial graphs: $G_1^f, \ldots, G_p^f$ in such a way so that every link of G^f appears exactly the same number of times (say k) in the set of graphs $\{G_1^f, \ldots, G_p^f\}$.

Fig. 2 shows two possible decompositions of $G^f = K_5$ (the complete graph on 5 vertices). Fig. 2(b) shows a decomposition of G^f into 2 ($p = 2$) hamiltonian

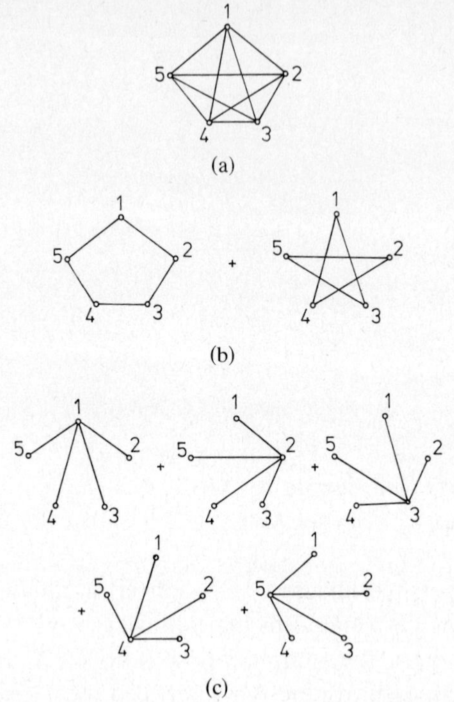

Fig. 2. (a) K_5; (b) decomposition into hamiltonian circuits; (c) decomposition into stars.

circuits H_1^f and H_2^f. Every link of G^f appears exactly once ($k=1$) in the set of graphs $\{H_1^f, H_2^f\}$. Fig. 2(c) shows a decomposition of G^f into 5 ($p=5$) star graphs $S_1^f, \ldots, S_5^f$. Every link of G^f appears exactly twice ($k=2$) in the set of graphs $\{S_i^f : i=1, \ldots, 5\}$.

Let the image number $M(G_1^f, G^d) = q$. In the example of Fig. 2(b)—if we also take $G^d = K_5$ (i.e. we consider a general QAP with 5 machines and 5 locations)—we have

$$q = |M(H_1^f, G^d)| = \frac{(5-1)!}{2} = 12,$$

i.e. q is the number of hamiltonian circuits in G^d.

In the example of Fig. 2(c) we have:

$$q = |M(S_1^f, G^d)| = 5,$$

i.e. q is the number of different stars in G^d.

Let us now assume that the optimum mapping of G_i^f into any one of its images in G^d can be computed and is of cost w_{ij} for image G_j^d. (For a given G_i^f and G_j^d a special case of a QAP is defined; and provided G_i^f and G_j^d are chosen

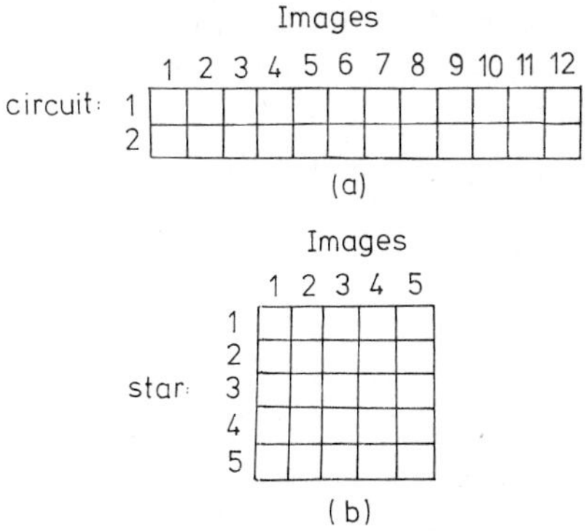

Fig. 3. $[w_{ij}]$ matrices with decomposition into (a) hamiltonian circuits and into (b) stars.

to be certain 'simple' graphs, e.g. trees, circuits, etc; polynomial-bounded algorithms for solving such special QAP's exist [4].) It is quite clear, that in the optimum solution to the general QAP every graph G_i^f into which G^f is decomposed, must be mapped into one and only one of its images in G^d. Thus, if V is the solution to the linear—and in general rectangular—assignment problem defined by $[w_{ij}]$, then V/k is a lower bound to the solution of the QAP. Hence, such a decomposition of G^f leads to a general procedure for calculating bounds.

Fig. 3 shows the matrices $[w_{ij}]$ resulting from the two decompositions shown in the example of Fig. 2.

3.1. *An equivalence relation*

Given a graph $G=(X, A)$, two vertices x_i and x_j are defined to be equivalent if and only if:

$$\{x_k : x_k \neq x_j, (x_i, x_k) \in A\} = \{x_k : x_k \neq x_i, (x_j, x_k) \in A\}.$$

For any graph, the above equivalence relation divides the vertices into one or more disjoint equivalence classes.

Consider now a decomposition of G^f into isomorphic partial graphs $G_1^f, \ldots, G_p^f$ and let us say that G_1^f has isomorphic number s and contains c equivalence classes with numbers of vertices $r_1, \ldots, r_c$. If $G^d = K_n$ the image number of G_1^f in G^d is:

$$q = \binom{n}{r_1} \cdot \binom{n-r_1}{r_2} \cdots \binom{n-\sum_{i=1}^{c-1} r_i}{r_c} \cdot \left[\frac{r_1! r_2! \cdots r_c!}{s}\right].$$

Since, in order to derive a bound, a linear assignment problem with p rows and q columns must be solved, this is only practical if p and q are reasonably small—certainly not much greater than n^2, say. Thus, in order to have a linear assignment probelm of reasonable dimensions, the graphs into which G^f is decomposed must have a maximum of $c = 2$ equivalence classes, and only the values of r_1, $r_2 = 1, 2, n-1, n-2$ could conceivably be useful for calculating a bound.

4. Enumeration of all practical decompositions

In this section we will enumerate all graphs with one or two equivalence classes each containing $1, 2, n-1$, or $n-2$ vertices. We illustrate these graphs (using $n = 7$ as an example) in Figs. 4, 5 and 6.

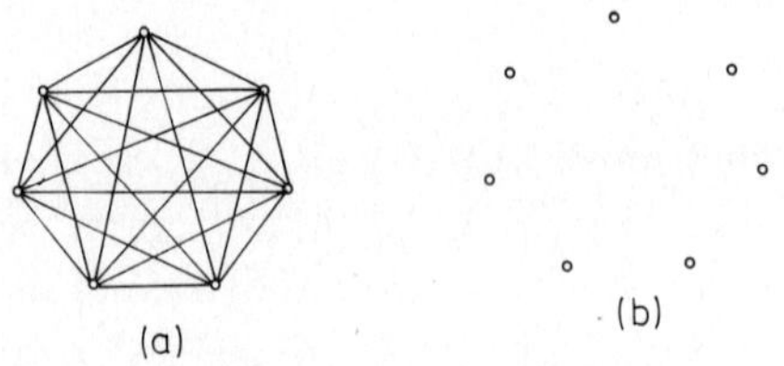

Fig. 4. Graphs with one equivalence class.

Fig. 4 shows the only two graphs with one equivalence class. Both are of no interest as far as the computation of bounds is concerned. The graph in Fig. 4(a) is the complete graph on n vertices and which, obviously, cannot be solved as a special case of the QAP since it is the general QAP itself. The graph in Fig. 4(b) is the trivial totally disconnected graph.

Fig. 5 shows the only two graphs with 2 equivalence classes with $r_1 = 1$ and $r_2 = n-1$. The graph in Fig. 5(a) is K_{n-1} with an additional isolated vertex, clearly much too difficult to be solved as a special case. The graph in Fig. 5(b) on the other hand, is the star graph which can easily be solved as a special case [4]. The bound resulting from the solution of the corresponding linear assignment problem is that proposed by Lawler [8] and Gilmore [7].

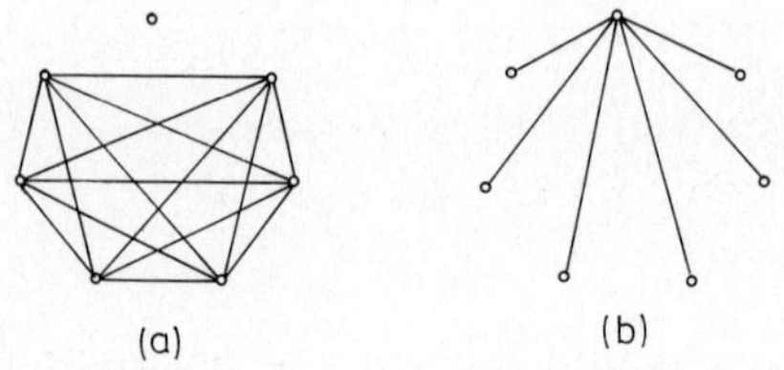

Fig. 5. Graphs with two equivalence classes and with $r_1 = 1$, $r_2 = n-1$.

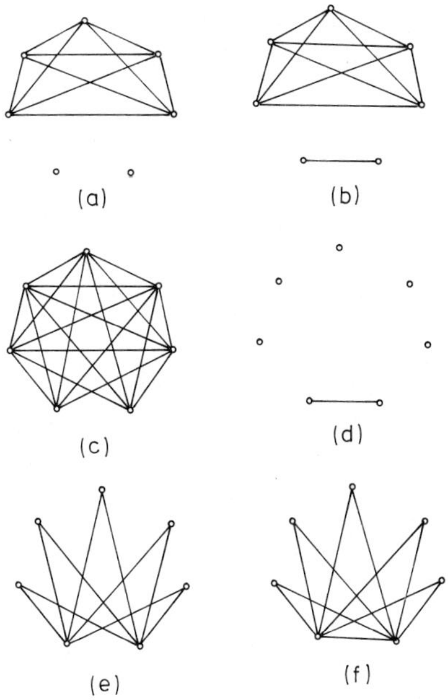

Fig. 6. Graphs with two equivalence classes and with $r_1 = 2$, $r_2 = n - 2$.

Fig. 6 shows the only 6 graphs with 2 equivalence classes with $r_1 = 2$ and $r_2 = n - 2$. The graphs in Figs. 6(a) to 6(c) are once more too complex to be special cases. The graph in Fig. 6(d) contains a single link. The bound resulting from the solution of the corresponding linear assignment problem is that proposed by Land [7] and Gavett and Plyter [5]. It should be noted here, that although in this case $p = q = \frac{1}{2}n(n-1)$, the coefficients of the resulting assignment problem are such that the value V of the solution can be obtained simply by ordering and without setting up the $\frac{1}{2}n(n-1)$ by $\frac{1}{2}n(n-1)$ assignment matrix. Thus, if the $\frac{1}{2}n(n-1)$ flows and distances are ordered in ascending and descending order, respectively, then V is simply the sum of the pairwise products of corresponding elements in the two lists. The graph in Fig. 6(e) corresponds to a 'double star' (i.e. two stars with different 'centers' but with the same 'end' vertices). The graph can easily be solved as a special case of the QAP, leading to a linear assignment problem of dimensions $p = q = \frac{1}{2}n(n-1)$. This bound requires considerably more computation time than the previous two bounds, but is considerably better in quality.[1] The graph in Fig. 6(f) is a trivial variation of the 'double star' of Fig. 6(e).

[1] A restricted number of tests indicate that the gap between the 'double star' bound and the optimum solution of general QAP's (with 12 vertices) is 45% of the gap between the 'single star' bound and the optimum solution.

5. Conclusion

We have expressed the QAP in terms of graph multiplication of a flow graph G^f with a distance graph G^d. By decomposing G^f into simpler graphs, we derive a general unified procedure to calculate bounds. This procedure generates the two previously known bounds as special cases and also generates some new bounds. By enumerating all practical decompositions we show that no better bounds within this class can be computed in reasonable time, other than those pointed out in the paper.

References

[1] V.J. Bowman, D.A. Pierce and F. Ramsey, A linear programming formulation of a special quadratic assignment problem, Carnegie-Mellon University, Managment Sciences Research Report No. 277 (1971).

[2] N. Christofides, Graph Theory—An Algorithmic Approach (Academic Press, London, 1975).

[3] N. Christofides and M. Gerrard, An improved algorithm for the QAP, Imperial College, Report MS 76-1 (1976).

[4] N. Christofides and N. Gerrard, Special cases of the QAP, Carnegie-Mellon University, W.P. 69-75-76, April 1976.

[5] J.W. Gavett and N.V. Plyter, The optimal assignment of facilities to locations by branch and bound, Operation Research 14 (1966) 210–232.

[6] P.C. Gilmore, Optimal and suboptimal algorithms for the quadratic assignment problem, J. SIAM 10 (1962) 305–313.

[7] A.H. Land, A problem of assignment with interrelated costs, Operational Res. Quart. 14 (1965) 185–199.

[8] E.L. Lawler, The quadratic assignment problem, Management Sci. 9 (1963) 586–594.

[9] W.L. Maxwell, The scheduling of single machine systems: a review, I.J. Prod. Res. 3 (1964) 177–199.

[10] J.M. Moore, Computer aided facilities design: an international survey, I.J. Prod. Res. 12 (1974) 21–44.

[11] C.E. Nugent, T.E. Vollman and J. Ruml, An experimental comparison of techniques for the assignment of facilities to locations, Operations Research 16 (1968) 150–173.

[12] J.F. Pierce and W.B. Crowston, Tree-search algorithms for the quadratic assignment problem, Naval Res. Logist. Quart. 18 (1971) 1–36.

[13] T. Pomentate, On minimization of backboard wiring functions, SIAM Review 9 (1967) 564–568.

[14] M. Scriabin and R.C. Vergin, Relative effectiveness of computer and manual methods for plant layout, Management Sci. 21 (1975) 564–568.

P. Hansen, ed., Studies on Graphs and Discrete Programming
© North-Holland Publishing Company (1981) 69–93

AN INVESTIGATION OF ALGORITHMS USED IN THE RESTRUCTURING OF LINEAR PROGRAMMING BASIS MATRICES PRIOR TO INVERSION

Kenneth DARBY-DOWMAN
Polytechnic of Central, London, 115 New Cavendish Street, London, W1M 8JS, England

Gautam MITRA
UNICOM Consultants, Brunel University, Uxbridge, Middx. UB8 3PH, England

Restructuring a basis matrix to create a sparse representation of the inverse is now established as an important algorithmic step in the solution of large scale LP problems. The most successful methods permute the basis matrix into block triangular form and Hellerman and Rarick were the first to propose practical algorithms. Since then the theory of block triangularisation has been studied extensively. The restructuring is achieved by first permuting the columns (or rows) of the basis matrix to produce a zero-free diagonal (maximum matching) and then finding the strong components of the directed graph associated with the permuted basis matrix. Of the many algorithms for the maximum matching problem the most established ones are those due to Hall and to Hopcroft and Karp. It is now accepted that the most efficient algorithm for finding strong components is due to Tarjan. In practice the triangular columns above and below the main block ('bump') can be isolated, thus allowing the block triangularisation algorithms to be applied only to the remaining bump. In this paper the implementation, including data structures, of several versions of algorithms used in the block triangularisation of basis matrices are described. A set of basis matrices from real-life LP problems are used to investigate and compare the behaviour of the algorithms.

1. Introduction

In modern linear programming (LP) codes the inverse of the basis matrix is not stored explicitly but is represented by the product form (PFI) or one of its variants. PFI is the product of a set of elementary matrices and for each such matrix only the non-zeros of the non-unit column (called an eta vector) and their positions need to be stored. An additional eta vector is added to the current set of vectors (called the eta file) at each iteration of the simplex method. As the eta file grows a point is reached at which the representation of the current inverse becomes 'inefficient'. Therefore a reinversion takes place from time to time in which the eta file is scrapped and the basis matrix is inverted from scratch to produce a sparser representation of the current inverse.

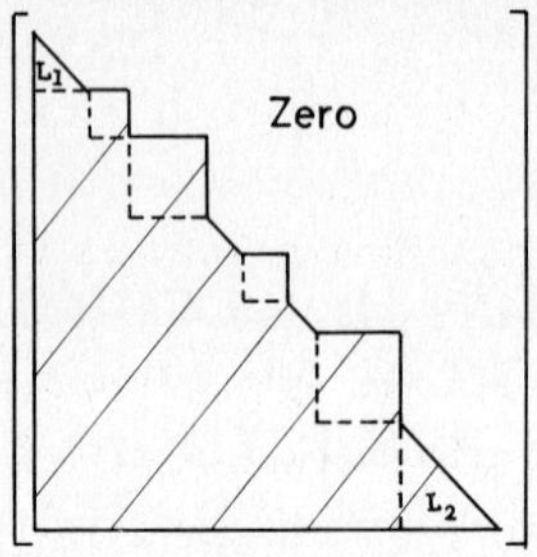

Fig. 1. A matrix in block triangular form.

Most basis matrices from large scale LP problems are very sparse (typically less than 1% dense). The sparsity of the new representation of the inverse depends on the non-zero structure of the basis matrix. By restructuring (i.e. permuting rows and columns) in a 'desirable' manner prior to inversion it is possible to obtain a much sparser representation of the inverse. A block triangular form, see Fig. 1, is such a desirable structure since non-zero build-up (fill-in) during inversion can be restricted to occur only within the blocks (the square diagonal sub-matrices).

The work described in this paper is applicable to the inversion of any sparse matrix and not just to sparse matrices arising in the field of linear programming. It should also be noted that restructuring to block triangular form is just one of many possible strategies. A review of algorithms for restructuring sparse matrices to other forms (e.g. minimum bandwidth) is given in [2].

The problem of restructuring matrices can be defined in graph theoretic terms and the next section gives some of the relationships between graphs and matrices. The methodology of block triangularisation and some practical considerations are given in Section 3. The block triangularisation procedure has two stages, finding a maximum matching followed by obtaining strong components. Section 4 describes several versions of algorithms to find a maximum matching. The implementation of a complete block triangularisation algorithm is described in Section 5 in the light of the data structures used. All the algorithms investigated are tested against a set of basis matrices generated during the solution of real life LP problems. Details of the experimental base and the results obtained are given in Section 6.

2. Graphs and matrices—terminology

A *directed graph*, G, has a set $V=\{1, 2, \ldots, m\}$, of vertices and a set, E, of edges each of which is an ordered pair (i, j) of vertices. Vertex l is *reachable* from vertex k if and only if there is a path in G from k to l. The *reachability matrix*, R, associated with G is an m-square (0, 1) matrix with element $(k, l) = 1$ if vertex l is reachable from vertex k. If every vertex in G is reachable from every other vertex then all the vertices in G are said to be *mutually reachable* and G is said to be *strongly connected.* The maximal strongly connected subgraphs of G are called the *strong components* of G.

Associated with an m-square matrix A with elements $\{a_{ij}\}$ is a directed graph consisting of a set of m vertices $(1, 2, \ldots, m)$ and a set of edges $\{(i, j)\}$ for which $a_{ij} \neq 0$. A symmetric permutation of A (PAP^{T} where P is a permutation matrix) does not affect the structure of the associated directed graph since the rows and columns of A are similarly reordered. The only change necessary to the graph is a relabelling of vertices and thus both graphs have an equivalent set of strong components.

Consider a matrix in block triangular form and its associated directed graph. It is clear from Fig. 1 that vertices associated with different blocks are not mutually reachable and hence are not in the same strong component. Vertices associated with any single block must be mutually reachable since otherwise the block could be reduced further into smaller blocks. (A more rigorous treatment of this point is contained in [1] and [3]). Thus a symmetric permutation of A to block triangular form is equivalent to finding the strong components of the directed graph associated with A.

3. Algorithms for block triangularisation

Preliminary discussion

To block triangularise a matrix A, permutation matrices P and Q are required such that PAQ is in block triangular form. P reorders the rows of A and Q reorders the columns. It can be shown [1, 3] that this can be done in 2 stages. First find a permutation matrix Q_1 such that each diagonal element of AQ_1 is a non-zero and then perform a symmetric permutation by finding P such that $P(AQ_1)P^{\mathrm{T}}$ is in block triangular form. The first stage is the well known problem of finding a maximum matching (or maximum transversal, system of distinct representatives etc.). The second stage is one of finding the strong components of the directed graph associated with the matrix AQ_1. Both problems have received extensive treatment over the past 20 years. It is, however, only recently that algorithms from graph theory have been applied to sparse matrix problems. A brief but comprehensive survey of sparse matrix research is given by Duff [2].

Most algorithms for finding a maximum matching are based on one devised by Marshall Hall [7] in 1956 (e.g. Ford and Fulkerson [5], Simmonard [13],

Yaspan [17] and Dulmage and Mendelsohn [4]). These algorithms are of complexity $O(m \cdot n)$ where m is the number of rows and n the number of non-zeros in the matrix. In 1973 Hopcroft and Karp [11] proposed an algorithm of complexity $O(m^{\frac{1}{2}} \cdot n)$. Both these complexity orders are, of course, obtained from a worst-case analysis and are not necessarily typical of the performance to be expected.

An early algorithm for finding the strong components of a directed graph was produced by Harary [8] in 1962. His approach involved computing the reachability matrix. The main disadvantage of this approach is that, apart from being expensive to compute, the reachability matrix is often quite dense even for very sparse matrices. Better algorithms have been developed in which the reachability matrix is not explicitly computed. These algorithms are based on the fact that vertices in the same strong component lie on a loop of the directed graph. The algorithms form these loops and as each loop is established the graph is condensed with the vertices of a loop being represented by a single vertex. The process is known as 'loop chasing' and the algorithms of Stewart [14, 15] and Sargent and Westerberg [12] are based on such an approach. Hellerman and Rarick's P^4 algorithm [10], designed specifically for restructuring LP basis matrices prior to inversion, is similar in that it exploits the fact that vertices in different strong components lie in different loops. More recently Tarjan [16] has developed an algorithm which uses a depth first search of the graph. It is extremely efficient for sparse matrices and is of complexity $O(m+n)$ compared with $O(m \log m + n)$ for the best of the other methods. Gustavson [6], has implemented the algorithm and concludes that it is 'near optimal' in both storage and operation count.

Practical considerations

The first stage of the block triangularisation procedure (that of finding a maximum matching) can be achieved by using either Hall's algorithm or that of Hopcroft and Karp. The efficiency of any algorithm depends very much on the quality of its implementation. Both these algorithms can be implemented in several ways. There is, however, very little published work on comparing such algorithms. The next section describes two versions of each algorithm as well as two versions of an algorithm which has taken certain features from each of the established algorithms.

It is now accepted that Tarjan's algorithm [16] for strong components is the most efficient of those currently available for carrying out the second stage of the block triangularisation procedure. Our implementation is very similar to that of Gustavson [6] and there is therefore no need to describe the algorithm or its implementation here.

The two stage block triangularisation method outlined earlier need not be applied to the complete basis matrix. Most LP bases contain a significant number of logical variables which, when the basis matrix has been restructured to block triangular form, contribute to the lower triangular sub-matrix labelled L_2 in Fig. 1. It is likely that many other columns of the basis matrix would be contained in L_2 and that the sub-matrix L_1 of Fig. 1 would also contain a significant number of columns. It is therefore possible to isolate L_1 and L_2 thus leaving a square sub-matrix (called the 'bump') to be block triangularised. Algorithms for isolating L_1 and L_2 are quite straight forward except that they require rowwise as well as columnwise access to the non-zeros of the basis matrix. The block triangularisation algorithms require only columnwise (or alternatively rowwise) storage of the matrix. It is therefore important to consider the data structures for the whole restructuring procedure rather than for each stage in isolation. Details of the implementation of the complete procedure, in the light of the data structures used, are given in Section 5.

4. Algorithms for maximum matching

In the context of column permuting a non-singular m-square matrix such that it has a non-zero diagonal, a maximum matching is a set of m non-zeros from the matrix such that there is one non-zero in each row and one in each column. Such a non-zero (specified by row and column number) is called an 'assignment'. An 'easy assignment' for a given column can be found if the column has a non-zero in a row that is not in the set of assignments already found. If an easy assignment is not possible then reassignments among the currently assigned columns have to be made. This is done by finding an 'augmenting path' which stretches the set of assignments by one. An augmenting path may be defined as $\{c_1, r_1, c_2, r_2, \ldots, c_k, r_k\}$ where each c_i is a column index and each r_i is a row index; c_1 is a currently unassigned column and r_k is a currently unassigned row; (c_i, r_i), $i = 1, 2, \ldots, k$, locates the position of a non-zero of the matrix and c_{i+1} is the current assignment in row r_i, $i = 1, 2, \ldots, (k-1)$. This augmenting path can be used to stretch the current set of assignments as illustrated below.

Old assignments		New assignments	
row	column	row	column
r_1	c_2	r_1	c_1
r_2	c_3	r_2	c_2
$\vdots$		$\vdots$	
r_{k-1}	c_k	r_{k-1}	c_{k-1}
		r_k	c_k

Hall's algorithm

Hall's algorithm processes the columns of the matrix sequentially. Consider column k:

"If there is an easy assignment in column k make such an assignment. Otherwise 'backtrack' through the columns already assigned to find an augmenting path from column k and make the necessary reassignments."

The procedure is repeated until all columns are assigned (this is possible since the matrix is assumed to be non-singular, see [1, 3]). The time-consuming task in this algorithm is that of finding an augmenting path each time a column is met which cannot be easily assigned. There is considerable flexibility in implementing such a procedure. The two basic approaches are:

(i) Find an augmenting path of shortest length (see implementation H1 in Section 5).

(ii) Find an augmenting path using a depth-first search of the columns already assigned (see implementation H2 in Section 5). This approach is also suggested by Gustavson [6].

In each case the process can be considered as the formation of a tree rooted at the column for which an assignment is to be made. An explicit search of the tree is not needed since just one path is required and the destination is known as soon as the tree is formed. The complete path can be simply derived from the structure of the tree. The complexity of Hall's algorithm to find and use an augmenting path is $O(n)$. The number of augmenting paths needed is $O(m)$ and thus the complexity of Hall's algorithm is $O(mn)$.

Hopcroft and Karp's algorithm

Hopcroft and Karp have attempted to improve on Hall's algorithm by finding several augmenting paths in a single pass of complexity $O(n)$. They have shown that the number of such passes needed is $O(m^{\frac{1}{2}})$ which leads to an overall algorithm complexity of $O(m^{\frac{1}{2}}n)$. The algorithm may be described as follows:

Step 1: Process the columns sequentially and find all the easy assignments.

Step 2: Construct the graph containing all the augmenting paths of shortest length from the set of all currently unassigned columns.

Step 3: Search the graph to find a maximal vertex disjoint set of augmenting paths. Use each path found to stretch the number of current assignments by one.

Step 4: Repeat Steps 2 and 3 until all columns are assigned.

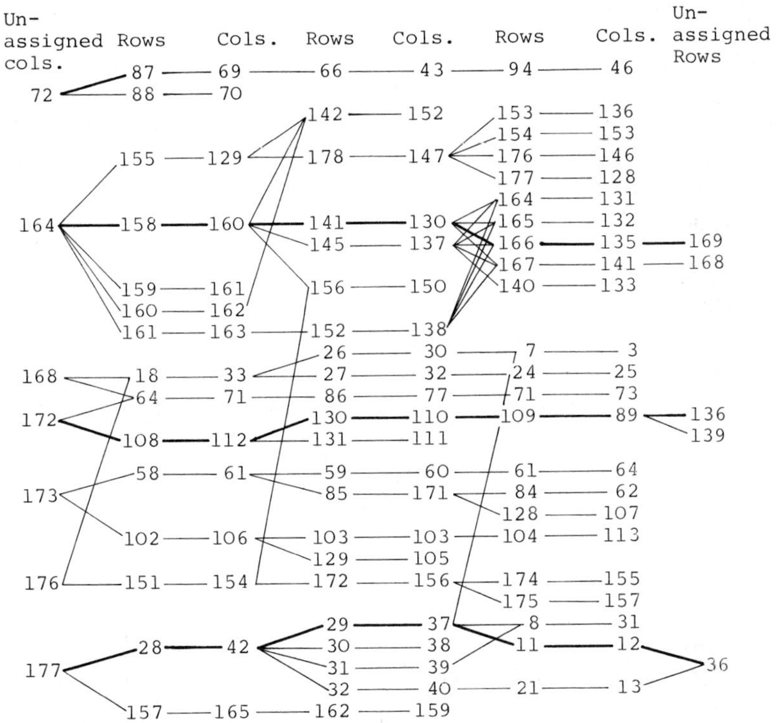

Fig. 2. An example of a graph formed by Hopcroft and Karp's algorithm.

Notes of Fig. 2

The heavy lines show a maximal vertex disjoint set of augmenting paths. The numbers shown in this graph are row and column numbers of non-zeros in the matrix. For example, column 72 has non-zeros in rows 87 and 88 (see the top left corner of the graph). Only the vertices corresponding to unassigned rows are shown in the last level of the graph.

In Step 1 all the easy assignments are made in one pass of the matrix. Consequently there are likely to be fewer non-easy assignments to be made, but of greater difficulty, than with Hall's algorithm. The search in Hall's algorithm is trivial (once the tree is formed). However in Hopcroft and Karp's algorithm, although fewer searches may be needed, each is of greater complexity. Thus it is by no means obvious which algorithm is the faster.

The flexibility of implementation of Hopcroft and Karp's algorithm lies in the type of search used to find a maximal vertex-disjoint set of augmenting paths in a graph. A depth-first search is efficient here but a decision has to be made as to whether the search is forwards (in the direction in which the graph was formed) (see implementation HK1 in Section 5) or backwards (see implementation HK2 in Section 5). Augmenting paths start with an unassigned column and end with an unassigned row. The number of unassigned rows in each graph will always be less than or equal to the number of unassigned columns. Fig. 2 shows an example of a graph formed during the execution of Hopcroft and Karp's algorithm and it can be seen that a backwards search, even though it is not as simple to implement as a forwards search, may prove to be generally more efficient.

A combined Hall–Hopcroft and Karp algorthm

In order to gain the attractive features of both Hall's and Hopcroft and Karp's algorithms the following algorithm is proposed:

Step 1: Process the columns sequentially and find all the easy assignments.

Step 2: For each non-assigned column: Backtrack through the columns already assigned to find an augmenting path and make the necessary reassignments.

Step 1 is taken from Hopcroft and Karp's algorithm and Step 2 from that of Hall. The choice of implementation depends, as in Hall's algorithm, on whether, in Step 2, an augmenting path is found of shortest length (see implementation HHK1) or by a depth-first search (see implementation HHK2).

5. Implementation of a block triangularisation algorithm

The stages of the algorithm can be illustrated as shown below

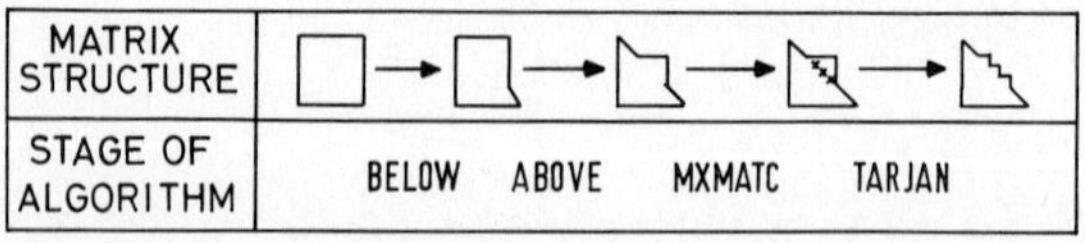

Let NNZA denote the number of non-zeros in the basis matrix and NRA the number of rows in the basis matrix.

Data structures for BELOW and ABOVE

It is assumed that initially the basis matrix is stored columnwise using an array IR0WN0 of length NNZA, which is a list of the row positions of the non-zeros, and an array ICPTR of length (NRA+1), where ICPTR(J) gives the position in IR0WN0 of the start of the data for column J.

The BELOW and ABOVE algorithms require that rows and columns of the matrix are 'deleted' and that row and column non-zero counts are updated. If only columnwise storage of the matrix is available then the locating of non-zeros within a row is slow since a complete scan of the matrix is needed for each such row. Rowwise storage can be achieved by the creation of two further arrays—IRPTR of length (NRA+1) and ICRPTR of length NNZA. IRPTR(*I*) gives the position in ICRPTR of the start of the data for row *I*. ICRPTR(*K*) refers to the *K*th non-zero of the basis matrix stored rowwise. It is a pointer to the position that this non-zero occupies in the array IR0WN0. To illustrate this data structure consider the following example.

$$
\begin{array}{cc}
 & \text{Column} \\
\text{Row} \begin{array}{c} \\ 1 \\ 2 \\ 3 \\ 4 \\ 5 \\ 6 \end{array} &
\left[\begin{array}{cccccc}
1 & 2 & 3 & 4 & 5 & 6 \\
X & & & & X & \\
X & & X & X & & X \\
 & X & & & & X \\
 & & & X & X & X \\
X & X & & & X & \\
 & X & & & &
\end{array}\right]
\end{array}
\qquad
\begin{array}{l}
X \text{ denotes a non-zero} \\
\text{NNZA} = 15 \\
\text{NRA} = 6
\end{array}
$$

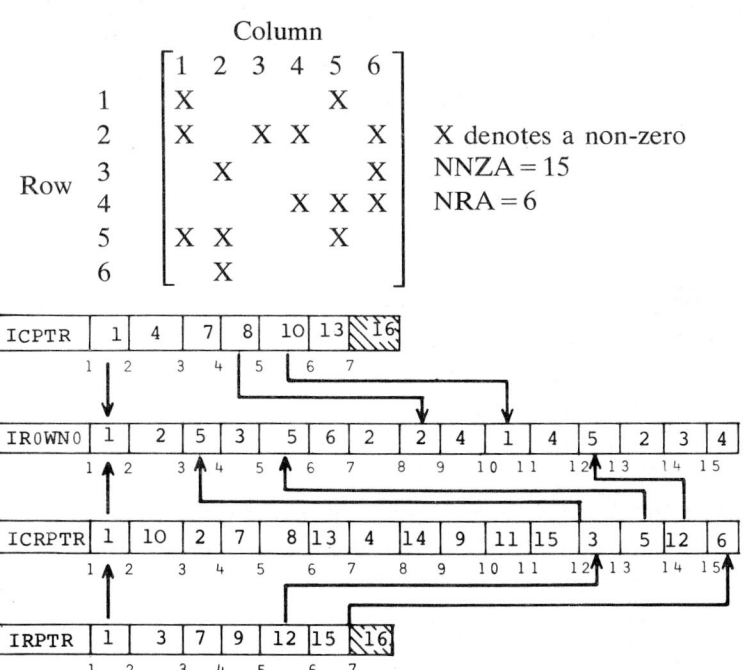

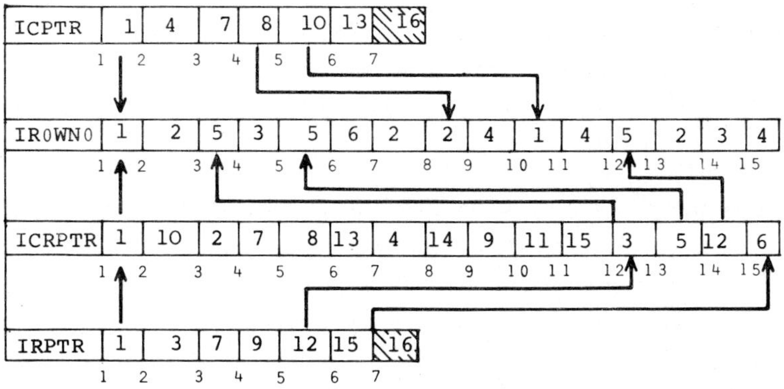

The non-zeros of column 4, for example, can be found as follows

$$
\left.\begin{array}{l}
\text{ICPTR}(4) = 8, \\
\text{ICPTR}(5) - 1 = 10 - 1 = 9
\end{array}\right\} \Rightarrow
\begin{array}{l}
\text{IR0WN0}(8) = 2, \\
\text{IR0WN0}(9) = 4.
\end{array}
$$

Thus column 4 has non-zeros in rows 2 and 4.

The non-zeros in row 5, for example, can be found as follows

$$
\left.\begin{array}{l}
\text{IRPTR}(5) = 12, \\
\text{IRPTR}(6) - 1 = 15 - 1 = 14
\end{array}\right\} \Rightarrow
\begin{array}{l}
\text{ICRPTR}(12) = 3, \\
\text{ICRPTR}(13) = 5, \\
\text{ICRPTR}(14) = 12.
\end{array}
$$

The column position of the non-zeros in row 5 can now be found by either a binary search over the array ICRPTR or, more conveniently, via an array JN0 of length NNZA where JN0(K) gives the column number of the (Kth) non-zero referred to by IR0WN0(K).

In this example,

the non-zero referred to by IR0WN0(3) is in column 1,
the non-zero referred to by IR0WN0(5) is in column 2,
the non-zero referred to by IR0WN0(12) is in column 5.

Thus row 5 has non-zeros in columns 1, 2 and 5.

It might appear more natural to have an array of column numbers instead of the indirect scheme using ICRPTR. However, when row and column deletions are made, a set of markers are needed to indicate the non-zeros of the original matrix that are no longer present in the matrix after deletion. With the suggested scheme this marking can be achieved by just negating elements in the array IR0WN0. Non-deleted non-zeros can then be found by examining just one array.

The rowwise storage can be created as the row counts are formed using the following algorithm.

Step 0: Initially set each row count to zero (IRCNT(I) = 0 for $I = 1, 2, \ldots,$ NRA). For each element of IR0WN0 (IR0WN0(L), $L = 1, 2, \ldots,$ NNZA) do Step 1.

Step 1: $I \leftarrow$ IR0WN0(L).
IRCNT(I) $\leftarrow$ IRCNT(I) + 1.
INUM(L) $\leftarrow$ IRCNT(I). Thus the Lth non-zero of the matrix (stored column by column) is the (INUM(L))th non-zero in its row (row I). The array JN0, mentioned earlier, (JN0(L) gives the column number of the Lth non-zero) can easily be created in this step if desired.

Step 2: The rowwise storage can now be constructed:
IRPTR(1) $\leftarrow$ 1,
IRPTR(I) $\leftarrow$ IRPTR($I-1$) + IRCNT($I-1$) for $I = 2, 3, \ldots,$ (NRA + 1)
and

$$\left.\begin{array}{l} I1 = \text{IR0WN0}(L) \\ \text{IP0SN} = \text{INUM}(L) + \text{IRPTR}(I1) - 1 \\ \text{ICRPTR(IP0SN)} = L \end{array}\right\} \text{for } L = 1, 2, \ldots, \text{NNZA}.$$

Once ICRPTR has been formed the array INUM can be discarded.

Brief description of BELOW and ABOVE

The methodology of the BELOW and ABOVE algorithms is well known and only brief descriptions are needed here in the light of the data structures used.

The BELOW algorithm is as follows:

Process the columns sequentially. For each column J, with a column count of unity (i.e. if ICCNT(J) = 1) do:

Step 1: $J1 \leftarrow J$.

Step 2: Find the row number, row I say, of the single undeleted non-zero in column $J1$. This is the only positive entry in the section of IR0WN0 relating to column $J1$.

Step 3: For each undeleted non-zero in row I do:
'Delete' the non-zero by negating its entry in IR0WN0. (Its position in IR0WN0 can be found via the arrays IRPTR and ICRPTR).
ICCNT($J2$) ← ICCNT($J2$) − 1 where $J2$ is the column number of this non-zero.
If ICCNT($J2$) = 1 place column $J2$ on a stack.

Step 4: If the stack is not empty: $J1$ ← the top element on the stack and go to Step 2.

The ABOVE algorithm is somewhat similar and can be described as follows:

Process the rows sequentially. For each row, I, with a row count of unity (i.e. if IRCNT(I) = 1) do:

Step 1: $I1 \leftarrow 1$.

Step 2: Find the column number, column J say, of the single undeleted non-zero in row $I1$. (This can be found by using IRPTR to point to ICRPTR which in turn points to IR0WN0 and the non-zero in question is the one relating to the only positive element of those pointed to in IR0WN0. The required column number can then easily be found using the array JN0 described earlier).

Step 3: For each undeleted non-zero in column J do:
'Delete' the non-zero by negating its entry in IR0WN0.
IRCNT($I2$) ← IRCNT($I2$) − 1 where $I2$ is the row number of this non-zero.
If IRCNT($I2$) = 1 place row $I2$ on a stack.

Step 4: If the stack is not empty: $I1$ ← the top element on the stack and go to Step 2.

Data structures for maximum matching (MXMATC)

In Section 4 the three alternative algorithms were described. Each algorithm has been implemented in two ways. The six implementations may be summarised as follows:

H1: Hall's algorithm with a shortest augmenting path (SAP) backtrack.
H2: Hall's algorithm with a depth-first search (DFS) backtrack.
HK1: Hopcroft and Karp's algorithm with a 'forward' search.
HK2: Hopcroft and Karp's algorithm with a 'backward' search.
HHK1: Combined Hall–Hopcroft and Karp algorithm with a SAP backtrack.
HHK2: Combined Hall–Hopcroft and Karp algorithm with a DFS backtrack.

Implementation H1 and HHK1 are very similar as are implementations H2 and HHK2 and only the details of the backtrack procedures are considered here.

SAP Backtrack—H1 and HHK1

A backtrack is necessary when a column (K say) cannot be easily assigned. Here an augmenting path of shortest length is obtained in order to find an assignment for the column. A tree, with column K as source, is formed level by level until an unassigned row is found. Only the nodes of the tree corresponding to row numbers need be stored since the nodes corresponding to column numbers can be found from the list of current assignments. The data structures used are described below.

Let the array IRLIST be a distinct list of row numbers. If IRFLAG(I) = K then row I is already in the tree (and hence in IRLIST) and should not be included again. Used in this manner, IRFLAG need only be initialised once to zero. IRTYPE(L) provides a pointer to the position in IRLIST of the row that preceded row (IRLIST(L)) in the tree. It therefore provides the linkage which enables a path to be easily constructed once the tree has been formed. The algorithm may be described as follows.

Step 1: IR ← Row number of the Lth non-zero in column K,
IRLIST(L) ← IR,
IRTYPE(L) ← 0,
IRFLAG(IR) ← K,
} for $L = 1, 2, \ldots,$ (number of non-zeros in column K).
$J \leftarrow 1$,
IRθW ← IRLIST(1).

Step 2: For each row number, I, in the column currently assigned to row IROW: If IRFLAG(I) $\neq K$ then
$L \leftarrow L+1$,
IRLIST(L) $\leftarrow I$,
IRTYPE(L) $\leftarrow J$,
IRFLAG(I) $\leftarrow K$.
If row I is currently unassigned go to Step 4.

Step 3: $J \leftarrow J+1$, IROW $\leftarrow$ IRLIST(J). Go to Step 2.

Step 4: An augmenting path has been found—use IRLIST and IRTYPE to update the assignments as illustrated below.

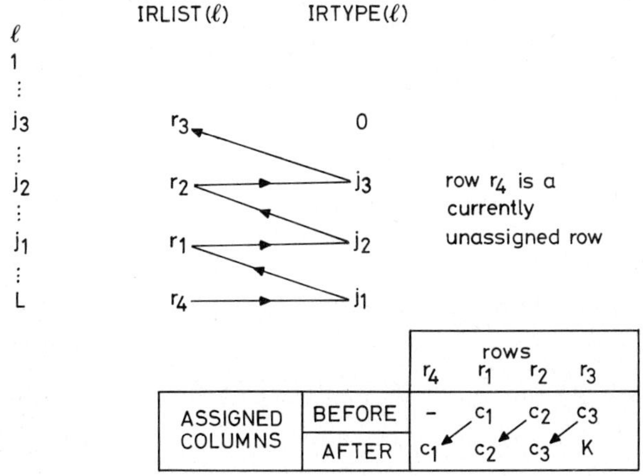

BFS Backtrack—H2 and HHK2

In these implementations a tree is formed using a depth first search approach instead of level by level. A stack (initially empty) is used to store the rows which are in a (partially constructed) augmenting path. (The columns need not be stored explicitly since these can be obtained from the list of current assignments). When the incomplete path cannot continue the top element of the stack is removed (Step 5) and the search for the path is continued from the next element on the stack. Suppose that column K cannot be easily assigned. The algorithm to find an augmenting path to enable the column to be assigned is described below.

Step 0: $I \leftarrow$ row number of the first non-zero in column K.

Step 1: Put I onto the stack.
$K1 \leftarrow$ number of the column that is currently assigned to row I.

Step 2: If there is an easy assignment in column $K1$ go to Step 6.

Step 3: $I \leftarrow$ row number of the next (first) non-zero in column $K1$. (If there is no 'next non-zero' go to Step 5).

Step 4: If I is on the stack go to Step 3. Otherwise go to Step 1.

Step 5: Remove the top element from the stack.
$I \leftarrow$ the row that is currently top of the stack.
$K1 \leftarrow$ the number of the column that is currently assigned to row I. Go to Step 3.

Step 6: An augmenting path has been formed. The necessary amendments to the current list of assignments and an assignment for column K can now be made.

The same procedure (Step 2) is used to test whether an augmenting path can be found as that used to obtain the initial set of easy assignments. It is only necessary to scan the non-zeros no more than once by setting a pointer to the next (unexamined) non-zero in each column that has been easily assigned. This is a suggestion put forward by Gustavson [6].

Hopcroft and Karp's graph formation

Hopcroft and Karp's algorithm requires that a graph is formed containing the set of all augmenting paths of shortest length, starting from the set of all currently unassigned columns. The graph is formed level by level and since it is only the set of shortest augmenting paths within the graph that is of interest, rows that have appeared in the graph at an earlier level should not appear again at a later level. This can easily be achieved by having an array IR0IND where IR0IND(I) is set equal to K at iteration K (the formation of the Kth graph) for all rows I that have appeared in the level just formed. Only the rows of the graph need to be stored—the columns can be obtained from the list of current assignments. The data structure used to represent the graph depends on the manner in which it is searched—forwards (in the same direction as the graph was created) or backwards.

*Data structure of the graph when a forward search is used—HK*1

This is the most natural form of search and hence requires the simplest data structure for the graph. Let JR0WS be an array containing a list of the rows in the graph with one entry for each edge in the graph leading into a row vertex. (Thus the row numbers in JR0WS are not necessarily distinct). Let JRPTR(J) be a pointer to the position in JR0WS at which the rows in the graph from column J start. When forming a new level of rows a list of distinct columns emanating from the rows in the previous level (columns currently assigned to these rows) have to be considered. This can be achieved by having an additional array IRT which is initialised to zero at the start of the formation of each graph. On the first occasion that a given row, row *I* say, of the previous level is examined to obtain potential rows in the next level IRT(*I*) is set equal to the sequence number in the graph of the column currently assigned to row *I*. Thus if IRT(*I*) $\neq 0$ when row *I* is being examined it means that the row has already acted as parent node to its assigned column and that the resultant rows have already been included in the new level of the graph. Examination of the graph in Fig. 2 shows the necessity of such a procedure. In the search for augmenting paths IRT is used to provide the linkages, via JRPTR and JROWS, necessary to form the paths.

Data structure of the graph when a backward search is used—HK2

A backwards search of the graph requires a more complicated data structure. Given a row of the graph its parent node must be identified. To achieve this the following arrays are used.

An array JR0WS is, as in the forwards search implementation, a list of the rows of the graph. JRFRST is an array such that JRFRST(*I*) points to the position in JR0WS at which row *I* first appears. An array JRLAST is used only during the construction of the graph and is such that JRLAST(*I*) points to the position in JR0WS at which row *I* last appeared in the current (partially constructed) graph. JFR0MR(*I*) gives the parent row of JR0WS(*I*) except when JR0WS(*I*) is a first level row in which case it gives the number of the unassigned column which leads into JR0WS(*I*). JRNEXT(*I*) points to the position in JR0WS at which the row numbered JR0WS(*I*) next appears. If there is no further appearance of the row in JR0WS then JRNEXT(*I*) = 0. The complete data structure is best illustrated by considering the following example of part of a typical graph constructed by Hopcroft and Karp's algorithm.

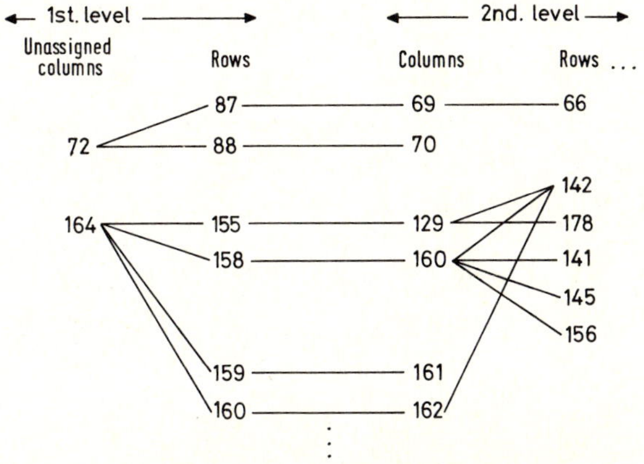

Consider the storage of rows in the second level as shown below:

subscript	JR0WS	JFR0MR	JRNEXT	
				JRFRST(66) $= (i_1)$
⋮	⋮			JRFRST(142) $= (i_1+1)$
(i_1)	66	87	0	JRFRST(178) $= (i_1+2)$
(i_1+1)	142	155	$\emptyset(i_1+3)$	JRFRST(141) $= (i_1+4)$
(i_1+2)	178	155	0	JRFRST(145) $= (i_1+5)$
(i_1+3)	142	158	$\emptyset(i_1+7)$	JRFRST(156) $= (i_1+6)$
(i_1+4)	141	158	0	
(i_1+5)	145	158	0	
(i_1+6)	156	158	0	JRLAST(66) $= i_1$
(i_1+7)	142	160	0	JRLAST(142) $= (i_1+7)$
				JRLAST(178) $= (i_1+2)$
⋮	⋮	⋮	⋮	JRLAST(141) $= (i_1+4)$
				JRLAST(145) $= (i_1+5)$
				JRLAST(156) $= (i_1+6)$

Hopcroft and Karp search

With the data structures defined above it is simple to perform a depth-first search of the graph to find a maximal set of augmenting paths. The set must be vertex-disjoint and therefore a marker must be set up to indicate whether a row has already been examined during the search of the current graph.

Storage requirements

The storage required by all the arrays used in each of the six implementations is summarised in the following table in which NR represents the number of rows and NNZ the number of non-zeros in the (sub-) matrix for which the maximum matching is being sought.

Implementation	Array storage requirement
H1	4NR
H2	5NR
HK1	8NR+NNZ+2
HK2	7NR+3NNZ
HHK1	5NR
HHK2	6NR

Only those arrays needed to contrast the approaches used in the six implementations have been described in this section. A more complete description is given in Darby-Dowman [1].

The final stage of block triangularisation—TARJAN

Full details of the implementation of Tarjan's algorithm used here can be found in Darby-Dowman [1] and Gustavson [6].

6. Experimentation and results

Several algorithms used in the restructuring of basis matrices have been described in the previous sections. This section presents and discusses the results of an investigation into their relative efficiency in practice. Randomly generated matrices have often been used in such an exercise. However basis matrices from real life LP problems are not random in structure. In this investigation, therefore, a set of six practical LP problems were collected and analysed. Their characteristics are shown in Table 1.

Table 1
Statistics of the sample of LP problems

Problem	No. of rows	No. of columns	No. of non-zeros	No. of non-zeros per column	Density
P1	315	458	2965	6.47	2.06%
P2	339	1066	8142	7.64	2.25%
P3	557	1917	13421	7.00	1.26%
P4	587	810	3724	4.60	0.78%
P5	822	1571	11127	7.08	0.86%
P6	1340	1675	14004	8.36	0.62%

In an attempt to obtain a representative cross-section, the problems were collected from several sources. Each problem was solved starting from a trivial basis using APEX 1 on a CDC 7600 machine. Basis files were produced at regular intervals and were used to generate a set of basis matrices for use in testing the restructuring algorithms. The characteristics of the sample of basis matrices are shown in Table 2.

Table 2
Statistics of the basis matrices

Problem	Basis matrix	No. of iterations from start	Basis status	No. of rows	No. of non-zeros	Density
P1	P1B1	601	INFEAS	315	1453	1.46%
	P1B2	1213	OPT		1662	1.67%
P2	P2B1	303	FEAS	339	1134	0.99%
	P2B2	401	OPT		1127	0.98%
P3	P3B1	500	INFEAS	557	1900	0.61%
	P3B2	1360	OPT		2574	0.83%
P4	P4B1	502	INFEAS	587	1686	0.49%
	P4B2	1069	OPT		2132	0.62%
P5	P5B1	3002	INFEAS	822	4453	0.66%
	P5B1	5002	FEAS		4605	0.68%
	P5B3	6664	OPT		4634	0.69%
P6	P6B1	801	INFEAS	1340	4848	0.27%
	P6B2	2142	OPT		7436	0.41%

Table 3
Basis matrix structure after isolation of the bump

Basis matrix	No. of rows	No. of logical variables in basis	No. of rows: Below the bump	No. of rows: Above the bump	No. of rows: In the bump	No. of non-zeros in the bump	Density of the bump
P1B1	315	74	103	34	178	831	2.62%
P1B2		57	81	18	216	1116	2.39%
P2B1	339	187	324	11	4	8	50%
P2B2		175	337	0	2	4	100%
P3B1	557	268	342	126	89	351	4.43%
P3B2		179	272	170	115	588	4.45%
P4B1	587	305	368	75	144	587	2.83%
P4B2		136	264	134	189	692	1.94%
P5B1		195	238	209	375	1781	1.27%
P5B2	822	157	190	197	435	2155	1.14%
P5B3		169	203	203	416	2053	1.19%
P6B1	1340	974	1159	106	75	219	3.89%
P6B2		753	1034	200	106	341	3.03%

Table 3 shows the structure of the basis matrices after the bump has been isolated. It can be seen that the application of the BELOW and ABOVE algorithms has considerably reduced the size of the matrix to be block triangularised. In particular the basis matrices associated with problem P2 have very small bump matrices and the structure is almost completely lower triangular. The question of whether isolation of the bump is worthwhile in terms of the overall speed of block triangularisation is considered later in this section.

Relative efficiency of maximum matching algorithms

All six versions of the maximum matching algorithms described earlier were applied to each bump matrix (with the exception of those of problem P2 which were too small to be of any use in the comparison).

The most obvious measure of efficiency is the amount of CP time that each implementation required. All algorithms were programmed in Fortran and compiled by the MNF compiler (version 5.3) on the CDC 6600 machine of the University of London running under the NOS/BE (version 1.2 level L454-F) operating system. The main difference between the algorithms is the way in which augmenting paths are found and it is in this area that further indicators of efficiency can be computed. Thus the following measures are used in addition to CP time.

(i) the total number of vertices corresponding to row numbers (row vertices) in the trees (graphs in the case of the Hopcroft and Karp implementations (HK1 and HK2)).

(ii) the total number of row vertices examined for entry into the trees (graphs). This includes the number of times a non-zero is examined in a row which is already in the tree (graph).

(iii) the average number of reassignments per augmenting path. Once an augmenting path has been found existing assignments have to be reassigned in order to obtain the new assignment.

(iv) the number of row vertices searched in the graphs to find the vertex disjoint sets of augmenting paths. This measure is only computed in the case of the Hopcroft and Karp implementation (HK1 and HK2) since an explicit search is not required by the other algorithms.

Table 4 shows the results of the application of Hall's algorithm. The depth first search version (H2) generally produces smaller trees than the shortest augmenting path version (H1), although generally more rows are examined for entry into the trees. The H2 version, however, has a tendency on occasions to produce very long augmenting paths as shown by the matrices from problem P5. H1 appears to consistently find short augmenting paths. On the basis of these results it is difficult to choose between the two versions.

Table 5 shows the results of the application of the combined Hall–Hopcroft and Karp algorithm. In this algorithm the columns that cannot be easily assigned are considered after all columns have been tested for easy assignments. This can result in considerably fewer columns that have to be assigned

Table 4
Relative efficiency of Hall's algorithm—implementations H1 and H2

Basis matrix	No. of rows in bump	No. of back-tracked assign-ments		For the set of all backtracked assignments: No. of row vertices in trees		For the set of all backtracked assignments: Total no. of row vertices examined for entry into trees		Average no. of reassignments per augmenting path		C.P. time (CDC 6600 millisecs)	
		H1	H2	H1	H2	H1	H2	H1	H2	H1	H2
P1B1	178	26	26	482	247	1006	1088	1.88	6.69	32	28
P1B2	216	35	35	703	339	1286	1353	1.97	7.37	41	34
P3B1	89	14	15	112	70	209	356	1.43	2.00	11	12
P3B2	115	28	28	411	173	1201	737	1.96	3.57	33	20
P4B1	144	17	18	315	155	776	575	2.41	6.22	26	17
P4B2	189	46	46	638	425	1437	1666	1.98	6.07	43	38
P5B1	375	71	70	1883	1500	5790	7278	2.17	12.10	145	130
P5B2	435	81	81	2200	2333	5726	9530	2.04	18.62	151	194
P5B3	416	74	71	2018	2094	5144	11418	1.96	16.90	137	205
P6B1	75	7	7	39	12	80	44	1.43	1.71	7	5
P6B2	106	17	17	137	31	253	114	1.47	1.83	11	9

Table 5
Relative efficiency of the combined Hall–Hopcroft and Karp algorithm—Implementations HHK1 and HHK2.

Basis matrix	No. of rows in bump	No. of back-tracked assign-ments	For the set of all backtracked assignments: No. of row vertices in trees		Total no. of row vertices examined for entry into trees		Average no. of reassignments per augmenting path		C.P. time (CDC 6600 millisecs)	
			HHK1	HHK2	HHK1	HHK2	HHK1	HHK2	HHK1	HHK2
P1B1	178	24	418	230	891	1050	2.00	6.71	29	28
P1B2	216	33	649	374	1246	1722	2.09	7.85	40	40
P3B1	89	10	66	26	99	85	1.20	2.40	8	7
P3B2	115	18	301	126	950	594	2.56	4.56	28	16
P4B1	144	14	284	133	684	561	2.79	6.29	23	18
P4B2	189	29	427	407	935	1565	2.10	9.66	30	38
P5B1	375	48	1787	1642	5808	8582	2.88	17.50	137	159
P5B2	435	59	1945	2206	4934	10283	2.39	22.31	123	202
P5B3	416	54	2229	2232	6552	10827	2.83	23.30	154	206
P6B1	75	6	31	13	57	48	1.33	2.17	5	5
P6B2	106	15	121	32	225	130	1.53	2.13	10	8

by backtracking although these columns are generally harder to assign. The other results are very similar to those of Hall's algorithm (Table 4).

Table 6 shows the results of Hopcroft and Karp's algorithm. Comparing the two versions of the algorithm, the backwards search version (HK2) is significantly more efficient than the forwards search version (HK1). However, on the matrices considered here, the algorithm is much less efficient than those discussed above. The attempt to find several augmenting paths in one pass of the matrix is achieved at the expense of constructing graphs which are relatively large and which then have to be searched. The time consuming nature of each pass outweighs the fact that fewer passes are needed.

Sub-bump structure

After a maximum matching has been found, the final stage of the block triangularisation procedure is to apply Tarjan's algorithm for strong components. Table 7 shows the sub-bump structure. It is interesting to note that there are several cases in which very large sub-bumps exist. In view of this it seems desirable to restructure the sub-bumps. Hellerman and Rarick [9, 10] have developed an algorithm to permute the sub-bumps in lower triangular form except for some (hopefully few) columns (called spikes). Their algorithm is intended to minimise the number of non-zeros above the diagonal.

Table 6
Relative efficiency of Hopcroft and Karp's algorithm—implementations HK1 and HK2.

Basis matrix	No. of rows in bump	No. of back-tracked assign-ments	No. of graphs formed		Average no. of assignments found per graph		No. of row vertices in graphs		Total no. of row vertices examined for entry into graphs		Average no. of reassignments per augmenting path		No. of row vertices searched		C.P. time (CDC 6600 millisecs)	
			HK1	HK2	HK1	HK2	HK1	HK2	HK1	HK2	HK1	HK2	HK1	HK2	HK1	HK2
P1B1	178	24	6	6	4.00	4.00	878	878	2102	2102	2.21	2.21	547	438	70	64
P1B2	216	33	5	5	6.60	6.60	967	967	2152	2152	2.09	2.09	705	569	76	69
P3B1	89	10	2	2	5.00	5.00	97	97	149	149	1.20	1.20	62	81	12	11
P3B2	115	18	6	6	3.00	3.00	606	606	1675	1675	2.56	2.56	456	244	51	47
P4B1	144	14	5	5	2.80	2.80	369	369	788	788	2.36	2.36	233	160	34	33
P4B2	189	29	6	6	4.83	4.83	751	751	1784	1784	2.21	2.21	530	309	67	59
P5B1	375	48	6	7	8.00	6.86	1888	2072	5354	5878	2.42	2.48	1112	791	160	160
P5B2	435	59	7	7	8.43	8.43	2604	2804	7297	7497	2.20	2.20	1667	1078	227	191
P5B3	416	54	9	9	6.00	6.00	3191	3148	9084	8916	2.52	2.50	2232	1223	277	228
P6B1	75	6	2	2	3.00	3.00	46	46	97	90	1.33	1.33	36	29	5	5
P6B2	106	15	3	3	5.00	5.00	179	180	326	327	1.53	1.53	117	134	15	16

Table 7
Sub-bump structure.

Basis matrix	Size of bump	No. of sub-bumps of size: 1	2	3	4	5	6	7	8	9	10	Plus single sub-bumps of size:	
P1B1	178	3	2					1				164	
P1B2	216		4									208	
P2B1	4		2										
P2B2	2		1										
P3B1	89	17	5	2	2		2	1	1			21	
P3B2	115	36	3	3	1	2	1	1			1	13	14
P4B1	144	40	3		5	6		2				34	
P4B2	189	54	11	1		3	1					89	
P5B1	375	35	1	1	5	3	1			1		285	
P5B2	435	10	3	1	2	2	1					392	
P5B3	416	16	1	3	3	1	1	1				365	
P6B1	75	9	12	1	1	1	1	2			1		
P6B2	106	3	28							3	2		

Overall efficiency

Table 8 shows the CP time required to complete each stage of the block triangularisation procedure. It can readily be seen that, because of the data structures required, the stages before that of finding a maximum matching are rather slow in comparison with the remaining stages. A question is therefore

Table 8
CP times for the complete block triangularisation procedure.

	Execution time in CDC 6600 CP millisecs					
Basis matrix	Set up rowwise storage	BELOW and ABOVE	Restore the bump matrix	Maximum matching H1	TARJAN	Total
P1B1	42	28	22	32	28	152
P1B2	46	25	26	41	32	170
P2B1	28	46	4	1	1	80
P2B2	35	50	6	1	1	93
P3B1	56	65	17	11	12	161
P3B2	72	83	20	33	18	226
P4B1	52	59	21	26	20	178
P4B2	62	68	24	43	27	224
P5B1	119	101	48	145	59	472
P5B2	120	98	54	151	68	491
P5B3	120	104	52	137	67	480
P6B1	139	197	30	7	9	382
P6B2	200	269	34	11	13	527

raised about the desirability of isolating the bump initially. It is possible to apply the maximum matching algorithm followed by the strong components algorithm to the basis matrix itself. Table 9 shows the results of adopting such a procedure and it can be seen that the speed of the complete block triangularisation has generally improved. The execution time of the maximum matching algorithm has not increased as much as had seemed likely. The possible assignments inside the bump are not unique but those outside the bump are unique. It seemed likely therefore that the maximum matching algorithm would find difficulty in making the assignments for those columns outside the bump. It appears that this was not the case. This result indicates that the structure of the unpermuted basis matrix was such that many of these assignments were made easily.

A possible improvement to the procedure would be to identify the column singletons of the basis matrix (this can easily be done since the basis matrix is stored columnwise) and make them 'permanent' assignments. Then a modified maximum matching algorithm could be applied to the basis matrix which would not allow permanent assignments to enter augmenting paths. Further permanent assignments of non-bump columns could be found in such an algorithm. Tarjans algorthim could then be modified to ignore rows/columns that are permanently assigned.

Table 9
CP times for block triangularisation.

	Execution time in CDC 6600 CP millisecs			
	Bump not isolated			Bump isolated
Basis matrix	Maximum matching H1	TARJAN	Total	Total
P1B1	48	45	93	152
P1B2	72	48	120	170
P2B1	48	39	87	80
P2B2	37	40	77	93
P3B1	66	67	133	161
P3B2	171	76	247	226
P4B1	94	66	160	178
P4B2	219	74	293	224
P5B1	360	129	489	472
P5B2	328	131	459	491
P5B3	370	132	502	480
P6B1	133	159	292	382
P6B2	163	200	363	527

Since reinversion of the basis matrix generally takes place every 40–70 iterations only a small proportion of basis matrix columns change between successive reinversions. The maximum matching found at a given reinversion may be retained and used as an initial approximation to the maximum matching required at the following reinversion. This matching is valid except for some or all of those columns of the basis matrix that have changed since the previous reinversion. Reassignments may then be made in the usual manner for those columns only. This suggestion has also been made by Bradford Stults (privage communication, 1979) and the referee of this paper.

7. Conclusions

Algorithms used in the block triangularisation of sparse basis matrices have been described and applied to basis matrices from real-life LP problems. The following points emerge from the investigation.

Hopcroft and Karp's algorithm for finding a maximum matching compares unfavourably with the algorithms based on Hall's method (Hall and combined Hall–Hopcroft and Karp). Of the latter algorithms there is no strong evidence to recommend one in preference to any other. The depth first search versions seem attractive but can on occasions be rather slow. The shortest augmenting path versions seem more predictable in their behaviour and can on occasions be considerably faster. More work is needed, especially with larger matrices, to explore further the unpredictable nature of the depth first search versions.

It has been shown that isolation of the bump before applying the block triangularisation algorithms is not always a worthwhile exercise. A restricted procedure (such as that described at the end of Section 6) which does not require a rowwise data structure for the matrix and which isolates many of the non-bump columns is worth considering.

References

[1] K. Darby-Dowman, The exploitation of sparsity in large scale linear programming problems—Data structures and restructuring algorithms for basis matrices, Ph.D. Thesis, Brunel University, Uxbridge, England, 1980.
[2] I.S. Duff, A survey of sparse matrix research, Proc. IEEE 65 (4) pp. 500–535. 1977.
[3] I.S. Duff, On permutations to block triangular form, J. Inst. Maths Applics. 19 (1977) 339–342.
[4] A.L. Dulmage and N.S. Mendelsohn, Two algorithms for bipartite graphs, J. SIAM 11 (1) (1963) 183–194.
[5] L.R. Ford and D.R. Fulkerson, A simple algorithm for finding maximal network flows and an application to the Hitchcock problem, Canad. J. Math. 9 (1957) 210–218.

[6] F. Gustavson, Finding the block lower triangular form of a sparse matrix in: J.R. Bunch and D.J. Rose, eds., Sparse matrix Computations (Academic Press, New York, 1976) 275–289.
[7] M. Hall, An algorithm for distinct representations, Amer. Math. Monthly 63 (1956) 716–717.
[8] F. Harary, A graph theoretic approach to matrix inversion by partitioning, Numer. Math. 4 (1962) 128–135.
[9] E. Hellerman and D. Rarick, Reinversion with the preassigned pivot procedure, Math. Programming 1 (1971) 195–216.
[10] E. Hellerman and D. Rarick, The partitioned preassigned pivot procedure. (P^4) in: D.J. Rose and R.A. Willoughby, eds., Sparse Matrices and Their Applications (Plenum, New York, 1972) 67–76.
[11] J.E. Hopcroft and R.M. Karp, An $n^{\frac{5}{2}}$ algorithm for maximum matchings in bipartite graphs, SIAM J. Comput. 2 (4) (1973) 225–231.
[12] R.W.H. Sargent and A.W. Westerberg, "Speed-up" in chemical engineering design, Trans. Inst. Chem. Eng. 42 (1964) 190–197.
[13] M. Simmonard, Linear Programming (Prentice-Hall, Englewood Cliffs, NJ, 1966).
[14] D.V. Steward, On an approach to techniques for the analysis of the structure of large systems of equations, SIAM Review 4 (4) (1962) 321–342.
[15] D.V. Steward, Partitioning and tearing systems of equations, J. SIAM Num. Anal. (Ser. B) 2 (2) (1965) 345–365.
[16] R.E. Tarjan, Depth first search and linear graph algorithms, SIAM J. Comput. 1 (1972) 146–160.
[17] A. Yaspan, On finding a maximal assignment, Operations Research 14 (1966) 646–651.

P. Hansen, ed., Studies on Graphs and Discrete Programming

ON THE NUMBER OF NONNEGATIVE INTEGER SOLUTIONS OF A SYSTEM OF LINEAR DIOPHANTINE EQUATIONS

M.A. FRUMKIN

Mathematical Institute of the Academy of Sciences, Moscow, U.S.S.R.

1. Introduction

In the paper we study the number $\nu(k)$ of integer points in the polyhedron

$$a_1x_1+\cdots+a_nx_n=k, \quad a_1,\ldots,a_n, k\in\mathbb{Z}^m, x_1,\ldots,x_n\in\mathbb{R}_+. \tag{1}$$

This number is equal to the number of nonnegative integer solutions of the following system of linear Diophantine equations

$$a_1x_1+\cdots+a_nx_n=k, \quad x_1,\ldots,x_n\in\mathbb{Z}. \tag{2}$$

Theorem 1.1. *Suppose that for all $k\in\mathbb{Z}^m$ the number $\nu(k)$ is finite. Then $\nu(k)$ may be expressed as a sum*

$$\nu(k)=\sum_{i=1}^{N}\chi_i(k)g_i(k),$$

where $\chi_i(k)$ is characteristic function of a cone with the vertex in 0 *and integer generators, $g_i(k)$ is a quasipolynomial, $i=1,\ldots,N$.*

Some results connected to the above mentioned are as follows.

Theorem 1.2 [1, 2]. *If $m=1$ and $a_1,\ldots,a_n\in\mathbb{Z}$, then $\nu(k)$ is a quasipolynomial of degree $n-1$. If k is large, then $\nu(k)>0$.*

Theorem 1.3 [6]. *If $n>\mathrm{rk}(a_1,\ldots,a_n)$ ans $s\in\mathbb{Z}_+$, then there exists a vector $k_s\in C(a_1,\ldots,a_n)\cap\mathbb{Z}(a_1,\ldots,a_n)$ such that $\nu(k)\geq s$ for all $k\in k_s+C(a_1,\ldots,a_n)\cap\mathbb{Z}(a_1,\ldots,a_n)$.* (Cf. *Corollary* 3.7 *of this paper.*)

Theorem 1.4 [1]. *Suppose that for all $k \in \mathbb{Z}^m$ the number $\nu(k)$ is finite. Then $\nu(sk)$, $s \in \mathbb{Z}_+$ is a quasipolynomial on s of degree $n - \mathrm{rk}(a_1, \ldots, a_n)$. The leading coefficient of the quasipolynomial is equal to the volume of the polyhedron* (1).

Theorem 1.5 [3, 4]. *If all vertices of polyhedra $\Gamma_1, \ldots, \Gamma_s$ are integer and $\Gamma = k_1\Gamma_1 + \cdots + k_s\Gamma_s$ is Minkowski's sum of $\Gamma_1, \ldots, \Gamma_s$, then $\nu(k_1, \ldots, k_s) = \nu(\Gamma)$ is a polynomial on $k_1, \ldots, k_s$.*

The following theorem is a result of this paper.

Theorem 1.6. *If the matrix $(a_1, \ldots, a_n)$ is totally unimodular (i.e. all its minors are $-1, 0, 1$), then in Theorem* 1.1 *the functions $g_i(k)$ are polynomials.*

G.H. Weidner [5] proved the following fact about the extremum of an integer programming problem.

Theorem 1.7. *Let us set*

$$M(k) = \max\{cx \mid ax = k,\ a,\ c,\ x \in \mathbb{Z}_+^n\}.$$

Then $M(k)$ is a sum of a linear function and a periodic one for large k.

In Section 4 we shall prove the following theorem.

Theorem 1.8. *Let us set*

$$M(k) = \max\{cx \mid a_1x_1 + \cdots + a_nx_n = k,\ k,\ a_1, \ldots, a_n \in \mathbb{Z}^m,\ x_1, \ldots, x_n \in \mathbb{Z}_+\}.$$

There are a partition $\mathbb{R}^m = \bigcup_{i=1}^N C_i$, where C_i, $i = 1, \ldots, N$ are cones with vertices in 0 *and integer generators, lattices $L_i \subset \mathbb{Z}^m$ and vectors $v_i \in \mathbb{Z}^m$ such that for all $b_1 \in v_i + L_i \cap C_i$, $b_2 \in C_i \cap \mathbb{Z}^m$ the equality $M(b_1 + b_2) = M(b_1) + M(b_2)$ holds.*

In the paper the following notations and notions are used.

By $\mathbb{Z}$, $\mathbb{Z}_+$, $\mathbb{R}$, $\mathbb{R}_+$, we denote integer, nonnegative integer, real and nonnegative real numbers respectively.

Let $a_1, \ldots, a_n \in \mathbb{Z}^m$. By $\mathbb{Z}(a_1, \ldots, a_n)$ and $C(a_1, \ldots, a_n)$ we denote a lattice and a cone generated by vectors $a_1, \ldots, a_n$, i.e.

$$\mathbb{Z}(a_1, \ldots, a_n) = \{x \mid x = a_1x_1 + \cdots + a_nx_n,\ x_i \in \mathbb{Z}\},$$
$$C(a_1, \ldots, a_n) = \{x \mid x = a_1x_1 + \cdots + a_nx_n,\ x_i \in \mathbb{R}_+\}.$$

Minkowski's sum of polyhedra $\Gamma_1, \ldots, \Gamma_s \subset \mathbb{R}^n$ with coefficients $k_1, \ldots, k_s$ is the polyhedron

$$\Gamma = k_1\Gamma_1 + \cdots + k_s\Gamma_s = \{k_1x_1 + \cdots + k_sx_s \mid x_i \in \Gamma_i\}.$$

By quasimonomial on variables $x_1, \ldots, x_n$ we mean the following function

$$cx_1^{r_1} \cdots x_n^{r_n} \exp(2\pi i(c_1x_1 + \cdots + c_nx_n)),$$

where $r_1, \ldots, r_n \in \mathbb{Z}_+$, $c, c_1, \ldots, c_n \in \mathbb{R}$ (i denotes $\sqrt{-1}$). The degree of this quasimonomial is the number $r = r_1 + \cdots + r_n$. By quasipolynomial we mean a sum of a finite number of quasimonomials. The degree of a quasipolynomial is the maximal degree of its quasimonomials.

By Q_r, P_r and X^m we denote the space of quasipolynomials of degree less or equal to r, the space of polynomials of degree r and the space of characteristic functions of convex polyhedral cones with integer generators respectively. We set $Q = \bigcup_{r=0}^{\infty} Q_r$, $X = \bigcup_{m=0}^{\infty} X^m$.

2. Auxiliary results on the geometry of polyhedra

Farkash's Lemma [7, D.3.6] implicates the following assertion.

Lemma 2.1. *A cone $C(a_1, \ldots, a_n)$, where $a_i \in \mathbb{Z}^m$, $a_i \neq 0$ does not contain any linear space except for 0 if and only if there exists a vector $v \in \mathbb{R}^m$ such that $va_1 > 0, \ldots, va_n > 0$.*

Proposition 2.2. *The system* (1) *has a bounded set of solutions for all $k \in \mathbb{R}^m$ if and only if the cone $C(a_1, \ldots, a_n)$ does not contain a nonzero linear space.*

Proof. ($\Leftarrow$) By Lemma 2.1 there exists $v \in \mathbb{R}^m$ such that $v(a_1, \ldots, a_n) = (\alpha_1, \ldots, \alpha_n) > 0$. Equality $\sum_{i=1}^n \alpha_i x_i = vk$, $x_i \geq 0$ shows that $0 \leq x_i \leq vk/\alpha_i < \infty$.

($\Rightarrow$) This implication is a result by N.N. Ivanov and V.N. Schevchenko [6, Theorem 2]. □

Now let us compare various ways of description of polyhedra families.

Proposition 2.3. *Suppose that all vertices of polyhedra $\Gamma_1, \ldots, \Gamma_s \subset \mathbb{R}^n$ are rational. Then there exist an integer matrix A and integer vectors b_i, $i = 1, \ldots, s$ such that the system $Ax \geq b_i$ determines the polyhedron Γ_i, $i = 1, \ldots, s$ and the system $Ax \geq k_ib_i + \cdots + k_sb_s$ determines the polyhedron $\Gamma = k_1\Gamma_1 + \cdots + k_s\Gamma_s \subset \mathbb{R}^n$.*

Proof. Let us consider embeddings $g_i : \mathbb{R}^n \to \mathbb{R}^{n+s}$, $g_i(x) = (x, 0, \ldots, 1, 0, \ldots, 0)$

(1 is on the $n+i$-th place). Let $\Delta_i = g_i(\Gamma_i)$ and Δ be the convex hull of the polyhedra Δ_i, i.e.

$$\Delta = \left\{ t_1 z_1 + \cdots + t_s z_s \mid z_i \in \Delta_i,\, t_i \geqslant 0,\, \sum_{i=1}^{s} t_i = 1 \right\}.$$

Let $Ax + Cy \geqslant b$ be any system of inequalities which determines Δ. Since all vertices of Δ are rational, A, C and b may be chosen integer.

It is obvious that the hyperplane $\sum_{i=1}^{s} y_i = 1$ contains Δ. Hence $\Delta \cap g_i(\mathbb{R}^n) = \Delta_i$. That is Δ_i is determined by the system

$$Ax + Cy \geqslant b, \qquad y_j = 0,\ i \neq j,\ y_i = 1$$

and $\Gamma_i \subset \mathbb{R}^n$ is determined by the system $Ax \geqslant b - c_i$, where c_i is the ith column of C. We set $b_i = b - c_i$.

The polyhedron $\Delta \cap \{y_1 = t_1, \ldots, y_s = t_s\} = t_1\Delta_1 + \cdots + t_s\Delta_s$ is determined by the system

$$Ax + Cy \geqslant b, \qquad y_1 = t_1, \ldots, y_s = t_s.$$

Thus, the polyhedron $\Gamma(t_1, \ldots, t_s) = t_1\Gamma_1 + \cdots + t_s\Gamma_s$ is determined by the system

$$Ax \geqslant b - t_1 c_1 - \cdots - t_s c_s = t_1 b_1 + \cdots + t_s b_s$$

in $\mathbb{R}^m$.

Now in the case when $\sum k_i = 1$ the proposition is proved. In the general case it is sufficient to use the equality

$$k_1\Gamma_1 + \cdots + k_s\Gamma_s = (k_1 + \cdots + k_s)\left(\frac{k_1}{\sum k_i}\Gamma_1 + \cdots + \frac{k_s}{\sum k_i}\Gamma_s \right)$$

and the following fact. If a polyhedron Γ is determined by the system $Ax \geqslant b$, then $t\Gamma$ is determined by the system $Ax \geqslant tb$. □

The following fact is well known.

Proposition 2.4. *Let $\Gamma \subset \mathbb{R}^n$ be a polyhedron. There exists an embedding $g : \mathbb{R}^n \to \mathbb{R}^N$ such that $g(\Gamma)$ is determined by the system $Ax = k$, $x \geqslant 0$ and $g(\mathbb{Z}^n) = \mathbb{Z}^N \cap g(\mathbb{R}^n)$.*

Now let the system $Ax = k$, $x \geqslant 0$ be one of a polyhedron $\Gamma(k)$. We shall show that there may be no polyhedra $\Gamma_1, \ldots, \Gamma_m$ such that $\Gamma(k) = k_1\Gamma_1 + \cdots + k_m\Gamma_m$, where $k = (k_1, \ldots, k_m)$.

Let us consider a positive matrix A such that $\dim \Gamma(k) \geqslant 1$ for some k. Then $\dim \Gamma_s \geqslant 1$ for some s. If $k_s \to \infty$, then $\Gamma(k)$ is bounded. On the other hand $k_s\Gamma_s$ and all the much so $k_1\Gamma_1 + \cdots + k_s\Gamma_s + \cdots + k_m\Gamma_m$ is unbounded.

Therefore Theorem 1.6 could not be reduced to Theorem 1.5. On the other hand not any polyhedron with integer vertices may be determined by a unimodular matrix. Hence Theorem 1.5 is not implied by Theorem 1.6.

Let us consider an integer vector a and a cone $C \subset \mathbb{R}^m$ with integer generators. We shall use a partition of the cone C_0 generated by a and C.

Let $cx = 0$ be an equation of a $m-1$-dimensional face of C. We call this face far (near) with respect to a if $ca > 0$ ($ca < 0$). Remaining faces we call intermediate. We call the cone generated by a and a far (near) face of C far (near) cone. Let C_i, $i = 1, \ldots, N$ be all far cones and $\tilde{C}_j$, $j = 1, \ldots, M$ be all near cones.

Lemma 2.5. (1) $C_0 = C \cup \bigcup_j \tilde{C}_j$.

(2) *If* C_0 *does not contain any linear space except for* 0, *then* $C_0 = \bigcup_i C_i$.

Proof. (1) By definition we have the inclusion $C_0 \supset C \cup \bigcup_j \tilde{C}_j$. We shall prove the opposite inclusion. Let $y \in C_0$ and $y \notin C$. Let us show that $y \in \tilde{C}_j$ for some j.

The vector y may be represented as $y = pa + qv$, where $p, q \geqslant 0$, $v \in C$. Since $y \in C$, $a \notin C$. Hence the segment $[pa, qv]$ has a common point w with the boundary of C. Let $\tilde{c}x = 0$ be an equation of the face of C that includes w. Since $a \notin C$, $\tilde{c}a < 0$ and this face $\tilde{C}$ is near. It is evident that y is a nonnegative linear combination of a and w. Hence y belongs to the near cone generated by $\tilde{C}$ and a.

(2) Suppose that $y \in C_0$ and y does not belong to any far cone. Let us consider a 2-dimensional plane generaded by y and a and the cones $K = V \cap C$ and $K_0 = V \cap C_0$. Let v_1 and v_2 be generators of K. Let $l_1 x = 0$ and $l_2 x = 0$ be equations of faces of C that contain v_1 and v_2. Let K_1 (K_2) be a cone generated by vectors v_1 and a (v_2 and a). Then $K_0 = K_1 \cup K_2$. If for example K_1 is a far cone then $K_0 = K_1$ and $l_1 a \geqslant 0$, $l_1 y \geqslant 0$ and y belongs to the far cone generated by a and by the face $l_1 x = 0$. Hence K_1 and K_2 are near cones i.e. $l_1 a \leqslant 0$ and $l_2 a \leqslant 0$, $-a \in K$. Therefore K_0 and C_0 include the line generated by a. Other possibilities may be considered similarly. □

Lemma 2.6. (1) *Let us set* $K_{i0} = C_i \cap C$, $K_{ij} = C_i \cap \tilde{C}_j$, $j \geqslant 1$. *Suppose that* $k \in K_{ij}$. *Then the condition* $k = as \in C$, $s \geqslant 0$ *is equivalent to* $l_1 k \leqslant \alpha s \leqslant l_2 k$, *where* $\alpha \in \mathbb{Z}$, $l_1, l_2 \in \mathbb{Z}^m$.

(2) *If* C_0 *does not contain a nonzero linear space, then* $C_0 = \bigcup_{i,j} K_{ij}$.

(3) *If the matrix formed by generators of the cone* C *and by* a *is totaliunimodular, then* $\alpha = 1$.

Proof. Let $c_i x = 0$ ($\tilde{c}_j x = 0$) be an equation of a $m-1$-dimensional face of C_i ($\tilde{C}_j$) that does not contain a.

If $k \in K_{ij}$, $j \geqslant 1$ we shall show that the condition $k - as \in C$, $s \geqslant 0$ is equivalent to conditions $c_i(k-as) \geqslant 0$ and $\tilde{c}_j(k-as) \geqslant 0$. If $k \in K_{i0}$, then we shall prove that the condition $k - as \in C$, $s \geqslant 0$ is equivalent to $c_i(k-as) \geqslant 0$, $s \geqslant 0$.

In the first case it is sufficient to prove the following assertion. If $k \in K_{ij}$ and $c_i(k-as) \geqslant 0$, $\tilde{c}_j(k-as) \geqslant 0$, then $s \geqslant 0$ and $k - as \in C$.

The condition $k \in K_{ij}$ may be rewritten in the following form:

$$\begin{cases} k = v_i + \alpha_i a & \text{where } c_i v_i = 0 \text{ and } c_r v_i \geqslant 0,\ \tilde{c}_l v_i \geqslant 0 \\ & \text{for all } r, l \text{ and } \alpha_i \geqslant 0, \\ k = \tilde{v}_j + \tilde{\alpha}_j a & \text{where } \tilde{c}_j \tilde{v}_j = 0 \text{ and } c_r \tilde{v}_j \geqslant 0,\ \tilde{c}_l \tilde{v}_j \geqslant 0 \\ & \text{for all } r, l \text{ and } \tilde{\alpha}_j \geqslant 0. \end{cases}$$

Hence the condition $c_i(k-as) \geqslant 0$ implies

$$0 \leqslant c_i(v_i + (\alpha_i - s)a) = (\alpha_i - s)c_i a,$$

i.e. $\alpha_i - s \geqslant 0$. Similarly the condition $c_j(k-as) \geqslant 0$ implies

$$0 \leqslant \tilde{c}_j(\tilde{v}_j + (\tilde{\alpha}_j - s)a) = (\tilde{\alpha}_j - s)\tilde{c}_j a,$$

i.e. $\tilde{\alpha}_j - s \leqslant 0$. In particular $s \geqslant 0$.

Let $c_r x = 0$ be an equation of a far face. Then $c_r(k-as) = c_r v_i + (\alpha_i - s)c_r a$. From inequalities $c_r v_i \geqslant 0$ and $c_r a \geqslant 0$ it follows that $c_r(k-as) \geqslant 0$. If $\tilde{c}_l x = 0$ is an equation of a near face we have just as above $\tilde{c}_l(k-as) = \tilde{c}_l(\tilde{v}_j + (\tilde{\alpha}_j - s)a) = \tilde{c}_l \tilde{v}_j + (\tilde{\alpha}_j - s)\tilde{c}_l a \geqslant 0$. If $c'x = 0$ is an equation of an intermediate face, then $c'(k-as) = c'v_i \geqslant 0$. As a result we have $k - as \in C$ and $s \geqslant 0$.

Let now $k \in K_{i0}$. In this case it is sufficient to prove the following assertion. If $k \in C \cap C_i$ and $c_i(k-as) \geqslant 0$, $s \geqslant 0$, then $k - as \in C$.

The condition $k \in K_{i0}$ may be rewritten in the following form:

$$\begin{cases} k = v_i + \alpha_i a & \text{where } c_i v_i = 0 \text{ and } c_r v_i \geqslant 0,\ \tilde{c}_l v_i \geqslant 0 \\ & \text{for all } r, l \text{ and } \alpha_i \geqslant 0, \\ c_r k \geqslant 0,\ \tilde{c}_l k \geqslant 0 & \text{for all } r, l. \end{cases}$$

The condition $c_i(k-as) \geqslant 0$, $s \geqslant 0$ implies

$$0 \leqslant c_i(k-as) = c_i(v_i + (\alpha_i - s)a) = (\alpha_i - s)c_i a,$$

Hence $0 \leqslant s \leqslant \alpha_i$.

Suppose that $c_r x = 0$ is an equation of a far face. Then

$$c_r(k-as) = c_r(v_i + (\alpha_i - s)a) = c_r v_i + (\alpha_i - s)c_r a \geqslant 0.$$

If $\tilde{c}_l x = 0$ is an equation of a near face, then $\tilde{c}_l a < 0$ and $\tilde{c}_l(k-as) = \tilde{c}_l k - s\tilde{c}_l a \geqslant 0$ for all $k \in C$. If $c'x = 0$ is an equation of an intermediate face then $c'(k-as) = c'k \geqslant 0$ for $k \in C$.

Now inequalities $c_i(k-as) \geqslant 0$ and $\tilde{c}_j(k-as) \geqslant 0$ may be rewritten in the form $\beta_1 c_j k \leqslant \alpha s \leqslant \beta_2 c_i k$ for some α, β_1, β_2. This completes the proof of the first part of the lemma.

(2) This assertion follows from Lemma 2.5:

$$C_0 = C \cup \bigcup_j \tilde{C}_j = \bigcup_i C_i = C \cup \bigcup_j \tilde{C}_j \cup \bigcup_i C_i$$

$$= \bigcup_i K_{i0} \cup \bigcup_i \bigcup_{j \geqslant 1} K_{ij}$$

(3) Since $\alpha = |\text{g.c.d.}\{c_i a, \tilde{c}_j a\}|$ it is sufficient to show that $c_i a = 1$ and $c_j a = -1$. Let $a_1, \ldots, a_n$ be all generators of a far face and $a_1, \ldots, a_{m-1}$ be a linearly independent system of generators. Each coordinate of c_i is a minor obtained by deletion of a row of the matrix $(a_1, \ldots, a_{m-1})$. Hence $c_i a = \det(a_1, \ldots, a_{m-1}, a)$ is a minor of totaliunimodular matrix $(a_1, \ldots, a_n, a)$ and $c_i a$ equals -1, 0 or 1. But $c_i x = 0$ is an equation of a far face with respect to a. This implies $c_i a = 1$. Similarly we have $\tilde{c}_j a = -1$. □

3. Proof of the main theorem

The proof of Theorem 1.1 is based on the following relation

$$\nu(k) = \sum_{s \in \mathbb{Z}_+} \nu_1(k - a_n s). \tag{3}$$

This formula expresses the number $\nu(k)$ of solutions of the system (2) by the number $\nu_1(k)$ of solutions of the system

$$a_1 x_1 + \cdots + a_{n-1} x_{n-1} = t, \quad x_i \in \mathbb{Z}_+.$$

This system has one variable less than the system (2). We shall prove by induction that first, $\nu_1(t)$ is a sum of productions of characteristic functions of cones and quasipolynomials and second, it is sufficient to consider only a finite segment of $\mathbb{Z}_+$ in (3). The formula (3) is used in Lemma 3.6. Lemma 3.2 is a basis of induction. Other lemmas of this section are some auxiliaries.

Lemma 3.1. *Let $L \subset \mathbb{Z}^m$ be a lattice of rank r. There exists a quasipolynomial g_L of degree 0 such that $g_L|_L = 1$, $g_L|_{\mathbb{Z}^n - L} = 0$.*

Proof. Let $C = (c_1, \ldots, c_m)$ be a basis of the lattice L, i.e.

$$L = \{k \mid k = Cv,\ v \in \mathbb{Z}^m\} = \{k \mid C^{-1} k \in \mathbb{Z}^M\}, \qquad d = \det C_1,$$

$$C^{-1} = \begin{bmatrix} c'_1 \\ \vdots \\ c'_m \end{bmatrix}, \qquad c'_1, \ldots, c'_m \in \mathbb{R}^m.$$

Let us prove that

$$g_L(k) = \sum_{i=1}^{m} \left(\frac{1}{d} \prod_{s=1}^{d} e^{2\pi i c_i' k s} \right)$$

is a characteristic function of L. If c_i' is a fractional number for some i, then its denominator divides d and

$$\sum_{s=1}^{d} e^{2\pi i c_i k s} = 0,$$

i.e. $g_L(k) = 0$. If $c_i' k \in \mathbb{Z}$ for all i, then

$$\sum_{s=1}^{d} e^{2\pi i c_i' k s} = 1$$

for all i and $g_L(k) = 1$. □

Lemma 3.2. *If the number m of equations in (2) equals n and $\det(a_1, \ldots, a_n) \neq 0$, then the number of nonnegative integer solutions of (2) equals $\chi(k)g(k)$, where $\chi(k)$ is characteristic function of $C(a_1, \ldots, a_n)$, $g(k)$ is a quasipolynomial of degree 0.*

Proof. The number $\nu(k)$ of nonnegative integer solutions of (2) equals

$$\nu(k) = \begin{cases} 0 & k \in \mathbb{Z}(a_1, \ldots, a_n) \cap C(a_1, \ldots, a_n), \\ 1 & \text{otherwise.} \end{cases}$$

Now Lemma 3.2 is implied by Lemma 3.1. □

Lemma 3.3. *Let $p(t) \in P_r$, $q(t) \in Q_{r-1}$, $t \in \mathbb{R}$. Then*

$$\sum_{u \leq dt \leq v} p(t) = P(u, v) + Q(u, v)$$

where $P \in P_{r+1}$, $Q \in Q_r$ and $\sum_{u \leq dt \leq v} q(t) \in Q_r$.

Proof. Let

$$g_d(t) = \frac{1}{d} \sum_{s=1}^{d} e^{2\pi i t s/d}$$

be a characteristic function of sublattice $d\mathbb{Z} \subset \mathbb{Z}$ and let $g(t)$ be a quasipolynomial. Then

$$\sum_{\substack{u \leq dt \leq v \\ t \in \mathbb{Z}}} g(t) = \sum_{\substack{u \leq t \leq v \\ t \in \mathbb{Z}}} g_d(t) g(t) = \sum_{\substack{u \leq t \leq v \\ t \in \mathbb{Z}}} \tilde{g}(t),$$

where $\tilde{g}(t)$ is a quasipolynomial. If $g(t)=p(t)\in P_r$ we have

$$g(t)=\frac{1}{d}\left(p(t)+\sum_{s=1}^{d-1} p(t)e^{2\pi is/d}\right).$$

Therefore in order to prove Lemma 3.3 it is sufficient to show that

$$\sum_{u\leq t\leq v} p(t)=P(u, v)\in P_{r+1}$$

and

$$\sum_{u\leq t\leq v} p(t)e^{2\pi ict}=Q(u, v)\in Q_r \quad \text{for } c\notin \mathbb{Z}.$$

It is sufficient to consider the case $p(t)=t^r$. If $S(y)$ and $T(y)$ satisfy the difference equations

$$S(y)-S(y-1)=y^r \tag{4}$$

$$T(y)-T(y-1)=y^r e^{2\pi icy}, \tag{5}$$

then $P(u, v)=S(v)-S(u-1)$ and $Q(u, v)=T(v)-T(u-1)$.

Since 1 is the root of the characteristic equation of (4) and $2\pi ic$ is not the root of (5), $S(y)$ and $T(y)$ may be represented in the following forms:

$$S(y)=a_{r+1}y^{r+1}+\cdots+a_0, \qquad T(y)=(b_ry^r+\cdots+b_0)e^{2\pi icy}. \quad \square$$

Lemma 3.4. *Suppose that* $p(x)\in P_r$, $q(x)\in Q_r$, $a, x\in \mathbb{Z}^m$. *Let us set* $W(x)=\{t\in\mathbb{Z} \mid l_1x+u_1\leq dt\leq l_2x+u_2\}$, *where* $l_1, l_2\in\mathbb{Z}^m$, $u_1, u_2\in\mathbb{Z}$. *Then*

$$\sum_{t\in W(x)} p(x-ta)=P(x)+Q(x)$$

where $P\in P_{r+1}$, $Q\in Q_r$, *and* $\sum_{t\in W(x)} q(x-ta)\in Q_{r-1}$.

Proof. Let $g(x)=x_1^{r_1}\cdots x_m^{r_m}e^{2\pi i(c_1x_1+\cdots+c_mx_m)}$ be a quasipolynomial of degree $r=r_1+\cdots+r_m$. Then

$$\sum_{t\in W(x)} g(x-ta)=\sum_{t\in W(x)} (x_1-ta_1)^{r_1}\cdots(x_m-ta_m)^{r_m}e^{2\pi ic(x-ta)}$$

$$=e^{2\pi icx}\sum_{0\leq s\leq r} p^{(s)}(x_1,\ldots,x_n)\sum_{t\in W(x)} t^{r-s}e^{-2\pi icat},$$

where $p^{(s)}(x_1,\ldots,x_n)\in P_s$. Now if we set $u=l_1x+u_1$ and $v=l_2x+u_2$ then assertions of Lemma 3.4 are implied by appropriate assertions of Lemma 3.3. $\square$

Lemma 3.5. *Let* C, C_0 *and* a *be as in Lemma* 2.5(2), $L\subset\mathbb{Z}^m$ *be a lattice,* M *be the lattice generated by* L *and* a, $g(k)=p(k)+q(k)$, $k\in L$, $p\in P_r$, $q\in Q_{r-1}$. *Let us set* $W=\{s\in\mathbb{Z}_+ \mid k-sa\in C\cap L\}$.

Then the equality

$$\sum_{s\in W} g(k-sa)=\sum_{i=1}^{N} \chi_i(k)(p_i(k)+q_i(k)),$$

where $\chi_i \in X$, $p_i \in P_{r+1}$, $q_i \in Q_r$ *holds for all* $k \in M \cap C_0$.

Proof. Let K_{ij} be cones of Lemma 2.6. By Lemma 2.6(2) $C_0 = \bigcup K_{ij}$. Therefore it is sufficient to prove that

$$\sum_{s\in W} g(k-sa)=p_{r+1}(k)+q_r(k)$$

where $p_{r+1} \in P_{r+1}$, $q_r \in Q_r$ for all $k \in M \cap K_{ij}$.

Lemma 2.6(1) implies that

$$\{s\in\mathbb{Z}_+ \mid k-sa\in C\}=\{s\in\mathbb{Z} \mid l_1k\leqslant \alpha s\leqslant l_2k\} \quad \text{for all } k\in K_{ij}.$$

Since $k\in M$, $k=l+sa$ for some $l\in L$, $s_0\in\mathbb{Z}$. Hence

$$\{s \mid k-sa\in L\}=\{s \mid (s_0-s)a\in L\}=\{s \mid s=-s_0+dt,\ t\in\mathbb{Z}\}$$

for some $d\in\mathbb{Z}$. Therefore

$$\begin{aligned} W&=\{s\in\mathbb{Z}_+ \mid k-sa\in C\cap L\}\\ &=\{s\in\mathbb{Z} \mid l_1k\leqslant\alpha s\leqslant l_2k\}\cap\{s \mid s=-s_0+dt,\ t\in\mathbb{Z}\}\\ &=\{t\in\mathbb{Z} \mid l_1k+s_0\alpha\leqslant\alpha\, dt\leqslant l_2k+s_0\alpha\}.\end{aligned}$$

Hence Lemma 3.5 is implied by Lemma 3.4.

Lemma 3.6. *Suppose that the number* $\nu_1(k)$ *of solutions of the system*

$$a_1x_1+\cdots+a_{h-1}x_{h-1}=k, \quad a_i, k\in\mathbb{Z}^m,\ x_i\in\mathbb{Z}_+$$

may be represented as $\sum_{i=1}^{N}\chi_i(k)(p_i(k)+q_i(k))$, *where* χ_i *is characteristic function of the cone* K_i, $p_i\in P_s$, $q_i\in Q_{s-1}$.

If the number $\nu(k)$ *of nonnegative integer solutions of the system*

$$a_1x_1+\cdots+a_hx_h=k, \quad a_h\in\mathbb{Z}^m,\ x_h\in\mathbb{Z}_+$$

is finite for all $k\in\mathbb{Z}(a_1,\ldots,a_h)=M$, *then it may be represented as*

$$\nu(k)=\sum_{i=1}^{N'}\chi_i'(k)(p_i'(k)+q_i'(k))$$

where $p_i'\in P_{s+1}$, $q_i'\in Q_s$, $\chi_i'\in X$.

Proof. Let us set

$$C=C(a_1,\ldots,a_{h-1}), \qquad C_0=C(a_1,\ldots,a_h), \qquad L=\mathbb{Z}(a_1,\ldots,a_h),$$
$$W=\{s\in\mathbb{Z}_+ \mid k-sa_h\in C\cap L\}, \qquad W_i=\{s\in\mathbb{Z}_+ \mid k-sa_h\in K_i\cap L\}.$$

By (3) we have $\nu(k)=\sum_{s\in W}\nu_1(k-sa_h)$. If $k-sa_h\in L$, then the equality

$$\nu_1(k-sa_0)=\sum_i \chi_i(k-sa_h)(p_i(k-sa_h)+q_i(k-sa_h))$$

implies

$$\nu(k)=\sum_{s\in W}\sum_{i=1}^{N}\chi_i(k-sa_h)g_i(k-sa_h)$$

$$=\sum_{i=1}^{N}\sum_{s\in W}\chi_i(k-sa_h)g_i(k-sa_h)$$

$$=\sum_{i=1}^{N}\sum_{s\in W_i}g_i(k-sa_h)\qquad\text{for all } k\in M,$$

where $g_i(k-sa_h)=p_i(k-sa_h)+q_i(k-sa_h)$.

From the finiteness of $\nu(k)$ and from Proposition 2.2 it follows that C_0 does not contain any linear space except for 0. Therefore Lemma 2.6 is applicable to the cone generated by K_i and a_h. Now Lemma 3.5 implies Lemma 3.6. □

Proof of the Theorem 1.1. Let us consider the system (2). Let V be the subspace generated by $a_1,\dots,a_n$, $r=\dim V$, $a_1,\dots,a_r$ be a linearly independent set of generators.

The number of nonnegative integer solutions of the system $a_1x_1+\cdots+a_rx_r=k$ is characteristic function of $C(a_1,\dots,a_r)\cap\mathbb{Z}(a_1,\dots,a_r)$.

By induction on the number of variables from Lemmas 3.2 and 3.6 it follows that $\nu(k)=\sum_i\chi_i(k)(p_i(k)+q_i(k))$, where $\chi_i\in X$, $p_i\in P_{n-r}$, $q_i\in Q_{n-r-1}$.

Let us choose $n-r$ integer vectors $v_1,\dots,v_{n-r}\in V$ such that $\operatorname{rk}L=n$, where $L=\mathbb{Z}(a_1,\dots,a_n,v_1,\dots,v_{n-r})$. Let χ and g be characteristic functions of V and L respectively. Then

$$\nu(k)=\chi(k)g(k)\sum\chi_i(k)(p_i(k)+q_i(k))\quad\text{for } k\in\mathbb{Z}^n.$$

Now Theorem 1.1 follows from Lemma 3.1 and the fact that the space X is closed with respect to the multiplication.

Corollary 3.7. *For the system (2) and for any integer s there exists a number $D(s)$ such that the number of nonnegative integer solutions of (2) is greater than s for all $k\in C(a_1,\dots,a_n)\cap\mathbb{Z}(a_1,\dots,a_n)$ such that $|k_i|>D(s)$, $i=1,\dots,m$.*

Proof. From the proofs of Lemma 3.5 and Theorem 1.1 it follows that $C=\bigcup_{i=1}^{N}K_i$ and that $\nu(k)=p_i(k)+q_i(k)$ for all $k\in\mathbb{Z}(a_1,\dots,a_n)\cap K_i$, where $p_i\in P_{s+1}$, $q_i\in Q_s$.

It is easy to see that leading coefficients of $p_i(k)$ are positive. On the other hand, a quasipolynomial has only imaginary exponents by definition. Therefore $p_i(k)>p_i(k)+s$ for large k. □

Remark 3.8. The assumption of finiteness of the number of nonnegative integer solutions of (2) is not essential. The following result may be proved.

Theorem 1.1*. *There exists a partition* $\mathbb{R}^m = \bigcup_{i=1}^N C_i$, *where* C_i *is a cone with integer generators and the vertex in* 0 *such that* $\nu(k)|_{C_i \cap \mathbb{Z}^m}$ *is either infinity or a quasipolynomial.*

Proof of Theorem 1.6. An analysis of the proof of Theorem 1.1 shows that a periodic component of $\nu(k)$ appears in two cases. First, characteristic function of a lattice $L \subset \mathbb{Z}^n$ is a periodic function (Lemma 3.2). Second, a periodic function appears if we sum a polynomial over a segment of an arithmetical progression with the difference $d > 1$ (Lemma 3.3).

If the matrix $(a_1, \ldots, a_n)$ is totaliunimodular, then we shall show that a periodic function does not appear neither by the first reason nor by the second one.

In the proof of Theorem 1.1, Lemma 3.2 is used as a basis of induction and it is applied to the system $a_1x_1 + \cdots + a_rx_r = k$ and the lattice $L = \mathbb{Z}(a_1, \ldots, a_r) \subset \mathbb{Z}^r$, where $r = \mathrm{rk}(a_1, \ldots, a_n)$. If $(a_1, \ldots, a_n)$ is totaliunimodular, then $L = \mathbb{Z}^r$ and characteristic function of L equals 1.

On the hth step of induction, $r < h \leqslant n$ (Lemma 3.6) the number $\nu(k)$ of nonnegative integer solutions of the system $a_1x_1 + \cdots + a_hx_h = k$ equals $\sum_{s \in W} \nu_1(k - sa_h)$, where $\nu_1(t)$ is the number of solutions of the system $a_1x_1 + \cdots + a_{h-1}x_{h-1} = t$,

$$W = \{s \in \mathbb{Z}_+ \mid k - sa_h \in C(a_1, \ldots, a_{h-1}) \cap \mathbb{Z}(a_1, \ldots, a_{h-1})\}.$$

Since the matrix $(a_1, \ldots, a_n)$ is totaliunimodular, $\mathbb{Z}(a_1, \ldots, a_{h-1}) = \mathbb{Z}^r$ and $k - sa_h \in \mathbb{Z}(a_1, \ldots, a_{h-1})$ for all integer s. By Lemma 2.6(1) there exists a partition $C(a_1, \ldots, a_h) = \bigcup_{i,j} K_{ij}$ such that $W = W_{ij}$ for $k \in K_{ij}$, where

$$W_{ij} = \{t \in \mathbb{Z} \mid l_{ij}k + s_{ij} \leqslant \alpha_{ij}t \leqslant l'_{ij}k + s'_{ij}\}.$$

By Lemma 2.6(3) $\alpha_{ij} = 1$. Now if $\nu_1(t)|_{C \cap \mathbb{Z}^r}$ is a polynomial, then Lemma 3.4 implies that $\nu(k)$ is a polynomial for all $k \in C \cap K_{ij} \cap \mathbb{Z}^r$. □

4. Proof of Theorem 1.8

Let us set $L(b) = \max\{cx \mid Ax = b, x > 0\}$, where $A = (a_1, \ldots, a_n)$ is a $m \times n$ integer matrix, $b \in \mathbb{Z}^m$, c, $x \in \mathbb{Z}^n$.

The following assertion is well known.

Lemma 4.1. $L(b_1 + b_2) \geqslant L(b_1) + L(b_2)$, $M(b_1 + b_2) \geqslant M(b_1) + M(b_2)$.

Lemma 4.2. *There exists a partition* $\mathbb{R}^m = \bigcup_{i=1}^N C_i$, *where* C_i *is a cone with integer generators and the vertex in* 0 *such that* $L(b_1+b_2) = L(b_1) + L(b_2)$ *for all* $b_1, b_2 \in C_i$, $i = 1, \ldots, N$.

Proof. We shall suppose that $m = \operatorname{rk} A$ and the polyhedron $Ax = b$, $x \geqslant 0$ is bounded for all $b \in \mathbb{Z}^m$.

Let V be an extremal vertex of the polyhedron $Ax = b$, $x \geqslant 0$ with respect to the function cx. If we rearrange coordinates appropriately then $V = (v_1, \ldots, v_m, 0, \ldots, 0)$ and the system of inequalities may be rewritten as follows: $A_1x_1 + A_2x_2 = b$, where A_1 is a $m \times m$ submatrix of A, $\det A_1 \neq 0$. Therefore $x_1 = A_1^{-1}(b - A_2x_2)$, $(v_1, \ldots, v_m) = A_1^{-1}b$, $L(b) = c_1A_1^{-1}b$ and $cx = c_1A_1^{-1}b + (c_2 - c_1A_1^{-1}A_2)x_2$. We have $c_2 - c_1A_1^{-1}A_2 \leqslant 0$. In the opposite case it would be possible to increase x_2 as well as cx and at the same time to keep the relation $x_1 = A_1^{-1}(b - A_2x_2) \geqslant 0$.

Hence for all $b_1, b_2 \in \mathbb{Z}^m$ such that $A_1^{-1}b_1 \geqslant 0$, $A_1^{-1}b_2 \geqslant 0$ the equalities $L(b_1) = c_1A_1^{-1}b_1$ and $L(b_2) = c_1A_1^{-1}b_2$ are valid. Therefore $L(b_1+b_2) = L(b_1) + L(b_2)$. Now we set $C_i = \{x \in \mathbb{Z}^m \mid x = A_1^{(i)}y,\ y \in \mathbb{Z}_+^m\}$ where $A_1^{(i)}$ pass through all $m \times m$ minors of A such that $c_2 - c_1A_1^{(i)-1}A_2^{(i)} \leqslant 0$. □

Theorem 1.8 is implied by the following lemma.

Lemma 4.3. *Let* $\mathbb{R}^m = \bigcup_{i=1}^N C_i$ *be the partition of Lemma* 4.2. *There are lattices* $L_i \subset \mathbb{Z}^m$ *and vectors* $v_i \in \mathbb{Z}^m$, $i = 1, \ldots, N$ *such that for all* $b_1 \in L_i \cap C_i$, $b_2 \in v_i + L_i \cap C_i$, $b_3 \in C_i \cap \mathbb{Z}^m$ *equalities* $M(b_1) = L(b_1)$ *and* $M(b_2+b_3) = M(b_2) + M(b_3)$ *are valid.*

Proof. Let C be a cone of the partition and $x = (x_1, x_2)$, $c = (c_1, c_2)$, $A = (A_1, A_2)$ be a representation of data of the problem with respect to an optimal vertex. As in Lemma 4.2 we have $c_2 - c_1A_1^{-1}A_2 \leqslant 0$.

If $b_1 \in L \cap C$, then $L(b_1) = c_1A_1^{-1}b_1 = c(A_1^{-1}b_1, 0)$. Since $(A_1^{-1}b_1, 0)$ is an integer vector, $M(b_1) \geqslant c(A_1^{-1}b_1, 0) = L(b_1)$ and hence $M(b_1) = L(b_1)$.

Now let $b_2 \in L \cap C$, $w \in \mathbb{Z}^m \cap C$ and $(x_1(b_2+w), x_2(b_2+w))$ be an optimal integer solution of the following problem:

$$\max\{c_1x_1 + c_2x_2 \mid A_1x_1 + A_2x_2 = b_2 + w,\ x_1, x_2 \geqslant 0\}. \tag{6}$$

Let a_{2i} be the ith column of A_2, $d_i = \min\{t \in \mathbb{Z}_+ \mid ta_i \in L\}$ and x_{2i} be the ith coordinate of x_2. Then the inequality $c_2 - c_1A_1^{-1}A_2 \leqslant 0$ implies that $x_{2i}(b_2+w) \leqslant d_i$. Therefore

$$w \in \{x \in \mathbb{R}^m \mid x = A_2(y_1, \ldots, y_{n-m}),\ 0 \leqslant y_i \leqslant d_i\} = W$$

and if $v \in \mathbb{Z}^m$, $v - w \in C$ for all $w \in W$, then $x_1(b_2+v) = A_1^{-1}(b_2+l)$ for some

$l \in L \cap C$. Thus, $x_1(b_2+v) \geqslant A_1^{-1}b_2$ and $(x_1(b_2+v)-A_1^{-1}b_2,\ x_2(b_2+v))$ is a feasible solution of the problem (6) with the constraint vector equal v. Hence $M(b_3) \geqslant M(b_2+b_3)-M(b_2)$ for all $b_3 \in v+\mathbb{Z}^m \cap C$ and $b_2 \in L \cap C$. Now Lemma 4.1 implies that $M(b_3)=M(b_2+b_3)-M(b_2)$. □

References

[1] E. Ehrhart, Sur un problème de géométrie diophantienne linéaire (Polyedres et réseaux), Z. Math. 226, 227 (1967).
[2] M. Hall, Combinatorics (Moscow, Mir, 1970).
[3] P. MacMullen, Metrical and combinatorial properties of convex polytopes, Proc. Intern. Congress. Math., Vancouver, Vol. 1 (1974) 431–435.
[4] D.N. Bernstein, On the number of integer points in integer polyhedra, Funct. Analysis and Appl. 10 (3) (1973) 72–73.
[5] H.G. Weidner, A periodicity lemma in linear Diophantine analysis, J. Number Theory 8 (1) (1976) 99–108.
[6] N.N. Ivanov and V.N. Shevchenko, On the structure of a semilattice with finite number of generators, Dokl. Akad. Nauk BSSR XIX (9) (1975) 773–774.
[7] K. Lancaster, Mathematical Economics (Moscow, Sov. Radio, 1972).

P. Hansen, ed., Studies on Graphs and Discrete Programming

THE ADMINISTRATION OF STANDARD LENGTH TELEPHONE CABLE REELS*

Túlio GONTIJO ROCHA, Ramiro de ARAUJO ALMEIDA,
Alberto de Oliveira MORENO
TELERJ, Telecomunicações do Rio de Janeiro S.A., Rio de Janeiro, Brazil

Nelson MACULAN
UFRJ, Universidade Federal do Rio de Janeiro, Rio de Janeiro, Brazil

When planning urban telephone network maintenance and expansion, one has to assign reels of cable to supply all lengths of cable needed.

In an ideal world, lengths needed, and the timing of their use, behave exactly as planned. The best solution for buying and using cable reels would then be: match cables exactly to the lengths demanded, and buy them accordingly; use them as needed.

In the real world, events may arise which turn desirable to buy cables in standard lengths. The administration of such a case is not at all simple: it involves the assignment of reels from which to cut the necessary lengths, in such a way as to minimize losses.

There are some model formulations to represent the problem of supplying lengths needed of some material from standard lengths. Unfortunately, most part of these models are posed from the manufacturer's point of view. That means: as factories do not keep significant finished materials inventory, preferring to produce when needed for supply, those models do not consider carrying costs.

In this paper, a model of the problem is presented, that considers both carrying and surplus costs. Some heuristics are proposed to give approximate solutions, and their performances are evaluated.

1. Introduction

This paper treats the problem of satisfying the demand for cables of specified length, occurring in consecutive work periods. These lengths must be cut from combinations of standard lengths and leftovers from previous periods.

This problem has been treated by many authors, under many names: minimization of scrap [1], trim problem [2], cutting stock problem [3–8, 12, 19], loading problem [11, 13, 14], 0–1 multiple-knapsack problem [15] and bin packing [16–18]. The approaches can be classified according to the solution technique: column generation [1–10], zero–one programming [8–15], combinatorial heuristics [16–18] and recently, subgradient optimization [19].

* An earlier version of this paper was presented at the Tenth International Symposium on Mathematical Programming, Montreal, August 27–31, 1979.

Mathematical models are here presented emphasizing the application to the supply of telephone cables [20]. Solution algorithms use heuristics naturally suggested.

Three models are presented in Section 2. They are characterized by: (i) demand must be satisfied from existing inventory, meaning that no backlogs are allowed; (ii) the problem must be feasible; (iii) different lengths can be assumed to exist in inventory. The first model seeks to minimize scraps. The second, tries to minimize the final number of cable reels, assuming that reels must be scrapped if they contain lengths below a specified minimum. The third model tries to combine both objectives, thus the resulting problem is of a vector optimization nature.

In Section 3 we describe a heuristic procedure designed to obtain approximate solutions for the bi-criterion problem. The acceptability of approximate, instead of optimal solutions, is justified on two grounds: first, because the problem is 'NP-hard in the strong sense' [31]; second, because this is a large scale problem.

We describe practical details in Section 4. This problem stemmed from a real problem that affected TELERJ—a government-owned company that operates telephone services in the State of Rio de Janeiro, Brazil. Results of the model implementation are presented in physical and economical terms. Section 5 is devoted to the perspectives of future research.

2. Model formulation

Define

λ_j = length of the jth incoming order, representing a stretch of cable to be laid; $j \in J = \{1, 2, \ldots n\}$.

β_i = length of cable remaining in the ith reel in the beginning of the period; $i \in I = \{1, 2, \ldots, m\}$.

λ^* = minimum length allowed in a reel (note: $\beta_i \geq \lambda^*$, all i, must be true)

$x_{ij} = 1$ if the jth stretch is to be supplied from the ith reel,
$= 0$ otherwise.

$u_i = 1$ if reel i has been used at least once,
$= 0$ otherwise.

$v_i = 1$ if reel i remains in stock at the end of the period, i.e., if length of cable remaining in the ith reel at the end of period is no less than λ^*,
$= 0$ otherwise.

Model I, formulated below, has the economic motivation of minimizing scraps: Find x_{ij}, u_i and min Z such that

$$Z = \sum_i \beta_i u_i, \tag{1}$$

$$\sum_j \lambda_j x_{ij} \leqslant \beta_i \quad \text{for all } i, \tag{2}$$

$$\sum_i x_{ij} = 1 \quad \text{for all } j, \tag{3}$$

$$-x_{ij} + u_i \geqslant 0 \quad \text{for all } i, j, \tag{4}$$

$$x_{ij} \in \{0, 1\} \quad \text{for all } i, j, \tag{5}$$

Some notes can be added:

(i) It can be shown that [20]: The objective function (1) is equivalent to

$$Z' = \sum_i s_i u_i \tag{6}$$

where s_i is the slack variable that can be introduced in (2), and means final length of reel i; this leads to the not so obvious conclusion that minimization of the total length of used reels is akin to minimization of leftovers. The linear form (1) is more tractable.

(ii) It could seem more appropriate to use the objective function

$$P(\lambda^*) = \sum_i s_i (1 - v_i)$$
$$= \text{sum of scraps at the end of the period.} \tag{7}$$

Besides being nonlinear, this objective function is really a worse representation of the practical problem than (1). In fact, there would be no difference between exhausting a reel and using a reel partially, without taking it out from inventory. In the long run, its use would thus lead to a huge inventory of partially used reels, which is certainly not desirable.

(iii) Eilon and Christofides [11] formulated a model similar to model I, with the objective function

$$\text{EC} = \sum_i u_i = \text{number of reels effectively used in the period.} \tag{8}$$

This objective function is not appropriate to our practical problem of minimizing scraps. Consider for example that we have n orders of length λ $(1 \leqslant \lambda^* < \lambda)$, to be satisfied from n reels of length λ and n of length $\lambda + 1$. A solution minimizing the objective function (8) will use *any* n reels, irrespective of their lengths. If (1) is used instead, only reels with length λ will be used, with zero scraps.

(iv) Still referring to Eilon and Christofides, an alternative to our definition of u_i is suggested:

$$\sum_j x_{ij} \leqslant L \cdot u_i, \qquad u_i \in (0, 1\}, L > j \quad \text{for all } i \tag{9}$$

in place of (4). There are n times less constraints in (9) than in (4). Yet, its use requires u_i to be 0–1.

(v) The use of the assignment variables x_{ij} in model I, suggests some relationship with the classical problem of assignment. It can indeed be shown that model I is equal to the 'generalized assignment problem' (see [21, 22]) but for the objective function, which is nonlinear in our case [20].

(vi) The models' characteristics (i) and (ii), outlined in the introduction, can be better understood as follows. As 'demand must be satisfied from existing inventory', it must be true that:

$$\exists I_1 \subset I \quad \text{such that} \quad \sum_{j \in J} \lambda_j \leqslant \sum_{i \in I_1} \beta_i \,.$$

In addition, as 'the problem must be feasible', it must be true that: $\exists P$ (partition of J) such that

–for each $p \in P, \exists i \in I_1$ such that $\sum_{j \in p} \lambda_j \leqslant \beta_i$, and

–for $p_1, p_2 \in P, \exists i_1, i_2 \in I_1$ such that (besides respecting the condition above for p_1, i_1 and p_2, i_2) $p_1 \neq p_2 \Rightarrow i_1 \neq i_2$.

Model II can be formulated as: find x_{ij}, v_i and min W such that

$$W = \sum_i v_i = \text{number of reels in inventory at the end of the period,} \tag{10}$$

$$\sum_j \frac{\lambda_j}{\beta_i} x_{ij} + v_i \geqslant \frac{\beta_i - \lambda^* + 1}{\beta_i} \quad \text{for all } i, \tag{11}$$

$$v_i \in \{0, 1\} \quad \text{for all } i \tag{12}$$

and constraints (2), (3), (5) of model I are satisfied. Models I and II could lead to different solutions. Model I will have a preference for using reels with lengthier cables, while model II will do the opposite. This is due to the flexibility that can be achieved by using reels with lengthier cables. Total demand being constant, using small length reels will obviously require more reels than using large length ones.

Model III below combines both objective functions. Define

$\mathbf{0}$ = line vector of m zeros,

$\mathbf{1}$ = line vector of m ones,

$\boldsymbol{\beta} = (\beta_1\, \beta_2 \cdots \beta_m)$,

$\boldsymbol{u} = (u_1\, u_2 \cdots u_m)'$,

$\boldsymbol{v} = (v_1\, v_2 \cdots v_m)'$.

We want to find x_{ij}, $\boldsymbol{u}$ and $\boldsymbol{v}$ that optimize

$$\begin{bmatrix} Z \\ W \end{bmatrix} = \begin{bmatrix} \boldsymbol{\beta} & \mathbf{0} \\ \mathbf{0} & \mathbf{1} \end{bmatrix} \begin{bmatrix} \boldsymbol{u} \\ \boldsymbol{v} \end{bmatrix} \tag{13}$$

subject to constraints (2), (3), (4), (5), (11), (12). We would like to minimize both Z and W, but this is not usually possible. Our optimization must then be understood in the sense of finding *efficient* points of the feasible set [23–26].

3. Heuristic procedure

The solution of model III is not practical, if we think of finding exactly the set of efficient points. This can be seen by studying the algorithms developed to date [25, 26] and by considering the scale of the actual problem. Dimensions of $m = 250$ and $n = 100$ are common, and would lead to intolerable high demands of computing time. This approach could, however, be useful for cables with low demand, or for very short periods of time.

On the other hand, the solution of models I and II, however possible [8–11, 13–15], would also require too much computer time.

These difficulties are not unexpected: the bin packing problem is a particular case of ours, if we restrict the reels to have the same lengths; being NP-hard in the strong sense, it is impossible to solve it using a pseudo-polynomial time algorithm, unless P = NP [27–31]; by implication, this must be true also in our case.

The use of approximate solutions was mandatory due to the urgent need of some practical results, as explained in Section 4. The above arguments, however, came as a help to ease our conscience.

The heuristics described below were developed to give an approximate solution to the problem, provided:

(i) The task of writing a program should not be too time consuming.

(ii) The computer program should be economic in terms of equipment requirements (CPU time, memory).

(iii) Solutions should be satisfactory in terms of attaining both objectives in (13).

The procedure decides which reels will be considered to satisfy demand. This is done by consulting a preestablished list, in which reels appear in order of priority. For each group of reels a subproblem is then solved: minimizing scraps by choosing orders that will be satisfied from this group of reels.

The priority list must be constructed considering the available length on each reel in inventory. The compromise between the two objectives is so achieved: the subproblem seeks to minimize losses, given priorities; the rule of priorities is responsible for the minimization of the number of reels at the end of the period.

We can visualize a reasonable priority rule from the discussion in Section 2: reels with shorter lengths should be tried first. This rule has also the advantage of not contributing to worsen the flexibility of use of short length reels.

The following notation will be used to describe the algorithm:

$\bar{i}$ = last reel selected;
$\bar{m}$ = number of reels of the same length as $\bar{i}$;
J' = set of orders not satisfied from reels that were selected in the previous steps;
J_1 = set of orders that, besides not yet satisfied, have length not larger than that of $\bar{i}$.

Description of the algorithm

Step 0: Rearrange I such that $\beta_{i+1} \geq \beta_i$, for any $i \in I - \{m\}$. Let $J' = J$, $\bar{i} = 0$, $x_{ij} = 0$, $i \in I$, $j \in J$.
Step 1: Form $J' = \{j \in J \mid \sum_i x_{ij} = 0\}$.
Step 2: If $J' = \emptyset$, then stop: a solution has been found.
Step 3: If $\bar{i} = m$, then stop: a feasible solution has not been achieved.
Step 4: Let $\bar{m} = \max\{i \in \mathbb{N} \mid i \leq m - \bar{i} \wedge \beta_{\bar{i}+i} = \beta_{\bar{i}+1}\}$.
Step 5: Let $\bar{i} = \bar{i} + \bar{m}$ and form the set $J_1 = \{j \in J' \mid \lambda_j \leq \beta_{\bar{i}}\}$.
Step 6: If $J_1 = \emptyset$, then go to Step 3.
Step 7: Solve the subproblem for $\bar{i}$, $\bar{m}$, J_1, and go to Step 1.

Description of the subproblem (minimization of losses)

First case: $\bar{m} = 1$

If $\bar{m} = 1$, the subproblem is a special case of the 0–1 knapsack problem [32–34]. We can see this, formulating model IV below, in which $\bar{i}$ and J_1 are assumed given.

$$\max \sum_{j \in J_1} \lambda_j x_{\bar{i}j} \tag{14}$$

subject to

$$\sum_{j \in J_1} \lambda_j x_{\bar{i}j} \leq \beta_{\bar{i}}, \tag{15}$$

$$x_{\bar{i}j} \in \{0, 1\}, \quad j \in J_1. \tag{16}$$

Note that the objective function coefficients (14) are equal to those in constraint (15). Model IV can be seen as a special case of the 0–1 knapsack problem, where this condition is not needed. This special case is called sum of subset problem [15]. Note also that model IV can be obtained from model I, after some manipulations [20].

The solution of model IV can be obtained using algorithms to solve the related knapsack problem (see [32–43]). However, due to the fact that the 0–1 KP is NP-hard (the language [29] that results from the use of an encoding scheme to the decision problem is NP-complete [27]), efforts have been dedicated to its reduction [44–47] or to obtaining approximate solutions [48–53]. (Note that the reduction methods are absolutely ineffective for the sum of subset problem: as can be seen in model IV, the ratios of the objective function and constraint coefficients for fixed j, all j, are equal to 1.)

Measured in terms of computing time, algorithms—among them pseudo-polynomial time algorithms—more efficient than the classical ones resulted from this approach.

This can be seen from the examples below (from [49, 51]):

(i) 0–1 knapsack problem:

–o(min$\{2^r, rM\}$) in [6, 54] and o(min$\{2^{r/2}, rM\}$) in [49], where r is the number of variables and M is the constraint's right hand side, for obtaining the optimal solution;

–o(kr^{k+1}) in [50], where r is as defined above and k is a parameter for the desired precision setting, such that $(P^* - \text{PMAX})/P^* < 1/(k+1)$, being P^* the value of the optimal solution and $PMAX$ the value of the approximate solution obtained;

(ii) sum of subset problem:

–o(max$\{2^r, r^2Q\}$) in [55] and o(max$\{2^{r/2}, r^2Q\}$) in [49], where r is as defined above and Q is the total number of partitions, for obtaining all the optimal solutions;

–o(r^k) in [48], where r and k are as defined above, for obtaining an approximate solution.

Second case: $\bar{m} > 1$

If $\bar{m} > 1$, two cases must be considered:

(i) If the reels suffice to satisfy all orders in J_1, the subproblem can be represented by model I; note also that, as reels have, in this case, the same length, the subproblem can be seen as the bin-packing problem.

(ii) If there are not enough reels, the subproblem can be seen as the 0–1-many-knapsack problem [13], 0–1 multiple-knapsack problem [15], or still as a kind of loading problem [11].

Heuristics for the subproblem solution

It must be added to the previous subsection discussion that, as we are here solving a subproblem, there's no reason for believing that it's better to obtain the optimal solution to each subproblem to be solved. In fact, a comparison of these two options was carried out [20] with competitive results. This reasoning justifies the simplification introduced by the following heuristics, although the comparison was made only after the heuristics implementation.

The simplification introduced is: irrespective of the value of $\bar{m}(\bar{i}$ or more), and of the reels' capability to satisfy all orders of J_1, the subproblem is solved as the bin packing problem. That means: start by partitioning J_1 in subsets eligible to use the $\bar{m}$ selected reels, and then select elements of the partition to use the reels in a best way—the number of elements selected will be equal to the least of $\bar{m}$ or the number of sets in the partition.

The rule for partition leads to the same solution as the First-Fit-Decreasing Algorithm (FFD) [16–18]. However, the construction of the solution is done in a different way: our rule defines each subset completely, before beginning to define another; the FFD's rule defines all subsets concurrently.

Two algorithms are presented for the subproblem, one with our rule for partition introduced and another with the FFD's.

Algorithm with our rule for partition

The rule for partition can be understood, if we note that:

(i) A subset will be formed by choosing orders not yet included in other subsets.

(ii) The order with greatest length will be chosen first.

(iii) An order will be added in sequence if it has the greatest length not greater than the length of the reel(s) selected (length $\beta_{\bar{i}}$) less the sum of lengths already added in this subset, until no order can be added.

(iv) Repeat the procedure until J_1 is exhausted.

The following notation will be used:

k = number of subsets in the partition,
k' = number of subsets selected,
$\bar{k}$ = index of subset being selected to use the selected reel(s),
K = set of generated subsets,
α_k = sum of lengths of orders in subset k,
y_j = subset to which order j has been assigned,

$\bar{j}$ = order selected to subset k, which is being generated,
J_1' = set of orders yet to be assigned,
J_2 = set of orders eligible to be members of subset k, which is being generated,
J_3 = auxiliary subset.

Description of the algorithm:

Step 0: Let $k=0$ and $J_1'=J_1$.
Step 1: Let $k=k+1$, $\alpha_k=0$ and $J_3=J_1'$.
Step 2: Find $\bar{j}\in J_3$ such that $\lambda_{\bar{j}}\geq\lambda_j$, $\forall j\in J_3$. Ties will be broken choosing any such $\bar{j}$.
Step 3: Let $y_{\bar{j}}=k$, $\alpha_k=\alpha_k+\lambda_{\bar{j}}$, $J_1'=J_1'-\{\bar{j}\}$.
Step 4: If $J_1'=\emptyset$, then go to Step 8.
Step 5: Let $J_2=\{j\in J_3 \mid \lambda_j\leq\beta_{\bar{i}}-\alpha_k\}$.
Step 6: If $J_2=\emptyset$, then go to Step 1.
Step 7: Let $J_3=J_2$ and go to Step 2.
Step 8: Let $K=\{1,\ldots,k\}$ and $k'=1$.
Step 9: If $k'>\min\{\bar{m},k\}$, then stop.
Step 10: Find $\bar{k}\in K$ such that $\alpha_{\bar{k}}\geq\alpha_k$, $\forall k\in K$. Ties will be broken choosing any such $\bar{k}$.
Step 11: Let $K=K-\{\bar{k}\}$, $J_3=\{j\in J_1 \mid y_j=\bar{k}\}$ and $x_{(\bar{i}-\bar{m}+k'),j}=1$, $\forall j\in J_3$.
Step 12: Let $k'=k'+1$ and go Step 9.

Algorithm with FFD's rule for partition

The rule for partition can be understood, if we note that:

(i) All subsets will be formed concurrently by sequential inclusion of orders in reels initially filled to level 0.

(ii) Orders will be included in subsets in the non-increasing order of their lengths.

(iii) An order will be included in the first reel where it fits; until J_1 is exhausted.

The following notation will be used, in addition to that used for the algorithm with our rule for partition ($\bar{j}$, J_1 and J_2 not used here, $\bar{k}$, K and J_3 used implicitly):

n' = number of elements in J_1,
j_l = lth order included in a subset, according to (ii) above,
l = index of order j_l, for which a reel is sought.

Description of the algorithm:

Step 0: Arrange the elements of J_1 in the order $j_1,\ldots j_{n'}$, such that $\lambda_{j_l}\geq\lambda_{j_{l+1}}$, $l\in\{1,\ldots,n'-1\}$, and let $k=0$, $k'=1$ and $l=1$.

Step 1: Let $k = k+1$ and $\alpha_k = 0$.
Step 2: If $\lambda_{j_l} \leq \beta_{\bar{i}} - \alpha_{k'}$, then let $y_{j_l} = k'$, $\alpha_{k'} = \alpha_{k'} + \lambda_{j_l}$ and go to Step 4. Else let $k' = k'+1$.
Step 3: If $k' \leq k$, then go to Step 2. Else go to Step 1.
Step 4: If $l < n'$, then let $l = l+1$, $k' = 1$ and go to Step 2.
Step 5 to 8: Same as Steps 9 to 12 of the algorithm with our rule for partition.

Worst-case performance bounds for the algorithms

A worst-case analysis was undertaken to evaluate the FFD algorithm [16–18].

Define L^* = minimum admissible number of elements of the partition and FFD(L) = number of elements in the partition (indicated by the algorithm). It can be proved that

$$\mathrm{FFD}(L) \leq \tfrac{11}{9} L^* + 4 \tag{17}$$

from which

$$\lim_{L^* \to \infty} \frac{\mathrm{FFD}(L)}{L^*} = \frac{11}{9} = 1.22. \tag{18}$$

We can obtain similar bounds for our algorithms. Define:

S = unused fraction of the total length contained in the reels in the partition;
S^* = unused fraction of the length contained in the partition with the minimum number of elements: k^*.

It can be proved that [20]

$$S \leq S^* + \tfrac{2}{9} + 4/k^* \tag{19}$$

which implies, if we let $k^* \to \infty$

$$S \leq S^* + \tfrac{2}{9} = S^* + 0.22. \tag{20}$$

It is reasonable to assume the existence of better bounds. This is due to the fact that we employ only the elements of the partition whose usage rate is greatest.

Note, however, that a probabilistic evaluation of the FFD algorithm [16] resulted in the limit 1.02, instead of 1.22 obtained in expression (18).

4. Application to a telephone company

This study was performed as part of a system for inventory control of telephone cables. Its main user is TELERJ—Telecomunicações do Rio de

Janeiro, a telephone company which operates in the state of Rio de Janeiro, Brazil.

Operation of telephone exchanges, in Brazil, is controlled by the government through a holding company—TELEBRÁS—of which TELERJ is a subsidiary. The manufacture of equipment, however, is entirely in the hands of private enterprises—cables, for example, are manufactured by a couple of firms, Pirelli being the largest.

When this study was undertaken, by request of the Operations Director, the situation was as follows: cables were not supplied in standard lengths; they were bought in sizes that followed closely the needs of specific projects, and were ordered well in advance with respect to the actual time they would be laid in the ground; this policy, as was envisaged at the time, would result in zero scrap; as some projects were postponed and actual distances did not match exactly those used in design, large amounts of scraps started to appear and inventory mounted to incredible heights; personnel in charge started to feel uneasy, and it was thought that the use of standard lengths would mitigate the loss in sunk capital. Cables would no longer be tied to projects, and would thus not accumulate when delays occurred. There was fear, however, that the amount of scraps would increase, unless procedures were developed to control the assignment of orders to reels. The O.R. staff was then called to provide a solution.

Telephone cables can be classified in types, characterized mainly by:

capacity—number of wires in the cable;

caliber—requirements of transmission quality dictate the caliber of wires in a cable;

coating—aluminium or lead alloys, used for protection.

For practical purposes, different types of cable were treated as completely different products.

Demand was characterized as a set of pairs (order, length) for each month. Only the next month is considered for assignment. Information about other months is useful only for ordering new reels.

The heuristic procedure was tested by simulation using historical data. Results indicated that percent scrap and amount of inventory would have been under acceptable limits, if it had been used in the past. The next step was using the program, with the same set of data, to choose the standard length to adopt for each type of cable. The final choice yielded a percent scrap around 1.5%, for the more commonly used types, and $\lambda^* = 30$ m. These simulation runs were held in March 1976, and were followed by measures designed to support the use of the procedure in a regular basis.

Implementation started in TELERJ in June 1977. There was a significant reduction in inventory, saving much in previously sunk capital: referring to

August, 1979 inventory levels, the reduction was 87%, saving about US$ 30 million; referring to the average inventory levels in the one-year period from September, 1978 to August, 1979, the reduction was 77%, saving about US$ 26 million. Note that the inventory reduction was obtained while demand increased up to four times the previous level. An area of 30 000 m^2 was liberated for other uses. The percent scrap has been approximately 1%, using $\lambda^* = 15$ m. As side effects, there were increases in the efficiency of cable-laying operations. More details can be found in [20].

5. Conclusions

Models of the problem, heuristics to give approximate solutions and results from its implementation in TELERJ were presented. It must be pointed out that: as discussed in Section 3, the solution of model III is not practical and the solution of model I and II would require too much computer time, considering the scale of the actual problem; thus, no effort was made to solve exactly the models, the heuristic approach being preferred. It is possible, by a slight modification in the heuristic procedure—namely selecting just one reel each time—to obtain only single knapsack type subproblems. This alternative was tested, using a dynamic programming procedure as subroutine. In terms of the overall solution, results were competitive with those of our heuristic procedure, for the cases examined.

As lines for future research we would indicate as most profitable:

(i) Analysis of the effects of a reduction in the work period (from one month to one week, for example), which could allow the use of exact algorithms [8–11, 13–15].

(ii) Worst-case and probabilistic analysis [16, 56] for the heuristics described.

(iii) Comparison between solutions obtained by heuristics and by models that use the concept of 'minimum saleable length' [20, 57–60].

Acknowledgements

We wish to acknowledge Dr. Sergio E. Girão Barroso for his help in reading this paper, correcting some mistakes, and giving many useful suggestions. We thank Dr. George L. Nemhauser and Dr. Gabriel R. Bitran for discussing with us, in different occasions, the models and heuristics presented here. We also thank Mr. Orlando Carlos G. de Souza for improving the paper's final version readability. Finally, we thank the referees, whose valuable comments led to improvements in the text. The responsibility for remaining errors we obviously assume.

References

[1] L.V. Kantorovich, Mathematical methods of organizing and planning production, Management Sci. 6 (1960) 363–422.

[2] K. Eiseman, The trim problem. Management Sci. 3 (1957) 279–284.

[3] P.C. Gilmore and R.E. Gomory, A linear programming approach to the cutting stock problem, Operations Research 9 (1961) 849–859.

[4] P.C. Gilmore and R.E. Gomory, A linear programming approach to the cutting stock problem—part II, Operations Research 11 (1963) 863–888.

[5] P.C. Gilmore and R.E. Gomory, Multistage cutting stock problems for two or more dimensions, Operations Research 13 (1965) 94–120.

[6] P.C. Gilmore and R.E. Gomory, The theory and computation of knapsack functions, Operations Research 14 (1966) 1045–1074.

[7] P.C. Gilmore, Cutting stock, linear programming, knapsacking, dynamic programming and integer programming, some interconnections, Annals of Discrete Math. 4 (1979) 217–235.

[8] J.F. Pierce, On the solution of integer programming cutting stock problems by combinatorial programming—part I, IBM Cambridge Center Technical Report No. 320–2001, 1966.

[9] J.F. Pierce, Application of combinatorial programming to a class of all-zero–one integer programming problems, Management Sci. 15 (1968) 191–209.

[10] J.F. Pierce and J.S. Lasky, Improved combinatorial programming algorithms for a class of all-zero-one integer programming problems, Management Sci. 19 (1973). 528–543

[11] S. Eilon and N. Christofides, The loading problem, Management Sci. 17 (1971) 259–268.

[12] N. Christofides and C. Whitlock, An algorithm for two-dimensional cutting problems, Operations Research 25 (1977) 30–44.

[13] G.P. Ingargiola and J.F. Korsh, An algorithm for the solution of 0–1 loading problems, Operations Research 23 (1975) 1110–1119.

[14] M.S. Hung and J.R. Brown, An algorithm for a class of loading problems, Naval Res. Logist. Quart. 25 (1978) 281–297.

[15] M.S. Hung and J.C. Fisk, An algorithm for 0–1 multiple-knapsack problems, Naval Res. Logist. Quart. 25 (1978) 571–579.

[16] D.S. Johnson, Near optimal bin packing algorithm, Ph.D. Dissertation, Massachusetts Institute of Technology, Cambridge, MA, 1973.

[17] D.S. Johnson, Fast algorithms for bin packing, J. Comput. System Sci. 8 (1974) 272–314.

[18] D.S. Johnson, A. Demers, J.D. Ullman, M.R. Garey and R.L. Graham, Worst-case performance bounds for simple one-dimensional packing algorithms, SIAM J. Comput. 3 (1974) 299–325.

[19] B.L. Golden, approaches to the cutting stock problem, AIEE Trans. 8 (1976) 265–274.

[20] T.G. Rocha, Aprovisionamento de Bobinas de Cabos Telefônicos: uma applicação de programação combinatória, M.Sc. Dissertation, COPP-UFRJ, Rio de Janeiro, RJ, Brazil, 1978.

[21] G.T. Ross and R.M. Soland, A branch and bound algorithm for the generalized assignment problem, Math. Programming 8 (1975) 91–103.

[22] G.T. Ross and A.A. Zoltners, Weighted assignment models and their application, Management Sci. 25 (1979) 683–696.

[23] B. Roy Problems and methods with multiple objective functions, Math. Programming 1 (1971) 239–266.

[24] S. Zionts, Integer linear programming with multiple objectives, Ann. Discrete Math. 1 (1977) 551–562.
[25] G.R. Bitran, Linear Multiple objective programs with zero–one variables, Math. Programming 13 (1977) 121–139.
[26] G.R. Bitran, Theory and algorithms for linear multiple objective programs with zero–one variables, Technical Report No. 150, Operations Research Center, MIT, 1978.
[27] R.M. Karp, Reducibility among combinatorial problems, in R.E. Miller, T.W. Thatcher, eds., Complexity of Computer Computations (Plenum, New York, 1972) 85–103.
[28] R.M. Karp, On the computational complexity of combinatorial problems, Networks 5 (1975) 45–68.
[29] A.V. Aho, J.E. Hopcroft and J.D. Ullman, The Design and Analysis of Computer Algorithms (Addison-Wesley, Reading, MA, 1974).
[30] S. Sahni and E. Horowitz, Combinatorial problems: reducibility and approximation, Operations Research 26 (1978) 718–759.
[31] M.R. Garey and D.S. Johnson, 'Strong' NP-completeness results: motivation, examples and implications, J. Assoc. Comput. Mach. 25 (1978) 499–508.
[32] R.S. Garfinkel and G.L. Nemhauser, Integer Programming (Wiley, New York, 1972).
[33] H.M. Salkin, Integer Programming (Addison-Wesley, Reading, MA, 1975).
[34] H.M. Salkin and C.A. DeKluyver, The knapsack problem: a survey, Naval Res. Logist. Quart. 22 (1975) 127–144.
[35] D. Fayard and G. Plateau, Resolution of the 0–1 knapsack problem: comparison of methods, Math. Programming 8 (1975) 272–307.
[36] R.M. Nauss, An efficient algorithm for the 0–1 knapsack problem, Management Sci. 23 (1976) 27–31.
[37] E. Balas and E. Zemel, Solving large zero–one knapsack problems, Management Sciences Report, 1977.
[38] J.C. Fisk, An initial bounding procedure for use with 0–1 single knapsack algorithm, Opsearch 14 (1977) 88–98.
[39] S. Martello and P. Toth, An upper bound for the zero–one knapsack problem and a branch and bound algorithm, European J. Oper. Res. 1 (1977) 169–175.
[40] S. Martello and P. Toth, Algorithm for the solution of the 0–1 single knapsack, Computing 21 (1978) 81–86.
[41] H. Müller-Merbach, An improved upper bound for the zero–one knapsack problem: a note on the paper by Martello and Toth, European J. Oper. Res. 2 (1978) 212–213.
[42] U. Suhl, An algorithm and efficient data structures for the binary knapsack problem, European J. Oper. Res. 2 (1978) 420–428.
[43] A.A. Zoltners, A direct descent binary knapsack problem, J. Assoc. Comput. Mach. 25 (1978) 304–311.
[44] G.P. Ingargiola and J.F. Korsh, Reduction algorithm for zero–one single knapsack problems, Management Sci. 20 (1973) 460–463.
[45] G.P. Ingargiola and J.F. Korsh, A general algorithm for one-dimensional knapsack problems, Operations Research 25 (1977) 752–759.
[46] D. Fayard and G. Plateau, Reduction algorithms for single and multiple constraints 0–1 linear programming problems, Proceedings of the Congress Methods of Mathematical Programming, Zakopane (Poland), 1977.
[47] R.S. Dembo and P.L. Hammer, Preprocessing knapsack problems, Research Report No. 26 of the School of Organization and Management, Yale University, 1978.
[48] D.S. Johnson, Approximation algorithms for combinatorial programming, Proceedings of the Fifth Annual ACM Symposium on Theory of Computing (1973) 38–49.
[49] E. Horowitz and S. Sahni, Computing partitions with applications to the knapsack problem. J. Assoc. Comput. Mach. 21 (1974) 277–292.
[50] S. Sahni, Approximate algorithms for the 0/1 knapsack problem, J. Assoc. Comput. Mach. 22 (1975) 115–124.

[51] O.H. Ibarra and C.E. Kim, Fast approximation algorithms for the knapsack and sum of subset problems, J. Assoc. Comput. Mach. 22 (1975) 463–468.
[52] A.K. Chandra, D.S. Hirschberg and C.K. Wong, Approximate algorithms for the knapsack problem and its generalizations, Report RC 5616, Research Center, Yorktown Heights, NY, 1975.
[53] E.L. Lawler, Fast approximation algorithm for knapsack problems, Research report of the Electronics Research Laboratory, University of California, Berkeley, 1977.
[54] G.L. Nemhauser and Z. Ullman, Discrete programming and capital allocation, Management Sci. 15 (1969) 494–505.
[55] D.R. Musser, Algorithms for polynomial factorization, Ph.D. Dissertation, University of Wisconsin, Madison, WI, 1971.
[56] R. M. Karp, The probabilistic analysis of some combinatorial search algorithms, in J.F. Traub, ed., Algorithms and Complexity: New Directions and Recent Results (Academic Press, New York, 1976) 1–19.
[57] L.P. Northcraft, Computerized inventory control in cable manufacture, The Western Electric Engineer 17 (1973) 56–60.
[58] L.P. Northcraft, Computerized cable inventory, Industrial Engineering 6 (1974) 45–49.
[59] L.P. Northcraft, A model for controlling cable inventory, The Western Electric Engineer 22 (1978) 26–30.
[60] R.S. Stainton, The cutting stock problem for the stockholder of steel reinforcement bars, Oper. Res. Quart. 28 (1977) 139–149.

P. Hansen, ed., Studies on Graphs and Discrete Programming

THRESHOLD NUMBERS AND THRESHOLD COMPLETIONS

P.L. HAMMER
Department of Combinatorics and Optimization, University of Waterloo, Waterloo, Ontario N2L 3G1, Canada

T. IBARAKI
Department of Applied Mathematics and Physics, Faculty of Engineering, Kyoto University, Kyoto, Japan

U.N. PELED
Department of Computer Science, Columbia University, New York, NY 10027, USA

The threshold number $t(f)$ of a positive Boolean function $f(x_1, \ldots, x_n)$ is the least number of linear inequalities whose solution set in 0–1 variables is the set of zeroes of f. These inequalities can be taken with nonnegative coefficients. If P is the collection of the prime implicants of f and $S \subseteq P$, then f_S denotes the Boolean sum of the prime implicants in S. S is called a threshold subcollection in f if there exists a threshold function g satisfying $f_S \leqslant g \leqslant f$. The threshold number $t(f)$ is equal to the least number of threshold subcollections in f that cover P. Thus $t(f) \leqslant |P| \leqslant \binom{n}{\lfloor n/2 \rfloor}$. A graph G_f with the vertex set P is defined, and it is shown that $t(f)$ is not less than its chromatic number, generalizing the results of Chvátal and Hammer. This is used to show that the maximum value of $t(f)$ is at least $\binom{n}{\lfloor n/2 \rfloor}/n$. When each prime implicant has the form x_ix_j, f is called graphic and corresponds naturally to a graph G with vertex set $\{x_1, \ldots, x_n\}$ and edge set P. In that case x_ix_j and x_kx_l are adjacent in G_f if and only if $x_ix_k, x_jx_l \notin P$ or $x_ix_l, x_jx_k \notin P$. Also for f graphic, a subset $S \subseteq P$ is a threshold subcollection in f if and only if G does not contain a closed walk alternating between S-edges and non-edges of G. This is proved both from the separation theorem for polytypes and by an $O(n^3)$ algorithm. When G_f is bipartite and S is one of the colours in a bicolouring of G_f, a further simplification of the condition is achieved.

1. Introduction

We consider here the following general type of problems: find a system with the fewest linear inequalities having a specified set F of solutions in 0–1 variables. If we can solve this problem efficiently for a given F, then for any 0–1 programming problem of the form 'maximize cx for $x \in F$', we have a most compact representation as a 0–1 linear programming problem. It is frequent in integer and 0–1 programming that F is already specified by linear inequalities, and it is well-known that the work involved in the solution of the problem often increases sharply with their number. Therefore finding the most compact representation may be a valuable preprocessing step in obtaining the solution of the problem.

It is convenient to discuss F in terms of the Boolean function $f:\{0,1\}^n \to \{0,1\}$ whose set of zeros is F, and we say that a system $Ax \leqslant b$ of linear inequalities *represents* f when for all 0–1 vectors $x=(x_1,\ldots,x_n)\in\{0,1\}^n$, which we call *points*, $Ax \leqslant b$ if and only if $f(x)=0$. Of course many systems can represent the same f, and the smallest possible number of inequalities in any such system is called *the threshold number of f*, and denoted by $t(f)$. This name originates from the fact that f is called a *threshold function* when $t(f)\leqslant 1$ (see Hu [2], Muroga [7]). In geometrical terms, $t(f)$ is the smallest number of hyperplanes that separate the zeros of f from the other points. In Boolean terms $t(f)$ is the smallest t such that f can be written as the Boolean sum of t threshold functions. Jeroslow [5] showed that every Boolean function f on n variables satisfies $t(f)\leqslant 2^{n-1}$ and that this upper bound is realized by the function $f=\sum_{i=1}^{n} x_i \pmod 2$.

In this work we shall assume that f is *positive*, which means that for any points x, y, $f(x)=0$ and $y\leqslant x$ imply $f(y)=0$. Here $y \leqslant x$ means $y_i \leqslant x_i$ for $i=1,\ldots,n$. Equivalently f is positive when it has a representation $Ax\leqslant b$ with a nonnegative A. There is a natural way to specify a positive Boolean function f. The *prime implicants* of f are the minimal points p (with respect to the partial order $\leqslant$) such that $f(p)=1$. If p is a prime implicant, the monomial $\prod_{p_i=1} x_i$ is also called a prime implicant by a harmless abuse of language. We denote by P the collection of all the prime implicants of f, and for $S\subseteq P$, f_S denotes the Boolean sum of the members of S. It is easy to see that $f_P=f$. A subcollection $S\subseteq P$ will be called *a threshold subcollection in f* when there exists a threshold function g such that $f_S\leqslant g\leqslant f$ (i.e. $f_S(x)\leqslant g(x)\leqslant f(x)$ for all points x). In geometrical terms this means that a single hyperplane can separate all the prime implicants in S from all the zeros of f. We shall prove below that for positive f, $t(f)$ is equal to the smallest number of threshold subcollections in f that cover P.

This result reduces the problem of computing $t(f)$ to a combinatorial one. Instead of looking at all possible systems representing f, one could follow the following program:

(1) compute P, the collection of prime implicants of f;

(2) generate all threshold subcollections in f;

(3) cover P with a minimum number of threshold subcollections in f.

Step (1) is standard, or it may be given already in the specification of f. Step (3) is an ordinary set covering problem. The main subject of this paper is Step (2). The difficulty in it is that no characterization is known for the threshold subcollections in f except for one using the concept of asummability, defined in Section 2.

We relax below the condition of asummability to that of 2-asummability, which is easier to work with. The solution of the corresponding set covering problem then provides a lower bound on $t(f)$. In particular, when for any two prime implicants, each one contains at least two variables that the other one lacks, we are able to show that the minimum cover contains at least $|P|$ threshold subcollections in f. This and the obvious bound $t(f) \leq |P|$ give $t(f) = |P|$. A construction from coding theory shows that under these conditions $t(f)$ can be at least $\binom{n}{\lfloor n/2 \rfloor}/n$, while Sperner theorem yields $t(f) < \binom{n}{\lfloor n/2 \rfloor}$ for any positive f and $n > 1$.

In order to bring $t(f)$ down to manageable size, we further restrict f to be *graphic*, which means that each prime implicant of f is of the form $x_i x_j$. Thus f corresponds naturally to a graph G whose vertices are $x_1, \ldots, x_n$ and whose edges are the prime implicants. Under these conditions we are able to give a good characterization of those sets of edges S that are thresholds in f: this is the case if and only if G does not contain a closed walk alternating between S-edges and non-edges of G. Moreover, we present an $O(n^3)$ algorithm that will find either the required threshold function g or the alternating closed walk.

Although it is easy to see that $f(f) < n$ for f graphic, Chvátal and Hammer [1] have shown that computing $t(f)$ is NP-hard even in that special case, so consequently it is NP-hard for positive f and for general f.

Most of what has been said above applies equally well, with suitable modifications, to the case of a *negative* f, i.e. such that $f(x) = 1$ and $y \leq x$ imply $f(y) = 1$. A negative f must have a representation of the form $Ax \geq b$ with a nonnegative A, and conversely. As is well-known, many important optimization problems, such as the set covering and set packing problems, have a corresponding negative or positive Boolean function. Thus our theory applies to the most compact representation of such problems.

2. Positive Boolean functions and bounds on $t(f)$

Proposition 1. *If $Ax \leq b$ represents a positive Boolean function f, then $A^+x \leq b$ represents the same f, where $A^+_{ij} = \max\{0, A_{ij}\}$.*

Proof. Obviously $A^+x \leq b$ implies $Ax \leq b$ for any point x. To show the converse, let $A_{ij} < 0$ be any negative entry in A, and define

$$A'_{rs} = \begin{cases} 0 & \text{if } r = i \quad \text{and} \quad s = j, \\ A_{rs} & \text{otherwise.} \end{cases}$$

It is enough to show that $Ax \leq b$ implies $A'x \leq b$ for any point x and proceed by induction on the number of negative entries in A. If $x_j = 0$, then $A'x = Ax$

and there is nothing to prove. If $x_j = 1$, define x' by

$$x'_s = \begin{cases} 0 & \text{if } s = j, \\ x_j & \text{otherwise.} \end{cases}$$

Since $x' \leqslant x$ and $f(x) = 0$, we have by the positivity of f that $f(x') = 0$ and thus $Ax' \leqslant b$. Hence

$$b_i \geqslant \sum_{s=1}^{n} A_{is}x'_s = \sum_{s=1}^{n} A'_{is}x_s,$$

whereas for $r \neq i$

$$b_r \geqslant \sum_{s=1}^{n} A_{rs}x_s = \sum_{s=1}^{n} A'_{rs}x_s.$$

Thus $A'x \leqslant b$. □

Corollary 1. *If f is a positive Boolean function, then $t(f)$ is equal to the smallest number of linear inequalities in any system with nonnegative coefficients that represents f, and also to the smallest number of positive threshold functions whose Boolean sum is f.*

Theorem 1. *Let f be a postive Boolean function and P its collection of prime implicants. Then f is the Boolean sum of t threshold functions if and only if P can be covered by t threshold subcollections in f.*

Proof. If $S_1, \ldots, S_t$ are threshold subcollections in f that cover P, then $P = S_1 \cup \cdots \cup S_t$ and there exist threshold functions $g^{(i)}$ such that $f_{S_i} \leqslant g^{(i)} \leqslant f$ for $i = 1, \ldots, t$. By taking the Boolean sum of these inequalities we obtain $f_P \leqslant \sum_{i=1}^{t} g^{(i)} \leqslant f$, and since $f_P = f$, $f = \sum_{i=1}^{t} g^{(i)}$. Conversely, assume that f is the Boolean sum of t threshold functions. By Corollary 1 f is the Boolean sum of t positive threshold functions $g^{(1)}, \ldots, g^{(t)}$. For $i = 1, \ldots, t$, let S_i be the subcollection of those prime implicants p of f such that $p(x) \leqslant g^{(i)}$. Then $f_{S_i} \leqslant g^{(i)} \leqslant f$ and the S_i are threshold subcollections in f. We shall prove that $S_1 \cup \cdots \cup S_t = P$, for otherwise there exists a prime implicant p of f which violates $p(x) \leqslant g^{(i)}$ for all $i = 1, \ldots, t$. This means that there are points $x^{(1)}, \ldots, x^{(t)}$ such that $p(x^{(i)}) = 1$ and $g^{(i)}(x^{(i)}) = 0$. But the first equality says that $p \leqslant x^{(i)}$ and since $g^{(i)}$ is positive, the second equality implies that $g^{(i)}(p) = 0$ for $i = 1, \ldots, t$, hence $f(p) = 0$, a contradiction. □

If we attempt to use Theorem 1, we are faced with the difficult Step (2) of the program in the Introduction, that is to characterize the threshold subcollections in f. Even to decide whether or not P itself is threshold in f, that is whether or not f is a threshold function, is the classical synthesis problem of threshold logic [2, 7]. There is one characterization of the threshold subcollections in f, analogous to the one characterization of threshold functions as follows.

We say that a subcollection $S \subseteq P$ is *k-summable in f* when there exist $2k$ points $a^{(i)}, b^{(i)}, i=1, \ldots, k$ (not necessarily distinct) such that $f_S(a^{(i)})=1$, $f(b^{(i)})=0$, $i=1, \ldots, k$ and $\sum_{i=1}^k a^{(i)} = \sum_{i=1}^k b^{(i)}$, where the sum denotes vector addition. If S is not k-summable in f, it is *k-asummable in f*. In geometrical terms summability means that some convex combination of the ones of f_S is equal to some convex combination of the zeros of f. Therefore if S is k-summable for some k, it cannot be a threshold subcollection in f, but if S is k-asummable for all k, then the convex hulls of the ones of f_S and of the zeros of f are disjoint and can be separated by a hyperplane (by the separation theorem for polytopes), hence S is a threshold subcollection in f.

As in the case of threshold logic, all characterizations of thresholdness in f somehow use a variant of linear programming. For example Peled [8] suggested heuristics to find $t(f)$ by finding a maximal threshold subcollection in f, then a maximal threshold subcollection in f from the rest of P and so on. The method recognizes a threshold subcollection in f (Step (2)) by using a perturbation technique in the simplex algorithm, and it finds the correct $t(f)$ in the extreme cases that $t(f)=1$ or $t(f)=|P|$.

Because of the difficulties associated with the use of the criterion of k-asummability in f for all k, we shall relax the latter to 2-asummability in f, thereby obtaining useful lower bounds on $t(f)$. As a first simplification, we shall prove that the minimal 2-summable collections in f must be pairs of prime implicants, which reduces the problem of recognizing 2-asummability in f to recognizing 2-asummability in f for pairs. To do this, define the *2-summability graph of f, G_f,* on the vertex set P, in which any two prime implicants p, q of f are adjacent precisely when $\{p, q\}$ is 2-summable in f. We note that G_f has no loops, since any single prime implicant is a threshold collection in f and hence 2-asummable in f.

Proposition 2. *Let f be a positive Boolean function, G_f its 2-summability graph, and S any collection of prime implicants of f. Then S is 2-asummable in f if and only if S is an independent set of vertices in G_f.*

Proof. If S is 2-asummable in f then so are all its subsets, and in particular all pairs of prime implicants from S, so S is independent in G_f. Conversely, if S is 2-summable in f, then there exist points $a^{(1)}, a^{(2)}, b^{(1)}, b^{(2)}$ such that $a^{(1)}+a^{(2)} = b^{(1)}+b^{(2)}$, $f(b^{(1)})=f(b^{(2)})=0$ and $f_S(a^{(1)})=f_S(a^{(2)})=1$. The last equalities show that $p \leqslant a^{(1)}$ and $q \leqslant a^{(2)}$ for some prime implicants p, q from S. Therefore

$f_{\{p,q\}}(a^{(1)}) = f_{\{p,q\}}(a^{(2)}) = 1$, so p and q are adjacent in G_f, and S is a dependent set of vertices of G_f. □

Corollary 2. *If f is a positive Boolean function with threshold number $t(f)$ and G_f its 2-summability graph with chromatic number $\chi(G_f)$, then $t(f) \geq \chi(G_f)$.*

As an application of Corollary 2 we find $t(f)$ for a class of positive functions f, which includes all symmetrical block designs with $k - \lambda \geq 2$ (where blocks correspond to prime implicants).

Proposition 3. Let *f be a positive Boolean function and P its collection of prime implicants. If for any $p, q \in P$ $(p \neq q)$ there are four distinct indices i, j, k, l such that $p_i = p_j = q_k = q_l = 1$ and $p_k = p_l = q_i = q_j = 0$, then $t(f) = |P|$.*

Proof. Since $t(f) \leq |P|$, we shall prove the opposite inequality, $t(f) \geq |P|$, by proving that G_f is a complete graph and using Corollary 2. Let p, q be any two distinct prime implicants of f and let i, j, k, l be as in Proposition 3. Consider the points x, y defined by

$$x_r = \begin{cases} 1 & r = i, k, \\ 0 & r = j, l, \\ p_r & \text{otherwise}, \end{cases} \qquad y_r = \begin{cases} 0 & r = i, k, \\ 1 & r = j, l, \\ q_r & \text{otherwise}. \end{cases}$$

Since there is just one index h such that $x_h = 1$ and $p_h = 0$ (namely $h = k$), for any point $z \leq x$ there is at most one index h such that $z_h = 1$ and $p_h = 0$, so by assumption z cannot be a prime implicant of f. Therefore $f(x) = 0$, and similarly $f(y) = 0$. Since $x + y = p + q$, $\{p, q\}$ is 2-summable in f. □

Theorem 2

$$\frac{1}{n}\binom{n}{\lfloor n/2 \rfloor} \leq \max_f t(f) < \binom{n}{\lfloor n/2 \rfloor},$$

where the maximum is taken over all positive Boolean functions f in $n > 1$ variables.

Proof. We know that $t(f) \leq |P|$. By Sperner theorem $|P| \leq \binom{n}{\lfloor n/2 \rfloor}$, with equality holding if and only if P consists of all products of $\lfloor \frac{n}{2} \rfloor$ variables. But in the latter case f is a threshold function and $t(f) = 1 < \binom{n}{\lfloor n/2 \rfloor}$. This establishes the upper bound. To establish the lower bound it is enough to show a function f satisfying the conditions of Proposition 3 and such that $|P| \geq \binom{n}{\lfloor n/2 \rfloor}/n$. Such examples are known from coding theory [4, 6], and we show one here for completeness. For any $w = 1, \ldots, n$, let C_w be the set of all $\binom{n}{w}$ points x

satisfying $\sum_{i=1}^{n} x_i = w$. Define a graph H_w with vertex set C_w in which x and y are adjacent when $\sum_{i=1}^{n} x_i(1-y_i)=1$, and consequently also $\sum_{i=1}^{n}(1-x_i)y_i=1$. Give the vertex x the colour $\sum_{i=1}^{n} ix_i \pmod n$. Then each colour is an independent set in H_w. Since there are n colours, some colour contains at least $\binom{n}{w}/n$ vertices, and if we regard them as prime implicants, the set P of these prime implicants satisfies the condition of Proposition 3. Therefore the largest P satisfying these conditions has at least

$$\max_{w=1,\ldots,n} \frac{1}{n}\binom{n}{w} = \frac{1}{n}\binom{n}{\lfloor n/2 \rfloor}$$

prime implicants. □

3. Graphic Boolean functions and asummability in f

From now on we specialize to the case that f is a graphic function associated with the graph G. Under this assumption we shall be able to perform Step (2) of the program discussed in the Introduction rather efficiently, based on a combinatorial characterization of the threshold subcollections in f that will be given below. We begin by strengthening the definition of summability in f.

Theorem 3. *Let S be a subcollection of prime implicants of the graphic Boolean function f. Then S is summable in f if and only if there exist a positive integer k and $2k$ points $a^{(r)}, b^{(r)}$, $r=1,\ldots,k$ satisfying*

(i) $f_S(a^{(r)})=1,\ f(b^{(r)})=0 \quad \text{for } r=1,\ldots,k,$

(ii) $\displaystyle\sum_{r=1}^{k} a^{(r)} = \sum_{r=1}^{k} b^{(r)},$

(iii) $\displaystyle\sum_{j=1}^{n} a_j^{(r)} = \sum_{j=1}^{n} b_j^{(r)} = 2 \quad \text{for } r=1,\ldots,h.$

Proof. The 'if' part is trivial, since (i) and (ii) alone constitute the definition of k-summability in f. For the 'only if' part we may assume (i) and (ii). Since $f_S(a^{(r)})=1$, there exist prime implicants $p^{(r)}$ in S such that $p^{(r)} \leqslant a^{(r)}$. Thus

(iv) $f_S(p^{(r)})=1 \quad \text{for } r=1,\ldots,k,$

(v) $\displaystyle\sum_{j=1}^{n} p_j^{(r)}=2 \quad \text{for } r=1,\ldots,k,$

and $\sum_{r=1}^{k} p^{(r)} \leqslant \sum_{r=1}^{k} b^{(r)}$. Therefore there exist points $q^{(r)}$ such that $q^{(r)} \leqslant b^{(r)}$ and

(vi) $\displaystyle\sum_{r=1}^{k} p^{(r)} = \sum_{r=1}^{k} q^{(r)}.$

Since f is positive, certainly

(vii) $f(q^{(r)})=0$ for $r=1,\ldots,k$.

Of all k, $p^{(r)}$ and $q^{(r)}$ satisfying (iv)–(vii) choose ones such that k is as small as possible. We then claim that

(viii) $\sum_{j=1}^{n} q_j^{(r)} \geqslant 2$ for $r=1,\ldots,k$.

For if $\sum_{j=1}^{n} q_j^{(r)}=0$ or 1 for some r, say $r=k$, then by (vi) there exists a single $p^{(r)}$, say $p^{(k)}$, satisfying $q^{(k)} \leqslant p^{(k)}$, and again by (vi) $\sum_{r=1}^{k-1} p^{(r)} \leqslant \sum_{r=1}^{k-1} q^{(r)}$. The same argument could then be repeated with $k-1$ replacing k, contradicting the minimality of k. This establishes (viii). Now from (vi), (v) and (viii) it follows that

$$2n=\sum_{r=1}^{k}\sum_{j=1}^{n} p_j^{(r)}=\sum_{r=1}^{k}\sum_{j=1}^{n} q_j^{(r)} \geqslant 2n,$$

proving that (viii) holds as an equality for $r=1,\ldots,k$. But then (iv), (vii), (vi), (v) and (viii) show that $p^{(r)}$ and $q^{(r)}$ satisfy (i)–(iii). □

As a consequence of Theorem 3 we can characterize, for a graphic f, which subcollections of two prime implicants are 2-summable in f, i.e. which are the edges of the 2-summability graph G_f. It turns out that G_f coincides with what Chvátal and Hammer [1] have denoted by G^*.

Proposition 4. *Let f be a graphic function and G_f its 2-summability graph. Then two prime implicants x_ix_j, x_kx_l of f are adjacent in G_f if and only if i, j, k, l are distinct and both x_ix_k and x_jx_l or both x_ix_l and x_jx_k are not prime implicants of f. Fig. 1 illustrates these situations in terms of the associated graph G, with dotted lines indicating non-eges of G.*

Proof. Suppose i, j, k, l are distinct and both x_ix_k and x_jx_l are not prime implicants of f (the other alternative is analogous). Letting $\delta^{(r)}$ denote the unit

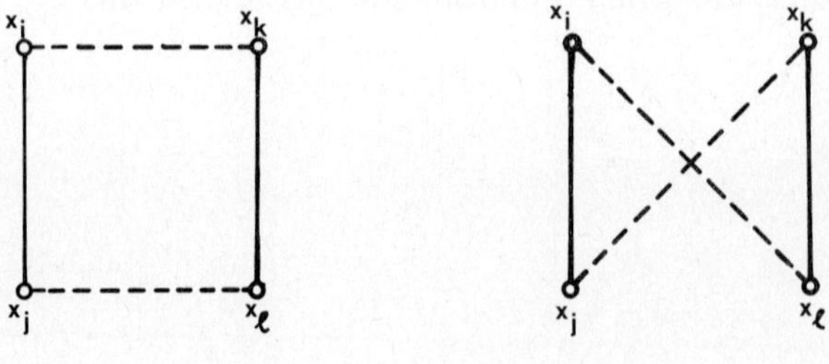

Fig. 1. x_ix_j and x_kx_l are adjacent in G_f.

vector with a 1 in position r and 0 elsewhere, we define the points $a^{(1)} = \delta^{(i)} + \delta^{(j)}$, $a^{(2)} = \delta^{(k)} + \delta^{(l)}$, $b^{(1)} = \delta^{(i)} + \delta^{(k)}$, $b^{(2)} = \delta^{(j)} + \delta^{(l)}$. Then $a^{(1)}$ and $a^{(2)}$ are prime implicants of f, hence $f_{\{x_ix_j, x_kx_l\}}(a^{(r)}) = 1$ for $r = 1, 2$. On the other hand $b^{(1)}$ and $b^{(2)}$ are not prime implicants, and since they have just two 1's, $f(b^{(r)}) = 0$ for $r = 1, 2$. Since $a^{(1)} + a^{(2)} = b^{(1)} + b^{(2)}$, $\{x_ix_j, x_kx_l\}$ is 2-summable in f, that is x_ix_j and x_kx_l are adjacent in G_f.

Conversely, assume that the prime implicants x_ix_j and x_kx_l are adjacent in G_f, i.e. $S = \{x_ix_j, x_kx_l\}$ is 2-summable in f. It is clear that i, j, k, l must all be distinct or else f_S would itself be a threshold function. By Theorem 3 there exist a k and $2k$ points $a^{(r)}, b^{(r)}, r = 1, \ldots, k$ satisfying conditions (i)–(iii) there, and moreover the smallest possible k satisfies $k \leqslant 2$ as is easily seen by following the proof there starting with $k = 2$. Since obviously 1-summability in f is a contradiction in terms, we may assume $k = 2$ and $a^{(1)} = \delta^{(i)} + \delta^{(j)}$, $a^{(2)} = \delta^{(k)} + \delta^{(l)}$. From condition (ii) we then obtain $b^{(1)} + b^{(2)} = \delta^{(i)} + \delta^{(j)} + \delta^{(k)} + \delta^{(l)}$, and since $b^{(1)}$ and $b^{(2)}$ are two distinct points, distinct from the $a^{(r)}$ and satisfying $\sum_{s=1}^{n} b_s^{(r)} = 2$, it follows necessarily that $b^{(1)} = \delta^{(i)} + \delta^{(k)}$, $b^{(2)} = \delta^{(j)} + \delta^{(l)}$ or $b^{(1)} = \delta^{(i)} + \delta^{(l)}$, $b^{(2)} = \delta^{(j)} + \delta^{(k)}$. Finally, since $f(b^{(r)}) = 0$, both x_ix_k and x_jx_l or both x_ix_l and x_jx_k are not prime implicants of f. □

By Propositions 2 and 4 it follows that if a graphic function f is associated with the graph G, then a subcollection S of prime implicants of f is 2-asummable in f if and only if the corresponding edges of G do not contain the configurations of Fig. 1. In particular f itself is 2-asummmable in f if and only if G does not contain the configurations of Fig. 1, i.e. G_f has no edges. In that case G is called a *threshold graph*, because in fact, Chvátal and Hammer [1] proved that for a graph f, G_f has no edges if and only if f is a threshold function. They also showed that $t(f)$ is NP-hard to compute even for a graphic f, and that $t(f) \geqslant \chi(G_f)$ for f graphic, a special case of our results above.

For a subcollection S of prime implicants of f, the condition of 2-asummability in f is necessary but not sufficient for being threshold in f, even for a graphic f, as the following example demonstrates.

$$f = x_1x_2 + x_2x_3 + x_3x_4 + x_4x_5 + x_5x_1 + x_2x_4 + x_3x_5,$$

$$f_S = x_1x_2 + x_5x_1 + x_3x_4.$$

Fig. 2 illustrates G and G_f for this example. S is independent in G_f and hence 2-asummable in f, but it is not a threshold subcollection in f, for if there were a single inequality $\sum_{i=1}^{5} a_ix_i \leqslant b$ representing a threshold function g such that $f_S \leqslant g \leqslant f$, then $g(1, 1, 0, 0, 0) \geqslant f_S(1, 1, 0, 0, 0) = 1$ and $g(0, 1, 0, 0, 1) \leqslant f(0, 1, 0, 0, 1) = 0$ would imply that $a_1 > a_5$, and similarly $a_5 > a_3 > a_1$, a contradiction. A study of this example shows that the nonexistence of g is due to the cycle with vertices $x_1, x_2, x_5, x_1, x_3, x_4$ in G, whose sides alternate between

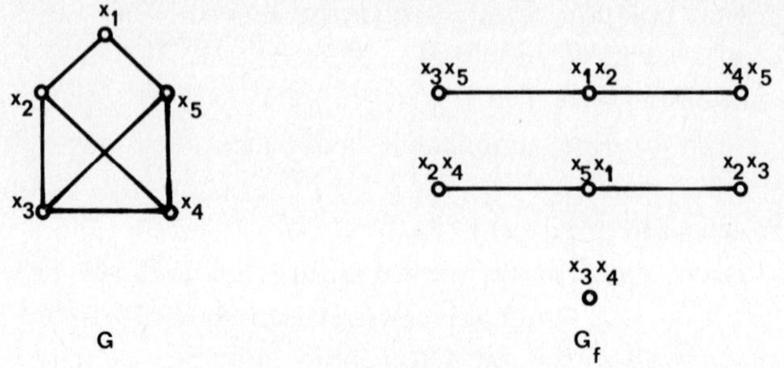

Fig. 2. An example.

edges in S and non-edges of G. The next theorem shows that such an obstruction is necessary and sufficient for S to be summable in f.

Theorem 4. *Let f be a graphic function associated with a graph G, and let S be a subcollection of prime implicants of f, i.e. edges of G. Then S is summable in f if and only if G contains a sequence of four or more vertices $x_{j_0}, x_{j_1}, \ldots, x_{j_{2k-1}}$, not necessarily distinct, such that for all r, $(x_{j_{2r-2}}, x_{j_{2r-1}})$ is an edge in S and $(x_{j_{2r-1}}, x_{j_{2r}})$ is a non-edge of G (indices are modulo $2k$). We call such a sequence* an alternating cycle relative to S.

Proof. If $x_{j_0}, \ldots, x_{j_{2k-1}}$ is an alternating cycle relative to S, then the $2k$ points $a^{(r)} = \delta^{(j_{2r-2})} + \delta^{(j_{2r-1})}$, $b^{(r)} = \delta^{(j_{2r-1})} + \delta^{(j_{2r})}$, $r = 1, \ldots, k$, satisfy $f_S(a^{(r)}) = 1$, $f(b^{(r)}) = 0$ and $\sum_{r=1}^{k} a^{(r)} = \sum_{r=1}^{k} b^{(r)}$, hence S is k-summable in f.

Conversely, if S is summable in f, then by Theorem 3 there exists an integer $k \geqslant 2$ and $2k$ points $a^{(r)}, b^{(r)}, r = 1, \ldots, k$ satisfying conditions (i)–(iii) there. Conditions (i) and (iii) mean that each $a^{(r)}$ corresponds to some edge in S and each $b^{(r)}$ to some non-edge of G, while condition (ii) says that each vertex is incident with as many S-edges corresponding to some $a^{(r)}$ as with non-edges corresponding to some $b^{(r)}$. Now consider the following process. Let $\delta^{(j_0)} + \delta^{(j_1)}$ be one of the $a^{(r)}$. Then we can find a j_2 such that $\delta^{(j_1)} + \delta^{(j_2)}$ is one of the $b^{(r)}$, then a j_3 such that $\delta^{(j_2)} + \delta^{(j_3)}$ is one of the $a^{(r)}$ and so on. Thus the corresponding vertices $x_{j_0}, x_{j_1}, x_{j_2}, \ldots$ are such that $(x_{j_0}, x_{j_1}), (x_{j_2}, x_{j_3}), \ldots$ are edges in S while $(x_{j_1}, x_{j_2}), (x_{j_3}, x_{j_4}), \ldots$ are non-edges of G. Since G is finite, there exists a vertex that is repeated twice (in fact infinitely often), $x_{j_k} = x_{j_l} = x_{j_m}$ with $k < l < m$. Obviously $l - k$, $m - l$ or $m - k$ must be even, say $l - k$ is even. Then the vertices $x_{j_k}, x_{j_{k+1}}, \ldots, x_{j_{l-1}}$ or $x_{j_l}, x_{j_{l-1}}, \ldots, x_{j_{k+1}}$ are an alternating cycle relative to S according as k and l are even or odd. □

Since being threshold in f is equivalent to being assumable in f, Theorem 4 has the following corollary:

Corollary 3. *Let f be a graphic function associated with a graph G, and let S be a subcollection of prime implicants of f, i.e. edges of G. Then S is a threshold subcollection in f if and only if G does not contain an alternating cycle relative to S.*

The result of Corollary 3 can be sharpened somewhat. If $G=(X, E)$ is a graph associated with a graphic function f and $S\subseteq E$, then by a *threshold completion of S in G* we mean a set T satisfying $S\subseteq T\subseteq E$ such that (X, T) is a threshold graph. The following result uses some of the ideas of the proof of Theorem 3.

Corollary 4. *If $G=(X, E)$ is a graph and $S\subseteq E$, then S has a threshold completion in G if and only if G does not have an alternating cycle relative to S.*

Proof. Let f be the graphic function associated with G. If T is a threshold completion of S in G, then the graphic function h associated with (X, T) is a threshold function and satisfies $f_S\leqslant h\leqslant f$, hence S is a threshold subcollection in f, and by Corollary 3 G does not have an alternating cycle relative to S. Conversely, if G does not have an alternating cycle relative to S, then by Corollary 3 there exists a threshold function g satisfying $f_S\leqslant g\leqslant f$. Let $\sum_{j=1}^{N} a_jx_j\leqslant b$ represent g. Then the graph $H=(X, T)$ such that $(x_i, x_j)\in T$ if and only if $a_i+a_j>b$ satisfies $S\subseteq T\subseteq E$. Moreover, H is a threshold graph by the theorem of Chvátal and Hammer. □

4. Algorithms to determine whether S has a threshold completion in G

Here we consider the following problem: given a graph $G=(X, E)$ and $S\subseteq E$, construct either a threshold completion of S in G or an alternating cycle in G relative to S showing that S does not have a threshold completion in G. Theorem 4 and Corollary 4 give such an algorithm in a very roundabout way as follows. Let f be the graphic function associated with G and consider the following system of linear inequalities:

$$\sum_{j=1}^{n} a_jx_j\leqslant b \quad \text{for all points } x \text{ such that } f(x)=0;$$

$$a_i+a_j>b \quad \text{for all edges } (x_i, x_j) \text{ in } S.$$

By linear programming methods either find a solution $a_1, \ldots, a_n$, b, in which case $T=\{(x_i, x_j): a_i+a_j>b\}$ is a threshold completion of S in G, or establish that no solution exists, in which case find points $a^{(r)}$ and $b^{(r)}$ satisfying the conditions of Theorem 3, and proceed to construct an alternating cycle in G relative to S as in the proof of Theorem 4.

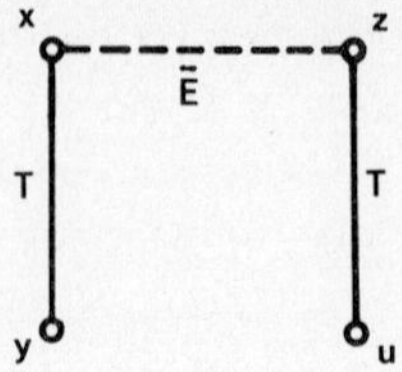

Fig. 3. Forcing $(y, u) \in T$.

This section gives direct and efficient combinatorial algorithms that use the Chvátal–Hammer characterization of threshold graphs rather than linear programming.

The first algorithm is very simple.

Step 0. Let $T = S$;

Step 1. Generate all subsets $Q = \{x, y, z, u\}$ of 4 vertices of G such that (x, y), $(z, u) \in T$ and $(x, z) \in \bar{E}$ (i.e. (x, z) is a non-edge of G). See Fig. 3. If $y, u \in \bar{E}$, stop, no threshold completion of S in G exists. Otherwise if y, u is not yet in T, adjoin it to T.

Step 2. If no new edges have been adjoined to T in Step 1, stop, T is a threshold completion of S in G. Otherwise repeat Step 1.

Clearly every step in the algorithm is dictated by the requirements from the threshold completion. If it stops in Step 2, (x, T) is indeed a threshold graph because it has no four vertices that induce the configurations of Fig. 1. If it stops in Step 1, an alternating cycle in G relative to S can easily be retrieved by retracing the steps of the algorithm. This algorithm can be refined so as to require $O(n^4)$ operations, namely when Step 1 is repeated, consider only those subsets Q such that (x, y) has just been adjoined to T in the previous execution of Step 1. The initial execution of Step 1 requires $O(n^4)$ operations. Since the final T has $O(n^2)$ edges, a total of $O(n^2)$ edges (x, y) are considered in all the subsequent executions of Step 1 combined. Since each (x, y) belongs to $O(n^2)$ subsets Q, the total number of operations in all the subsequent executions of Step 1 is $O(n^4)$.

We present now a more structured algorithm requiring $O(n^3)$ operations.

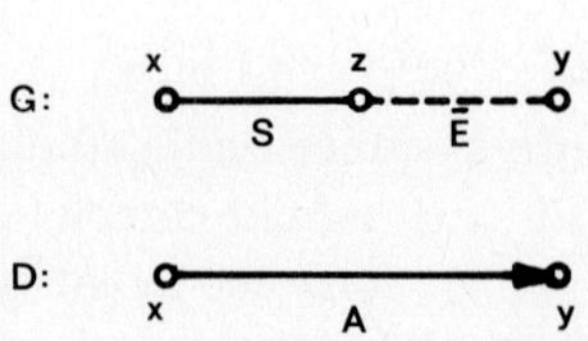

Fig. 4. Illustrating how the digraph D is constructed from G.

Some notation is useful for its presentation. For each vertex $x \in X$ of G, denote

$$N_S(x)=\{y \in X:(x, y) \in S\} \quad \text{and} \quad N_{\bar{E}}(x)=\{y \in X:(x, y) \in \bar{E}\}.$$

Define a digraph $D=(X, A)$ whose set of directed arcs is

$$A=\{\overrightarrow{(x, y)} \in X \times X: N_S(x) \cap N_{\bar{E}}(y) \neq \emptyset\},$$

see Fig. 4 for an illustration. By a *source* (*sink*) of a digraph is meant as usual a vertex with no incoming (outgoing) arcs incident to it. We can now describe the algorithm.

Begin by putting $D_0=D$ and $X_0=X$. If D_k has been constructed and $X_k \neq \emptyset$, then put

$$\begin{aligned}
&U_k = \text{set of non-isolated sources of } D_k,\\
&V_k = \text{set of non-isolated sinks of } D_k,\\
&W_k = \text{set of isolated vertices of } D_k,\\
&X_{k+1} = X_k - (U_k \cup V_k \cup W_k),\\
&D_{k+1} = \text{subdigraph of } D_k \text{ induced by } X_{k+1}.
\end{aligned}$$

This process can be continued until D_{k+1} becomes empty, say for $k=K$, unless $X_{k+1}=X_k$ happens earlier. But in the latter case it is easy to find a directed circuit in D_k, hence in D, hence an alternating cycle in G relative to S. We assume this does not happen and proceed to construct a threshold completion of S in G. For each k, we wish to partition W_k into two parts U_k' and V_k' to be adjoined to U_k and V_k respectively. For this end consider any connected component in the subgraph of $(X, S \cup \bar{E})$ induced by W_k. Such a component cannot contain both S-edges and $\bar{E}$-edges, since otherwise it would contain vertices x, y, z such that $(x, y) \in S$ and $(y, z) \in \bar{E}$, and consequently $\overrightarrow{(x, z)} \in A$, contradicting the independence of W_k in D. If the component contains S-edges, we put all of its vertices in U_k', and if it contains $\bar{E}$-edges or no edges at all, we put all of its vertices in V_k'. Thus we have obtained a partition $X=\bigcup_{k=0}^K (U_k^* \cup V_k^*)$, where $U_k^*=U_k \cup U_k'$ and $V_k^*=V_k \cup V_k'$. We can now specify a set T of edges for which $U^*=\bigcup_{k=0}^K U_k^*$ is complete and $V^*= \bigcup_{k=0}^K V_k$ is independent, and such that for each $x \in V_k^*$,

$$N_T(x)=\bigcup_{i=0}^{k} \bigcup_{y \in V_i^*} N_S(y).$$

Theorem 5. *The set T constructed above is a threshold completion of S in G.*

Proof. Obviously if $x \in V_k^*$ and $y \in V_l^*$ with $k \leq l$, then $N_T(x) \subseteq N_T(y)$, i.e. the

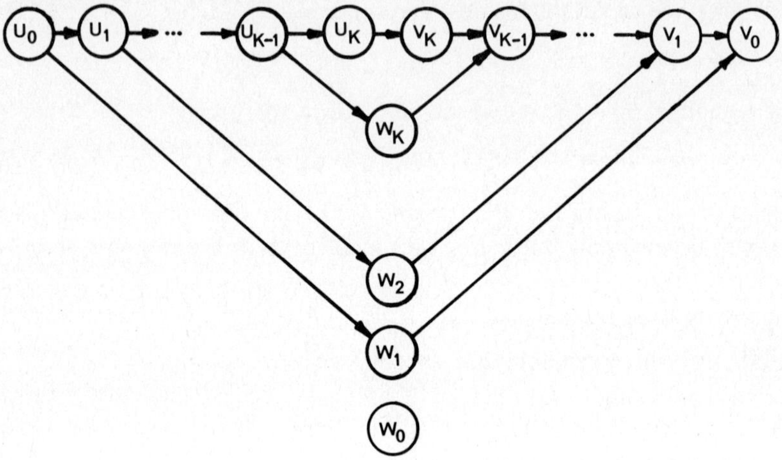

Fig. 5. The arcs of D.

neighbourhoods $N_T(x)$, $x \in V^*$ are totally ordered by inclusion. This, together with the fact that T makes U^* complete and V^* independent, easily implies that all the neighbourhoods $N_T(x)$, $x \in X$ are totally ordered by inclusion. This means that (X, T) does not contain the configurations of Fig. 1 and hence is a threshold graph by the theorem of Chvátal and Hammer. What remains to be proved is that $S \subseteq T \subseteq E$.

The main tool in the proof are the following observations, which are easily established by induction on k:

(1) For every vertex $x \in U_k$ $(k > 0)$ there exists some vertex y such that $\overrightarrow{(y, x)} \in A$, and necessarily $y \in U_{k-1}$.

(2) For every vertex $x \in V_k$ $(k > 0)$ there exists some vertex y such that $\overrightarrow{(x, y)} \in A$, and necessarily $y \in V_{k-1}$.

(3) For every vertex $x \in W_k$ $(k > 0)$ there exist vertices y and z such that $\overrightarrow{(y, x)}$, $\overrightarrow{(x, z)} \in A$, and necessarily $y \in U_{k-1}$, $z \in V_{k-1}$.

Fig. 5 illustrates schematically the arcs of D according to these observations (note that $U_k \neq \emptyset$ if and only if $V_k \neq \emptyset$). It follows that for each $x \in U_k$, the longest directed path into x has length exactly k (i.e. contains k arcs), namely from U_0 to $U_1 \cdots$ to U_k; and there exists a directed path from x of length greater than k, namely from U_k to V_k (along one of the existing directed paths) to $V_{k-1} \cdots$ to V_0. Similarly for each $x \in V_k$, the longest directed path from x has length exactly k and there exists a directed path into x of length greater than k. For each $x \in W_k$, the longest directed paths into x and drom x have length exactly k.

In order to prove that $S \subseteq T \subseteq E$, it suffices to show that
(a) no $\bar{E}$-edge joins two vertices of U^*,
(b) no S-edge joins two vertices of V^*, and
(c) if $x \in V_i^*$, $y \in V_k^*$ and $i \leqslant k$, then $N_S(x) \cap N_{\bar{E}}(y) = \emptyset$.

To prove (a), assume that $x, y \in U^*$ and yet $(x, y) \in \bar{E}$. Without limiting generality we may assume $x \in U_k^*$, $y \in U_i^*$ and $i \leqslant k$. Consider first the case $i < k$. There exists in D a directed path $x = y_0, y_1, \ldots, y_k$ of length k from x. This means that there exist vertices $z_0, z_1, \ldots, z_{k-1}$ such that $(y_j, z_j) \in S$ and $(z_j, y_{j+1}) \in \bar{E}$ for $j = 0, \ldots, k-1$. See Fig. 6. But then $z_{k-1}, z_{k-2}, \ldots, z_0, y$ is a directed path of length k into y, which contradicts the fact that the longest directed path into y has length i and $i < k$. Now consider the case $i = k$. Since both x and y belong to $U_k^* = U_k \cup U_k'$ and $(x, y) \in \bar{E}$, at least one of them must be in U_k (by definition of U_k') say $x \in U_k$. Then using a directed path of length greater than k from x, the previous argument applies again. This proves (a), and similarly (b). To prove (c), assume that $x \in V_i^*$, $y \in V_k^*$, $i \leqslant k$, $(x, z) \in S$ and $(y, z) \in \bar{E}$. Then $\overrightarrow{(x, y)} \in A$. Since D has a directed path of length k from y, it has a directed path of length $k+1$ from x, which is again a contradiction. □

Proposition 5. *The above algorithm requires at most* $O(n^3)$ *operations.*

Proof. First, the construction of D from G requires $O(n^3)$ operations, because to determine the arcs of D we examine all triples of vertices of G. Second, once D is constructed and stored in incoming and outgoing vertex-arc incidence lists, we can construct T in $O(n^2)$ operations. For we can construct U_0, V_0 and W_0 in $O(n)$ operations, then remove their vertices and the incident arcs from the lists, and construct U_1, V_1, and W_1 in $O(n)$ operations, and so on. Therefore U_k, V_k and W_k, $k = 0, \ldots, K$ are constructed in $O(Kn) = O(n^2)$ operations altogether. To construct U_k' and V_k' we only have to determine which pairs of vertices of W_k are connected by S-edges, requiring $O(|W_k|^2)$ operations, and hence U_k^* and $V_k^*, k = 0, \ldots, K$ are constructed in $O(n^2)$ operations altogether. The edges of T that make U^* complete arc constructed in $O(|U^*|^2) = O(n^2)$ operations, and so are the edges that connect V_k^* to U^* for all k combined. □

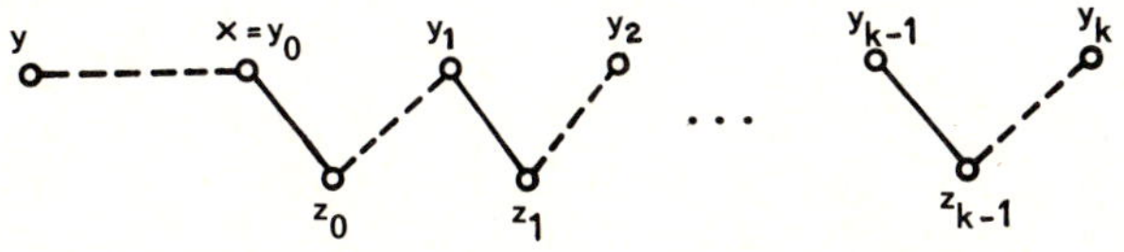

Fig. 6. Illustrating the proof of Theorem 5. Solid lines indicate S-edges, dotted lines $\bar{E}$-edges.

5. Variation and bipartite G_f

If S is a subset of the edges of a graph G, then by Corollary 4 S has a threshold completion in G if and only if G does not have an alternating cycle relative to S. By definition the vertices of an alternating cycle need not be distinct. In this section we present a family of configurations with distinct vertices, which are special alternating cycles relative to S, and must be present in G whenever G has any alternating cycle relative to S. In the case that the corresponding 2-summability graph G_f is bipartite, a much simpler subfamily still has the same property.

By an *alternating path* (relative to S) we mean a sequence of distinct vertices $v_0, v_1, \ldots, v_p$ $(p \geq 0)$ such that $(v_{2i}, v_{2i+1}) \in S$, $(v_{2i-1}, v_{2i}) \in \bar{E}$ for all i or $(v_{2i}, v_{2i+1}) \in \bar{E}$, $(v_{2i-1}, v_{2i}) \in S$ for all i. If $p \geq 3$ and the vertices are distinct except that $v_0 = v_p$, then we have an *even* or an *odd alternating polygon* (relative to S) according as p is even or odd. Note that an even alternating polygon is a special alternating cycle, while an odd alternating polygon is not, because its two edges incident with v_0 are both S-edges or both $\bar{E}$-edges. A *double flower* (relative to S) consists of two odd alternating polygons $u_0, \ldots, u_p$ and $v_0, \ldots, v_q$, and an alternating path $w_0, \ldots, w_r$, where all the vertices are distinct except that $u_0 = u_p = w_0$ and $v_0 = v_q = w_r$, such that $(u_0, u_1), (u_0, u_{p-1}) \in S$, $(w_0, w_1) \in \bar{E}$ or $(u_0, u_1), (u_0, u_{p-1}) \in \bar{E}$, $(w_0, w_1) \in S$, and similarly at vertex v_0. Note that the double flower defines an alternating cycle relative to S $u_0, u_1, \ldots, u_p, w_1, \ldots, w_r, v_1, \ldots, v_q, w_{r-1}, \ldots, w_0$ which traverses the alternating path twice in opposite directions. Fig. 7 illustrates these configurations.

Proposition 6. *Let S be a subset of the edges of a graph $G = (X, E)$. If G has an alternating cycle relative to S, then G has an even alternating polygon or a double flower relative to S.*

Proof. Let $A = v_0, v_1, \ldots, v_{2t-1}$ be the sequence of vertices of the alternating

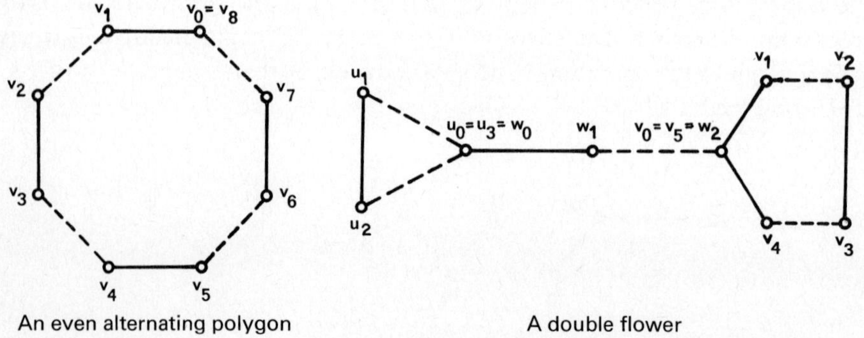

Fig. 7. Two special alternating cycles relative to S. (Solid lines indicate S-edges, dotted ones $\bar{E}$-edges.)

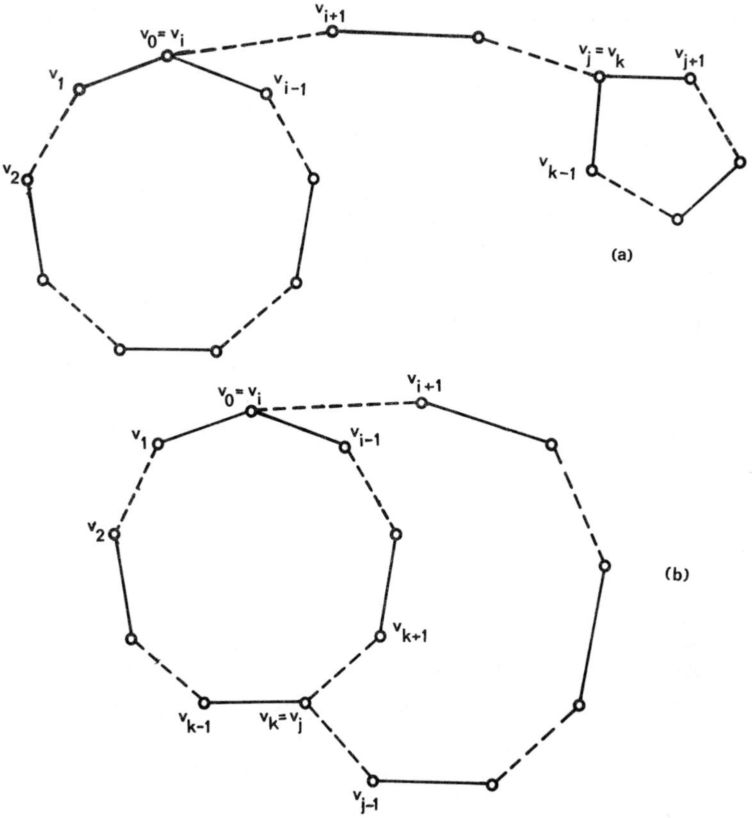

Fig. 8. Illustrating the proof of Proposition 6.

cycle. If no vertex is repeated in A, then A followed by v_0 is an even alternating polygon. Assuming there are repetitions in A, we may perform a convenient cyclic permutation of the vertices of A and reverse its order if necessary that so v_0 becomes the first vertex of A that is repeated, and $(v_0, v_1) \in S$, $(v_{2t-1}, v_0) \in \bar{E}$. Let v_i be the first vertex that is equal to v_0, $i > 0$. We may assume that i is odd, for otherwise $v_0, v_1, \ldots, v_i = v_0$ is an even alternating polygon. Thus $(v_i, v_{i+1}) \in \bar{E}$. Since $(v_{2t-1}, v_0) \in \bar{E}$, some vertex v_j in the subsequence $v_{i+1}, v_{i+2}, \ldots, v_{2t-1}$ must coincide with some previous vertex v_k, $k < j$, for otherwise we have a double flower having the odd alternating polygons $v_0, v_1, \ldots, v_i = v_0$ and $v_i, v_{i+1}, \ldots, v_{2t-1}, v_0 = v_i$ and a degenerate path of length 0 v_0. Let then v_j be the first vertex in the subsequence $v_{i+1}, v_{i+2}, \ldots, v_{2t-1}$ that coincides with some vertex v_k, $k < j$. Again we may assume that $j - k$ is odd, for otherwise $v_j, v_{j+1}, \ldots, v_k = v_j$ is an even alternating polygon. Thus if $k \geqslant i$ we have a double flower with the odd alternating polygons $v_0, v_1, \ldots, v_i = v_0$ and $v_j, v_{j+1}, \ldots, v_k = v_j$ and the alternating path $v_i, v_{i+1}, \ldots, v_j$ (see Fig. 8(a)). On the other hand if $k < i$, then the fact that

(v_k, v_{k+1}) and (v_{j-1}, v_j) are both in S or both in $\bar{E}$ shows that one of (v_{k-1}, v_k) and (v_{j-1}, v_j) is in S and the other one is in $\bar{E}$ (depending on the parity of k). Therefore $v_0, v_1, \ldots, v_k = v_j, v_{j-1}, \ldots, v_i = v_0$ is an even alternating polygon (see Fig. 8(b) for the case of k odd). □

Corollary 5. *Let S be a subset of the edges of a graph G. Then S has a threshold completion in G if and only if G has neither an even alternating polygon nor a double flower relative to S.*

Let us mention in this context a result of Younger [9]. From the results of Chvátal and Hammer [1] it is known that any threshold graph must be a *split graph*, i.e. its vertices can be partitioned into a clique and an independent set of vertices. Therefore if $G = (X, E)$, $S \subseteq E$ and S has a threshold completion in G, then S has certainly a *split completion in* G, i.e. there exists a set T satisfying $S \subseteq T \subseteq E$ such that (X, T) is a split graph. The above-mentioned result of Younger is, in our terminology, that S has a split completion in G if and only if G does not contain any double flower (relative to S) in which the two odd alternating polygons and the alternating path are allowed to overlap each other (but there may be even alternating polygons). By an argument similar to the proof of Proposition 6 it is not hard to show that if indeed there is overlap, then the configuration must contain an even alternating polygon as a sub-configuration. Therefore if S does not have a split completion in G, then G contains a double flower or an even alternating polygon, and hence by Corollary 5 S does not have a threshold completion in G. This confirms that Younger's condition is indeed a relaxation of the one of Corollary 5.

By definition of the 2-summability graph G_f of G, if S is a subset of edges of G that is independent in G_f, then the smallest possible even alternating polygon relative to S is the *alternating hexagon*, i.e. an even alternating polygon with $p = 6$, illustrated in Fig. 9(a). Also the smallest possible double flower relative to S is the *alternating pentagon*, i.e. a double flower with $p = q = 3$ and $r = 0$, illustrated in Fig. 9(b). It turns out that in a certain

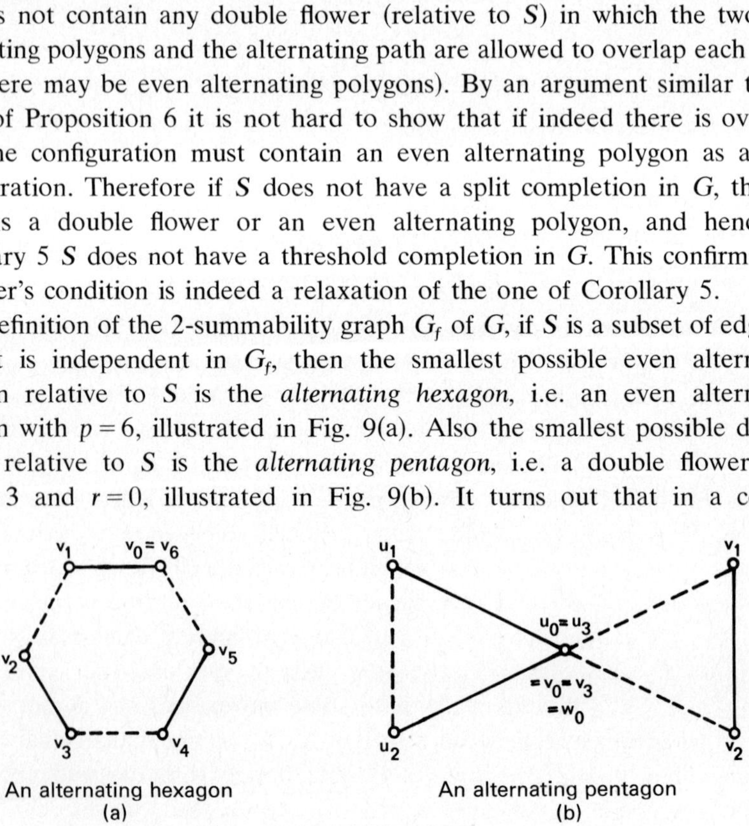

Fig. 9. The smallest alternating cycles relative to an independent set.

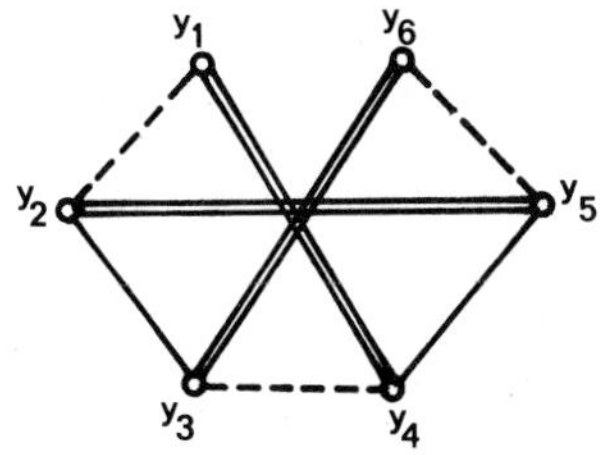

Fig. 10. Illustrating Case 1 in Theorem 6. (Single lines indicate S_1-edges, double lines S_2-edges and dotted lines $\bar{E}$-edges.)

important special case, the conclusion of Proposition 6 can be sharpened to read that G has an alternating hexagon or pentagon rather than just an even alternating polygon or a double flower relative to S.

Theorem 6. *Let the set E of edges of the graph $G=(X, E)$ be partitioned into two sets S_1 and S_2, each independent in the associated 2-summability graph G_f (i.e. G_f is bipartite). If G has an even alternating polygon or a double flower relative to S_1, then G has an alternating hexagon or pentagon relative to S_1.*

Proof. Let $y_0, y_1, \ldots, y_{2t}=y_0$ be the sequence of vertices of an alternating cycles relative to S_1, such that $(y_{2i}, y_{2i+1}) \in S_1$ and $(y_{2i-1}, y_{2i}) \in \bar{E}$ (indices mod $2t$), with t as small as possible. We have to prove that $t=3$, so we assume $t \geqslant 4$ and derive a contradiction. We distinguish two cases.

Case 1: It is possible to make a cyclic permutation of the indices $y_j \to y_{j+2k}$ ($j=0, \ldots, 2t-1$) after which $y_1, \ldots, y_6$ are all distinct. This only means that along the alternating cycle one can find consecutive edges of the form $\bar{E}, S_1, \bar{E}, S_1, \bar{E}$ with six distinct vertices (see Fig. 10). Since (y_2, y_3) and (y_4, y_5) are both in the independent S_1 and $(y_3, y_4) \in \bar{E}$, (y_2, y_5) must be in E. Moreover, it must belong to S_2, for otherwise $y_0, y_1, y_2, y_5, y_6, \ldots, y_{2t}=y_0$ would be a smaller alternating cycle relative to S_1. By a similar minimality argument (y_1, y_4) must be in E, and it must be in S_2 because it is adjacent in G_f to (y_2, y_3) and S_1 is independent in G_f. Similarly (y_3, y_6) must be in S_2. Since the independent S_2 contains (y_1, y_4) and (y_3, y_6), and $(y_3, y_4) \in \bar{E}$, (y_1, y_6) must be in E. Moreover, it must be in S_1 because it is adjacent in G_f to (y_2, y_5). At this point we see that G contains the alternating hexagon $y_6, y_1, y_2, y_3, y_4, y_5, y_6$ relative to S_1.

Case 2: It is impossible to make the above cyclic permutation. By Proposition 6, since t is minimal, the alternating cycle is an even alternating polygon or a double flower. Case 2 rules out the former. Moreover, none of the two odd alternating polygons in the double flower can have more than three vertices, since otherwise it would contain two $\bar{E}$-edges without a common vertex, and the alternating path or the other alternating odd polygon would furnish a third $\bar{E}$-edge and reduce us back to Case 1. Thus each of the two alternating odd

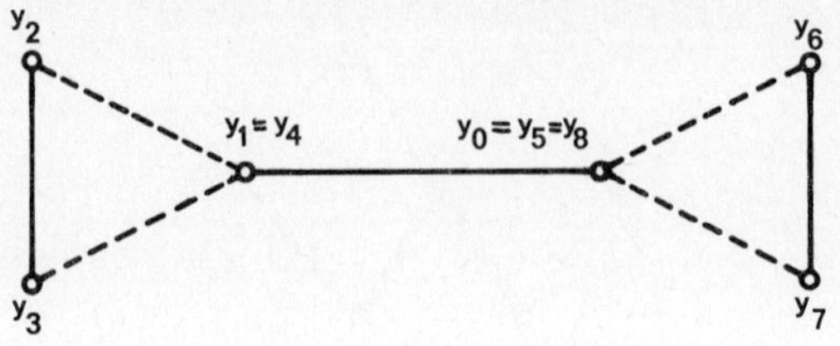

Fig. 11. Illustrating Case 2 in Theorem 6.

polygons is a triangle. For a similar reason the alternating path cannot contain any $\bar{E}$-edges, and since $t \geqslant 4$, it must consist of precisely one S_1-edge. The situation is thus as illustrated in Fig. 11. Since (y_2, y_3) and (y_0, y_1) are in S_1 and $(y_1, y_2) \in \bar{E}$, (y_0, y_3) must be in E. Similarly (y_1, y_6) must be in E. Since these two edges are adjacent in G_f, one of them must be in S_1, say $(y_0, y_3) \in S_1$. Then $y_0, y_1, y_3, y_0, y_6, y_7, y_0$ is an alternating pentagon relative to S_1. □

Corollary 6. *Let the edges of G be partitioned into S_1 and S_2, each independent in the associated 2-summability graph G_f. Then S_1 has a threshold completion in G if and only if G does not contain an alternating hexagon or pentagon relative to S_1.*

The condition of Corollary 6 is extremely convenient to work with. Suppose that G_f is bipartite and one wishes to show that $t(f)=2$. This means to partition the edges of G into two sets, each having a threshold completion in G. Obviously each must be independent in G_f. Thus one may partition the edges of G into two sets S_1 and S_2 that are independent in G_f, and then search for an alternating hexagon or pentagon relative to S_1 and relative to S_2 (this search is clearly polynomially bounded). If none exists, the algorithm is successful. If one exists, the sets S_1 and S_2 are modified so as to destroy it. This can be done by exchanging the S_1- and S_2-edges in a connected component of G_f that contains one of the edges of the alternating hexagon or pentagon. The difficulty here is that if G_f has k connected components, the sets S_1 and S_2 can be selected in 2^k ways (or rather 2^{k-1} if we disregard a total exchange of S_1 with S_2). In a companion paper [3] we prove, based on Corollary 6, that $\chi(G_f)=2$ implies $t(f)=2$ in the following two cases: (1) that the vertices of G can be partitioned into a clique and an independent set (i.e. G is a split graph); (2) that G_f contains at most two non-singleton connected components and any number of isolated vertices.

Acknowledgements

We thank Professor T. Kameda of the University of Waterloo for his discussion on coding theory. P.H. gratefully acknowledges the support of NRC grant A8552. T.I's stay in Waterloo was made possible by a Canada Council Grant for Exchange of Scientists between Japan and Canada.

References

[1] V. Chvátal and P.L. Hammer, Aggregation of inequalities in integer programming, Ann. Discrete Math. 1 (1977) 145–162.

[2] S.T. Hu, Threshold Logic (Univ. of California Press, Berkeley and Los Angeles, 1965).

[3] T. Ibaraki and U.N. Peled, Sufficient conditions for graphs to have threshold number two, in: P. Hansen, ed., Studies on Graphs and Discrete Programming (North-Holland, Amsterdam, 1981), in this Volume.

[4] N. Ikeno, G. Nakamura and K. Naemura, A general method of generating constant-weight codes (in Japanese), Technical Report of Automaton and Information Theory Research Group, Institute of Electronics and Communication Engineers of Japan, No. AIT 71-15, April 1971.

[5] R.G. Jeroslow, On defining sets of vertices of the hypercube by linear inequalities, Discrete Math. 11, (1975) 119–124.

[6] W.H. Kautz and B. Elspas, Single-error-correcting codes for constant-weight data words, IEEE Trans. Information Theory IT-11 (1965) 132–141.

[7] S. Muroga, Threshold Logic and its Applications (Wiley-Interscience, New York, 1971).

[8] U.N. Peled, Regular Boolean functions and their polytopes, Thesis, University of Waterloo, Dept. of Combinatorics and Optimization, 1976, pp. 6.3–6.11.

[9] D. Younger, private communication.

P. Hansen, ed., Studies on Graphs and Discrete Programming

CONSTRAINED LOCATION AND THE WEBER–RAWLS PROBLEM*

P. HANSEN

Institut d'Economie Scientifique et de Gestion, Lille, France and Faculté Universitaire Catholique de Mons, Belgium

D. PEETERS

Unité de Géographie Economique, Université Catholique de Louvain, Louvain-la-Neuve, Belgium

J.-F. THISSE

SPUR, Université Catholique de Louvain, Louvain-la-Neuve, Belgium

The Weber problem with maximum distance constraints is considered. Efficient algorithms, based on Goldman's concept of visibility, are proposed. Specifically, an $O(m)$ algorithm is given for the rectilinear norm case, where m is the number of users; an $O(m^2)$ algorithm for determining the feasible region in the Euclidean norm case and an efficient algorithm for solving the corresponding location problem are presented. These algorithms are further used to provide the set of Pareto-optimal locations of the Weber–Rawls problem in which both efficiency and equity criteria are minimized.

1. Introduction

Weber's problem consists in locating one facility in $\mathbf{R}^2$ in order to minimize the total transportation cost with respect to a given set of points corresponding to the users' locations. Usually, the transportation cost associated with every user is assumed to be a non-decreasing and convex function of the distance covered. The cost function is deduced from the user's utility and the distance from a norm on $\mathbf{R}^2$. Among the main contributions dealing with the Euclidean case are those of Kuhn and Kuenne [25], Cooper [5], Kuhn [23, 24], Katz [20], Cordelier and Fiorot [6], Ostresh [35]; with the rectilinear case those of Francis [12], Wesolowsky and Love [42], Huriot and Perreur [18], Wendell and Hurter [40]; with the l_p-norm those of Love and Morris [28], Morris and Verdini [32]; finally, the mixed-norm case has been recently considered by Planchart and Hurter [36], Hansen, Perreur and Thisse [18]. In all these papers, it is

* Paper presented at the Tenth International Symposium on Mathematical Programming, Montréal, August 1979 and at the Colloque Analyse Spatiale et Services Publics de l'Association de Science Régionale de Langue Française, Louvain-la-Neuve, September 1979.

supposed that any point of $\mathbf{R}^2$ is a feasible location. The more realistic case where only some regions of $\mathbf{R}^2$ can accommodate the facility has been studied by Hurter, Schaefer and Wendell [19], Schaefer and Hurter [38], Hansen, Peeters and Thisse [15, 16]. A related, but somewhat different, problem where some regions of $\mathbf{R}^2$ are forbidden not only for locating the facility but also for transportation has recently been investigated by Katz and Cooper [21].

The constrained location problem considered by Schaefer and Hurter [38] is to establish a facility within a maximum distance from each user. This corresponds to a constraint particularly relevant to the location of an emergency service; see, e.g. [33]. The first purpose of the present paper is to study this problem further; specifically, an $O(m)$ algorithm is proposed for the rectilinear case (where m is the number of users), an $O(m^2)$ algorithm for determining the feasible region in the Euclidean case and an efficient algorithm for solving the corresponding location problem. Computational results are reported on.

The minimization of the total transportation cost supported by all users can be viewed as a criterion of efficiency. Alternatively, one might prefer a criterion of equity: the minimization of the largest transportation cost supported by a user; see [31]. We call this the Rawls problem in reference to the leximin criterion proposed by Rawls [37] and used in social choice theory, [22, 7]. References to the Rawls problem include [4, 13, 9, 10] for the Euclidean case, [10, 14, 41] for the rectilinear case, and [8] for the l_p-norm. Our second purpose is to examine a bicriterion location problem in which efficiency and equity are simultaneously taken into account. We give it the name of Weber–Rawls problem. The algorithms developed for the constrained location problem are shown to provide easily the set of undominated, or Pareto-optimal locations associated with that problem, both in the Euclidean and rectilinear cases. This last case has been considered previously by McGinnis and White [31].

2. The Weber problem

The *unconstrained Weber problem* (in short UWP) is defined as follows:

(i) There is a single facility to be located and any point $x \in \mathbf{R}^2$ is a feasible location.

(ii) There are m users of the facility which are located respectively in points $d_1, \ldots, d_m$ of $\mathbf{R}^2$.

(iii) The distance between the facility located in x and the user located in d_i is expressed by $\|x - d_i\|$, where $\|\cdot\|$ is a l_p-norm defined on $\mathbf{R}^2$.

(iv) The transportation cost between the facility located in x and the user located in d_i is given by a non-decreasing and convex function of the distance covered; it is denoted $C_i(\|x - d_i\|)$.

(v) The facility must be located in a point of $\mathbf{R}^2$ where the total transportation cost, i.e. $\sum_{i=1}^{m} C_i(\|x-d_i\|)$ is minimized.

Formally, the UWP is therefore defined by

$$\min_{x \in \mathbf{R}^2} F(x) = \sum_{i=1}^{m} C_i(\|x-d_i\|). \tag{1}$$

In the *constrained Weber problem* (in short CWP), assumption (i) is replaced by the following one:

(i′) There is a single facility to be located and any point $s \in S \subset \mathbf{R}^2$ is a feasible location, where S is closed in $\mathbf{R}^2$. The CWP can then be written as

$$\min_{s \in S} F(s) = \sum_{i=1}^{m} C_i(\|s-d_i\|). \tag{2}$$

For our problem, the set of feasible locations is defined as the intersection of the disks $B(d_i\,;R_i) = \{s \in \mathbf{R}^2;\ \|s-d_i\| \le R_i\}$ with center d_i and radius $R_i > 0$, i.e.

$$S = \bigcap_{i=1}^{m} B(d_i;\ R_i). \tag{3}$$

Clearly, this set is compact, convex and its boundary is formed by the union of a finite number of simple arcs.

A point $s \in S$ is said to be *visible* from $t \notin S$ if and only if the linear segment $]s, t]$ contains no point of S. Given the properties of S, the following result is then known to hold [19]:

Proposition 1. *Let x^* be a solution to the UWP. Then x^* does not belong to S or there exists a solution s^* to the CWP visible from x^*.*

Let $V(x^*)$ be the set of points of S which are visible from a solution x^* to the UWP. As S is convex, $V(x^*)$ is connected and closed. Accordingly, since $V(x^*)$ is a subset of the boundary of S, a continuous bijection $s(\theta)$ from $[0, 1]$ on $V(x^*)$ exists. Given the convexity of $F(x)$ on $\mathbf{R}^2$, we then have [15]:

Proposition 2. *$F[s(\theta)]$ is quasi-convex on $[0, 1]$ and any local minimizer of $F[s(\theta)]$ is a global minimizer.*

We say that a point s of $S = \bigcap_{i=1}^{m} B(d_i;\ R_i)$ is *angulous* if and only if s belongs to the boundary of two disks which are not included one into the other.

Let ND be the set of points of $\mathbf{R}^2$ where function $F(x)$ is not differentiable. In the rectilinear case, ND is formed by the straight lines passing through points d_i and parallel to the axes; for $p \in\,]1, 2]$, ND is equal to $\{d_1, \ldots, d_m\}$. In view of the above two propositions, we may now state a general algorithm for the CWP where S is given by (3):

(a) *Feasible region.* Determine $S=\bigcap_{i=1}^{m} B(d_i;R_i)$. If $S=\emptyset$ end, the CWP having no feasible solution. If $S=\{s\}$ end, s being the sole feasible solution to the CWP.

(b) *Unconstrained solution.* Solve the UWP associated with the CWP. Denote by x^* an optimal solution to the UWP. If $x^*\in S$ end, the optimal unconstrained solution being feasible.

(c) *Visible boundary.* Determine the set $V(x^*)$ of feasible locations visible from x^*. Let $s(0)$ and $s(1)$ be the extremities of $V(x^*)$. Check if $V(x^*)$ contains zero, one or more angulous points or points of ND different from $s(0)$ and $s(1)$.

(d) *Constrained solution.* (i). If $V(x^*)$ contains no angulous points and no points of ND different from $s(0)$ and $s(1)$, determine a constrained optimal solution s^* by a unimodal search along the simple arc $V(x^*)$.

(ii). If $V(x^*)$ contains exactly one angulous point or one point of ND different from $s(0)$ and $s(1)$, say $\bar{s}$, compute the directional derivatives at $\bar{s}$ towards $s(0)$ and $s(1)$ along $V(x^*)$. If both derivatives are positive, $s^*=\bar{s}$. Otherwise, search s^* as in (i) along the sub-arc of $V(x^*)$ corresponding to the negative derivative.

(iii). If $V(x^*)$ contains at least two angulous points or points of ND different from $s(0)$ and $s(1)$, evaluate the objective function at each of these points. Select the point $\bar{s}$ of minimum value and proceed as in (ii) to find s^*.

3. The rectilinear norm case

In this section, we specialize the general scheme given in Section 2 to the rectilinear norm and linear transportation costs case. We shall prove that the time and space complexity of the resulting algorithm is $O(m)$, where m is the number of points d_i. To this end, each step of the algorithm will be analyzed successively in the sequel.

(a) *Feasible region.* Suppose that $\mathbf{R}^2$ is endowed with a pair of orthogonal axes. Then the *disks* associated with the rectilinear norm are squares whose sides form a $\frac{1}{4}\pi$-angle with one axis.

Let $I_k=\bigcap_{i=1}^{k} B(d_i;R_i)$ be the intersection of the k first *disks*. Clearly, I_k may be: (1) empty; (2) a point; (3) a segment with a slope of $\frac{1}{4}\pi$ with one axis; (4) a rectangle whose sides also form $\frac{1}{4}\pi$-angles with the axes. Looking for the intersection of I_k and $B(d_{k+1};R_{k+1})$, no work is needed in the first case. In the second one, I_{k+1} is empty if the distance betwen d_{k+1} and the single point of I_k is greater than R_{k+1}, and equal to I_k otherwise. In the third case, I_{k+1} can be found by computing at most 2 intersections of two straight lines and by performing at most 7 comparisons. Finally, if I_k is a rectangle, at most 6 intersections of two straight lines and 23 comparisons allow to determine I_{k+1}.

Thus, in any case, a constant number of operations is sufficient to obtain I_{k+1} from I_k and $B(d_{k+1}; R_{k+1})$. A direct consequence is that the intersection of m rectilinear *disks* can be found in $O(m)$ time. Regarding the space requirement, we need three vectors of length m (first and second coordinates of points d_i, radii R_i). The current intersection I_k can be described by its vertices (at most 4 ordered pairs of numbers) and a variable indicating the nature of I_k. Thus the space complexity is also $O(m)$.

(b) *Unconstrained solution.* The objective function now writes as

$$\min_{x \in \mathbf{R}^2} F(x) = \sum_{i=1}^{m} w_i(|x^1 - d_i^1| + |x^2 - d_i^2|) \tag{4}$$

where the w_i are positive constants.

Given the separability of $F(x)$, it is known that (4) amounts to the resolution of two one-dimensional problems defined by

$$\min_{x^j \in \mathbf{R}} F^j(x^j) = \sum_{i=1}^{m} w_i \, |x^j - d_i^j| \tag{5}$$

with $j = 1, 2$. Dropping the upper index for notational simplicity, we know that a point $d_{i^*} \in \{d_1, \ldots, d_m\}$ which satisfies the inequalities

$$\sum_{\{i\,;d_i < d_{i^*}\}} w_i < \frac{1}{2}\left(\sum_{i=1}^{m} w_i\right), \tag{6a}$$

$$\sum_{\{i\,;d_i \leqslant d_{i^*}\}} w_i \geqslant \frac{1}{2}\left(\sum_{i=1}^{m} w_i\right) \tag{6b}$$

is an optimal solution to (5) [12]. If the inequality is strict in (6b), then the optimal solution d_{i^*} is unique; otherwise, all points of the segment $[d_{i^*}, d_{i^{**}}]$, where $d_{i^{**}} = \min\{d_i\,; d_i > d_{i^*}\}$, are optimal. A straightforward resolution method consists in sorting points d_i into non-decreasing order and in scanning the so-obtained sequence until conditions (6a) and (6b) hold. This procedure has an $O(m \log m)$ time complexity. We propose a better procedure regarding the worst-case performance. Firstly, we determine the median $\tilde{d}$ of points $d_1, \ldots, d_m$. Secondly, we split up the set of indices $I = \{1, \ldots, m\}$ in three subsets $I^<, I^=$ and $I^>$: $I^<$ (resp. $I^=$ and $I^>$) is the subset of indices such that $d_i < \tilde{d}$ (resp. $d_i = \tilde{d}$ and $d_i > \tilde{d}$). Summing the coefficients w_i with indices in $I^<$ (in $I^=$) yields $w^<(w^=)$. If $w^< \geqslant \frac{1}{2}(\sum_{i=1}^{m} w_i)$, it follows from (6b) that the index of the solution to (5) belongs to $I^<$. If $w^< < \frac{1}{2}(\sum_{i=1}^{m} w_i)$ and if $w^< + w^= \geqslant \frac{1}{2}(\sum_{i=1}^{m} w_i)$, then $\tilde{d}$ is an optimal solution. If $w^{\leqslant} = w^< + w^= < \frac{1}{2}(\sum_{i=1}^{m} w_i)$, we have to consider $I^>$ to obtain an optimal solution. Thirdly, if $d_{i^*} \neq \tilde{d}$, the above procedure is applied either to $I^<$ or to $I^>$, $w^< + w^=$ being added to the coefficient of $\min\{d_i\,; i \in I^>\}$ in the latter case. Given that the number of elements in $I^<$ and $I^>$ is not greater than $\frac{1}{2}m$, the search is carried on in sets of cardinality $m, \frac{1}{2}m, \frac{1}{4}m, \ldots$ This leads to a number of operations proportional to m. We now state the procedure in Pidgin ALGOL (see [1] for notational conventions).

procedure RECTI1(m, d, w, ind, opt1, opt2)
/* **input:** m = number of points;
$d[1:m]$ = coordinates of the points;
$w[1:m]$ = associated coefficients;
output: if ind = 1 {opt1} is the solution;
if ind = 2 the segment [opt1, opt2] is the solution */
begin

1 $w^* \leftarrow \left(\sum_{i=1}^{m} w_i\right) * 0.5;$
2 $I \leftarrow \{1, \ldots, m\};$
3 $w^I \leftarrow 0.0;$
4 ITER(I, m, d, w, w^*, w^I, ind, opt1, opt2);
end;
procedure ITER(I, m, d, w, w^*, w^I, ind, opt1, opt2);
begin
5 $\tilde{d} \leftarrow$ MEDIAN(I, d); /* $\tilde{d}$ is the median of the family d_i, $i \in I$) */
6 PARTITION(I, d, $\tilde{d}$, $I^<$, $I^=$, $I^>$); /* $I^< = \{i \in I: d_i < \tilde{d}\}$
$I^= = \{i \in I: d_i = \tilde{d}\}$
$I^> = \{i \in I: d_i > \tilde{d}\}$ */

7 $w^< \leftarrow w^I + \sum_{i \in I^<} w_i;$
8 **if** $w^< > w^*$ **then** ITER($I^<$, m, d, w, w^*, w^I, ind, opt1, opt2)
else if $w^< = w^*$ **then** $d_{i^*} \leftarrow \max_{i \in I^<} d_i;$
ind $\leftarrow$ 2; opt1 $\leftarrow d_{i^*}$; opt2 $\leftarrow \tilde{d}$
else $w^{\leqslant} \leftarrow w^< + \sum_{i \in I^=} w_i;$
if $w^{\leqslant} > w^*$ **then** ind $\leftarrow$ 1; opt1 $\leftarrow \tilde{d}$
else if $w^{\leqslant} = w^*$ **then** $d_{i^{**}} \leftarrow \min_{i \in I^>} d_i;$
ind $\leftarrow$ 2; opt1 $\leftarrow \tilde{d}$; opt2 $\leftarrow d_{i^{**}}$
else ITER($I^>$, m, d, w, w^*, $w^{\leqslant}$, ind, opt1, opt2);
end;

Let us now study the time complexity of RECTI1. Step 1 requires $m+1$ operations. The set of indices I can be represented throughout the procedure by pointers on vectors $d[1:m]$ and $w[1:m]$, so that Step 2 is O(1). Step 3 is also O(1). The heart of the matter is undoubtedly procedure ITER. Suppose we call it with $|I| = k$ and denote $T(k)$ its worst-case time complexity. Linear time median finding algorithms are known; see e.g. [1, 39]. Partitionning I in $I^{<}, I^{=}$ and $I^{>}$ is achieved in $O(k)$ by permutations of items in vectors $d[1:m]$ and $w[1:m]$ and moves of pointers. At most $\lfloor \frac{1}{2}k \rfloor$ operations are needed in Step 7. The computation of $w^{\leq}$ in Step 8 needs at most $k+1$ operations. Moreover, three cases may arise: (1) a unique optimal solution is found and 2 operations are required; (2) a segment of optimal solutions is obtained and $O(\lfloor \frac{1}{2}k \rfloor)$ operations are needed; (3) no optimal solutions is obtained and we call ITER again. Now, $|I^{<}| \leq \frac{1}{2}k$ and $|I^{>}| \leq \frac{1}{2}k$ so that we obtain the following recurrence relationship on the time complexity:

$$T(k) \leq T(\tfrac{1}{2}k) + ak$$

where a is a positive constant. Using [1, Theorem 2.1], we deduce:

Proposition 3. *Procedure* RECTI1 *solves the UWP* (5) *in* $O(m)$ *time. Its space requirement is* $O(m)$.

The second part of the proposition is straightforward.

The technique of balancing used in RECTI1 is similar to that employed by Balas and Zemel [2], Fayard and Plateau [11] and Lawler [26] to solve the linear relaxation of the binary knapsack problem in linear time.

(c) *Visible boundary.* This step of the algorithm is skipped unless the feasible region S is a rectangle. Assume that the optimal solution x^* to the UWP is unique. The supporting lines including the sides of S divide $\mathbf{R}^2$ into nine regions. At most 4 computations and comparisons are sufficient to determine the region containing x^*. If x^* does not belong to S, the sides of S visible from x^* are then immediately known and the step is achieved in $O(1)$. The same result holds when the optimal solution to the UWP is not unique; the proof is left to the patient reader.

(d) *Constrained solution.* Consider the cases where the visible boundary $V(x^*)$ is composed of one or two adjacent sides of S, the other cases being trivial.

The straight lines passing through at least one point d_i and parallel to the axes form a grid which divides the plane into disjoint cells. Given that $F(x)$ is linear on each cell $\mathscr{C}$, the gradient is constant over $\mathscr{C}$. Denoting by $x^1 = d_e^1$ and $x^2 = d_n^2$ the straight lines containing the east and north sides of $\mathscr{C}$, the gradient (g_1, g_2) in $\mathscr{C}$ is given by

$$g_1 = -\sum_{i=1}^{m} w_i + 2 \sum_{\{i\,;\, d_i^1 < d_e^1\}} w_i, \qquad g_2 = -\sum_{i=1}^{m} w_i + 2 \sum_{\{i\,;\, d_i^2 < d_n^2\}} w_i.$$

For any $x \in \mathscr{C}$ and any direction (α, β), the directional derivative of F in x along the direction (α, β), denoted f, is equal to $\alpha g_1 + \beta g_2$.

Assume that $V(x^*) = [s(0), \bar{s}] \cup [\bar{s}, s(1)]$, where $\bar{s}$ is an angulous point.

Clearly, $2m$ comparisons are sufficient to determine the cell containing $\bar{s}$ and $3m+4$ additions and multiplications to compute g_1 over that cell. Computing the directional derivatives along the directions associated with $[s(0), \bar{s}]$ and $[\bar{s}, s(1)]$ at $\bar{s}$ requires 4 subtractions, 4 multiplications and 2 additions. Accordingly, determining whether $\bar{s}$ is an optimal solution or which segment contains an optimal solution can be performed in $O(m)$. Let us examine the latter case. We establish that finding an optimal solution in, say, $[s(0), \bar{s}]$ can be done in $O(m)$ running time. Since the gradient (g_1, g_2) in a cell $\mathscr{C}$ is constant, $f = (\bar{s}^1 - s(0)^1) \times g_1 + (\bar{s}^2 - s(0)^2) \times g_2$ is also constant in $\mathscr{C}$. Consequently, the finite set formed by $s(0)$, by $\bar{s}$ and by the intersection points of $[s(0), \bar{s}]$ with the straight lines of the grid must contain an optimal solution. Characterize each of these points by its parametric coordinate $\theta \in \mathscr{H}$ along $[s(0), \bar{s}]$. The procedure is as follows. First, select the median $\tilde{\theta}$ of $\mathscr{H}$. Second, compute the left and right directional derivatives of F at $\tilde{\theta}$, denoted by f^{l} and f^{r}. If both $f^{\mathrm{l}} \leqslant 0$ and $f^{\mathrm{r}} \geqslant 0$, $\tilde{\theta}$ is an optimal solution. If $f^{\mathrm{l}} > 0$, we have to explore the part of $\mathscr{H}$ situated on the left of $\tilde{\theta}$. If $f^{\mathrm{r}} < 0$, the optimal solution must be searched on the right of $\tilde{\theta}$. As in RECTI1, the above procedure is applied to the resulting subset of $\mathscr{H}$.

The case where $V(x^*) = [s(0), s(1)]$ can be dealt with as $[s(0), \bar{s}]$ is in the previous paragraph.

The procedure is described below in Pidgin ALGOL (for notational simplicity, we put $a = s(0)$ and $b = \bar{s}$).

procedure RECTI2(*m*, *d*, *w*, *a*, *b*, *opt*);
/* **input:** m = number of points;
$d[1:m]$ = coordinates of the points with $d_i = (d_i^1, d_i^2)$;
$w[1:m]$ = associated coefficients;
$a = (a^1, a^2)$ and $b = (b^1, b^2)$ = extremities of the segment $[a, b]$;
output: opt = a solution to CWP */
begin
1 wsum $\leftarrow \sum_{i=1}^{m} w_i$; $\alpha \leftarrow b^1 - a^1$; $\beta \leftarrow b^2 - a^2$;
2 $f_a^1 \leftarrow$ **if** $\alpha > 0$ **then** $-\mathrm{wsum} + 2 \sum_{\{i;\, d_i^1 \leqslant a^1\}} w_i$
else $-\mathrm{wsum} + 2 \sum_{\{i;\, d_i^1 < a^1\}} w_i$;
3 $f_a^2 \leftarrow$ **if** $\beta > 0$ **then** $-\mathrm{wsum} + 2 \sum_{\{i;\, d_i^2 \leqslant a^2\}} w_i$
else $-\mathrm{wsum} + 2 \sum_{\{i;\, d_i^2 < a^2\}} w_i$;
4 $f_a^{\mathrm{r}} \leftarrow \alpha * f_a^1 + \beta * f_a^2$;
5 **if** $f_a^{\mathrm{r}} \geqslant 0$ **then** opt $\leftarrow a$
6 **else** $f_b^1 \leftarrow$ **if** $\alpha > 0$ **then** $-\mathrm{wsum} + 2 \sum_{\{i;\, d_i^1 < b^1\}} w_i$
else $-\mathrm{wsum} + 2 \sum_{\{i;\, d_i^1 \leqslant b^1\}} w_i$;

7 $\quad f_b^2 \leftarrow$ **if** $\beta > 0$ **then** $-\text{wsum} + 2 \sum_{\{i;\, d_i^2 < b^2\}} w_i$
$\quad\quad$ **else** $-\text{wsum} + 2 \sum_{\{i;\, d_i^2 \leqslant b^2\}} w_i$;
8 $\quad f_b^\ell \leftarrow \alpha * f_b^1 + \beta * f_b^2$;
9 $\quad$ **if** $f_b^\ell \leqslant 0$ **then** opt $\leftarrow b$
10 $\quad$ **else** $I^1 \leftarrow \{i: \min(a^1, b^1) < d_i^1 < \max(a^1, b^1)\}$;
$\quad\quad I^2 \leftarrow \{i: \min(a^2, b^2) < d_i^2 < \max(a^2, b^2)\}$;
11 $\quad\quad$ **for** all $i \in I^1$ **do** $\theta^i \leftarrow (d_i^1 - a^1)/\alpha$;
12 $\quad\quad$ **for** all $i \in I^2$ **do** $\theta^i \leftarrow (d_i^2 - a^2)/\alpha$;
13 $\quad\quad$ merge lists $(\theta^i: i \in I^1)$ and $(\theta^i: i \in I^2)$
$\quad\quad$ to form the family $\mathscr{H}$;
14 $\quad\quad$ RESULT$(m, d, w, a, b, I^1, I^2, \mathscr{H}, f_a^1, f_a^2, \text{opt})$;
end
procedure RESULT$(m, d, w, a, b, I^1, I^2, \mathscr{H}, f_a^1, f_a^2, \text{opt})$;
begin
15 $\tilde{\theta} \leftarrow$ MEDIAN$(\mathscr{H})$; $\tilde{\theta}^1 \leftarrow a^1 + \tilde{\theta} * \alpha$; $\tilde{\theta}^2 \leftarrow a^2 + \tilde{\theta} * \beta$;
$\quad$ /* $(\tilde{\theta}^1, \tilde{\theta}^2)$ are the coordinates of the median
$\quad$ intersection point */
16 PARTITION1$(\mathscr{H}, \tilde{\theta}, \mathscr{H}^<, \mathscr{H}^=, \mathscr{H}^>)$;
17 PARTITION2$(I^1, d^1, \tilde{\theta}^1, I^{1<}, I^{1=}, I^{1>})$;
18 PARTITION2$(I^2, d^2, \tilde{\theta}^2, I^{2<}, I^{2=}, I^{2>})$;
19 $f^{1\ell} \leftarrow f_a^1 + \sum_{i \in I^{1<}} w_i$;
20 $f^{2\ell} \leftarrow f_a^2 + \sum_{i \in I^{2<}} w_i$;
21 **if** $\alpha * f^{1\ell} + \beta * f^{2\ell} > 0$ **then** RESULT$(m, d, w, a, b, I^{1<}, I^{2<}, \mathscr{H}^<, f_a^1, f_a^2, \text{opt})$
22 **else** $f^{1r} \leftarrow f^{1\ell} + \sum_{i \in I^{1=}} w_i$;
23 $f^{2r} \leftarrow f^{2\ell} + \sum_{i \in I^{2=}} w_i$;
24 **if** $f^{1r} * \alpha + f^{2r} * \beta \geqslant 0$ **then** opt $\leftarrow (\tilde{\theta}^1, \tilde{\theta}^2)$
25 **else** RESULT$(m, d, w, a, b, I^{1>}, I^{2>}, \mathscr{H}^>, f^{1r}, f^{2r}, \text{opt})$;
end;

We now analyze the procedure. Steps 1, 2, 3, 6 and 7 are O(m) while Steps 4, 5, 8 and 9 are O(1). For each set in line 10 we have at most m interchanges of items and moves of pointers. Since $|I^1| \leqslant m$ and $|I^2| \leqslant m$ lines 11 and 12 requires at most m divisions each, and Step 13 can be done in O(1) by a simple transfer of pointer. We denote by $T(k)$ the worst case running time of procedure RESULT. Let us suppose $|\mathscr{H}| = k$, $|I^1| = k_1$, $|I^2| = k_2$ with $k_1 + k_2 = k$. Finding the median in Step 15 takes O(k); the rest is O(1). Step 16 is achieved in O(k) through at most k_1 comparisons, interchanges and moves of pointers. So Steps 17 and 18 are respectively O(k_1) and O(k_2). Steps 19 and 22 need at most k_1 additions while Steps 20 and 23 together need k_2 additions. Line 24 is done in O(1). Finally, procedure RESULT is recursively invoked in Steps 21 and 23 with $\max\{|\mathscr{H}^{<}|, |\mathscr{H}^{>}|\} \leq \frac{1}{2}k$. Accordingly,

$$\max\{|I^{1<}|, |I^{1>}|\} \leq \tfrac{1}{2}k \quad \text{and} \quad \max\{|I^{2<}|, |I^{2>}|\} \leq \tfrac{1}{2}k.$$

Hence we obtain the recurrence relationship $T(k) = T(\frac{1}{2}k) + ak$, where a is a positive constant. Consequently, procedure RECTI2 is $O(m)$. It is easily seen that the space requirement is $O(m)$ too. We have thus proved:

Proposition 4. *The algorithm defined by Steps* (a), (b), (c), (d) *solves the CWP with rectilinear norm and linear transportation costs in* $O(m)$ *time and space complexity.*

Computational experience has been obtained with an algorithm, simpler than that one just described, which sorts the sets $\{d_1^1, \ldots, d_m^1\}$ and $\{d_1^2, \ldots, d_m^2\}$. As this algorithm has time complexity $O(m \log m)$, its performance might be slightly less good than those with RECTI1 and RECTI2. However, computation times are very low.

The algorithm has been programmed in FORTRAN IV and implemented on an IBM 370/158 computer. We have generated a certain amount of problems and eliminated as follows the trivial ones. The coordinates (d_i^1, d_i^2) and their associated coefficients w_i, with $i = 1, \ldots, m$, have been chosen randomly according to the normal bivariate distribution $N(0, \sigma^2 I_2)$ with $\sigma = 10$, and to the N(20, 5) normal distribution respectively. We have used a $N(\mu, \sigma^2)$ generator to produce the radii, with $\mu = 27.5, 30, 32.5, 35, 37.5, 40, 42.5$ and $\sigma = 5, 6, 7,$

Table 1
Computational results for the rectilinear norm case

m	1V	2V	EXTR	SIDE	CPU in sec.
10	9	1	1	9	0.12
20	7	3	0	10	0.16
30	9	1	0	10	0.20
40	7	3	2	8	0.24
50	7	3	2	8	0.29
60	8	2	0	10	0.33
70	4	6	4	6	0.37
80	6	4	1	9	0.42
90	5	5	2	8	0.46
100	7	3	1	9	0.50
150	6	4	2	8	0.72
200	8	2	1	9	0.95
300	7	3	3	7	1.40

8. For each m, we have dropped the cases of infeasibility and those where the unconstrained solution was feasible, to keep only the ten first occurrences of a complete running of the procedure. The results are reported on in Table 1. In all cases the feasible region has been found to be a rectangle and the unconstrained solution to be a single point. Columns 1V and 2V give the number of problems for which the visible part of the feasible set consists in respectively one or two sides. In columns EXTR and SIDES we report on the number of constrained solutions located at an extreme point or along a side.

Finally, the average CPU time (virtual CPU time given by the IBM 370/158 operating in CMS; we worked at a low priority level during a week-end) of the ten problems can be found in the last column. Obviously, despite its non-optimal worst-case performance, the algorithm is extremely powerful.

4. The Euclidean norm case

We now examine the case of the Euclidean norm. Both because of the search for the unconstrained solution and the exploration of the visible part of S, we cannot expect to obtain a polynomially bounded algorithm. However, the specialized version of the general scheme given in Section 2 proves to be very efficient. Specifically, an $O(m^2)$ procedure for determining the feasible region, based on the result below, is obtained.

Proposition 5. *The intersection I_k of k disks contains at most $2k-2$ angulous points.*

Proof. The disks are ranked in a way such that $R_i \leqslant R_{i+1}$, $i = 1, \ldots, k-1$. The proposition is well known when $k = 2$. Suppose that it is true for $k = j-1$ and show that it holds for $k = j$. Assume, on the contrary, that $I_j = I_{j-1} \cap B(d_j; R_j)$ contains at least three angulous points $\bar{s}_1$, $\bar{s}_2$ and $\bar{s}_3$ belonging to the boundary of $B(d_j; R_j)$. Two cases may arise. First, the arcs delimitated respectively by $\bar{s}_1$ and $\bar{s}_2$ and by $\bar{s}_2$ and $\bar{s}_3$, denoted arc$(\bar{s}_1, \bar{s}_2)$ and arc$(\bar{s}_2, \bar{s}_3)$, are included in I_j. This would mean that $\bar{s}_2$ is not an angulous point, a contradiction. Second, the arc delimitated by $\bar{s}_1$ and $\bar{s}_2$, say, is not included in I_j. Then there exists at least one disk $B(d_h; R_h)$, with $h < j$, which intersects $B(d_j; R_j)$ in some interior point(s) of arc$(\bar{s}_1, \bar{s}_2)$. Since $B(d_h; R_h)$ cannot intersect the linear segment $[\bar{s}_1, \bar{s}_2]$ included in I_j, $B(d_h; R_h)$ must intersect arc$(\bar{s}_1, \bar{s}_2)$ in two points. But this is possible only if $R_h > R_j$, a contradiction.

To reduce tediousness, we limit ourselves to a brief presentation of Step (a), which corresponds to the most interesting step concerning complexity. The procedure is as follows. First, we sort in $O(m \log m)$ the disks into a non-decreasing order of the radii. Second, for each $k \in \{2, \ldots, m\}$, we select the arcs of I_{k-1} delimited by two adjacent angulous points, which are included in $B(d_k; R_k)$; the number of operations is proportional to $2k-4$. Third, we compute the intersection points of $B(d_k; R_k)$ with the arcs of I_{k-1} cutting $B(d_k; R_k)$; at most two arcs of I_{k-1} are to be considered and each intersection is $O(1)$. It can then be easily deduced that Step (a) can be performed in $O(m^2)$ time. The complete procedure is available from the authors upon request.

In Step (c), one might think that it should be sufficient to consider the disks not containing the unconstrained solution to form the visible part of S. This is not so, as shown by the following counter-example: $d_1 = (0, 2)$, $d_2 = (3.5, 2)$, $d_3 = (3.5, 0)$; $w_1 = 1$, $w_2 = 99.9$, $w_3 = 1$; $R_1 = 2$, $R_2 = 2$, $R_3 = 2.2$ (see Fig. 1).

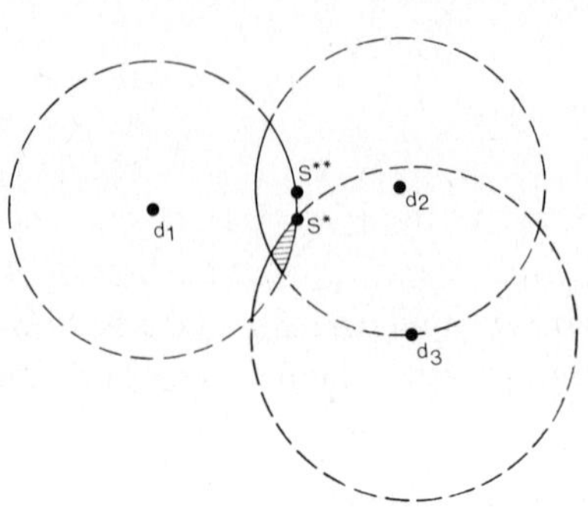

Fig. 1. Counter-example.

Clearly, the unconstrained solution is d_2. The feasible region is the shaded area; the constrained solution is $s^* = (1.9516, 1.5629)$ and the optimal value of the objective function is 164.93. Since $d_2 \in B(d_3; R_3)$, dropping the constraints $s \in B(d_2; R_2)$ and $s \in B(d_3; R_3)$ would yield the solution $s^{**} = (2, 2)$ with value 154.35, but s^{**} is not feasible.

The algorithm for the Euclidean problem has been coded in FORTRAN IV. We have used Ostresh's algorithm [34] to solve the UWP. Particular subroutines have been written to find the intersection of the disks and to determine its visible boundary. For the computation of the directional derivatives and the linear search along an arc of I_m, we have expressed the objective function in polar coordinates and used the Newton–Raphson method. Computational experiments have been carried out on the set of problems generated in the rectilinear case. Again, for each m, we have only kept the ten first complete runnings. We first report in columns CALC, MAX and INT of Table 2 the average total number of computed angulous points, the average length of the vectors describing I_m and the average number of angulous points of I_m respectively. Looking at column INT, we notice that the number of angulous

Table 2
Computational results for the Euclidean norm case

m	CALC	MAX	INT	VIS	ARC	VERT	F	F'	N–R	CPU (in sec.)
10	7.2	4	3.3	2	10	0	1	2	3.3	0.19
20	12	6.8	3.9	2.1	7	3	1	2.1	2.8	0.25
30	12	7	3.8	2.4	7	3	1	2.7	2.4	0.33
40	14.2	7.5	3.7	2.2	7	3	1	2.4	2.6	0.40
50	14.8	8.1	3.6	2.1	8	2	1	1.8	2.8	0.47
60	15.6	8.6	4.2	2.3	5	5	1	2.4	1.7	0.54
70	17.4	9.3	3.8	2.2	7	3	1	2.1	2.8	0.61
80	14.8	7.6	4.1	2.4	4	6	1	2.7	1.5	0.67
90	16.2	8.7	3.9	2.2	6	4	1	2.2	2.1	0.75
100	16.8	9.1	3.4	2.2	8	2	1	2.4	3	0.86
150	20.9	11.3	4.6	2.3	8	2	1	2.6	2.8	1.22
200	20	10.9	4.2	2.1	8	2	1	1.9	2.8	1.57
300	23.8	12.6	4.2	2.1	8	2	1	2.1	2.4	2.30

points is far from its theoretical maximum $2m-2$. In VIS we give the average number of visible arcs of I_m, which varies slightly. The number of constrained solutions located along an arc or at an angulous point are given in columns ARC and VERT. The average number of evaluations of F, of directional derivatives and the average number of iterations in the Newton–Raphson search are reported on in columns F, F' and N–R. Finally, the average CPU times on an IBM 370/158 are reproduced in the last column (working conditions are identical with those of the rectilinear case). The low computational times and their quasi-linear dependance on m suggest that the algorithm is very efficient.

Before concluding the section, we compare the performance of our algorithm with Schaefer and Hurter's one [38]. Five small data sets are produced in that paper, on which we have run our code. The results are summarized in Table 3.

Table 3
Comparison with Schaefer–Hurter's algorithm

Date set	Number of points	H–S CDC 6400	H–P–T IBM 370/158
I	3	8.6	0.15
II	3	8.9	0.31
III	3	2.35	0.12
IV	4	10.8	0.11
V	8	17.9	0.14

Schaefer and Hurter's computational times in seconds on a CDC 6400 are reported in the third column and our results in the last one. Even making a generous allowance for the difference of computer, our algorithm clearly outperforms that of Schaefer and Hurter. Noteworthy, the abnormal result for data set II is due to the fact that 200 iterations are necessary in the Ostresh's algorithm for solving the UWP.

5. The Weber–Rawls problem

The *unconstrained Rawls problem* (in short URP) is defined by the assumptions (i)–(iv) of the UWP and by:

(v′) The facility must be located in a point of $\mathbf{R}^2$ where the maximum transportation cost supported by a user, i.e. $\max_{i=1,\dots,m} C_i(\|x-d_i\|)$, is minimized. Formally, the URP is given by

$$\min_{x\in\mathbf{R}^2} G(x) = \max_{i=1,\dots,m} C_i(\|x-d_i\|). \tag{7}$$

The *constrained Rawls problem* (in short CRP) obtains when assumption (i) is replaced by (i′). Thus, the CRP may be written

$$\min_{s\in S} G(s) = \max_{i=1,\dots,m} C_i(\|s-d_i\|). \tag{8}$$

Properties equivalent to those of the CWP can be easily shown to hold for the CRP. Consequently, algorithms similar to those presented above can be used for solving the CRP.

We now consider an extension of both the UWP and the URP: we make the assumptions (i)–(iv) as before and take as objectives both the minimization of total transportation cost and of maximum transportation cost; the resulting problem is called *Weber–Rawls problem* (in short WRP). As the objectives of the WRP are usually conflicting, its solution is provided by the set of Pareto-optimal—or undominated—locations.

Given that $F(x)$ and $G(x)$ are convex functions, it is known that the set of efficient locations can be obtained from the minimization of all the convex combinations $\theta F(x)+(1-\theta)G(x)$ with $\theta\in[0,1]$. Solutions to these sub-problems are related to θ through the following result, H denoting the convex hull of $\{d_1,\dots,d_m\}$.

Proposition 6. *The correspondance*

$$\mu(\theta)=\left\{x\in H;\ \theta F(x^*)+(1-\theta)G(x^*)=\min_{x\in\mathbf{R}^2}[\theta F(x)+(1-\theta)G(x)]\right\}$$

from $[0,1]$ *to* H *is upper hemi-continuous from* $[0,1]$ *to* H.

Proof. First, given that H is compact, the correspondance Γ from $[0, 1]$ to H such that $\Gamma(\theta) = H$ for any $\theta \in [0, 1]$ is continuous [3, p. 115]. Furthermore, we deduce from a theorem of Wendell and Hurter [40] that H contains a minimizer of $\theta F(x) + (1-\theta)G(x)$ for any $\theta \in [0, 1]$. Finally, since $\theta F(x) + (1-\theta)G(x)$ is continuous with respect to (θ, x) in $[0, 1] \times \mathbf{R}^2$, the theorem of the maximum applies [3, p. 123] and the proposition follows.

Interestingly, when only one minimizer exists for each $\theta \in [0, 1]$, the above proposition tells us that the set of Pareto-optimal locations is a continuous curve in H.

Rather than developing a specialized algorithm to minimize $\theta F(x) + (1-\theta)G(x)$ for given values of θ, we choose to minimize $F(x)$ for different

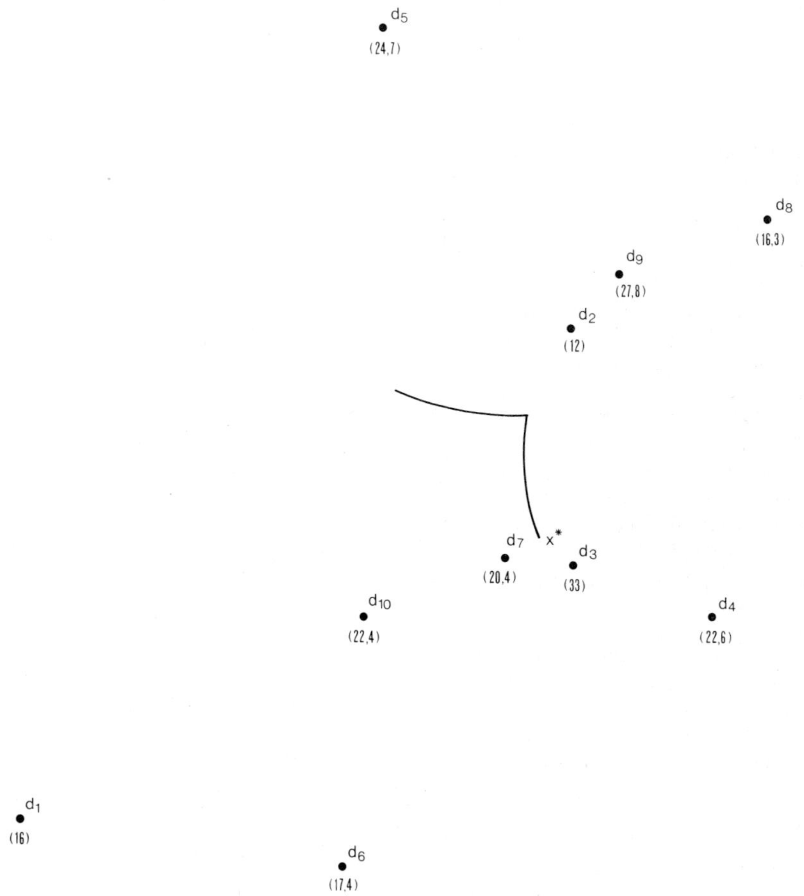

Fig. 2. Pareto-optimal locations.

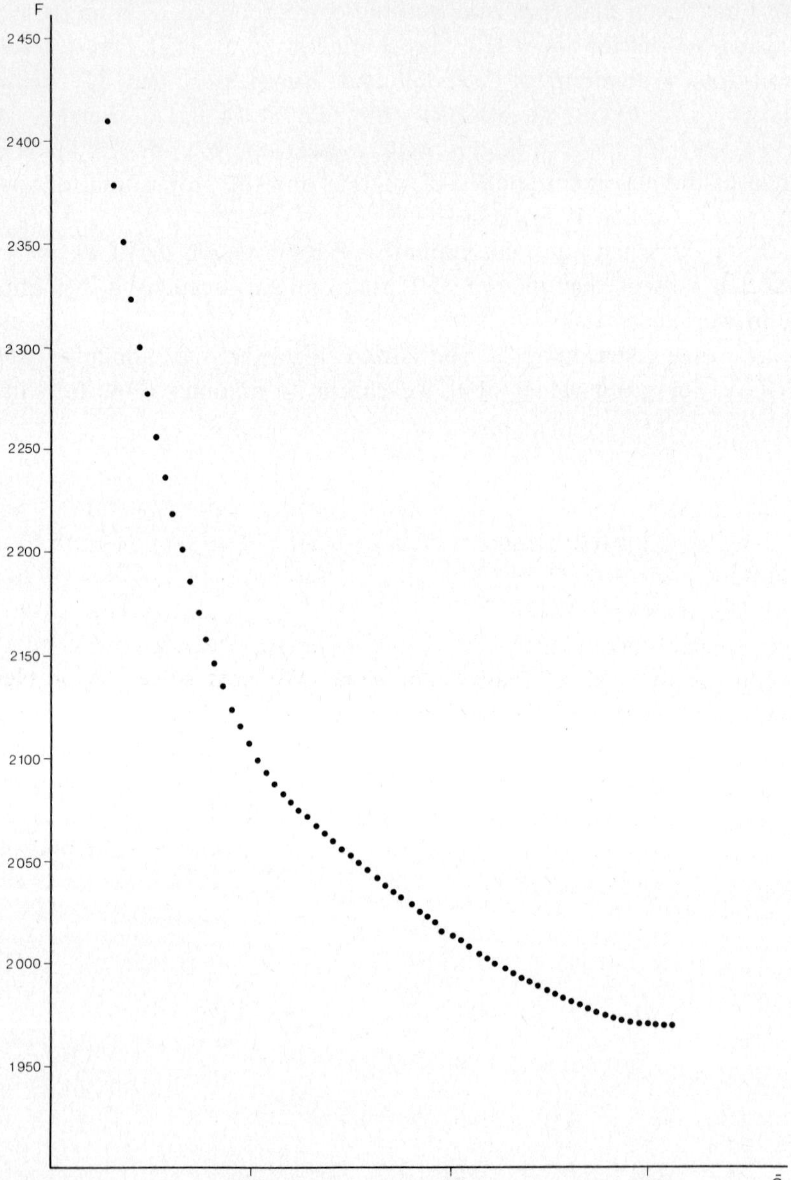

Fig. 3. Values of Pareto-optimal locations.

fixed levels of $G(x)$, using to this end the general scheme described in Section 2. More precisely, any given level $\bar{G}$ of $G(x)$ defines a maximum distance D_i associated with d_i such that $C_i(D_i) = \bar{G}$, and, therefore, a set of disks $B(d_i; D_i)$. We thus obtain a CWP, where S is equal to the (possibly empty) intersection of the disks. A parametrization on $\bar{G}$ then yields the solution to the WRP.

Table 4
Computational results for the Euclidean WRP

m	$\overline{\text{ITER}}$	$\overline{\text{CPU}}$	$\widetilde{\text{ITER}}$	$\widetilde{\text{CPU}}$	ITER_m	CPU_m
10	54.8	2.66	52	2.45	104	4.59
20	50.5	4.28	36	2.89	162	13.58
30	49.5	5.46	31	3.48	108	11.81
40	50.9	7.59	27	4.08	142	21.76
50	73.3	13.36	69	12.64	164	29.46
60	43.8	9.17	41	8.80	84	18.35
70	55.5	12.79	42	8.74	114	27.61
80	56.8	16.46	46	13.32	95	28.88
90	44.0	14.52	37	11.39	85	28.28
100	54.1	19.59	39	14.93	101	34.73

In the following, we give an example of the procedure in the Euclidean case. There are 10 points represented on Fig. 2 with the corresponding coefficients in brackets. We solve the UWP and find a point x^*. The intial value of $\bar{G}$ is provided by $\max_i w_i \|x^* - d_i\|$. We choose $\text{STEP} = \sum_{i=1}^m w_i/100$ and update $\bar{G}$ by $\bar{G} := \bar{G} - \text{STEP}$. For each point d_i, we compute the corresponding radius $R_i = \bar{G}/w_i$ of the disk of feasible locations. We next solve the problem so defined as in Section 3. This procedure is iterated until the intersection of the shrinking disks is empty. So doing, we scan the set of Pareto-optimal locations linking the solution to the UWP and the solution to the URP, for which the last computed point is an approximation depending on the magnitude of STEP. This set is represented in heavy lines on Fig. 2. Fig. 3 gives the optimal value of function F for each level of $\bar{G}$.

Table 5
Computational results for the rectilinear WRP

m	$\overline{\text{ITER}}$	$\overline{\text{CPU}}$	$\widetilde{\text{ITER}}$	$\widetilde{\text{CPU}}$	ITER_m	CPU_m
10	75.2	0.72	64	0.63	133	1.21
20	64.0	0.86	49	0.68	167	2.01
30	57.3	1.01	20	0.46	172	2.64
40	49.6	1.09	41	0.94	93	1.89
50	67.8	1.70	52	1.40	161	3.35
60	63.7	1.91	50	1.54	133	3.67
70	72.2	2.40	56	1.93	168	5.16
80	64.2	2.46	57	2.14	149	5.27
90	57.8	2.40	40	1.75	126	4.86
100	77.5	3.43	73	3.34	119	4.99

A slight modification of the algorithms presented in Sections 3 and 4 was made to solve the WRP in both the Euclidean and rectilinear cases. For each value of m that can be found in the first columns of Tables 4 and 5, we have randomly generated ten problems in the way explained in Section 3 and solved them. We have chosen $\text{STEP}=\sum_{i=1}^{m} w_i/10m$. The characteristics we were looking for, are the number of iterations (ITER) and the CPU time. Since both widely varied, we have computed their mean ($\overline{\text{ITER}}$ and $\overline{\text{CPU}}$), their median ($\widetilde{\text{ITER}}$ and $\widetilde{\text{CPU}}$) and their maximum value (ITER_m and CPU_m). Results are reported on in Tables 4 and 5.

6. Concluding remarks

(i) The algorithm proposed in this paper proves to be very efficient for solving the CWP when the set of feasible locations is defined by the intersection of a finite set of disks. This agrees with previous good results obtained when the set of possible locations is equal to the union of a finite number of convex polygons [16] and emphasizes the usefulness of the visibility concept for constrained location problems.

(ii) Fixed transporation costs—i.e. costs independent of the distance covered and to be added to $C_i(\|s-d_i\|)$ when s differs from d_i—can easily be taken into account. Indeed, it is sufficient to compare the solution s^* and the solutions d_i, which belong to the set of possible locations taking in both cases the fixed costs into account.

(iii) Norms l_p with $p\in\,]1,2[$ appear to be useful for estimating the distance within some transportation networks; see [27, 29]. The general scheme of Section 2 could be specialized to solve some CWP involving such norms. We conjecture that Proposition 5 still holds in these cases.

Acknowledgment

The authors are grateful to A.H.G. Rinnooy Kan for a helpful suggestion and to P. Hanjoul for comments on a first draft of the paper.

Note added in proof

In the period between the submission of this paper and its publication in this Volume, we have heard about a paper on a related subject by C.D.T. Watson-Gandy and entitled "Multifacility constrained Weber problems".

References

[1] A.V. Aho, J.E. Hopcroft and J.D. Ullman, The Design and Analysis of Computer Algorithms (Addison-Wesley, Reading, MA, 1976).

[2] E. Balas and E. Zemel, Solving large zero–one knapsack problem, Oper. Res. 28 (1980) 1130–1145.

[3] C. Berge, Espaces topologiques. Fonctions multivoques (Dunod, Paris, 1966).

[4] L.M. Blumenthal and G.E. Wahlin, On the spherical surface of smallest radius enclosing a bounded subset of n-dimensional Euclidean space, Amer. Math. Soc. Bull. 47 (1941) 771–777.

[5] L. Cooper, Location–allocation problems, Oper. Res. 11 (1963) 331–343.

[6] F. Cordelier and J.Ch. Fiorot, On the Fermat–Weber problem with convex cost functions, Math. Programming 14 (1978) 295–311.

[7] C. d'Aspremont and L. Gevers, Equity and the informational basis of collective choice, Rev. Econom. Stud. 44 (1977) 199–209.

[8] Z. Drezner and G.O. Wezolowsky, Single facility l_p-distance minimax location, SIAM J. Algebraic and Discrete Methods 1 (1980) 315–321.

[9] D.J. Elzinga and D.W. Hearn, The minimum covering sphere problem, Management Sci. 19 (1972) 96–104.

[10] D.J. Elzinga and D.W. Hearn, Geometrical solutions for some minimax location problems, Transportation Sci. 6 (1972) 379–394.

[11] D. Fayard and G. Plateau, Reduction algorithms for single and multiple constraints 0–1 linear programming problems, Publication 92 of the Laboratoire de Calcul de l'Université des Sciences et Techniques de Lille 1, and Proceedings of the Congress 'Methods of Mathematical Programming', Zakopane, Poland (1977).

[12] R.L. Francis, A note on the optimum locations of new machines in existing plant layout, J. Indust. Eng. 14 (1963) 57–59.

[13] R.L. Francis, Some aspects of a minimax location problem, Oper. Res. 15 (1967) 1163–1168.

[14] R.L. Francis, A geometrical solution procedure for a rectilinear distance minimax location problem, AIIE Trans. 4 (1972) 328–332.

[15] P. Hansen, D. Peeters and J.-F. Thisse, An algorithm for a constrained Weber problem, Management Sci. (forthcoming).

[16] P. Hansen, D. Peeters and J.-F. Thisse, Some localization theorems for a constrained Weber problem, J. Regional Sci. 21 (1981) 103–115.

[17] P. Hansen, J. Perreur and J.-F. Thisse, Location theory, dominance and convexity: some further results, Oper. Res. 28 (1980) 1241–1250.

[18] J.M. Huriot and J. Perreur, Modèles de localisation et distance rectilinéaire, Revue d'Economie Politique 83 (1973) 640–662.

[19] A.P. Hurter, M. K. Schaefer and R.E. Wendell, Solutions of constrained location problems, Management Sci. 22 (1975) 51–56.

[20] I.N. Katz, On the convergence of a numerical scheme for solving locational equilibrium problems, SIAM J. Appl. Math. 17 (1969) 1224–1231.

[21] I.N. Katz and L. Cooper, Facility location in the presence of forbidden regions I: Formulation and the case of the Euclidean distance with one forbidden circle, European J. Oper. Res. 6 (1981) 166–173.

[22] S.C. Kolm, Justice et équité (C.N.R.S. Paris, 1973).

[23] H.W. Kuhn, On a pair of dual non-linear programs, in: J. Abadie, ed., Non-Linear Programming (North-Holland, Amsterdam, 1967) 37–54.

[24] H.W. Kuhn, A note on Fermat's problem, Math. Programming 4 (1973) 98–107.

[25] H.W. Kuhn and R.E. Kuenne, An efficient algorithm for the numerical solution of the generalized Weber problem in spatial economics, J. Regional Sci. 4 (1962) 21–33.

[26] E.L. Lawler, Fast approximation algorithms for knapsack problems, Math. Oper. Res. 4 (1979) 339–356.

[27] R.F. Love and J.G. Morris, Modelling inter-city road distances by mathematical functions, Oper. Res. Quart. 23 (1972) 61–71.

[28] R.F. Love and J.G. Morris, Solving constrained multi-facility location problems involving *lp*-distances using convex programming, Oper. Res. 23 (1975) 581–587.

[29] R.F. Love and J.G. Morris, Mathematical models of road travel distances, Management Sci. 25 (1979) 130–139.

[30] L.F. McGinnis and J.A. White, A single facility rectilinear location problem with multiple criteria, Transportation Sci. 12 (1978) 217–231.

[31] R.L. Morrill and J. Symons, Efficiency and equity aspects of optimum location, Geographical Analysis 9 (1977) 215–225.

[32] J.G. Morris and W.A. Verdini, Minisum l_p-distance location problems solved via a perturbated problem and Weiszfeld's algorithm, Oper. Res. 27 (1979) 1180–1188.

[33] National Board of Fire Underwriters, Standard schedule for grading cities and towns of the U.S. with reference to their fire defenses and physical conditions (1956).

[34] L.M. Ostresh, Weber: Exact solution to the one source location problem, in: G. Rushton, M.F. Goodchild and L.M. Ostresh, ed., Computer Programs for Location–Allocation Problems (Department of Geography, University of Iowa, 1973) 1–14.

[35] L.M. Ostresh, On the convergence of a classe of iterative methods for solving the Weber location problem, Oper. Res. 26 (1978) 597–609.

[36] A. Planchart and A.P. Hurter, An efficient algorithm for the solution of the Weber problem with mixed norms, SIAM J. Control 13 (1975) 650–665.

[37] J. Rawls, A Theory of Justice (Harvard University Press, Cambridge, 1971).

[38] M.K. Schaefer and A.P. Hurter, An algorithm for the solution of a location problem with metric constraints, Naval Res. Logist. Quart. 21 (1974) 625–636.

[39] A. Schönhage, M. Paterson and N. Pippenger, Finding the median, J. Comput. System Sci. 13 (1976) 184–199.

[40] R.E. Wendell and A.P. Hurter, Location theory, dominance and convexity, Oper. Res. 21 (1973) 314–320.

[41] G.O. Wesolowsky, Rectangular distance location under the minimax optimality criterion, Transportation Sci. 6 (1972) 103–113.

[42] G.O. Wesolowsky and R.F. Love, The Optimal location of new facilities using rectangular distances, Oper. Res. 19 (1971) 124–130.

P. Hansen, ed., Studies on Graphs and Discrete Programming

A BOUNDING TECHNIQUE FOR INTEGER LINEAR PROGRAMMING WITH BINARY VARIABLES*

Frederick S. HILLIER,
Department of Operations Research, Stanford University, Stanford CA 94305, USA

Nancy Eileen JACQMIN,
Department of Mathematical Sciences, Virginia Commonwealth University, Richmond, VA 23284, USA

We present a bounding technique for use in implicit enumeration algorithms for solving the integer linear programming problem with binary variables. This technique enables fixing the values of some of the variables from the outset, and then provides 'joint' bounds on some of the other variables so that many combinations of their binary values can be eliminated from further consideration. After illustrating the approach, we give some indications that the bounds may be used efficiently and then present some preliminary computational experience. Finally, a class of problems particularly well-suited to this bounding procedure is specified.

1. Introduction

This paper presents a bounding technique for use in implicit enumeration algorithms for solving the integer linear programming problem with binary variables. This problem may be stated as follows,

$$\text{(P)} \qquad \begin{aligned} &\text{maximize} && x_0 = \boldsymbol{c}\boldsymbol{x}, \\ &\text{subject to} && A\boldsymbol{x} \leqslant \boldsymbol{b}, \\ & && x_j = 0 \text{ or } 1 \quad (j = 1, \ldots, n) \end{aligned}$$

where $\boldsymbol{c}$ is a $1 \times n$ vector, $\boldsymbol{x}$ is an $n \times 1$ vector, $\boldsymbol{b}$ is an $m \times 1$ vector and A is an $m \times n$ matrix.

For ease of exposition, we make the following assumptions: (1) the linear programming relaxation of (P),

$$(\text{P}_\text{R}) \qquad \begin{aligned} &\text{maximize} && x_0 = \boldsymbol{c}\boldsymbol{x}, \\ &\text{subject to} && A\boldsymbol{x} \leqslant \boldsymbol{b}, \\ & && 0 \leqslant x_j \leqslant 1 \quad (j = 1, 2, \ldots, n) \end{aligned}$$

* This research was supported by the Office of Naval Research and the National Science Foundation.

possesses a unique optimal solution, (2) the optimal solution to (P_R) is not binary, and (3) a good feasible solution to (P) is available. However, if (1) does not hold, a perturbation can be made in (P_R) to attain this condition (see [7, pp. 56–60] for details). If (2) does not hold, then the optimal solution to (P_R) is also the optimal solution to (P) and the algorithm would not be required. Assumption (3) may be met by using one of the many heuristic procedures in the literature [2, 5, 9, 10]. An alternative to assumption (3) requires that a lower bound on the optimal objective function value of a related problem (P'), defined in Section 3, be available.

In what follows we show that 'joint' bounds can be obtained on the values of a subset of the variables by adapting the approach used by the bound-and-scan algorithm [4] in which the variables are restricted to be non-negative integers rather than binary. After illustrating this approach we give some indications that these bounds may be used efficiently and then present some preliminary computational experience. Finally, a class of problems particularly well-suited to this bounding procedure is specified.

2. Conceptual outline of the bounding technique

This technique is motivated by the following considerations. In an implicit enumeration algorithm the goal is to quickly eliminate large subsets of the 2^n possible combinations of values for the variables by various tests. Having a lower bound on the objective function value of the optimal solution provides one such test since constraining feasible points to yield a better objective function value may greatly decrease the size of the feasible region. Also, this additional constraint assures that any new strictly better feasible solution found will decrease the size of the feasible region even further. Moreover, we will be able to eliminate even more of the 2^n combinations if we can obtain a representation for a group of variables that (1) allows us to fix the values of some of the variables in an optimal solution to (P) and (2) yields 'joint' bounds on the other variables as well as a systematic generation of all relevant combinations of their values. Further, if the number of variables in this group is large as a function of n and we use the 'joint' bounds in a computationally frugal way, additional time savings will be achieved as a result of having only a small number of variables to deal with in another manner. The technique presented here seeks to take advantage of these ideas.

Consider the region determined by the constraints that are binding[1] in the

[1] Here we exclude constraints that are binding due to degeneracy. Hence, the binding constraints are those constraints out of $\sum_{j=1}^{n} a_{ij}x_j \leq b_i$ $(i = 1, 2, \ldots, m)$ whose slack variables are nonbasic in the optimal solution to (P_R), and those nonnegativity constraints that correspond, after the change of variables, to nonbasic variables in this solution.

optimal solution to (P_R) after substituting $x_j' = 1 - x_j$ for those variables that were nonbasic at their upper bound one in this solution. (This change of variables is made automatically by the simplex method when it is streamlined by the standard upper bound technique [6, pp. 682–685], the effect of which is to have all the nonbasic variables at level zero.) In addition to satisfying these constraints, we require that the value of the objective function not be less than the bound generated by the known feasible solution to (P). The optimal solution to (P) clearly lies in this region. After identifying the extreme points of this region, each feasible point within the region can be represented as a convex combination of these extreme points. This representation provides a particularly convenient way of identifying the relevant combinations of values for the variables that were nonbasic in the optimal solution to (P_R) since it immediately reveals which of these variables can be fixed definitely and then provides 'joint' bounds on the rest of them.

We anticipate that the most important application of this bounding technique will be in accelerating branch-and-bound and other implicit enumeration algorithms for solving (P). (See [7] for one such implementation.) In particular, each time a relevant set of values for these nonbasic variables is determined, one can exit from this technique to the underlying algorithm to see if this partial set of values for the x_j can yield a completion which satisfies all the constraints of (P) as well as the constraint on the objective function value. If such a completion is found, it yields a new lower bound on the optimal objective function value of (P) and the corresponding constraint would be revised accordingly. Either way, one would next seek to identify another relevant set of values for this same group of variables and then repeat the testing for whether an improved completion exists. Hence, each iteration corresponds to a relevant set of values for these variables and the algorithm terminates when all such sets have been considered.

3. Notation and initialization

This section introduces the notation and terminology that will be used in what follows. An $n \times 1$ vector $\boldsymbol{x}$ is a *feasible* solution to (P) if it is in the region defined by the constraints; it is an *optimal* solution if it is feasible and maximizes the objective function value among all feasible solutions. The value of x_0 for an optimal solution will be called the *optimal objective function value.* If values have been assigned to only certain of the variables (components of $\boldsymbol{x}$), this specification of values will be called a *partial solution* for these variables. Given a partial solution, a *completion* is a solution resulting from specification of values for the remaining variables. An *eligible partial solution* is one that does not violate any known bounds on the variables involved, i.e., one that, based on our present information, could yield a completion that is an optimal solution.

The optimal solution to (P_R) may be included in the input stream or (P_R) may be solved during initialization if the underlying algorithm is simplex-based. In either case, we will denote this solution as $\boldsymbol{x}^* = [x_1^*, \ldots, x_n^*]^T$ and its objective function value as $x_0^* = \boldsymbol{c}\boldsymbol{x}^*$. Also we define

$$J_B = \{j \mid x_j^* \text{ is basic}\},$$
$$J_0 = \{j \mid x_j^* = 0 \text{ and } x_j^* \text{ is nonbasic}\},$$
$$J_1 = \{j \mid x_j^* = 1 \text{ and } 1 - x_j^* \text{ is nonbasic}\},$$
$$J_N = J_0 \cup J_1,$$
$$K_B = \{i \mid i \in \{1, \ldots, m\} \text{ and the } i\text{th constraint is binding at } \boldsymbol{x}^*\},$$
$$K_N = \{1, \ldots, m\} - K_B.$$

We wish to make the following change of variables to define $\boldsymbol{x}' \in \mathbb{R}^n$ for any $\boldsymbol{x} \in \mathbb{R}^n$:

$$(*) \qquad x_j' = \begin{cases} 1 - x_j, & \text{if } j \in J_1, \\ x_j, & \text{otherwise.} \end{cases}$$

This change of variables directly induces the problem

$$(\mathrm{P}') \qquad \begin{aligned} &\text{maximize} && x_0' = \boldsymbol{c}'\boldsymbol{x}', \\ &\text{subject to} && A'\boldsymbol{x}' \leqslant \boldsymbol{b}', \\ &&& x_j' = 0 \text{ or } 1 \quad (j = 1, 2, \ldots, n) \end{aligned}$$

from (P). Note that an optimal solution to (P'), $\boldsymbol{x}'^{\text{opt}}$, yields an optimal solution to (P) via $(*)$. Henceforth we will direct our attention to solving (P').

For notational convenience it will be assumed that the original indexing of the x_j and b_i was such that if $n_B = |J_B|$, then

$$J_B = \{1, \ldots, n_B\}, \qquad K_B = \{1, \ldots, n_B\},$$
$$J_N = \{n_B + 1, \ldots, n\}, \qquad K_N = \{n_B + 1, \ldots, m\}.$$

Finally, denote the initial good feasible solution to (P) by $\boldsymbol{x}^{(F)} = [x_1^F, \ldots, x_n^F]^T$, let $\boldsymbol{x}'^{(F)}$ be obtained from $\boldsymbol{x}^{(F)}$ via $(*)$, and let $x_0'^{(F)} = \boldsymbol{c}'\boldsymbol{x}'^{(F)}$. Note that $x_0'^{(F)}$ is a lower bound on the optimal objective function value of (P').

4. Details of the bounding technique

We now are ready to develop the bounding technique. Define the $n \times n$ matrix A^* whose ith row is

$$\mathbf{A}_i^* = \begin{cases} \mathbf{A}_i', & i = 1, \ldots, n_B, \\ -\boldsymbol{e}_i, & i = n_B + 1, \ldots, n, \end{cases}$$

where $\boldsymbol{e}_i$ is the n-dimensional unit row vector with unity assigned to component i. Also define the n-dimensional column vector $\boldsymbol{b}^*$ whose component i is

$$b_i^* = \begin{cases} b_i', & i=1,\ldots,n_{\mathrm{B}}, \\ 0, & i=n_{\mathrm{B}}+1,\ldots,n. \end{cases}$$

Now consider the set of points satisfying

$$\text{(i)} \quad \mathrm{A}^*\boldsymbol{x} \leq \boldsymbol{b}^*,$$
$$\text{(ii)} \quad \boldsymbol{c}'\boldsymbol{x} \geq x_0'^{(\mathrm{F})}.$$

Since (i) is the set of constraints that are binding on $\boldsymbol{x}'^*$ (induced by $\boldsymbol{x}^*$ via the change of variables) and $x_0'^{(\mathrm{F})}$ is a lower bound on the optimal objective function value of (P′), any optimal solution to (P′) must be in this set. If $(\boldsymbol{c}'\boldsymbol{x}'^* - \boldsymbol{c}'\boldsymbol{x}'^{(\mathrm{F})})$ is small, this set will contain few binary points and the search for an optimal binary point may not be difficult.[2]

We now seek to find the $(n+1)$ extreme points of the n-simplex defined by (i) and (ii). The first n extreme points are determined by the system of equations

$$\mathrm{A}^{(i)}\boldsymbol{x}^{(i)} = \boldsymbol{b}^{(i)}, \quad (i=1,\ldots,n)$$

where $\mathrm{A}^{(i)}$ is obtained by replacing row i of A^* by $\boldsymbol{c}'$, and $\boldsymbol{b}^{(i)}$ is the vector obtained by replacing component i of $\boldsymbol{b}^*$ by $x_0'^{(\mathrm{F})}$. The $(n+1)$st extreme point is $\boldsymbol{x}^{(0)} = \boldsymbol{x}'^*$ [4, p. 646]. It can be shown that $\mathrm{A}^{(i)}$ is nonsingular $(i=1,\ldots,n)$ [4, p. 464] and, hence, we have explicitly

$$\boldsymbol{x}^{(i)} = (\mathrm{A}^{(i)})^{-1}\boldsymbol{b}^{(i)}, \quad (i=1,\ldots,n).$$

We wish to represent the feasible points as convex combinations of these extreme points. To this end define

$$M = \begin{bmatrix} \boldsymbol{x}^{(0)} & \boldsymbol{x}^{(1)} & \cdots & \boldsymbol{x}^{(n)} \\ 1 & 1 & \cdots & 1 \end{bmatrix}$$

and

$$\boldsymbol{\rho} = [\rho_0, \ldots, \rho_n]^{\mathrm{T}}.$$

Then setting

$$M\boldsymbol{\rho} = \begin{bmatrix} \boldsymbol{x} \\ 1 \end{bmatrix}$$

yields such a representation for $\boldsymbol{x}$ via $\boldsymbol{\rho}$. $\boldsymbol{x}$ is in the simplex determined by (i) and (ii) if and only if $\boldsymbol{\rho} \geq 0$. For an arbitrary $\boldsymbol{x}$, $\boldsymbol{\rho}$ exists by virtue of the fact that

[2] If all of the components of $\boldsymbol{c}'$ are integer, this set can be made even smaller by adding one to $x_0'^{(\mathrm{F})}$ above and in the definition of ρ_0. (Similarly, one would be added to x_0^{B} in the definition of ρ_0^{B} for the theorem presented subsequently.) In this case, the absence of any binary points in the set implies that $\boldsymbol{x}'^{(\mathrm{F})}$ (or $\boldsymbol{x}'^{(\mathrm{B})}$ subsequently) is optimal.

M is nonsingular [4, pp. 648–649], so

$$\boldsymbol{\rho} = M^{-1}\begin{bmatrix} \boldsymbol{x} \\ 1 \end{bmatrix}$$

provides the relevant vector $\boldsymbol{\rho}$.

Fortunately, this representation can be obtained directly without explicitly performing a matrix inversion, since $\mathbf{A}_i^* \boldsymbol{x}^{(j)} = b_i^*$ for $j \neq i$ (by definition of $A^{(i)}$ and $b^{(i)}$, $i = 1, \ldots, n$), so

$$\begin{aligned} b_i^* - \mathbf{A}_i^* \boldsymbol{x} &= b_i^* - \sum_{j=0}^{n} \rho_j \mathbf{A}_i^* \boldsymbol{x}^{(j)} \\ &= b_i^* - (1-\rho_i) b_i^* - \rho_i \mathbf{A}_i^* \boldsymbol{x}^{(i)} \\ &= \rho_i (b_i^* - \mathbf{A}_i^* \boldsymbol{x}^{(i)}). \end{aligned}$$

Therefore,

$$\rho_i = \frac{b_i^* - \mathbf{A}_i^* \boldsymbol{x}}{b_i^* - \mathbf{A}_i^* \boldsymbol{x}^{(i)}}, \quad (i = 1, \ldots, n)$$

and, similarly,

$$\rho_0 = \frac{\mathbf{c}'\boldsymbol{x} - x_0'^{(F)}}{\mathbf{c}'\boldsymbol{x}^{(0)} - x_0'^{(F)}}.$$

Solving (P′) has now been reduced to finding those $\boldsymbol{x}$ with $\boldsymbol{\rho} \geq 0$ which satisfy the constraints of (P′) that are not binding on $\boldsymbol{x}'^*$ and identifying the one which maximizes $\mathbf{c}'\boldsymbol{x}$.

It can be shown that each element of $\{x_j \mid j \in J_N\}$ has the property that it has value zero at all of the extreme points of our n-simplex except one, where it has a strictly positive value. In particular, $x_j^{(j)} > 0$ for each $j \in J_N$. This follows from a previously proven lemma [4, p. 646], which for completeness is restated here.

Lemma. *Under the assumptions of Section* 1, $\boldsymbol{x}^{(i)}$ $(i = 0, \ldots, n)$ *satisfies constraints* (i) *and* (ii). *Furthermore,* $\mathbf{A}_i^* \boldsymbol{x}^{(i)} < b_i^*$ *for* $i = 1, \ldots, n$ *and* $\mathbf{c}'\boldsymbol{x}^{(0)} > x_0'^{(F)}$.

Hence, for $j \in J_N$,

$$\rho_j = \frac{b_j^* - \mathbf{A}_j^* \boldsymbol{x}}{b_j^* - \mathbf{A}_j^* \boldsymbol{x}^{(j)}} = \frac{x_j}{x_j^{(j)}}.$$

This result, combined with the fact that ρ_0 increases as the best known objective function value of (P′) increases, yields the following theorem [4, p. 652].

Theorem. *A necessary condition for* $[x_{n_B+1}, \ldots, x_n]^T$ *to be an eligible partial solution to* (P′) *for the variables whose indices are in* J_N *is that*

$$\sum_{j=n_B+1}^{n} \frac{x_j}{x_j^{(j)}} \leq 1 - \rho_0^B,$$

where ρ_0^{B} *is the value of* ρ_0 *for the current best feasible solution to* (P′), $\boldsymbol{x}^{\mathrm{B}} = [x_1^{\mathrm{B}}, \ldots, x_n^{\mathrm{B}}]^{\mathrm{T}}$ (*with objective function value* x_0^{B}).

This leads immediately to the following corollary.

Corollary. *If* $x_j^{(j)} < 1$ *where* $j \in J_{\mathrm{N}}$, *then* $x_j' = 0$ *in any optimal solution to* (P′).

Proof. $x_j^{(j)} < 1$ implies that $1/x_j^{(j)} > 1$, since $x_j^{(j)} > 0$. Hence, $\rho_j > 1$ when $x_j = 1$ and this partial solution for x_j is not eligible.

Thus we may set $x_j' \equiv 0$ for those j where $x_j^{(j)} < 1$. For notational convenience let this occur for $j = n_{\mathrm{B}} + 1, \ldots, n_0$, and our necessary condition is now reduced to

$$\sum_{j=n_0+1}^{n} \frac{x_j}{x_j^{(j)}} \leq 1 - \rho_0^{\mathrm{B}}.$$

Thus, by calculating $1/x_j^{(j)}$ for $j = n_0 + 1, \ldots, n$ once at the outset and updating ρ_0 whenever a new improved feasible solution to (P′) is found, only additions and one comparison are needed to check whether any particular partial solution for these variables is an eligible one. Hence, it is simple to generate all the eligible partial solutions one at a time, stopping each time to explore its completions using the underlying algorithm.

We now describe one possible enumeration scheme for systematically generating all of the relevant partial solutions on the computer. The scheme is initiated by starting at $[x_{n_{\mathrm{B}}+1}, \ldots, x_n]^{\mathrm{T}} = \mathbf{0}$. Each subsequent partial solution is obtained by first searching the set $\{x_{n_0+1}, \ldots, x_n\}$ from left to right for the first zero value, x_q, and then setting $x_{n_0+1} = x_{n_0+2} = \cdots = x_{q-1} = 0$ and $x_q = 1$, leaving $x_{q+1}, \ldots, x_n$ at their current values. The necessary condition for eligibility is then checked for this partial solution. If the condition is satisfied, some underlying algorithm can be used to explore its completions. When upon searching the set $\{x_{n_0+1}, \ldots, x_n\}$ no zero-valued variable is found, the relevant partial solutions have been exhausted and the algorithm terminates.

5. Example

Consider the following allocation problem of Trauth and Woolsey [11],

$$\begin{aligned} \text{maximize} \quad & 20x_1 + 18x_2 + 17x_3 + 15x_4 + 15x_5 + 10x_6 + 5x_7 + 3x_8 + x_9 + x_{10}, \\ \text{subject to} \quad & 30x_1 + 25x_2 + 20x_3 + 18x_4 + 17x_5 + 11x_6 + 5x_7 + 2x_8 + x_9 + x_{10} \leq 55, \\ & x_j = 0 \text{ or } 1 \quad (j = 1, 2, \ldots, 10). \end{aligned} \tag{P}$$

The optimal solution to ($\mathrm{P_R}$) is $\boldsymbol{x}^* = [0, 0, 0.9, 0, 1, 1, 1, 1, 1, 1]$ with objective function value $x_0^* = 50.3$. We have, as defined in Section 3:

$$J_B = \{3\},$$
$$J_1 = \{5, 6, 7, 8, 9, 10\},$$
$$J_N = \{1, 2, 4, 5, 6, 7, 8, 9, 10\},$$
$$K_B = \{1\},$$
$$K_N = \emptyset.$$

Hillier's heuristic procedure [5] gives an initial feasible solution $\boldsymbol{x}^{(F)} = (0, 0, 0, 1, 1, 1, 1, 1, 1, 1)$. The $x_j^{(j)}$, $j \in J_N$, are:

$$x_1^{(1)} = 0.05455, \qquad x_7^{(7)} = 0.4,$$
$$x_2^{(2)} = 0.09231, \qquad x_8^{(8)} = 0.23077,$$
$$x_4^{(4)} = 1.0, \qquad x_9^{(9)} = 2.0,$$
$$x_5^{(5)} = 0.54545, \qquad x_{10}^{(10)} = 2.0.$$
$$x_6^{(6)} = 0.46154,$$

The corollary to the theorem of Section 4 yields

$$x_1' = x_2' = x_5' = x_6' = x_7' = x_8' = 0$$

in any optimal solution to (P′) (derived from (P) via the change of variables). At this point, the eligible partial solutions consist of these values together with binary assignments for x_4', x_9', and x_{10}'. Since $\rho_0^B = 0$ initially, the bounding inequality given by the theorem of Section 4 is

$$x_4' + 0.5x_9' + 0.5x_{10}' \leq 1.$$

Of the $2^3 = 8$ partial solutions still remaining under consideration, 3 of them violate this inequality. Hence, of the $2^9 = 512$ possible partial solutions for the variables with indices in J_N, all but 5 are eliminated by the bounding technique. ($\boldsymbol{x}^{(F)}$ is the optimal solution to (P).)

6. Computational considerations

The bounding technique presented in Section 4 has not yet been fully tested for computational efficiency. However, we discuss below some indications that it will be efficient, consider which kinds of problems can best take advantage of it, and briefly summarize the available computational experience.

It was mentioned in Section 4 that a small value of $(\boldsymbol{c}'\boldsymbol{x}'^* - \boldsymbol{c}'\boldsymbol{x}'^{(F)})$ is a key to the efficiency of the technique. Several heuristic procedures for obtaining $\boldsymbol{x}'^{(F)}$ have been tested and shown to give good results [13], so a relatively small value of the above seems to be a reasonable expectation. Such a value also tends to yield relatively small values for the $x_j^{(j)}$, which greatly reduces the number of partial solutions that are eligible. For example, if $1 \leqslant x_j^{(j)} < 2$ for all $j = n_0 + 1, \ldots, n$, then the number of eligible partial solutions grows only linearly with $(n - n_0)$, whereas this number could grow essentially exponentially if the $x_j^{(j)}$ are sufficiently large.

In identifying the extreme points $\boldsymbol{x}^{(1)}, \ldots, \boldsymbol{x}^{(n)}$ it is not really necessary to find the $(A^{(i)})^{-1}$ or even to construct the $A^{(i)}$ (or, in fact, A^*) in a simplex-based code. Beginning with the inverse of the final basis matrix for $(\mathrm{P_R})$ (recalling that the change of variables does not affect the basic columns of A), the extreme points may be computed by performing a succession of pivots. First append the constraint $\boldsymbol{c}'\boldsymbol{x}' \geqslant x_0'^{(\mathrm{F})}$ to (P'_R) (induced by $(\mathrm{P_R})$ via the change of variables), introducing its slack variable as the additional basic variable in the optimal solution. Now successively introduce each nonbasic variable $(x_j : j \in J_\mathrm{N})$ into the basis while removing the variable that was introduced into the preceding basis. The first n components of the resulting n basic solutions are the desired extreme points.

An explicit representation for the ρ_i has already been given, obviating the need for finding M^{-1} or even explicitly constructing the matrix M.

Hence, though the theory employed in arriving at this approach is rather involved, the only additional storage (beyond that of the underlying algorithm) is ρ_0^B, $1/x_j^{(j)}$ $(j = n_0 + 1, \ldots, n)$, n_B, and a vector indicating which j are in J_1. (The latter is needed to deal with the required change of variables.) The added computation involves the change of variables, the addition of the constraint $\boldsymbol{c}'\boldsymbol{x}' \geqslant x_0'^{(\mathrm{F})}$ to (P'_R), n pivots to find $\boldsymbol{x}^{(1)}, \ldots, \boldsymbol{x}^{(n)}$, the computation of $1/x_j^{(j)}$ $(j = n_0 + 1, \ldots, n)$, the updating of ρ_0^B, the enumeration scheme for generating relevant partial solutions and testing each one for eligibility with the bounding inequality.

These modest computational and storage requirements suggest that we should usually be able to efficiently investigate a relatively large number of variables with this bounding technique. In fact, a similar technique embedded in Hillier's bound-and-scan algorithm for the general integer linear programming problem has proven to be quite efficient under computational testing [4].

We know that n_B, the number of variables that are not handled by this technique, is bounded above by m, the number of functional constraints. This result follows from Weingartner's proof that at most m of the original variables can be basic in a solution to (P) [12, pp. 35–37]. Therefore, the bounding technique seems particularly promising for problems where m is small, since relatively few variables would then need to be handled by the underlying algorithm.

Preliminary computational results for this technique using an algorithm developed by Jacqmin [7] and implemented on the IBM 370/168 are promising. In solving problems of Balas [1], Petersen [8] and Trauth and Woolsey [11], a maximum of 25% of the possible partial solutions for the variables with indices in J_N survived the bounding process defined by the theorem of Section 4. For problems where the ratio of constraints to variables was small ($\leqslant 0.10$), this maximum was 8%.

In summary, in a simplex-based implicit enumeration procedure to solve (P), the use of this bounding technqiue for problems where the number of constraints is small in comparison to the number of variables seems very promising computationally. It will be fully implemented and tested in the near future and the results reported.

References

[1] E. Balas, An additive algorithm for solving linear programs with zero–one variables, Oper. Res. 13 (1965) 517–546.

[2] E. Balas and C.H. Martin, Pivot and complement—A heuristic for 0–1 programming, Management Sci. 26 (1980) 86–96.

[3] R.S. Garfinkel, and G.L. Nemhauser, Integer Programming (Wiley, New York, 1972).

[4] F.S. Hillier, A bound-and-scan algorithm for pure integer linear progamming with general variables, Oper. Res. 17 (1969) 638–679.

[5] F.S. Hillier, Efficient heuristic procedures for integer linear programming with an interior, Oper. Res. 17 (1969) 600–637.

[6] F.S. Hillier, and G.J. Lieberman, Operations Research, third edition (Holden-Day, San Francisco, CA, 1980).

[7] N.E. Jacqmin, Binary integer linear programming: A hybrid implicit enumeration approach, Ph.D. Dissertation as well as Technical Report No. 97 (ONR Contract N00014-76-C-0418) and Technical Report No. 80-30 (NSF Grant MCS76-81259), Department of Operations Research, Stanford University (1980).

[8] C.C. Petersen, Computational experience with variants of the Balas algorithm applied to the selection of R & D projects, Management Sci. 13 (1967) 736–750.

[9] S. Senju, and Y. Toyoda, An approach to linear programming with 0–1 variables, Management Sci. 15 (1968) B196–B207.

[10] Y. Toyoda, A simplified algorithm for obtaining approximate solutions to zero–one programming problems, Management Sci. 21 (1975) 1417–1427.

[11] C.A. Trauth, Jr. and R.E. Woolsey, Integer linear programming: Study in computational efficiency, Management Sci. 15 (1969) 481–493.

[12] H. Weingartner, Mathematical Programming and the Analysis of Capital Budgeting (Prentice-Hall, Englewood Cliffs, NJ, 1963).

[13] S.H. Zanakis, Heuristic 0–1 linear programming: An experimental comparison of three methods, Management Sci. 24 (1977) 91–104.

P. Hansen, ed., Studies on Graphs and Discrete Programming

SCHOOL TIMETABLES

A.J.W. HILTON
Department of Mathematics, University of Reading, England

1. Introduction

For the purposes of this paper, a *school timetable* is an $n\times n$ matrix $T=(t_{ij})$ in which each cell (i, j) is filled with a number (possibly zero) of distinct symbols drawn from a set $\{\sigma_1,\ldots,\sigma_b\}$ of b symbols and with the property that each symbol occurs exactly once in each row and exactly once in each column. Then t_{ij} is the set of symbols in the cell (i, j). We may think of the rows of the matrix as corresponding to the teachers, the columns to the classes and the symbols to the periods of the week. Thus in this model we suppose that there are the same number of teachers as there are of classes and that every teacher and every class is occupied for each of the given periods. Of course one could simulate free periods by the introduction of dummy teachers and dummy classes. We assume that there are no difficulties with regard to classrooms.

In this paper we prove two theorems about school timetables. The first may be of practical use when drawing up a school timetable as it shows how one may build a timetable in stages. The second deals with the problem of constructing a timetable when there are a number of preassigned symbols around which the rest of the timetable must be constructed; this is a well-known problem upon which much has been written, but, so far as the author knows, the theorem here has not appeared in print. Finally we combine these two theorems to give a more general result.

2. Reconstructing timetables

2.1. *The reduction modulo (P, Q, S) of a timetable T*

A *composition* C of a positive integer n is a sequence $(c_1,\ldots,c_r)$ of positive integers such that $c_1+c_2+\cdots+c_r=n$.

Let an $n\times n$ timetable $T=(t_{ij})$ on the symbols $1,\ldots,b$ be given and let $P=(p_1,\ldots,p_u)$, $Q=(q_1,\ldots,q_v)$ and $S=(s_1,\ldots,s_w)$ be three given compositions of n, n and b respectively. We now define the reduction modulo (P, Q, S)

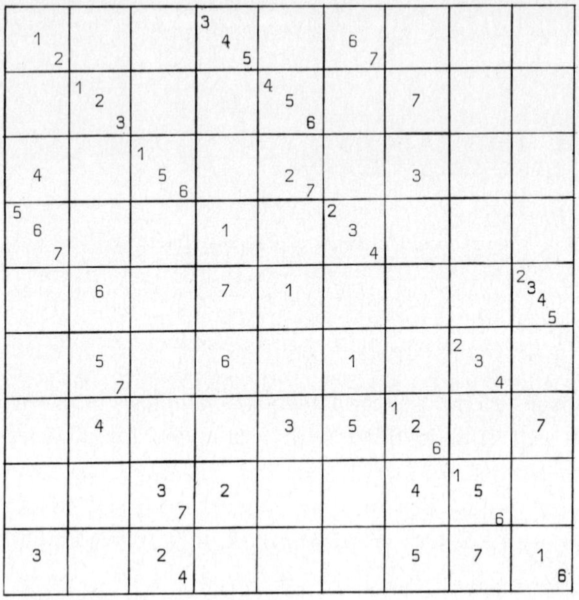

1 2			3 4 5		6 7			
	1 2 3			4 5 6		7		
4		1 5 6		2 7		3		
5 6 7			1		2 3 4			
	6		7	1				2 3 4 5
	5 7		6		1		2 3 4	
	4			3	5	1 2 6		7
		3 7	2			4	1 5 6	
3		2 4				5	7	1 6

Fig. 1.

of T as follows. Intuitively it is the matrix obtained from T by amalgamating rows $p_1+\cdots+p_{i-1}+1,\ldots,p_1+\cdots+p_i$ columns $q_1+\cdots+q_{j-1}+1,\ldots,q_1+\cdots+q_j$ and symbols $\sigma_1+\cdots+\sigma_{k-1}+1,\ldots,\sigma_1+\cdots+\sigma_k$ for $1\leqslant i\leqslant u$, $1\leqslant j\leqslant v$ and $1\leqslant k\leqslant w$.

More precisely, for $1\leqslant\lambda\leqslant u$, $1\leqslant\mu\leqslant v$ and $1\leqslant\xi\leqslant b$, let $x(\lambda,\mu,\xi)$ be the number of times that symbol ξ occurs in the set of cells

$$\{(i,j): p_1+\cdots+p_{\lambda-1}+1\leqslant i\leqslant p_1+\cdots+p_\lambda,\\ q_1+\cdots+q_{\mu-1}+1\leqslant j\leqslant q_1+\cdots+q_\mu\}$$

and, for $1\leqslant\nu\leqslant w$, let

$$x_\nu(\lambda,\mu)=x(\lambda,\mu,s_1+\cdots+s_{\nu-1}+1)+\cdots+x(\lambda,\mu,s_1+\cdots+s_\nu).$$

Then the *reduction modulo* (P,Q,S) of T is a $u\times v$ matrix whose cells are filled from a set of w symbols (say $\{\tau_1,\ldots,\tau_w\}$) and in which cell (λ,μ) contains τ_ν $x_\nu(\lambda,\mu)$ times.

Given a reduction modulo (P,Q,S) of a school timetable T, clearly this itself may be further reduced by further amalgamations of rows, columns or symbols.

We now illustrate this reduction process with an example. Let the given timetable be as in Fig. 1.

Then $n=9$ and $b=7$. Let u, v and w be 4, 3 and 3 respectively, and let $(p_1,p_2,p_3,p_4)=(2,2,3,2)$, $(q_1,q_2,q_3)=(3,4,2)$ and $(s_1,s_2,s_3)=(2,1,4)$. Let I

be the composition consisting of a sequence (of length appropriate to the context) of 1's.

The reduction modulo (P, Q, I) of T is given in Fig. 2.

1 1 2 2 3	3 4 4 5 5 6 6 7 7	
1 4 5 5 6 6 7	1 2 2 3 3 4 7	
4 5 6 7	1 1 1 2 3 5 6 6 7	2 2 3 3 4 4 5 7
2 3 3 4 7	2 4 5	1 1 5 6 6 7

Fig. 2.

In this diagram, and also in the ones in Figs. 1 and 3, there is not intended to be any significance in the way the symbols are arranged in each cell.

Finally we reduce modulo (I, I, S); this means that we replace 1 and 2 by $\alpha(=\tau_1)$, 3 by $\beta(=\tau_2)$ and 4, 5, 6 and 7 by $\delta(=\tau_3)$. We obtain the diagram in Fig. 3. This is then the reduction modulo (P, Q, S) of T.

α α α α β	β δ δ δ δ δ δ δ δ	
α δ δ δ δ δ δ	α α α β β δ δ	
δ δ δ δ	α α α α β δ δ δ δ	α α β β δ δ δ δ
α β β δ δ	α δ δ	α α δ δ δ δ

Fig. 3.

2.2. *Outline timetables*

We now define an outline timetable. Let D be a $u \times v$ matrix filled with w symbols $\tau_1, \ldots, \tau_w$, each cell of which may contain any number (possibly 0) of symbols and may contain repetitions of symbols. For $1 \leq \lambda \leq u$, $1 \leq \mu \leq v$ and $1 \leq \nu \leq w$ let ρ_λ be the number including repetitions of symbols which occur in row λ, let c_μ be the number including repetitions of symbols which occur in column μ and let d_ν be the number of times τ_ν appears in D. Then D is called an *outline timetable* if, for some integers n and b the following properties are obeyed for each λ, μ, ν such that $1 \leq \lambda \leq u$, $1 \leq \mu \leq v$ and $1 \leq \nu \leq w$:

(i) b divides each ρ_λ and c_μ;

(ii) n divides each d_ν;

(iii) the number of times τ_ν appears in row λ is $(\rho_\lambda d_\nu)/nb$;

(iv) the number of times τ_ν appears in column μ is $(c_\mu d_\nu)/nb$.

Proposition 1. *The reduction modulo (P, Q, S) of T is an outline timetable and has the further properties*

(v) $(\rho_1, \ldots, \rho_u) = (bp_1, \ldots, bp_u)$,

(vi) $(c_1, \ldots, c_v) = (bq_1, \ldots, bq_v)$;

(vii) $(d_1, \ldots, d_w) = (ns_1, \ldots, ns_w)$;

(viii) $\sum_{\lambda=1}^{u} \rho_\lambda = \sum_{\mu=1}^{v} c_\mu = \sum_{\nu=1}^{w} d_\nu = nb$.

Proposition 1 is easy to verify.

2.3. *Forming timetables from outline timetables*

In this section we show that any outline timetable could have been formed from some timetable T by reduction modulo (P, Q, S) for some suitable compositions P, Q and S. The main tool is a theorem of de Werra which we now explain.

2.3.1. *de Werra's theorem*

The graph theory terminology we employ here is standard if it is used without explanation and may be found in [5] or [11].

Let G be a graph with vertex set V and edge set E. Let G contain multiple edges but no loops. An *edge-colouring* of G with colours $1, \ldots, k$ is a partition of E into k mutually disjoint subsets $C_1, \ldots, C_k$. Thus $C_i \cap C_j = \emptyset$ $(1 \leq i < j \leq k)$ and $C_1 \cup \cdots \cup C_k = E$. An edge has colour i if it belongs to C_i. Note that we do not make the usual requirement that two edges having the same colour do not have a vertex in common.

Given an edge-colouring of G, for each $v \in V$ let $C_i(v)$ be the set of edges on v of colour i, and, for each $u, v \in V$, $u \neq v$ let $C_i(u, v)$ be the set of edges joining u and v of colour i.

An edge-colouring of G is called *equitable* if, for all $v \in V$,

(a) $$\max_{1 \leq i < j \leq k} \big| |C_i(v)| - |C_j(v)| \big| \leq 1,$$

and it is called *balanced* if, in addition, for all $u, v \in V$, $u \neq v$,

(b) $$\max_{1 \leq i < j \leq k} \big| |C_i(u, v)| - |C_j(u, v)| \big| \leq 1.$$

Thus an edge-colouring is balanced if the colours occur as uniformly as possible at each vertex and if the colours are shared out as uniformly as possible on each multiple edge. D. de Werra [8–10] proved the following important theorem. (There is a short proof in [1] and [6].)

Proposition 2 (de Werra). *For each $k \geq 1$, any finite bipartite graph has a balanced edge-colouring with k colours.*

2.3.2. Application of de Werra's theorem to forming timetables from outline timetables

We now state the main result of Section 2.

Theorem 1. *To each outline timetable D there is a timetable T and compositions P, Q and S such that D is the reduction of T modulo (P, Q, S).*

Proof. First observe that if x denotes the number of entries in D, then

$$x = \sum_{\lambda=1}^{u} \rho_\lambda = \sum_{\nu=1}^{w} d_\nu.$$

Then from (iii) we deduce that

$$x = \sum_{\lambda=1}^{u} \sum_{\nu=1}^{w} \frac{1}{nb} \rho_\lambda d_\nu$$
$$= \frac{1}{nb}\left(\sum_{\lambda=1}^{u} \rho_\lambda\right)\left(\sum_{\nu=1}^{w} d_\nu\right) = \frac{x^2}{nb}.$$

Therefore $x = nb$.

Suppose now that $u < n$. Then we show that D can be obtained from a $(u+1)\times v$ outline timetable D' by amalgamating the cells of two rows (so that any pair of cells in these two rows which are in the same column are identified) or, in other words, by reduction modulo (P^*, I, I), where P^* is a composition with one term 2, the rest all ones. Repeated applications of this and of the analogous statement relating to the columns will show that D can be obtained from an $n \times n$ outline timetable on w symbols.

Since $u \neq n$, b divides $\rho_1, \ldots, \rho_u$ and $\sum_{\lambda=1}^{u} \rho_\lambda / b = n$ there is at least one λ for which $\rho_\lambda / b > 1$. We may assume without loss of generality that $\rho_u / b \geq 2$. We wish to form an outline timetable D' by splitting the last row of D into two new rows. To do this we construct a bipartite graph G with vertex sets $\{\gamma_1, \ldots, \gamma_v\}$ and $\{\tau_1, \ldots, \tau_w\}$ where the vertex γ_μ is joined to the vertex τ_ν by y edges if and only if the symbol τ_ν occurs y times in cell (u, μ) of D. Then the degree of γ_μ is the number, including repetitions, of symbols in the cell (u, μ) and the degree of the vertex τ_ν is the number of times τ_ν occurs in row u of D, namely $(\rho_u d_\nu)/nb$. This is illustrated in Fig. 4.

We now give G an equitable edge-colouring with ρ_u/b colours, say $1, \ldots, \rho_u/b$. Then each vertex τ_ν has exactly

$$\frac{b}{\rho_u}\left(\frac{1}{nb} \rho_u d_\nu\right) = \frac{1}{n} d_\nu$$

edges of colour 1 on it. Now split row u of D into two rows u^* (to be row $u+1$ of D') and u^{**} (to be row u of D') by placing a symbol τ_ν in cell (u^*, μ) x times

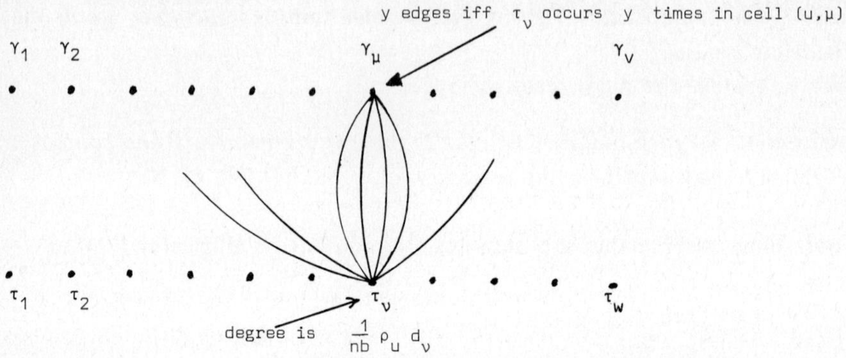

Fig. 4. The graph G.

if and only if there are x edges of colour 1 joining the vertices γ_μ and τ_ν, and by placing τ_ν in cell (u^{**}, μ) y times if and only if there are y edges of colours different from 1 joining the vertices γ_μ and τ_ν. [Note that we could equally well split row u into ρ_u/b rows directly.]

We now check that D' is an outline timetable. Let $\rho'_\lambda = \rho_\lambda$ $(1 \leq \lambda < u)$, $\rho'_u = \rho_u - b$ and $\rho'_{u+1} = b$. Let $c'_\mu = c_\mu$ $(1 \leq \mu \leq v)$, $\tau'_\nu = \tau_\nu$ and $d'_\nu = d_\nu$ $(1 \leq \nu \leq w)$. Then clearly b divides each ρ'_λ and c'_μ and n divides each d'_ν. Thus conditions (i) and (ii) applied to D' are satisfied. Clearly condition (iv) applied to D' is satisfied. Each symbol τ_ν occurs

$$\frac{1}{nb} d_\nu \rho_u - \frac{d_\nu}{n} = \frac{1}{nb} d_\nu (\rho_u - b) = \frac{1}{nb} \rho'_u d'_\nu$$

times in row u of D' and

$$\frac{d_\nu}{n} = \frac{1}{nb} \rho'_{u+1} d'_\nu$$

times in row $u+1$. Thus for these rows condition (iii) applied to D' is satisfied. For all the other rows, condition (iii) applied to D' is clearly satified. Thus D' is an outline timetable.

When an $n \times n$ outline timetable D^* on w symbols is obtained and if $w < b$ then some symbol occurs at least $2n$ times in D^*. We may suppose without loss of generality that τ_w is such a symbol. Then we construct similarly a bipartite graph H with vertex sets $\{\delta_1, \ldots, \delta_n\}$, $\{\gamma_1, \ldots, \gamma_n\}$ in which δ_λ is joined to γ_μ by z edges if and only if cell (λ, μ) contains symbol τ_w exactly z times. The degree of each vertex is d_w/n. We then colour the edges of G equitably with d_w/n colours, say $1, \ldots, d_w/n$. We replace τ_w by τ_w^* and τ_w^{**} according to the rule: τ_w^* is placed y times in cell (λ, μ) if and only if δ_λ is joined to γ_μ by y edges of colour 1, and the remaining occurrences of τ_w are replaced by τ_w^{**}. It is easy to check that an $n \times n$ outline timetable on $w+1$ symbols is obtained.

Iterating this we obtain the required $n \times n$ timetable on b symbols. This proves Theorem 1.

We should remark that we did not actually use balanced edge-colourings in the proof (although they are needed for various analogues and generalizations) and that it was convenient, but not necessary, to use equitable edge-colourings in forming D' in the first part of the proof.

Theorem 1 may be of some practical use when drawing up a school timetable. One might, for example, as a first step, draw up an outline timetable in which all teachers of French are counted together, all of Mathematics are counted together, etc., all classes in Year 1 are counted together, etc., and the periods in Day 1 are counted together, etc. When this outline timetable is deemed to be suitable, it might then be developed into a proper timetable.

3. Extending partial timetables

Suppose we are given a part of a timetable involving only r out of the n teachers and s out of the n classes in which there is a period allocated for some or all of their meetings, can the partial timetable be completed to form a timetable (in our technical sense). In this section we provide some answers to these kinds of questions.

First let us define a partial timetable. An $r \times s$ *partial timetable* on b symbols is an $r \times s$ matrix R in which each cell is filled with a number (possibly zero) of distinct symbols drawn from a set $\{\sigma_1, \ldots, \sigma_b\}$ of b symbols and such that no symbol occurs more than once in any row or column. The partial timetable R is said to be *completed* to form an $n \times n$ timetable T on the same set of b symbols if R may be obtained from T by deleting a set of $n-r$ rows and set of $n-s$ columns, and deleting some of the symbols from the remaining cells.

Theorem 2. *A partial timetable R of size $r \times s$ on the symbols $\sigma_1, \ldots, \sigma_b$ can be completed to form an $n \times n$ timetable T on the same symbols in such a way that no further symbols are placed in any cell of R if and only if*

$$N(\nu) \geq r+s-n,$$

where $N(\nu)$ denotes the number of times the symbol σ_ν occurs in R.

Proof. *Necessity.* Suppose that R can be completed to form T with no further symbols being placed in any cell of R and suppose that T is subdivided as indicated in Fig. 5. Any symbol σ_ν occurs $n-r$ times in B and at most $n-s$ times in A; since σ_ν occurs n times in T it follows that σ_ν occurs at least $n-(n-r)-(n-s)=r+s-n$ times in R.

We give two proofs of the sufficiency; the first is an adaptation of the proof

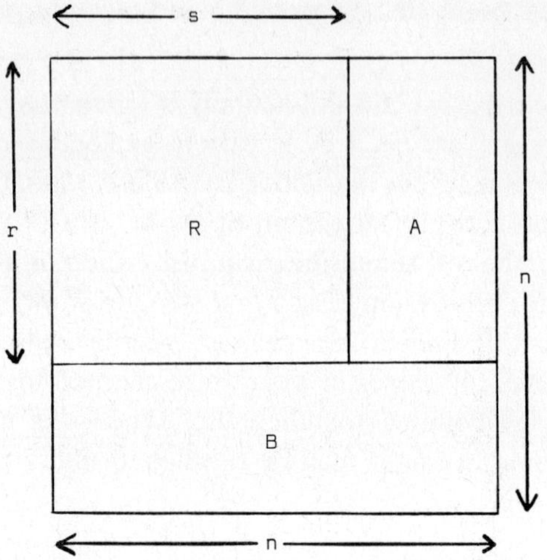

Fig. 5.

of a theorem of Ryser [7] on latin squares, whilst the second is a deduction from Theorem 1.

Sufficiency (first proof). Let R be given and suppose that $N(\nu) \geq r+s-n$ for each symbol σ_ν $(1 \leq \nu \leq b)$. Form a bipartite graph G whose vertex sets are $\{\rho_1, \ldots, \rho_r\}$ and $\{\sigma_1, \ldots, \sigma_b\}$ in which ρ_λ is joined to σ_ν if and only if σ_ν does not occur in row λ of R. Then the degree of each vertex σ_ν is $r - N(\nu) \leq r-(r+s-n) = n-s$. Now give G an equitable edge-colouring with $n-s$ colours $C_{s+1}, \ldots, C_n$. Then place symbol σ_ν in cell (λ, μ) if and only if there is an edge joining ρ_λ and σ_ν coloured C_μ. It is clear that each symbol occurs exactly once in each of the first r rows and at most once in the first s columns. Since no vertex σ_ν has more than one edge of any given colour on it, it follows that no symbol occurs more than once in any of the columns $s+1, \ldots, n$.

The final $n-r$ rows can be added now by an analogous argument to form a timetable.

Sufficiency (second proof). Let R be given and suppose that $N(\nu) \geq r+s-n$ $(1 \leq \nu \leq b)$. Form a matrix R^* by adjoining one further row and one further column, as follows: for $1 \leq i \leq r$, cell $(i, s+1)$ contains all symbols not already occurring in row i; for $1 \leq j \leq s$, cell $(r+1, j)$ contains all symbols not already occurring in column j; finally each symbol σ is placed in cell $(r+1, s+1)$ the number of times necessary in order that σ may occur $n-r$ times altogether in the final row; the condition $N(\nu) \geq r+s-n$ $(\forall \nu)$ ensures that no symbol occurs more than $n-r$ times in the final row. It is easy to check that R^* is an outline timetable and sufficiency now follows from Theorem 1.

This completes the proof of Theorem 2.

Corollary 2.1. *An $r \times n$ partial timetable with $r<n$ on b symbols can be completed to an $n \times n$ timetable on the same set of symbols by adjoining one row at a time.*

Corollary 2.2. *An $r \times (n-r)$ partial timetable R on b symbols can be completed to form an $n \times n$ timetable on the same symbols without placing any further symbols in any cell of R.*

If $N(\nu)<r+s-n$ for some values of ν, can we always complete R if we are willing to permit the addition of some more symbols to the cells of R? We can clearly add in any symbol σ_ν so that it occurs altogether $\min(r, s)$ times in R. Then, by Theorem 2, R can be completed. But, more generally, if T' is any partial $n \times n$ timetable on b $(\leqslant n)$ symbols, then additional symbols can be placed in the cells of T' so that T' is completed to form a timetable T. For if a symbol σ_ν occurs $l<n$ times in T', then clearly a set of $n-l$ cells can be found such that, if σ_ν is placed in each of them, then σ_ν occurs once in each row and once in each column. Doing this for each symbol separately will yield a timetable T which completes the partial timetable T'.

4. A common generalization

In this section we state and prove a common generalization of Theorems 1 and 2.

Theorem 3. (a) *Let b and n be given. Let R be an $r \times s$ partial timetable on the symbols $\tau_1, \ldots, \tau_w$, where $w \leqslant b$. Let $\rho_1, \ldots, \rho_r$, $c_1, \ldots, c_s$ and $d_1, \ldots, d_w$ be such that, for each λ, μ, ν satisfying $1 \leqslant \lambda \leqslant r$, $1 \leqslant \mu \leqslant s$ and $1 \leqslant \nu \leqslant w$,*

(i) *b divides ρ_λ and c_μ;*

(ii) *n divides d_ν;*

(iii)′ *the number of times τ_ν appears in row λ is $\leqslant (\rho_\lambda d_\nu)/nb$;*

(iv)′ *the number of times τ_ν appears in column μ is $\leqslant c_\mu d_\nu/nb$;*

(viii)′ $\sum_{\lambda=1}^{r} \rho_\lambda < nb$; $\sum_{\mu=1}^{s} c_\mu < nb$; $\sum_{\nu=1}^{w} d_\nu = nb$.

Then R is the reduction modulo (P, Q, S) of a $(\rho_1+\cdots+\rho_r)/b \times (c_1+\cdots+c_s)/b$ submatrix of a timetable T on b symbols with

$$P=\left(\frac{\rho_1}{b}, \ldots, \frac{\rho_r}{b}\right), \quad Q=\left(\frac{c_1}{b}, \ldots, \frac{c_s}{b}\right) \quad \text{and} \quad S=\left(\frac{d_1}{n}, \ldots, \frac{d_w}{n}\right)$$

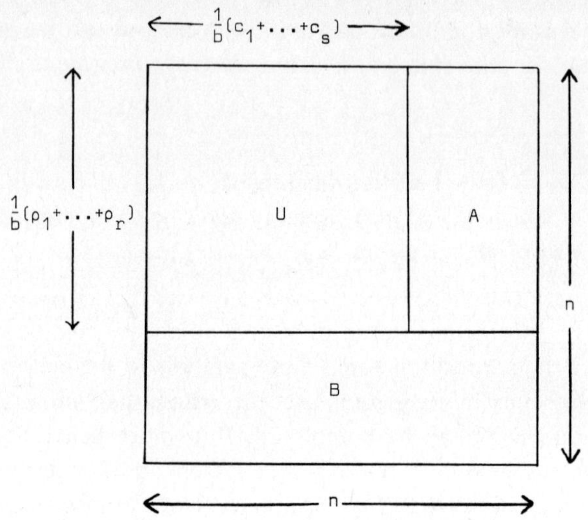

Fig. 6.

if and only if, for $1 \leqslant \nu \leqslant w$,

$$N(\nu) \geqslant \frac{d_\nu}{n}\left\{\frac{1}{b}(\rho_1+\cdots+\rho_r)+\frac{1}{b}(c_1+\cdots+c_s)-n\right\}, \tag{1}$$

where $N(\nu)$ *is the number of times* τ_ν *appears in* R.

(b) *The result remains true if* (iii)′ *is replaced by* (iii) *and* $\sum_{\lambda=1}^{r} \rho_\lambda < nb$ *in* (viii)′ *is replaced by* $\sum_{\lambda=1}^{r} \rho_\lambda = nb$.

(c) *The result remains true if* (iv)′ *is replaced by* (iv) *and* $\sum_{\mu=1}^{s} c_\mu < nb$ *in* (viii)′ *is replaced by* $\sum_{\mu=1}^{s} c_\mu = nb$.

Proof. *Necessity.* Suppose R is the reduction modulo (P, Q, S) of a $(\rho_1+\cdots+\rho_r)/b \times (c_1+\cdots+c_s)/b$ submatrix U of a timetable T on symbols $\sigma_1, \ldots, \sigma_b$. Let T be partitioned as indicated in Fig. 6. By the same argument as in the proof of Theorem 2, each symbol σ appears at least

$$\frac{1}{b}(\rho_1+\cdots+\rho_r)+\frac{1}{b}(c_1+\cdots+c_s)-n$$

times in U. But the symbol τ_ν is obtained by amalgamating d_ν/n of the symbols σ. Therefore the symbol τ_ν will appear at least

$$\frac{d_\nu}{n}\left\{\frac{1}{b}(\rho_1+\cdots+\rho_r)+\frac{1}{b}(c_1+\cdots+c_s)-n\right\}$$

times in R.

Sufficiency. (a) Define ρ_{r+1} and c_{s+1} by the equations

$$\frac{\rho_{r+1}}{b}=n-\frac{1}{b}(\rho_1+\cdots+\rho_r), \qquad \frac{c_{s+1}}{b}=n-\frac{1}{b}(c_1+\cdots+c_s).$$

From R form an $(r+1)\times(s+1)$ matrix R^* on the symbols $\tau_1,\ldots,\tau_w$ by adjoining a final row and a final column as follows:

(I) for $1\leq\lambda\leq r$, $1\leq\nu\leq w$ the symbol τ_ν is placed in the cell $(\lambda, s+1)$ the number of times necessary to ensure that τ_ν appears exactly $(\rho_\lambda d_\nu)/nb$ times in row λ of R^*;

(II) similarly, for $1\leq\mu\leq s$, $1\leq\nu\leq w$ the symbol τ_ν is placed in the cell $(r+1, \mu)$ the number of times necessary to ensure that τ_ν appears exactly $(c_\mu d_\nu)/nb$ times in column μ;

(III) finally the symbol τ_ν is placed in the cell $(r+1, s+1)$ the number of times necessary to ensure that τ_ν appears exactly $(c_{s+1}d_\nu)/nb$ times in column $s+1$.

First we note that (I) and (II) can obviously be satisfied in view of (iii)′ and (iv)′. To see that (III) can be satisfied, we show that (1) implies that τ_ν does not occur more than $(c_{s+1}d_\nu)/nb$ times in column $s+1$. This is because if (I) is satisfied for each of the first r rows of R^*, then in these first r rows τ_ν will appear exactly

$$\frac{1}{nb}d_\nu(\rho_1+\cdots+\rho_r)$$

times. Therefore (1) implies that the number of times that τ_ν appears in the first r rows of column $s+1$ is

$$\leq\frac{1}{nb}d_\nu(\rho_1+\cdots+\rho_r)-\frac{d_\nu}{n}\left\{\frac{1}{b}(\rho_1+\cdots+\rho_r)+\frac{1}{b}(c_1+\cdots+c_s)-n\right\}$$

$$=\frac{d_\nu}{n}\left(n-\frac{1}{b}(c_1+\cdots+c_s)\right)=\frac{1}{nb}d_\nu c_{s+1}.$$

Therefore condition (III) can be satisfied.

We now wish to show that R^* is an outline timetable. To show that (I) is obeyed we need only observe that b divides $\rho_1,\ldots,\rho_r$ and $c_1,\ldots,c_s$ by hypothesis, and that b divides ρ_{r+1} and c_{s+1} by the definitions of ρ_{r+1} and c_{s+1} respectively. Condition (ii) is obeyed by hypothesis. Condition (iv) is obeyed for $1\leq\mu\leq s$ by (II) and for $\mu=s+1$ it is obeyed by (III). Condition (iii) is obeyed for $1\leq\lambda\leq r$ by construction (I); for $\lambda=r+1$ observe that τ_ν appears $d_\nu(c_1+\cdots+c_{s+1})/nb$ times altogether in R^*, and that $d_\nu(\rho_1+\cdots+\rho_r)/nb$ of these occurrences are in the first r rows. Therefore the number of times τ_ν

appears in row $r+1$ is

$$\frac{1}{nb}d_\nu\{(c_1+\cdots+c_{s+1})-(\rho_1+\cdots+\rho_r)\}$$

$$=\frac{1}{nb}d_\nu(nb-(\rho_1+\cdots+\rho_r))=\frac{1}{nb}d_\nu\rho_{r+1},$$

as required. Therefore R^* is an outline timetable.

There is therefore, by Theorem 1, a timetable T and compositions P, Q, S such that R^* is the reduction of T modulo (P, Q, S). It is easy to see that P, Q and S must be $(\rho_1/b, \ldots, \rho_{r+1}/b)$, $(c_1/b, \ldots, c_{s+1}/b)$ and $(d_1/n, \ldots, d_w/n)$ respectively. Case (a) now follows.

Cases (b) *and* (c). These follow by an analogous (but simpler) argument to that of Case (a). This concludes the proof of Theorem 3.

5. Final remarks and acknowledgements

Theorem 2 is an analogue of the well-known theorem of Ryser [7] on latin squares. The kind of arguments used in this paper were developed by L.D. Andersen and the author together in [1–4]; the author wishes to acknowledge his debt to L.D. Andersen. Finally, we thank the referee for finding some errors.

References

[1] L.D. Andersen and A.J.W. Hilton, Generalized latin rectangles I: construction and decomposition, Discrete Math. 31 (1980) 125–152.
[2] L.D. Andersen and A.J.W. Hilton, Generalized latin rectangles II: embedding, Discrete Math. 31 (1980) 235–260.
[3] L.D. Andersen and A.J.W. Hilton, Generalized latin rectangles, in: R.J. Wilson, ed., Proceedings on the One-day Conference on Combinatorics at the Open University, 1978, Research Notes in Mathematics 34, 1–17.
[4] L.D. Andersen and A.J.W. Hilton, Quelques théorèmes sur carrés latins generalisés (où sur graphes complets équitablement colorés), Coll. Math. Discrètes: Codes et Hypergraphes, Bruxelles 1978, Cahiers C.E.R.O. 20 (3–4) (1978) 307–313.
[5] F. Harary, Graph Theory (Addison-Wesley, Reading, MA, 1969).
[6] A.J.W. Hilton, The reconstruction of latin squares with applications to school timetabling and to experimental design, Proceedings of CP79 at the University of East Anglia, Combinatorial Optimization II, Mathematical Programming Study 13 (1980) 68–77.
[7] H.J. Ryser, Combinatorial Mathematics, Carus Mathematical Monographs (Wiley, New York, 1963).
[8] D. de Werra, Balanced schedules, INFOR 9 (1971) 230–237.
[9] D. de Werra, A few remarks on chromatic scheduling, in: B. Roy, ed., Combinatorial Programming: Methods and Applications (S. Reidel, Dordrecht, The Netherlands, 1975) 337–342.
[10] D. de Werra, On a particular conference scheduling problem, INFOR 13 (1975) 308–315.
[11] R.J. Wilson, Introduction to Graph Theory (Oliver and Boyd, Edinburgh, 1972).

P. Hansen, ed., Studies on Graphs and Discrete Programming

HOW TO COLOR CLAW-FREE PERFECT GRAPHS*

Wen-Lian HSU

Northwestern University, Chicago, IL 60201, USA

An $O(n^4)$ minimum coloring algorithm on claw-free perfect graphs is presented. The algorithm proceeds recursively as follows. Let x be any vertex of G. Having colored the subgraph $G\backslash x$ using no more than $\omega(G)$ (the size of a maximum clique in G) colors, the algorithm shows how to color G using no more than $\omega(G)$ colors. The algorithm demonstrates that the size of a minimum coloring is equal to the size of a maximum clique for graphs containing no claws, odd holes or odd anti-holes, hence it provides an alternative proof that the strong perfect graph conjecture is true for claw-free graphs.

1. Introduction

Let $G=(V,E)$ be a *simple graph*, i.e. a finite, undirected, loopless graph without multiple edges. Let V and E denote the vertex and edge sets of G, respectively. Denote the cardinality of V by n. A *clique* is a complete subgraph. A *coloring* of G is an assignment of colors to V such that no two adjacent vertices have the same color. Let

$$\omega(G)=|\text{maximum clique in } G|$$

and

$$\theta(G)=|\text{minimum coloring of } G|.$$

A *claw* is a complete bipartite graph $K_{1,3}$. A simple graph G is claw-free if it does not contain $K_{1,3}$ as a vertex induced subgraph. Claw-free graphs have recently received some attentions (see [2, 6–8, 10]). An interesting subclass of claw-free graphs is the class of line graphs.

A graph G is *perfect* if $\omega(G)=\theta(G)$ for every induced subgraph H of G. An *odd hole* is an odd cycle in which no two non-consecutive vertices are adjacent. An *odd anti-hole* is the complementary graph of an odd hole. It is easy to see that odd holes and odd anti-holes are not perfect. Berge's (see [1]) *strong perfect graph conjecture* (SPGC) asserts that a graph G is perfect if and only if G contains no odd holes or odd anti-holes. The SPGC has been shown to be true for claw-free graphs (see [2, 6, 7, 9]).

* This work was supported by National Science Foundation grant ENG 75-00568 to Cornell University.

We present an $O(n^4)$ coloring algorithm which shows $\theta(G)=\omega(G)$ under the assumption that G does not have claws, odd holes or odd anti-holes, hence it provides an alternative proof that the SPGC is true for claw-free graphs. It should be noted that the vertex coloring problem on claw-free graphs is NP-hard since the edge coloring problem on general graphs (known to be NP-hard [4]) can be reduced to it in polynomial time. In the preparation of this paper, we learn that a polynomial algorithm for the vertex coloring problem on perfect graphs has been established by Grötschel, Lovász and Schrijver [2]. However, their ellipsoid algorithm is not graphical and is unlikely to be computationally practical.

Our coloring algorithm makes use of color adjustment along certain paths. In the following we will define the paths that we are interested in and then study some path properties of a claw-free perfect graph G.

Let f be a function mapping each vertex of G to a color. f is said to be a *coloring* of G if for any two vertices x and y, $(x, y)\in E\Rightarrow f(x)\neq f(y)$. Now consider a coloring f of G. Let i, j be two colors in f and consider the subgraph G_{ij} induced on vertices colored i or j. An *(i, j)-path* in G is a simple path with vertices colored alternately in i and j such that no two non-consecutive vertices are adjacent. Note that such a path may be a simple cycle.

Proposition 1. *Each component of G_{ij} is either a single vertex or an (i, j)-path.*

Proof. It suffices to show that each vertex in a component of G_{ij} has at most two neighbors. Suppose not and assume without loss of generality that y has at least three neighbors and $f(y)=i$. Then since f is a coloring, all the neighbors of y are colored j and form an independent set. But then y and its neighbors contain at least one claw centered at y. □

By switching of colors in an (i, j)-component, we mean changing all i's to j and all j's to i. Note that such a switching will still result in a coloring of G. Hence if x and y do not have the same color and are in distinct $(f(x), f(y))$-components, then we can switch the colors in one component so that x and y will have the same color. To facilitate the description of our algorithm we need the following

Proposition 2. *Let P be an acyclic (i, j)-path with an odd number of edges in a claw-free perfect graph G. Let x and y be the endpoints of P. Let v be a vertex not on the path and suppose (v, x), $(v, y)\in E$. Let x_1 (y_1) be the neighbor of x (y) on P.*

(1) *If P is of length 3, i.e. $(x_1, y_1)\in E$, then v must be adjacent to at least one of x_1 and y_1.*

(2) *If P is of length greater than* 3, *then v must be adjacent to exactly one of* x_1 *and* y_1 *and no other intermediate vertices on P.*

Proof. Since vx (along P) yv is an odd cycle in a perfect graph, we must have a trianglc. Thus either $(x_1, v) \in E$ or $(y_1, v) \in E$. If P is of length greater than 3, then v cannot be adjacent to any intermediate vertex between x_1 and y_1 on P; since otherwise we would get a claw; furthermore, if both (x_1, v) and (y_1, v) are in E then we would get a shorter odd hole vx_1 (along P) y_1v. □

2. Main idea of the algorithm

Let G be a claw-free perfect graph. Let u be any vertex of G, G^0 the subgraph induced on its neighbors $N(u)$ and $\bar{G}^0$ its complement. We use the fact that the number of colors in a minimum cardinality coloring is equal to the size of a maximum clique, $\omega(G)$, which can be found easily since

Lemma 1 [6]. *$\bar{G}^0$ is bipartite.*

The algorithm colors the vertices successively to maintain a coloring that uses no more than $\omega(G)$ colors. Thus a vertex is assigned an arbitrary color different from those of its already colored neighbors so long as these neighbors have used fewer than $\omega(G)$ colors. However if these neighbors have already used $\omega(G)$ colors then we switch colors along appropriate (i, j)-paths. These paths are identified from the solution of an edge-matching problem as described below.

Suppose that in a coloring of $G \setminus u$, $N(u)$ uses $\omega(G)$ colors. Since G is claw-free, at most two vertices in $N(u)$ can have the same color. Let M be the set of edges in $\bar{G}^0$ between two vertices with the same color. Then M is an edge-matching in $\bar{G}^0$. Furthermore we have

Theorem 1. *If $N(u)$ used $\omega(G)$* colors, *then M is not a maximum matching.*

Proof. Let K be the set of colors that color two vertices of $N(u)$. Let

$$V_1 = \{v \mid v \in N(u), f(v) \in K\}.$$

Then $|V_1| = 2|K|$ and $|N(u) \setminus V_1| = \omega(G) - |K|$. Hence

$$|N(u)| = |V_1| + |N(u) \setminus V_1| = 2|K| + \omega(G) - |K|$$
$$= \omega(G) + |K|.$$

Since $\bar{G}^0$ is bipartite and the size of a maximum packing P in $\bar{G}^0$ is $\leqslant \omega(G) - 1$, by Konig's Theorem, the size of a maximum matching in $\bar{G}^0$ is equal to

$$\begin{aligned}|N(u)|-|P| &\geq (\omega(G)+|K|)-(\omega(G)-1)\\ &=|K|+1.\end{aligned}$$

But $|M|=|K|$. Hence M is not maximum. □

3. Color adjustment procedure

Given that M is not a maximum matching we can find an augmenting path P with respect to M. We now show how this path identifies (i, j)-paths for color switching so that the number of colors for $N(u)$ is reduced by one.

Theorem 2. *If P consists of a single edge (x, y) where $f(x)=j$ and $f(y)=k$, then by switching colors along the (j, k)-path in G containing y and assigning vertex u the color k, we obtain a coloring for G that uses $\omega(G)$ colors.*

Proof. It suffices to show that the (j, k)-path in G containing j does not contain x. Since $(x, y) \notin E(G)$, a (j, k)-path joining x to y must contain at least two intermediate vertices. Since $(x, y) \notin M$, none of the intermediate vertices are in $N(u)$. But this implies that the cycle formed by the (j, k)-path together with the two edges (u, x) and (u, y) is an odd-hole. □

For longer augmenting path, we can reduce its length to 1 using the following induction:

Theorem 3. *Let $P=x_0y_1x_1y_2 \cdots x_ty_{t+1} \cdots y_ax_ay_0$ be an augmenting path with respect to M with $a \geq 1$ and suppose that P cannot be shortened by the existence of an edge (x_l, y_m), $l \geq 1$, $m > l+1$. Then by switching colors along some (i, j)-paths we can obtain a different coloring of $G \backslash u$ with a corresponding matching M' and augmenting path P' such that P' is shorter than P.*

Proof. We will give a constructive proof of this theorem using three types of color switches as shown in Figs. 1, 2 and 3. For each color switch the figure gives the relevant part of $\bar{G}^0$ with matching edges indicated by heavy lines. Furthermore, our reduction procedure always starts from the ends of the path P (i.e. x_0 and y_0). Consider two cases:

Case 1: $a=1$. It can be shown that there do not exist both an (i, j)-path joining x_0 and y_1 and an (i, k)-path joining x_1 and y_0. If there is no (i, j)-path joining x_0 and y_1, then by switching colors along the (i, j)-path containing y_1

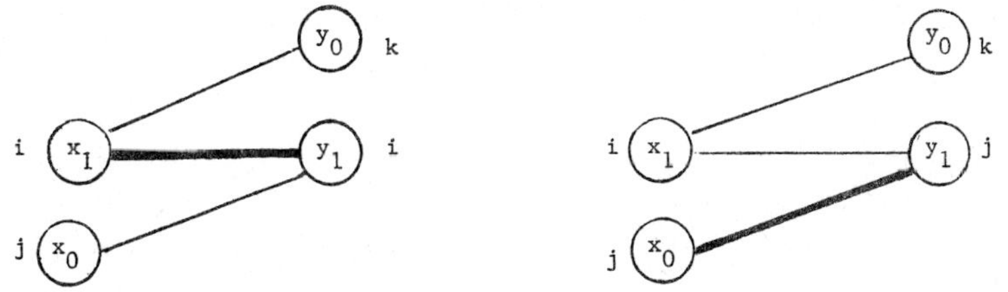

Fig. 1. Type I color switching.

and revising the matching accordingly, we obtain the augmenting path consisting of the single edge (y_0, x_1) (see Fig. 1). Similar result can be obtained if there is no (i, k)-path joining x_1 and y_0. This is called a Type I switch. (It is easy to check that the Type I switch can be applied to shorten augmenting paths of length >3 whenever we have a similar color setting). Hence suppose both paths exist. By Proposition 2 and the fact that u cannot be adjacent to any vertices colored i or j other than x_0, x_1, y_1, the (i, j)-path from y_1 to x_0 in $G\backslash u$ must go from y_1 to x_1 and then to x_0. Similarly the (i, k)-path from x_1 to y_0 must go from x_1 to y_1 and then to y_0. Therefore suppose there exists an (i, j)-path $y_1 Q_{(i,j)} x_1 x_0$ and an (i, k)-path $x_1 Q_{(i,k)} y_1 y_0$ in G. Consider two subcases:

(i) $|Q_{(i,j)}| = 1$, i.e. $Q_{(i,j)} = \{x^*\}$ with $f(x^*) = j$. Then $x_0 y_0 y_1 x^* x_1 x_0$ is an odd cycle. Hence $(x^*, y_0) \in E(G)$. Let the path $x_1 Q_{(i,k)} y_1 y_0$ be $x_1 y_3' y_4'$ (along $Q_{(i,k)}$)$y_1 y_0$ (note y_4' could be y_1). Since x_0 is adjacent to x_1, y_0, the end points of an (i, k)-path, and $(x_0, y_1) \notin E(G)$, we have, by Proposition 2, $(x_0, y_3') \in E(G)$ and $(x_0, y_4') \notin E(G)$. Also, bearing in mind that u cannot be adjacent to any vertices colored i, j or k other than x_0, x_1, y_0, y_1, we can easily derive the following:

$$\{x_1, (x^*, y_3', u)\} \Rightarrow (x^*, y_3') \in E(G),$$
$$\{y_3', (x_0, x^*, y_4')\} \Rightarrow (x^*, y_4') \in E(G),$$
$$\{x^*, (x_1, y_1, y_4')\} \Rightarrow y_4' = y_1.$$

Hence we get an odd anti-hole $y_3' y_0 x_1 y_1 x_0 x^* u y_3'$.

(ii) $|Q_{(i,j)}| > 1$. Let the path $y_1 Q_{(i,j)} x_1 x_0$ be $y_1 x_3' x_4'$ (along $Q_{(i,j)}$)$x_1 x_0$ (note $x_4' \neq x_1$). Since $x_0 y_0 y_1 Q_{(i,j)} x_1 x_0$ is an odd cycle, we have $(x_3', y_0) \in E(G)$. By Proposition 4.7, $(x_4', y_0) \notin E$. Let the path $x_1 Q_{(i,k)} y_1 y_0$ be $x_1 y_3'$ (along $Q_{(i,k)}$)$y_4' y_5' y_1 y_0$. We have $(x_0, y_3') \in E(G)$. Now

$$\{y_1, (x_3', y_5', u)\} \Rightarrow (x_3', y_5') \in E(G),$$
$$\{x_3', (y_0, x_4', y_5')\} \Rightarrow (x_4', y_5') \in E(G).$$

Thus $x_4' = y_4'$. Now $(x_3', y_3') \notin E(G)$ otherwise we would have the claw $\{x_3', (y_0, y_3', y_4')\}$. Hence $x_0 y_3'$ (along $Q_{(i,k)}$)$y_4' x_3' y_0 x_0$ is an odd hole.

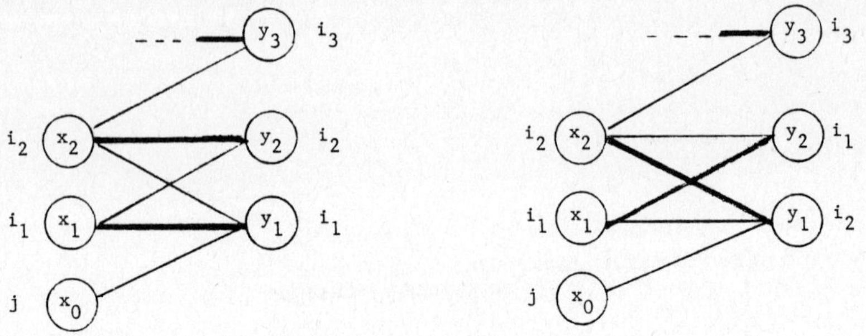

Fig. 2. Type II color switching.

We have derived contradictions in both cases. Hence either $Q_{(i,j)}$ or $Q_{(i,k)}$ does not exist and we can apply Type I switch to get a shorter augmenting path.

Case 2: $a>1$. Type II and Type III color switches will be applied in this case. The vertex coloring on P is given by $f(x_0)=j$, $f(y_0)=k$ and $f(x_l)=f(y_l)=i_l$, $1\leq l\leq a$. Consider two subcases:

(i) Either $(y_1, x_2)\in E(\bar{G})$ or $(x_a, y_{a-1})\in E(\bar{G})$. Without loss of generality, assume $(y_1, x_2)\in E(\bar{G})$. It can be shown that the (i_1, i_2)-path containing y_1, y_2 contains neither x_1 nor x_2. Thus by switching colors along this (i_1, i_2)-path and revising the matching accordingly, we obtain the shorter augmenting path through $x_0y_1x_2y_3$ (see Fig. 2). This is called a Type II switch. Similar switch can be applied to $y_ax_ay_ax_{a-1}y_{a-1}$ when $(x_a, y_{a-1})\in E(\bar{G})$.

Now suppose x_1, x_2, y_1, y_2 are on the same (i_i, i_2)-path. By Proposition 2 and the fact that u cannot be adjacent to vertices colored i_1 or i_2 other than x_1, x_2, y_1, y_2, the (i_i, i_2)-path containing x_1, x_2, y_1, y_2 must have either a part $x_1x_2Q_1y_2y_1$ or a part $x_2x_1Q_2y_1y_2$, where Q_i, $i=1, 2$, denote connecting (i_1, i_2)-paths. In the former case we get an odd hole $y_3x_1x_2Q_1y_2y_3$ since y_3 cannot be adjacent to vertices colored i_1 other than x_1, y_1; in the latter case we get an odd hole $x_0x_1Q_2y_1y_2x_0$ since x_0 cannot be adjacent to vertices colored i_2 other than x_2, y_2. Hence such a path cannot exist.

(ii) Assume that $(y_1, x_2), (x_a, y_{a-1})\in E(G)$ so that the Type II switch cannot be applied; furthermore, it is assumed that there is an (i_1, j)-path $y_1Q_{(i_1,j)}x_1x_0$ and an (i_a, k)-path $x_aQ_{(i_0,k)}y_ay_0$ so that the Type I switch is also not permissible. Then it can be shown that one of the following color switches can be performed:

(a) $f(x_0)=f(y_1)=i_2$, $f(y_2)=j$, $f(x_2)=i_1$, and switch colors on the (i_1, j)-path $Q_{(i_1,j)}$.

(b) $f(y_0)=f(x_a)=i_{a-1}$, $f(x_{a-1})=k$, $f(y_{a-1})=i_a$, and switch colors on the (i_a, k)-path $Q_{(i_a,k)}$ as shown in Fig. 3. This is called a Type III switch. Either switch yields an augmenting path that has four fewer edges than P. (Note that the new paths start from x_2 in (a) and y_{a-1} in (b)). In the following we will prove either (a) or (b) can be performed.

Let y_1' be the successor of y_1, on $Q_{(i_1,j)}$, x_a' the successor of x_a on $Q_{(i_a,k)}$. Since $y_2y_1Q_{(i_1,j)}x_1x_0y_2$ and $x_{a-1}x_aQ_{(i_a,k)}y_ay_0x_{a-1}$ are odd cycles, we must have (y_2, y_1'), $(x_{a-1}, x_a') \in E(G)$. By Proposition 2, y_2 cannot be adjacent to any other vertex of $Q_{(i_1,j)}$ and x_{a-1} cannot be adjacent to any other vertex of $Q_{(i_a,k)}$.

We now claim that if $(x_2, y_1') \notin E(G)$ then the color switch (a) can be performed and, if $(y_{a-1}, x_a') \notin E(G)$ then the color switch (b) can be performed. Finally we will show that one of the above two conditions must hold.

Proof of the claim. We will only prove for the case $(x_2, y_1') \notin E(G)$ since a symmetric argument can be applied to the case $(y_{a-1}, x_a') \notin E(G)$.

If $(x_2, y_1') \notin E(G)$, then x_2 cannot be adjacent to any vertex of $Q_{(i_1,j)}$. We then switch the colors as described in (a): $f(y_1) = i_2$, $f(y_2) = j$, $f(x_2) = i_1$, $f(x_0) = i_2$, $f(x_1) = j$ and switch colors i_1, j on $Q_{(i_1,j)}$. Note that one relation holds for every vertex: a vertex gets a new color b if and only if it is previously adjacent to two vertices with color b, where b is a color taken from $\{i_1, i_2, j\}$. It is straightforward to check that the new coloring is a proper one. End of proof of the claim.

Now we are ready for the final part of this theorem. Suppose (x_2, y_1'), $(y_{a-1}, x_a') \in E(G)$, since $(x_2, y_1') \in E(G)$, x_2 is adjacent to y_1, y_1', x_1, x_0 on the (i_1, j)-path y_1y_1' (along $Q_{(i_1,j)}$)x_1x_0. By Proposition 2, $(y_1', x_1) \in E(G)$. We now get a vertex $y_1' \notin N(u)$ which is adjacent to x_1, y_1, x_2, y_2 in G. In Fig. 4 we show the adjacency in $\bar{G}$. If $(y_1', y_l) \in E(\bar{G})$ for some y_l, $2 < l \leq a$, let y_{l^*} be such a vertex with the lowest index l^*. The only triangle y_1' can create in the odd cycle

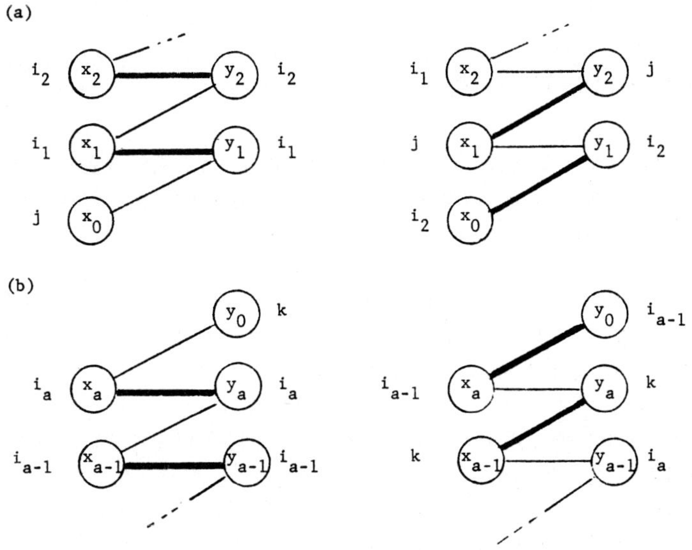

Fig. 3. Type III color switching.

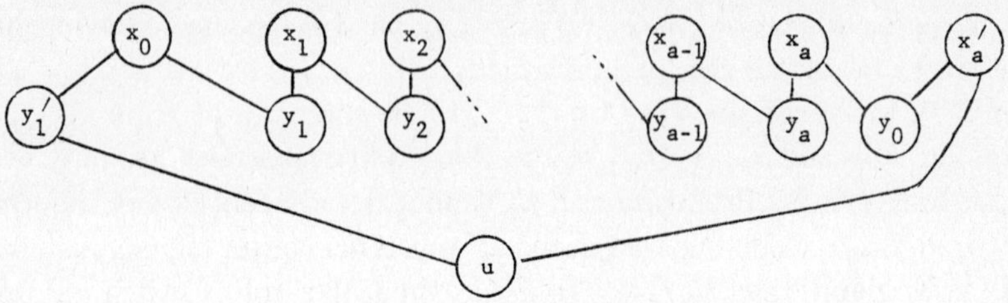

Fig. 4. An odd cycle in $\bar{G}$.

$y'_1x_0y_1x_1y_2x_2 \cdots x_{l^*-1}y_{l^*}y'_1$ in $\bar{G}$ is $y'_1x_{l^*-1}y_{l^*}$. But then we get a claw $\{x_1, (y'_1, x_{l^*-1}, y_{l^*})\}$ in G. Hence y'_1 cannot be adjacent to any y_l with $1 \leq l \leq a$. Similarly y'_1 cannot be adjacent to y_0 in $\bar{G}$. Therefore y'_1 is not adjacent to any y_l, $0 \leq l \leq a$. By a similar argument, x'_a is not adjacent to any x_l, $0 \leq l \leq a$. But (y'_1, u), $(x'_a, u) \in E(\bar{G})$ and $\{x_1, (u, x'_a, y'_1)\} \Rightarrow (x'_a, y'_1) \in E(G)$. Hence in $\bar{G}$, we get an odd cycle $uy'_1x_0y_1x_1y_2x_2 \cdots y_ax_ay_0x'_au$ containing no triangles. This implies the existence of an odd hole, a contradiction. Hence this case can never happen. □

Theorem 3 shows that our path reduction procedure will work if we apply three types of color switches and approach from both ends of the path. It is also easy to see that in the proof of those theorems, we only have to assume that G contains no claws, odd holes or odd anti-holes. Hence the algorithm again demonstrates that the SPGC is true for claw-free graphs.

4. Summary of the algorithm

In summary, our algorithm proceeds by coloring the vertices one by one using no more than $\omega(G)$ colors such that no adjacent vertices get the same color. If, at some iteration, the algorithm picks a vertex x whose neighbor has already used $\omega(G)$ colors, then we can switch the existing colors by considering the augmenting path described in Section 3. Having found an augmenting path, we shorten the path by applying three types of reductions until its length is reduced to 1, in which case we can easily get a spare color for x. At each shortening stage, we only have to test for the existence of some (i, j)-paths in order to switch the colors. After coloring x properly, the algorithm picks another uncolored vertex and repeats this procedure. Hence the running time of this algorithm is determined by

(1) The number of iterations, which is $O(n)$.

(2) Finding an augmenting path in the complement graph of a neighbor of a

vertex, which can be done in $O(n^{2.5})$ time by the bipartite matching algorithm of Hopcroft and Karp [2].

(3) Shortening an augmenting path, which is done by testing for the existence of some (i, j)-paths in $\bar{G}^0$, which can be done in $O(n^2)$ time. The number of shortenings is at most $O(n)$.

At each iteration we might have to do both 2 and 3, which can be bounded by $O(n^3)$. Hence the entire algorithm is bounded by $O(n^4)$.

References

[1] C. Berge, Graphes et Hypergraphes (Dunod, Paris, 1970).
[2] R. Giles, and L.E. Trotter, On stable set polyhedra for $K_{1,3}$-free graphs, Universität Bonn, West Germany, 1979, Report 79136.
[3] M. Grötschel, L. Lovász and A. Schrijver, The ellipsoid method and its consequences in combinatorial optimization, Report 80.151, Bonn Universitat, West Germany, 1980.
[4] I. Holyer, The NP-completeness of edge-coloring, unpublished manuscript, 1979.
[5] J.E. Hopcroft and R.M. Karp, An $n^{\frac{5}{2}}$ algorithm for maximum matching in bipartite graphs, SIAM J. Comput. 2 (1973) 225–231.
[6] W.-L. Hsu and G.L. Nemhauser, Algorithms for minimum coverings by cliques and maximum cliques in claw-free perfect graphs, to appear in *Discrete Math.* (1981).
[7] E. Koch, Ramifications of matching theory, Ph.D. Dissertation, University of Waterloo, Waterloo, Ontario, 1979.
[8] G.J. Minty, On maximal independent sets of vertices in claw-free graphs, J. Combinatorial Theory (B) 28 (1980) 284–304.
[9] K.R. Parathasarathy and G. Ravindra, The strong perfect graph conjecture is true for $K_{1,3}$-free graphs, J. Combinatorial Theory (B) 21 (1976) 212–223.
[10] N. Sbihi, Algorithmes de recherche d'un stable de cardinalité maximum dans un graphe sans étoile, Discrete Math. 29 (1980) 53–76.

P. Hansen, ed., Studies on Graphs and Discrete Programming

PROPERTIES OF SOME EXTREMAL PROBLEMS OF PERMUTATION CYCLES

Ming S. HUNG
Graduate School of Management, Kent State Univ., Kent, OH 44242, USA

Allan D. WAREN and Walter O. ROM
College of Business Administration, Cleveland State Univ., Cleveland, OH 44115, USA

Given a set of n real numbers $A=\{a_i \mid i=1,\ldots,n\}$ and a permutation α of the integers from 1 to n, define the function $f(\alpha, A)=\frac{1}{2}\sum_{i=1}^{n}(a_{\alpha_i}-a_{\alpha_{i+1}})^2$. We determine the permutations α^* and α^{**} which maximize and minimize this function. Next, given the integer k, we find the subsets B_k^* and B_k^{**} which maximize and minimize $f(\alpha^*, B_k^*)$ and $f(\alpha^{**}, B_k^{**})$ respectively, over all subsets B_k of A with cardinality k. Then we determine the values for k, $2\leq k\leq n$ which maximize and minimize the values of $f(\alpha^*, B_k^*)$ and $f(\alpha^{**}, B_k^{**})$ respectively.

These results are then applied to a special assignment problem to determine if the diameter of two property of the assignment can be achieved by an adjacent extreme point method, i.e. can the problem be solved in two steps. It is shown that in general this is not possible.

1. Introduction

Given a set $A_n=\{a_i \mid i=1,\ldots,n\}$ of real numbers and a permutation

$$\alpha=(\alpha_1, \alpha_2, \ldots, \alpha_n)$$

of the integers 1 to n inclusive, define

$$f^{\pm}(\alpha, A_n)\equiv\tfrac{1}{2}\sum_{i=1}^{n}(a_{\alpha_i}\pm a_{\alpha_{i+1}})^2 \quad \text{where } a_{\alpha_{n+1}}\equiv a_{\alpha_1}$$

$$=f^{\pm}(\alpha_1, \alpha_2, \ldots, \alpha_n, A_n).$$

The following properties are easily established:

Property 1

$$f^{\pm}(\alpha_1, \alpha_2, \ldots, \alpha_k, \ldots, \alpha_n, A_n)=f^{\pm}(\alpha_k, \alpha_{k+1}, \ldots, \alpha_n, \alpha_1, \alpha_2, \ldots, \alpha_{k-1}, A_n)$$

Property 2

$$f^{\pm}(\alpha_1, \alpha_2, \ldots, \alpha_k, \ldots, \alpha_n, A_n)=f^{\pm}(\alpha_n, \alpha_{n-1}, \ldots, \alpha_{k+1}, \alpha_k, \ldots, \alpha_2, \alpha_1, A_n)$$

That is, the functions $f^{\pm}(\alpha, A_n)$ are invariant under cyclic rearrangement of the permutation α, being independent of starting element (Property 1) or direction through the permutation (Property 2). As an immediate consequence, we have

Property 3

$$f^{\pm}(\alpha_1, \alpha_2, \alpha_3, \ldots, \alpha_{n-1}, \alpha_n, A_n) = f^{\pm}(\alpha_1, \alpha_n, \alpha_{n-1}, \ldots, \alpha_3, \alpha_2, A_n)$$

which will be of specific use later.

Two permutations $\alpha, \beta \in S_n$, the set of all possible permutations of the integers from 1 to n, are cyclically related (C related) if there exists an integer k, $1<k<n$ such that

$$(\alpha_1, \alpha_2, \ldots, \alpha_k, \ldots, \alpha_n) = (\beta_k, \beta_{k+1}, \ldots, \beta_n, \beta_1, \ldots, \beta_{k-1})$$

or

$$(\alpha_1, \alpha_2, \ldots, \alpha_k, \ldots, \alpha_n) = (\beta_k, \beta_{k-1}, \ldots, \beta_1, \beta_n, \beta_{n-1}, \ldots, \beta_{k+1}).$$

Clearly, C is an equivalence relation and divides the set S_n (of order $n!$) into $(n-1)!/2$ equivalence classes each with $2n$ elements. Furthermore, if α and β are C related, then $f^{\pm}(\alpha, A_n) = f^{\pm}(\beta, A_n)$.

In terms of the functions $f^{\pm}(\alpha, A_n)$ it is sufficient to consider only one permutation from each equivalence class. To fix the initial element of each permutation, we assume henceforth that

$$a_{\alpha_1} \leqslant a_{\alpha_j}, \quad j = 1, \ldots, n$$

and, for all k such that $a_{\alpha_1} = a_{\alpha_k}$, then $\alpha_1 < \alpha_k$. In order to determine the 'direction' of the permutation we also require

$$a_{\alpha_n} \geqslant a_{\alpha_2}$$

and, if $a_{\alpha_n} = a_{\alpha_2}$, then $\alpha_n > \alpha_2$ (refer to Property 3).

Property 4. Given the set A_n, if α^* is the permutation which satisfies

$$f^{+}(\alpha^*, A_n) = \max_{\alpha \in S_n} f^{+}(\alpha, A_n)$$

and β^* satisfies

$$f^{+}(\beta^*, A_n) = \min_{\beta \in S_n} f^{+}(\beta, A_n),$$

then

$$f^{-}(\alpha^*, A_n) = \min_{\alpha \in S_n} f^{-}(\alpha, A_n) \quad \text{and} \quad f^{-}(\beta^*, A_n) = \max_{\beta \in S_n} f^{-}(\alpha, A_n).$$

Proof

$$f^{\pm}(\alpha, A_n) = \tfrac{1}{2} \sum_{i=1}^{n} (a_{\alpha_i} \pm a_{\alpha_{i+1}})^2$$

$$= \sum_{i=1}^{n} a_i^2 \pm \sum_{i=1}^{n} a_{\alpha_i} a_{\alpha_{i+1}} \qquad (\text{recall } a_{\alpha_{n+1}} \equiv a_{\alpha_1}).$$

Since $\sum_{i=1}^{n} a_i^2$ is independent of α the results follow immediately. □

Note that for the function

$$q(\alpha, A_n) = \sum_{i=1}^{n} a_{\alpha_i} a_{\alpha_{i+1}}$$

α^* and β^* (as defined in Property 4) are also the maximizing and minimizing permutations respectively.

As a result of Property 4, in terms of extremizing permutations there is no need to consider both $f^+(\alpha, A_n)$ and $f^-(\alpha, A_n)$. In the balance of this paper, only

$$f(\alpha, A_n) \equiv f^-(\alpha, A_n)$$

will be considered.

Property 5

$$f(\alpha, kA_n + l) \equiv \tfrac{1}{2} \sum [(ka_{\alpha_i} + l) - (ka_{\alpha_{i+1}} + l)]^2 = k^2 f(\alpha, A_n).$$

This property shows that extremal permutations are invariant under linear transformations of the given set of numbers A_n. If this set consists of rational numbers then it can be transformed to a set of integers with smallest element equal to zero.

In Section 2 of this paper, we examine the specific permutations which minimize and maximize $f(\alpha, A_n)$. We denote such an extremizing permutation by an * or ** (i.e. α^* in $f(\alpha^*, A_n)$ is the permutation of the integers 1 to n which maximizes f and α^{**} is the minimizing permutation). Then in Section 3 we examine the problem of selecting a subset B_k of A_n, of cardinality k, such that the extremal values of $f(\alpha, B_k)$ will be extremized over all possible sets B_k. Denoting these sets B_k^* and B_k^{**}, we then determine the values of k which extremizes $f(\alpha, B_k)$.

An application of these results is presented in Section 4. This application is to the computational complexity of a special class of the assignment problems. The earlier results are used to identify those extreme points of the assignment polytope which are adjacent to a given extreme point. We rank extreme points, in particular 'worst' and 'best' points, in terms of the assignment problem objective function value. Then we examine the question of how many iterations are required to reach the optimal solution. This question is of interest in relation to the results of Balinski and Russakoff [1] who showed the diameter of the assignment polytope is two. This result means that given any pair of distinct extreme points, either the points are adjacent to each other or there exists a third extreme point which is adjacent to both. The usual strategy (e.g. the simplex method) for solving the assignment problem is to move from one extreme point to an adjacent, improving extreme point until the optimal point is reached. If the initial point is not adjacent to the optimum, the Balinski–Russakoff result guarantees that there exists an intermediate point adjacent to both and hence implies it may be possible to solve the problem in two steps. However, these intermediate points may have an objective function value worse than that of the initial point and hence none of the usual solution strategies could find it.

Another application of the material from Sections 2 and 3 is for solving a special travelling salesman problem. This and some concluding remarks are presented in Section 5.

2. Extremal permutations

In the following, for convenience, we subscript the elements of all sets, such as A_n, so that

$$a_1 \leqslant a_2 \leqslant \cdots \leqslant a_n$$

In addition, for all permutations α, we can assume that

$$a_{\alpha_1} = a_1 \quad \text{and} \quad a_{\alpha_2} \leqslant a_{\alpha_n}$$

in order to determine a specific permutation from each C-equivalence class. The set of these specific permutations we denote by E_n.

For notational simplicity we represent the function $f(\alpha, A_n)$ by $f(\alpha)$, unless the specific set A and number of elements n is particularly germane. Furthermore, since $f(\alpha) \equiv 0$ if $a_1 = a_n$, we assume $a_1 < a_n$, i.e. that the set consists of at least two distinct elements.

Theorem 2.1. *If* $\alpha^* \in E_n$ *satisfies*

$$f(\alpha^*) \geqslant f(\alpha) \quad \forall \alpha \in S_n,$$

then there exists $\bar{\alpha} \in E_n$ *such that* $f(\bar{\alpha}) = f(\alpha^*)$ *and* $\bar{\alpha}_n = n$.

Proof. If $\alpha_n^* = n$, then $\bar{\alpha} = \alpha^*$ and the theorem is true. Assume $\alpha_n^* \neq n$, hence there is a j, $2 \leqslant j \leqslant n-1$ such that $\alpha_j^* = n$. Let:

$$\alpha^* = (1, \alpha_2^*, \ldots, \alpha_{j-1}^*, n, \alpha_{j+1}^*, \ldots, \alpha_{n-1}^*, \alpha_n^*),$$

$$\beta = (1, \alpha_2^*, \ldots, \alpha_{j-1}^*, \alpha_n^*, \alpha_{n-1}^*, \ldots, \alpha_{j+1}^*, n).$$

Then it can easily be shown that

$$f(\beta) - f(\alpha^*) = (a_n - a_{\alpha_n^*})(a_{\alpha_{j-1}^*} - a_1).$$

Since $a_n \geqslant a_{\alpha_n^*}$ and $a_1 \leqslant a_{\alpha_{j-1}^*}$, then $f(\beta) - f(\alpha^*) \geqslant 0$. But $f(\alpha^*) \geqslant f(\alpha)$ for all α, hence $f(\beta) = f(\alpha^*)$ and $\bar{\alpha} = \beta$ satisfies the theorem. $\square$

Corollary 2.1. *If the elements of* A_n *are distinct, then for a maximizing permutation* $\alpha^* \in E_n$, $\alpha_n^* = n$.

Proof. From the proof of Theorem 2.1

$$(a_n - a_{\alpha_n^*})(a_{\alpha_{j-1}^*} - a_1) = 0,$$

so either $a_n = a_{\alpha_n^*}$ or $a_{\alpha_{j-1}^*} = a_1$. If $a_n = a_{\alpha_n^*}$, then the uniqueness of elements implies $\alpha_n^* = n$ proving the corollary. If $a_{\alpha_{j-1}^*} = a_1$, then $\alpha_{j-1}^* = 1$ and $\alpha_j^* = n = \alpha_2^*$. But for $\alpha \in E_n$, $a_{\alpha_2} < a_{\alpha_n}$, hence $a_{\alpha_2^*} = a_n = a_{\alpha_n^*}$. However, since all elements of a are unique, $a_n > a_j$ for $j = 1, \ldots, n-1$. Thus $a_{\alpha_{j-1}^*} \neq a_1$. □

Theorem 2.2. *If* $\alpha^* \in E_n$ *satisfies*

$$f(\alpha^*) \geq f(\alpha) \quad \forall \alpha \in S_n$$

then $\exists \bar{\alpha} \in E_n$ *such that* $f(\bar{\alpha}) = f(\alpha^*)$ *and*

$$\bar{\alpha} = (1, n-1, 3, n-3, \ldots, n-4, 4, n-2, 2, n).$$

Furthermore, if the elements of A_n *are unique, then* $\alpha^* = \bar{\alpha}$.

Proof. From Theorem 2.1, there exists $\beta \in E_n$ such that $f(\beta) = f(\alpha^*)$ and $\beta_1 = 1$, $\beta_n = n$. Assume $\beta_2 \neq n-1$. Hence there exists j such that $\beta_j = n-1$, $3 < j < n-1$. Consider

$$\beta = (1, \beta_2, \ldots, \beta_{j-1}, n-1, \beta_{j+1}, \ldots, \beta_{n-1}, n),$$
$$\beta^1 = (1, n-1, \beta_{j-1}, \ldots, \beta_3, \beta_2, \beta_{j+1}, \ldots, \beta_{n-1}, n).$$

Then

$$f(\beta^1) - f(\beta) = (a_{\beta_2} - a_{n-1})(a_1 - a_{\beta_{j+1}}).$$

Since $\beta_2 \neq n$, $a_{\beta_2} - a_{n-1} \leq 0$ and since $\beta_{j+1} \neq 1$,

$$a_1 - a_{\beta_{j+1}} \leq 0 \quad \text{and} \quad f(\beta^1) - f(\beta) = f(\beta^1) - f(\alpha^*) \geq 0.$$

But maximality of α^* implies $f(\beta^1) = f(\alpha^*)$.

A similar argument can be used to establish $a_{\beta_{n-1}} = a_2$. Then repeated applications will prove that $f(\bar{\alpha}) = f(\alpha^*)$. If the elements of A_n are unique, then from above and for $f(\beta^1) = f(\alpha^*)$

$$(a_{\beta_2} - a_{n-1}) = 0 \quad \text{or} \quad (a_1 - a_{\beta_{j+1}}) = 0.$$

But $\beta_{j+1} \neq 1$ i.e. $a_1 \neq a_{\beta_{j+1}}$ and hence $\beta_2 = n-1$.

Again repeated application establishes that $\beta = \alpha^*$. □

One of the specific forms of the maximizing permutation α^* is given as follows. Note that $a_1 \leq a_2 \leq \cdots \leq a_n$.

$$\alpha^* = \begin{cases} (1, n-1, 3, n-3, \ldots, x, y, \ldots, 4, n-2, 2, n) & n \text{ even}, \\ (1, n-1, 3, n-3, \ldots, r, s, t, \ldots, 4, n-2, 2, n) & n \text{ odd} \end{cases}$$

where for n even

$$(x, y) = \begin{cases} (\tfrac{1}{2}n+1, \tfrac{1}{2}n) & \text{if } \tfrac{1}{2}n \text{ even}, \\ (\tfrac{1}{2}n, \tfrac{1}{2}n+1) & \text{if } \tfrac{1}{2}n \text{ odd} \end{cases} \tag{2.1}$$

and for n odd

$$(r, s, t) = \begin{cases} (\tfrac{1}{2}(n-1), \tfrac{1}{2}(n+1), \tfrac{1}{2}(n+3)) & \text{if } \tfrac{1}{2}(n+1) \text{ even}, \\ (\tfrac{1}{2}(n+3), \tfrac{1}{2}(n+1), \tfrac{1}{2}(n-1)) & \text{if } \tfrac{1}{2}(n+1) \text{ odd}. \end{cases}$$

Theorem 2.3. *Let $\alpha^{**} \in E_n$ satisfy $f(\alpha^{**}) \leq f(\alpha), \forall \alpha \in S_n$. Then*

$$\alpha^{**} = \begin{cases} (1, 2, 4, \ldots, n-2, n, n-1, \ldots, 5, 3) & n \text{ even}, \\ (1, 2, 4, \ldots, n-1, n, n-2, \ldots, 5, 3) & n \text{ odd}. \end{cases} \tag{2.2}$$

Proof. The proof parallels those in Theorems 2.1 and 2.2. □

If the given set of numbers A_n is that of natural numbers, i.e. $A_n \equiv Z_n = \{1, 2, 3, \ldots, n\}$, then the following is straightforward.

Lemma 2.1

$$f(\alpha^*) = \begin{cases} \tfrac{1}{6}(n^3 - 4n + 6) & n \text{ even}, \\ \tfrac{1}{6}(n^3 - 4n + 3) & n \text{ odd} \end{cases}$$

and $f(\alpha^{**}) = 2n - 3$. □

3. Extremal permutations of subsets of $\{a_1, \ldots, a_n\}$

3.1. *Permutations of k elements, $2 \leq k \leq n$*

Let $B_k = \{a_{i_1}, a_{i_2}, \ldots, a_{i_k}\}$ be an ordered subset of $A_n = \{a_1, \ldots, a_n\}$ $n \geq k \geq 2$ (that is $a_i \leq a_j$ for $i \leq j$ for both B_k and A_n). The number of distinct subsets of size k is $\binom{n}{k}$ or less (because of possible repetitions of some elements in A_n). For each subset B_k it is easy to find the maximizing and the minimizing permutation, as shown by Theorem 2.2 and 2.3. Let $f(\alpha^*, B_k)$ and $f(\alpha^{**}, B_k)$ be respectively the maximum and the minimum of f achievable by B_k.

The objective is to find B_k^* and B_k^{**} such that

$$f(\alpha^*, B_k^*) \geq f(\alpha^*, B_k) \quad \forall B_k \subset A_n,$$
$$f(\alpha^{**}, B_k^{**}) \leq f(\alpha^{**}, B_k) \quad \forall B_k \subset A_n.$$

Theorem 3.1. *If k is even, then B_k^* consists of the $\frac{1}{2}k$ largest and $\frac{1}{2}k$ smallest elements of $A_n = \{a_1, \ldots, a_n\}$.*

Proof. (i) Suppose $\frac{1}{2}k$ is even. Let $\bar{B}$ consist of the $\frac{1}{2}k$ smallest and $\frac{1}{2}k$ largest elements of A_n, i.e.

$$\bar{B}_k = \{a_1, \ldots, a_{k/2}, a_{n-k/2+1}, a_{n-1}, a_n\} = \{b_1, \ldots, b_k\},$$

then by Theorem 2.2, $\max_\alpha f(\alpha, B_k)$ is given by:

$$\alpha^* = (1, k-1, 3, k-3, \ldots, \tfrac{1}{2}k+1, \tfrac{1}{2}k, \ldots, 4, k-2, 2, k).$$

If B_k is any other ordered subset of A_n, f is also maximized by α^*.

Let $B_k = \{a_{i_1}, \ldots, a_{i_k}\}$. We have

$$a_l \leq a_{i_l} \quad \text{for } l = 1, \ldots, \tfrac{1}{2}k,$$
$$a_l \geq a_{l-(n-k)} \quad \text{for } l = n-\tfrac{1}{2}k+1, \ldots, n$$

with strict inequality holding, for at least one.

Then

$$2(f(\alpha^*, \bar{B}_k) - f(\alpha^*, B_k)) = (a_1 - a_{n-1})^2 + (a_{n-1} + a_3)^2 + \cdots + (a_2 - a_n)^2 - (a_{i_1} - a_{i_{k-1}})^2 - (a_{i_{k-1}} - a_{i_3})^2 - \cdots - (a_{i_2} - a_{i_k})^2.$$

Clearly

$$(a_1 - a_{n-1})^2 \geq (a_{i_1} - a_{i_{k-1}})^2,$$
$$(a_{n-1} - a_3)^2 \geq (a_{i_{k-1}} + a_{i_3})^2, \quad \text{etc.}$$

Thus $f(\alpha^*, \bar{B}_k) > f(\alpha^*, B_k)$, and $\bar{B}_k = B_k^*$

(ii) If $\frac{1}{2}k$ odd: same as above with

$$\alpha^* = (1, k-1, 3, \ldots, \tfrac{1}{2}k, \tfrac{1}{2}k+1, \ldots, 2, k). \quad \square$$

Theorem 3.2. *If k is odd, then B_k^* consists of $\frac{1}{2}(k-1)$ largest and $\frac{1}{2}(k-1)$ smallest elements of $A_n = \{a_1, \ldots, a_n\}$, with the additional element, denoted as a_j^*, satisfying, for every $a_u \leq a_{\bar{j}} \leq a_v$,*

$$|a_j^* - \tfrac{1}{2}(a_u + a_v)| \geq |a_j - \tfrac{1}{2}(a_u + a_v)|$$

where $u = \frac{1}{2}(k-1)$, $v = n - \frac{1}{2}(k+3)$. (That is, a_u is the uth smallest element chosen for B_k^* and a_v is the uth largest element.)

Proof. The proof consists of verifying that B_k^*, as defined above, is optimal.

I.e. let

$$B_k^* = \{a_1, \ldots, a_u, a_{j^*}, a_v, \ldots, a_n\} = \{b_1, \ldots, b_k\}.$$

(i) $\frac{1}{2}(k+1)$ odd:

$$\alpha^* = (1, k-1, \ldots, \tfrac{1}{2}(k+3), \tfrac{1}{2}(k+1), \tfrac{1}{2}(k-1), \ldots, 2, k) \tag{3.1}$$

The proof requires two steps. First, suppose $\bar{B}_k$ is the same as B_k^* except that the middle element a_j^* is replaced by $a_{\bar{\imath}}$, with $\bar{\imath} \neq j^*$. Then

$$\begin{aligned} f(\alpha^*, B_k^*) - f(\alpha^*, \bar{B}_k) &= \tfrac{1}{2}[(a_v - a_{j^*})^2 + (a_{j^*} - a_u)^2 - (a_v - a_{\bar{\imath}})^2 - (a_{\bar{\imath}} - a_u)^2] \\ &= (a_{j^*} - a_{\bar{\imath}}^2) - (a_v + a_u)(a_{j^*} - a_{\bar{\imath}}) \\ &= (a_{j^*} - \tfrac{1}{2}(a_v + a_u))^2 - (a_{\bar{\imath}} - \tfrac{1}{2}(a_v + a_u))^2. \end{aligned}$$

So $f(\alpha^*, B_k^*) \geq f(\alpha^*, \bar{\bar{B}}_k)$ be definition of a_{j^*}. Second, suppose

$$\bar{B}_k = \{a_{i_1}, \ldots, a_{i_u}, a_{\bar{\imath}}, a_{i_v}, \ldots, a_{i_k}\}$$

where $a_{i_1} \leqslant a_{i_2} \leqslant \cdots \leqslant a_{i_u} \leqslant a_{\bar{\imath}} \leqslant a_{i_v} \leqslant \cdots \leqslant a_{i_k}$ since $\bar{B}_n$ is ordered. Then

$$a_l \leqslant a_{i_l} \qquad \text{for } l = 1, \ldots, u,$$

$$a_l \geqslant a_{i_{l-(n-k)}} \qquad \text{for } l = v, \ldots, n$$

with at least one strict inequality holding. From the proof in Theorem 3.1, $f(\alpha^*, \bar{B}_k) > f(\alpha^*, \bar{\bar{B}}_k)$. Hence, $f(\alpha^*, B_k^*) \geq f(\alpha^*, \bar{B}_k) > f(\alpha^*, \bar{B}_k)$.

(ii) $\tfrac{1}{2}(k+1)$ even. Proof similar to (i) above. □

The minimizing permutations are more difficult to establish than the maximizing permutations. In fact, only necessary condition is possible,

Theorem 3.3. *If B_k^{**} minimizes $f(\alpha^{**}, B_k)$, $B_k \subset A_n$, then B_k^{**} must consist of k consecutive elements of A_n. In other words, $B_k^{**} = \{a_l, a_{l+1}, \ldots, a_{l+k-1}\}$ for some l, $1 \leqslant l \leqslant n-k+1$.*

Proof. Consider the case:

$$\bar{B}_k = \{a_1, \ldots, a_{t-1}, a_{t+1}, \ldots, a_{k+1}\} = \{b_1, \ldots, b_k\}$$

and

$$B_k^{**} = \{a_1, \ldots, a_k\}$$

with k even and t odd. The minimizing permutation for both is:

$$\alpha^{**} = (1, 2, \ldots, k-2, k, k-1, \ldots, 5, 3).$$

so

$$\begin{aligned} &f(\alpha^{**}, B_k^{**}) - f(\alpha^{**}, \bar{B}_k) \\ &\quad = \tfrac{1}{2}[(a_{t-2} - a_t)^2 + (a_t - a_{t+2})^2 + (a_{k-1} - a_k)^2 + (a_{t+1} - a_{t-1})^2 \\ &\qquad - (a_{t-2} - a_{t+1})^2 - (a_k - a_{k+1})^2 - (a_{k+1} - a_{k-1})^2 - (a_{t+2} - a_{t-1})^2] \\ &\quad = (a_{t-1} + a_{t-2})\,(a_{t+2} - a_{t+1}) + (a_t - a_{t-2})\,(a_t - a_{t+2}) + (a_k - a_{k+1})\,(a_{k+1} - a_{k-1}). \end{aligned}$$

Since

$$a_t - a_{t-2} \geqslant a_{t-1} - a_{t-2} \geqslant 0, \qquad -(a_t - a_{t+2}) \geqslant a_{t+2} - a_{t+1} \geqslant 0,$$

we have:

$$-(a_t - a_{t-2})(a_t - a_{t+2}) \geq (a_{t-1} - a_{t-2})(a_{t+2} - a_{t+1})$$

so the sum of the first two terms of $f(\alpha^{**}, B_k^{**}) - f(\alpha^{**}, \bar{B}_k)$ is negative, and its last term is negative because $a_k \leq a_{k+1}$ and $a_{k+1} \geq a_{k-1}$. Therefore $f(\alpha^{**}, B_k^{**}) \leq f(\alpha^{**}, \bar{B}_k)$. A similar proof applies when t is even or when k is odd. □

When A_n is specialized to a set of natural numbers, closed form expressions B_k^* and B_k^{**} can again be obtained.

Let $A_n = Z_n = \{1, 2, \ldots, n\}$.

Theorem 3.4. *If B_k^* maximizes $f(\alpha^*, B_k)$, $B_k \subset Z_n$, then for k even*

$$B_k^* = \{1, 2, 3, \ldots, \tfrac{1}{2}k, n - \tfrac{1}{2}k + 1, n - \tfrac{1}{2}k + 2, \ldots, n-1, n\}$$

and for k odd

$$B_k^* = \{1, 2, 3, \ldots, \tfrac{1}{2}(k-1), \tfrac{1}{2}(k+1), n - \tfrac{1}{2}(k-3), \ldots, n-1, n\}.$$

Furthermore,

$$f(\alpha^*, B_k^*) = \begin{cases} \tfrac{1}{2}(k-1)(k^2+k-3) + n^2k - nk^2 + 1 & \text{for } k \text{ even}, \\ \tfrac{1}{3}(k-1)(k^2-2k+3) + n^2k + 2nk - nk^2 - n^2 - 3n + 2 & \text{for } k \text{ odd}. \end{cases}$$

Theorem 3.5. *If B_k^{**} minimizes $f(\alpha^{**}, B_k)$, $B_k \subset Z_n$, then B_k^{**} consists of any subset of k consecutive elements, i.e., $B_k^{**} = \{l, l+1, \ldots, l+k-1\}$, $n-k+1 \geq l \geq 1$ and $f(\alpha^{**}, B_k^{**}) = 2k-3$.*

3.2. *Extrema of all cardinalities*

The next question to be addressed is what is the maximum $f(\alpha^*, B_k^*)$ and what is the minimum $f(\alpha^{**}, B_k^{**})$ as k varies between 2 and n. The minimum will be studied first because it is easier.

Theorem 3.6. $f(\alpha^{**}, B_k^{**}) \geq f(\alpha^{**}, B_{k-1}^{**})$ *for $k = 3, \ldots, n$.*

Proof. The B_k^{**} that yields minimum $f(\alpha^{**}, B_k)$ is known and has the following form:

$$B_k^{**} = \{a_{i_1}, \ldots, a_{i_k}\}$$

where $a_{i_1} \leq a_{i_2} \leq \cdots \leq a_{i_k}$ and are a subset of consecutive elements from $\{a_1, \ldots, a_n\}$, are required by Theorem 3.3. Suppose k is odd. Then

$$a^{**} = \{1, 2, 4, \ldots, k-1, k, k-2, \ldots, 5, 3\}.$$

Let $\bar{B}_{k-1}=\{a_{i_1},\ldots,a_{i_{k-1}}\}$. Then it is straightforward to show that

$$f(\alpha^{**},B_k^{**})\geqslant f(\alpha^{**},\bar{B}_{k-1})\geqslant f(\alpha^{**},B_{k-1}^{**})$$

the last inequality arises from the fact that $\bar{B}_{k-1}$ may not be minimum among all subsets of size $k-1$, as shown in Corollary 3.3. A similar argument holds for k even. The above theorem immediately suggests the following. □

Corollary 3.6. *The minimum of $f(\alpha^{**},B_k^{**})$, for $k=2,\ldots,n$ is achieved by a pair $B_2=\{a_{i_1},a_{i_2}\}$ for which $|a_{i_1}-a_{i_2}|\leqslant|a_{i_u}-a_{i_v}|$ for $u,v=1,\ldots,n$ and $u\neq v$.* □

For the set $Z_n=\{1,2,\ldots,n\}$, the overall minimum of $f(\alpha^{**},B_k^{**})$ is achieved by $B_2^{**}=\{i,i+1\}$ for $i=1,\ldots,n-1$.

To determine the over all maximum of $f(\alpha^*,B_k^*)$, $k=2,\ldots,n$, a difficulty similar to that for determining the minimum $f_k(\alpha^*,B_k)$ for a given k is encountered. That is, actual computation must be carried out before the result is known. However, necessary conditions are available.

Theorem 3.7. *If $f(\alpha^*,B_{k^*}^*)=\max_{2\leqslant k\leqslant n}f(\alpha^*,B_k^*)$, then k^* must be even. Moreover $f(\alpha^*,B_k^*)\leqslant f(\alpha^*,B_{k+1}^*)$ and $f(\alpha^*,B_k^*)\leqslant f(\alpha^*,B_{k-1}^*)$ for k odd.*

Proof. Let k be odd and $\frac{1}{2}(k+1)$ odd. Then $f(\alpha^*,B_k)$ is maximized by

$$B_k^*=\{a_1,a_2,\ldots,a_u,a_j^*,a_v,\ldots,a_{n-1},a_n\}=\{b_1,\ldots,b_k\}$$

with

$$\alpha^*=(1,k-1,\ldots,\tfrac{1}{2}(k+3),\tfrac{1}{2}(k+1),\tfrac{1}{2}(k-1),\ldots,2,k)$$

where $v=n-\frac{1}{2}(k+3)$, $u=\frac{1}{2}(k-1)$, and a_{j^*} is specified in Theorem 3.2. Consider

$$\bar{B}_{k-1}=\{a_1,a_2,\ldots,a_u,a_v,\ldots,a_n\}.$$

Then $\bar{B}_{k-1}=B_{k-1}^*$ and,

$$\begin{aligned}f(\alpha^*,B_k^*)-f(\alpha^*,\bar{B}_{k-1})&=\tfrac{1}{2}[(a_v-a_{j^*})^2+(2_{j^*}-a_u)^2-(a_v-a_u)^2]\\&=(a_{j^*}-a_v)(a_{j^*}-a_u)\leqslant 0\end{aligned}$$

because $a_{j^*}\geqslant a_u$ and $a_{j^*}\leqslant a_v$.

On the other hand, if $k\neq n$, then another element may be inserted to B_k^*. Suppose $a_{j^*}=a_{u+1}$, then by Theorem 3. We have:

$$B_{k+1}^*=\{a_1,\ldots,a_u,a_{j^*},a_{v-1},a_v,\ldots,a_n\}$$

and

$$\begin{aligned}f(\alpha^*,B_k^*)-f(\alpha^*,B_{k+1}^*)&=\tfrac{1}{2}[(a_{j^*}-a_u)^2-(a_{j^*}-a_{v-1})^2-(a_{v-1}-a_u)^2]\\&=-(a_{v-1}-a_u)(a_{v-1}-a_{j^*})\leqslant 0.\end{aligned}$$

Similarly, if $j^* = v-1$, $B_{k+1} = \{a_1, \ldots, a_{n+1}, \ldots, a_{j^*}, a_v, \ldots, a_n\}$. Therefore $f_k(\alpha^*, B_k^*)$ for k odd is worse than $f(\alpha^*, B_{k-1}^*)$ and $f(\alpha^*, B_{k+1}^*)$. □

Comparisons between $f(\alpha^*, B_k^*)$ of even k values do not suggest any good behavior in the function f. But the situation is different on the natural number series, $Z_n = \{1, \ldots, n\}$. Using the values given in Theorem 3.4, straightforward comparisons yield the following result.

Theorem 3.8. *For the set $Z_n = \{1, 2, \ldots, n\}$, and $B_k \subset Z_n$:*
(i) *n even $f(\alpha^*, Z_n) = f(\alpha^*, B_{n-2}) > f(\alpha^*, B_k)$ for $k = 2, \ldots, n-3, n-1$;*
(ii) *n odd $f(\alpha^*, B_{n-1}) > f(\alpha^*, B_k)$ for $k = 2, \ldots, n-2, n$.* □

4. Application to a special assignment problem

4.1.

The assignment problem is a special case of a linear program, and is also a permutation problem. As a linear program it is written:

$$\text{Minimize} \quad \sum_i \sum_j c_{ij} x_{ij},$$

$$\sum_i x_{ij} = 1,$$

$$\sum_j x_{ij} = 1,$$

$$x_{ij} \geqslant 0.$$

Every basic solution to this problem has $2n-1$ basic variables, with $n-1$ of them degenerate. Thus, each basis consists of n non-degenerate variables. These can be represented by a permutation $\alpha = (\alpha_1, \ldots, \alpha_n)$ of the integers $(1, \ldots, n)$ where $x_{i\alpha_i} = 1$, $i = 1, \ldots, n$, and $x_{ij} = 0$ for $j \neq \alpha_i$. It is known (see e.g., Balinski and Russakoff [1]) that there is a one-to-one correspondence between extreme points of the linear program and permutations. Thus any extreme point solution x corresponds to a permutation $\alpha(x)$.

Property 4.1 (Balinski and Russakoff [1]). Two extreme points x, y of the assignment problem are adjacent if and only if $\alpha(x) = \alpha(y)z$ where z is a permutation containing one cycle of length two or more.

Property 4.2 (Balinski and Russakoff [1]). The diameter of the assignment polytope is two.

Thus in principle, an adjacent extreme point following method should be able to solve any assignment problem in two steps, since for any starting point it is either adjacent to the optimum or there exists an extreme point adjacent to the starting point and the optimum. One reason this minimum step number is in general not achieved using the Simplex Method is that it proceeds from a basis to an adjacent basis and for the assignment problem each extreme point is represented by $2^{n-1}n^{n-1}$ bases [1]. Even if we are considering a basis-free adjacent extreme point method (or not counting degenerate pivots in the simplex method) it is not clear that this diameter of two property can be achieved. The reason is that, although there exists an extreme point adjacent to both the current point and the optimum, we cannot assume that any of these points are the most improving, i.e. it might be impossible to identify one of the intermediate extreme points. This difficulty will be addressed in the remainder of this section.

4.2. *A special assignment problem*

In order to pursue the questions raised above we will consider an assignment problem with a special structure [3]: Let

$$c_{ij} = a_i \cdot a_j, \qquad i = 1, \ldots, n, \quad j = 1, \ldots, n.$$

Without loss of generality, we will assume:

$$a_1 \leqslant a_2 \leqslant \cdots \leqslant a_n$$

Thus

$$\sum_{i,j} c_{ij} x_{ij} = \sum_{j=1}^{n} a_i a_{j_i} x_{ij_i}$$

where $x_{ij_i} = 1$, all other $x_{ij} = 0$. At this point we will pursue how assignment solutions are represented as permutations. The solution

$$x = (x_{13} = x_{26} = x_{35} = x_{42} = x_{51} = x_{64} = 1, \text{ all other } x_{ij} = 0)$$

can be represented by the permutation

$$\alpha(x) = \begin{pmatrix} 1 & 2 & 3 & 4 & 5 & 6 \\ 3 & 6 & 5 & 2 & 1 & 4 \end{pmatrix}.$$

Alternatively, every permutation can be written as a product of independent cycles, [2]. In this example:

$$\alpha(x) = \begin{pmatrix} 1 & 3 & 5 \\ 3 & 5 & 1 \end{pmatrix} \begin{pmatrix} 2 & 6 & 4 \\ 6 & 4 & 2 \end{pmatrix},$$

or more simply: $\alpha(x) = (1\ 3\ 5)\ (2\ 6\ 4)$. The value of the objective function is

$$a_1 a_3 + a_3 a_5 + a_5 a_1 + a_2 a_6 + a_6 a_4 + a_4 a_2.$$

Thus, in general, any extreme point solution to the assignment problem, written as a permutation, can be written as a product of independent cycles: $\alpha=(\alpha^1)\cdots(\alpha^p)$, where $\alpha^i=(\alpha_1^i,\alpha_2^i,\ldots,\alpha_{l_i}^i)$. For the special objective function we are considering the value of such a solution would be:

$$c(x)=\sum_{i=1}^{p}\left[\left(\sum_{j=1}^{l_i-1} a_{\alpha_j^i}a_{\alpha_{j+1}^i}\right)+a_{\alpha_1^i}a_{\alpha_{l_i}^i}\right].$$

We can now prove the following result.

Theorem 4.1. *For our special assignment problem the worst solution (i.e. the maximizing solution) is* $x^*=\{x_{ij}\mid x_{ii}=1,\ i=1,\ldots,n$ *and* $x_{ij}=0$ *for* $i\neq j\}$, *with value* $c(x^*)=\sum_{i=1}^n a_i^2$.

Proof. Let $\bar{x}$ be any other solution. Then

$$c(x^*)-c(\bar{x})=\sum_{i=1}^{n} a_i^2-\sum_{i=1}^{p}\left[\sum_{j=1}^{l_i-1} a_{\alpha_j^i}a_{\alpha_{j+1}^i}+a_{\alpha_1^i}a_{\alpha_{l_i}^i}\right],$$

where $\alpha=(\alpha^1)\cdots(\alpha^p)$ is the permutation representing $\bar{x}$.

Consider the terms of $c(x^*)-c(\bar{x})$ corresponding to the first cycle of $\bar{x}$:

$$\sum_{j=1}^{l_1} a_{\alpha_j}^2-\left(\sum_{j=1}^{l_1-1} a_{\alpha}a_{{}_j^1\alpha_{j+1}^1}+a_{\alpha_1^1}a_{\alpha^1}\right)$$
$$=\frac{1}{2}\left[\sum_{j=1}^{l_1-1}(a_{\alpha_j^1}-a_{\alpha_{j+1}^1})^2+(a_{\alpha_1^1}-a_{\alpha_{l_i}^1})^2\right]\geqslant 0.$$

Similarly for all other cycles of $\bar{x}$. Thus, $c(x^*)-c(\bar{x})\geqslant 0$. $\square$

In order to determine the minimizing solution we will use the following lemma:

Lemma 4.1. *Let* $g(\alpha)=\frac{1}{2}[(a_{\alpha_1}-a_{\alpha_2})^2+(a_{\alpha_3}-a_{\alpha_4})^2+\cdots+(a_{\alpha_{n-1}}-a_{\alpha_n})^2]$, *where, for n odd the last term of* $g(\alpha)$ *is taken to be* $(a_{\alpha_n}-a_{\alpha_n})^2=0$. *Then for* $\alpha^*=(1,n,2,n-1,\ldots)$ *we have* $g(\alpha^*)\geqslant g(\alpha)$, *for any other permutation* α.

Proof. Let α^1 be any permutation. Then

$$g(\alpha^1)=\tfrac{1}{2}[(a_{\alpha_1^1}-a_{\alpha_2^1})^2+(a_{\alpha_3^1}-a_{\alpha_4^1})^2+\cdots].$$

Since $g(\cdot)$ remains unchanged if terms in the sum are interchanged or if terms inside the square are interchanged, we can assume $\alpha_1^1=1$. Thus:

$$g(\alpha^1)=\tfrac{1}{2}[(a_1-a_k)^2+\cdots+(a_t-a_n)^2+\cdots].$$

Let a^2 be identical to a^1 except, that the positions of a_k and a_n are interchanged. Then

$$g(\alpha^2)-g(\alpha^1)=\tfrac{1}{2}(a_1-a_n)^2+\tfrac{1}{2}(a_t-a_k)^2-\tfrac{1}{2}[(a_1-a_k)^2+(a_t-a_n)^2]$$
$$=(a_1-a_t)(a_k-a_n)\geq 0.$$

Thus α^1 does not maximize $g(\cdot)$. Now if $\alpha^2=(1, n, 2, k, \ldots)$ with $k\neq n-1$, then a similar exchange between a_k and a_{n-1} results in α^3 with $g(\alpha^3)\geq g(\alpha^2)$. Repetitive exchanges will produce the desired result. □

We can now prove the following result:

Theorem 4.2. *For the special assignment problem the best (i.e. minimizing) solution is* $x^{**}=\{x_{ij} \mid x_{i,n-i+1}=1,$ *and* $x_{ij}=0, j\neq n-i+1\}$.

Proof. By direct substitution: $c(x^*)-c(x^{**})=g(\alpha^*)$. Hence, the largest difference between values of solutions occurs for x^* and x^{**}. Hence, x^{**} must be the minimizing solution. □

If $a_1, \ldots, a_n$ are the natural numbers, $a_i=i$, then the foregoing results can be specialized as follows:

Corollary 4.1. *For the special assignment problem, if* $a_i=i$, $i=1, \ldots, n$ *then* x^* *and* x^{**} *are unique and*

$$c(x^*)=\sum_{i=1}^{n} i^2=\tfrac{1}{6}[n(n+1)(2n+1)],$$

$$c(x^{**})=\sum_{i=1}^{n} i(n+1-i)=\tfrac{1}{6}[n(n+1)(n+2)].$$

4.3. *Characterizing adjacent extreme points on the assignment polytope*

If x and y are two solutions to an assignment problem and α, β are the corresponding permutations then, by Property 4.1, x and y are adjacent if and only if $\alpha=\beta z$ with z containing exactly one cycle of length two or more. (Cycles of length one indicate where x and y coincide.) This approach allows us to determine the value of extreme points adjacent to a given extreme point. In particular, if x is a solution adjacent to x^* (the worst solution), i.e., $\alpha(x)=\alpha^{-1}(x^*)z$ for some $z=(j_1, j_2, \ldots, j_k)$, then $\alpha(x)=z$, since $\alpha(x^*)$ is the identity and

$$c(x^*)-c(x)=\sum_{t=1}^{k} a_{j_t}^2-\left(\sum_{t=1}^{k-1} a_{j_t}a_{j_{t+1}}-a_{j_k}a_{j_1}\right).$$

Alternatively we can write:

$$c(x^*)-c(x)=\tfrac{1}{2}\sum_{t=1}^{k}(a_{j_t}-a_{j_{t+1}})^2 \quad \text{with } a_{j_{k+1}}\equiv a_{j_1}.$$

We can now use the results of the previous sections to address the question of whether the diameter of two property of the assignment problem can be achieved, i.e. can an adjacent extreme point method solve the problem in two steps. Let us take the special case $a_i = i$, $i=1,\ldots,n$. Then using Theorem 3.8 and formula (2.1), $c(x^*)-c(x)$ is maximized, for $n=6$, by $z=(1,5,3,4,2,6)$. (Although this is not unique.) Thus the best extreme point adjacent to x^*, call it $\bar{x}$ is represented by:

$$\alpha(\bar{x})=\begin{pmatrix}1 & 2 & 3 & 4 & 5 & 6\\ 5 & 6 & 4 & 2 & 3 & 1\end{pmatrix}.$$

But $\bar{x}$ differs from x^{**} by two cycles:

$$\alpha(x^{**})=\begin{pmatrix}1 & 2 & 3 & 4 & 5 & 6\\ 6 & 5 & 4 & 3 & 2 & 1\end{pmatrix}=\begin{pmatrix}1 & 2 & 3 & 4 & 5 & 6\\ 5 & 6 & 4 & 2 & 3 & 1\end{pmatrix}\begin{pmatrix}1 & 2\\ 2 & 1\end{pmatrix}\begin{pmatrix}4 & 5\\ 5 & 4\end{pmatrix}.$$

This is also true of the other cycle maximizing $c(x^*)-c(x)$. Thus even though the diameter of the polytope is two, it is not possible to get to the best extreme point from the worst by choosing the maximum increase.

As n increases, the number of cycles between $\bar{x}$ and x^{**} increases, but each cycle is always a pair of consecutive elements. In fact, a cursory examination of the formulas in Theorem 3.8 shows that for $n \geqslant 4$, the number of cycles between $\bar{x}$ and x^{**} is $\frac{1}{2}n-1$ for n even and $\frac{1}{2}(n-3)$ for n odd. From this, the diameter of two property is achieved for $n=4, 5$. For $n=2$, 3, x^* and x^{**} are adjacent.

We also note that x^* and x^{**} differ by $\frac{1}{2}n$ cycles for n even and by $\frac{1}{2}(n-1)$ for n odd. Thus the extreme point giving the maximum improvement reduces the number of cycles to the optimum by one. It is conjectured that for an arbitrary starting point and general linear objective function the best adjacent extreme point will reduce the number of cycles to the optimum by at least one.

5. Conclusions

Several problems in extremal permutations of n real numbers have been presented and solved. These problems were encountered when we studied the assignment problem. A fairly thorough but by no means complete search in combinatorics textbooks did not reveal any result to these problems. We thus feel the results presented here may be a useful addition to the theory of combinatorics.

An important and still elusive problem in extremal combinatorics is the travelling salesman problem in which a person is to make a tour of n cities with the minimum total distance. If the distance between city i and city j is measured as $(a_i - a_j)^2$, then Theorems 2.2 and 2.3 show the maximal and minimal solutions in closed form.

As mentioned above, we discovered these permutation problems in the course of studying the computational complexity of the assignment problem. The interest in identifying improving extreme points adjacent to a given point is pragmatic. Most algorithms for solving programming problems or any optimization problems employ the so-called 'primal method' in which one starts with a feasible solution and then tries to move to an adjacent improving solution. The well-known Simplex Method in linear programming is a good example. Therefore we feel that the study of computational complexity cannot be complete without taking into account this 'greedy' behavior, in which one tries to move to the most improving adjacent extreme point.

We showed that the greedy approach fails to solve the special assignment problems in two steps. We did not explicitly consider the role of the basis, which is rather important in the Simplex Method. In essence, we consider only non-degenerate pivots and ignore the required degenerate pivots.

References

[1] M.L. Balinski and A.D. Russakoff, The assignment polytope, SIAM Review (1974).
[2] I. Tomescu, Introduction to Combinatorics (Collets, London, 1975).
[3] R. Silver, An algorithm for the assignment problem, Comm. Assoc. Comput. Mach. 3 (1966) 603–606.

P. Hansen, ed., Studies on Graphs and Discrete Programming
© North-Holland Publishing Company (1981) 215–240

MINIMAL TEST SET FOR DIAGNOSING A TREE SYSTEM*

T. IBARAKI

Department of Applied Mathematics and Physics, Faculty of Engineering, Kyoto University, Kyoto, Japan 606

T. KAMEDA

Department of Electrical Engineering, University of Waterloo, Waterloo, Ontario, Canada N2L 3G1

S. TOIDA

Department of Systems Design, University of Waterloo, Waterloo, Ontario, Canada N2L 3G1

Let $G=(V, E)$ be an undirected tree representing a system in such a way that each edge corresponds to a system component. Each edge (i.e., component) may become faulty. Assuming that at most one edge becomes faulty at any time, we want to detect and locate a faulty edge if one exists, by applying tests to G. A test performed between a pair of vertices u and v can detect the existence of a faulty edge on the path between u and v (but cannot locate it if the path has more than one edge). Let V_L and V_M denote the set of vertices with degree one and the set of vertices with degree two in G, respectively. Our main result shows that a minimal test set (i.e., a test set detecting and locating a faulty edge with the minimum number of tests) has $\lceil \frac{1}{2}|V_L|+\frac{1}{2}|V_M|\rceil$ tests if $|V_M|\geq|V_L|$ and $\lceil \frac{2}{3}|V_L|+\frac{1}{3}|V_M|\rceil$ tests if $|V_M|<|V_L|$, except for one special tree with seven vertices, where $|\cdot|$ denotes the cardinality of the set therein and $\lceil\cdot\rceil$ denotes the smallest integer not smaller than the number therein. Moreover, such a minimal test set can be constructed in $O(|V|^2)$ time.

1. Introduction

A graph is a convenient tool to represent a system in which many components are interconnected. In this paper, we assume that each edge of the graph represents a system component (the transformation from a graph in which each vertex represents a component to such a graph is usually straightforward). In a real system, each edge (i.e., component) is subject to malfunctioning, and we want to detect and locate a faulty edge, if there is one, by applying tests to the system. This is called the *system diagnosis* and discussed in a number of papers

* This work was partially supported by the Natural Science and Engineering Research Council of Canada under grant numbers A4315 and A7396.

[2–11], where various models are used. We employ here a model which is closely related to the one initiated by Mayeda and Ramamoorthy [6, 7, 9]. Recently [3, 4] have examined the complexity of diagnosis problems of this model when systems are represented by *directed* graphs. Most problems turn out to be NP-complete, strongly suggesting their computational intractability [1], if general directed graphs are to be dealt with. If systems are represented by directed trees, however, there is a polynomial time algorithm to generate a minimal (in terms of the number of tests) test set which detects and locates a faulty edge [4].

This paper considers the case of *undirected trees.* An undirected graph would be more suitable than a directed graph for representing a system consisting of bilateral components, e.g., information networks in which signals are sent along lines in both directions.

To be more specific, let $G=(V, E)$ be an *undirected tree*, where V and E are the sets of vertices and (undirected) edges, respectively. A *test* t denoted by $[u, v]$ is performed between two vertices u and v, where u and v are called the *test points* of t, by injecting a test signal at u and monitoring it at v. We do not differentiate test $[u, v]$ from $[v, u]$, and assume that any vertex can be used as a test point. If a faulty edge lies on the path between u and v, denoted by either $u \rightarrow v$ or $v \rightarrow u$ (the path is unique up to its direction because G is a tree), t can detect the existence of a faulty edge on the path but cannot locate it (if the path has more than one edge). Therefore several tests are in general required to locate a faulty edge. Throughout this paper, we adopt the *single fault assumption* that at most one edge may become faulty at any time.

A vertex $v \in V$ is called a *leaf* if its degree is one, an *intermediate vertex* if its degree is two and a *junction vertex* if its degree is more than two. Let V_L, V_M and V_J denote the sets of leaves, intermediate vertices and junction vertices, respectively. For a given G, a set of tests T is called a *minimal test set* if (1) it can detect and locate any single faulty edge in G, and (2) it contains the minimum number of tests among those having property (1). Our main result shows that a minimal test set has $\lceil \frac{1}{2}|V_M|+\frac{1}{2}|V_L|\rceil$ tests if $|V_M| \geq |V_L|$, and $\lceil \frac{2}{3}|V_L|+\frac{1}{3}|V_M|\rceil$ tests if $|V_L|>|V_M|$, except for one special tree with seven vertices, where $|\cdot|$ denotes the cardinality of the set therein and $\lceil \cdot \rceil$ denotes the smallest integer not smaller than the number therein. Moreover, such a minimal test set can be constructed in $O(|V|^2)$ time.

In Section 2, some definitions and basic lemmas are given. Based on these lemmas, the case of $|V_M|=|V_L|$ is treated in Section 3, which forms the basis for developing the rest of results. Section 4 discusses the case where $|V_M|>|V_L|$. Sections 5–7 consider the case where $|V_L|>|V_M|$ by separately treating the cases of $|V_M|=0$ and $|V_M|=1$. Finally in Section 8, the minimality of a test set with the above number of tests is proved.

2. Preliminary lemmas

Let $G=(V, E)$ be a tree. Without loss of generality, we assume that G is *connected* since, if not, each component can be treated independently. We also assume that $|V| \geq 3$ to avoid some trivial cases. A test $t=[u, v]$ is said to *cover* the set of edges (together with their end vertices) lying on the path $u \rightarrow v$. $C(t)$ denotes the set of edges covered by t. By definition of a test, a test set T can *detect* any faulty edge if:

(a) Each edge $e \in E$ is covered by at least one test $t \in T$ (in this case we say that T *covers* E).

Furthermore t can *locate* a faulty edge if

(b) $T(e_1) \neq T(e_2)$ holds for any two distinct edges e_1 and e_2, where $T(e)$ denotes the set of the tests in T which cover e.

Obviously there exists a test set satisfying both (a) and (b), e.g., $T=\{[u, v] \mid u, v \in V \text{ and } u \neq v\}$. Therefore our goal is to find a test set T with the minimum cardinality among those satisfying (a) and (b). As in the case of directed graphs [7], the condition (b) above can be restated as follows.

(b′) For any two distinct edges e_1 and e_2, there is a test $t \in T$ which covers exactly one of e_1 and e_2.

In the above case, t is said to *distinguish* e_1 and e_2. A set of edges is said to be *distinguished* by a test set T if any two edges in it are distinguished by a test in T. Two sets of edges, E_1 and E_2, are said to be *mutually distinguished* by T if any $e_1 \in E_1$ and $e_2 \in E_2$ are distinguished by a test in T. Using these terminologies, we can say that a minimal test set is a set of tests T which covers and distinguishes E with the minimal $|T|$.

We begin with a few lemmas characterizing a minimal test set.

Lemma 1. *There is a minimal test set T such that $n_T(v)=0$ for all $v \in V_J$, where $n_T(v)$ is the number of tests in T which use v as their test points.*

Proof. Given a minimal test set T' which has a test $t=[u, v]$ with $u \in V_J$, we modify t to $t'=[u', v]$ satisfying $|C(t')|>|C(t)|$ such that the resulting set is also a minimal test set. The property $|C(t')|>|C(t)|$ prevents this operation from repeating indefinitely, and hence a minimal test set T with the stated property must result eventually.

Now let $u_0, u_1, \ldots, u_k$ $(k \geq 2)$ be the vertices adjacent to u and let $e_i=(u_i, u)$, $i=0,1, \ldots, k$, where t covers e_0. Since T' covers E, there exist tests $t_i=[w_i, x_i]$ $(i=1,2, \ldots, k)$ covering e_i respectively (all t_i are not necessarily distinct). This is illustrated in Fig. 1. Note that each O——O generally represents a path (possibly with length 0) unless stated as an edge. We use this convention throughout this paper. Denote the last vertex common to both

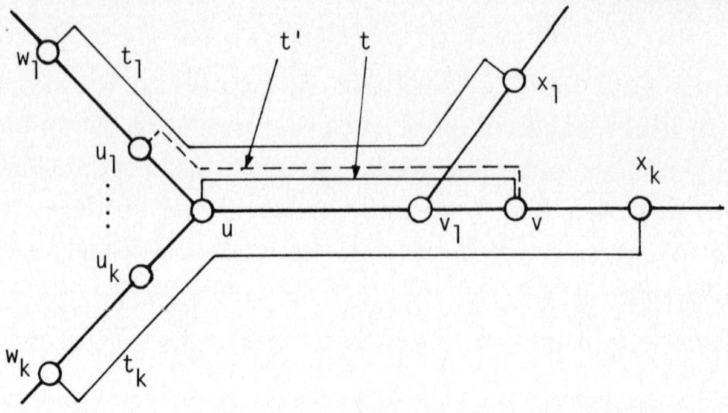

Fig. 1. Illustration for the proof of Lemma 1. Each O——O generally represents a path (possibly of length 0) unless it is specified as an edge.

$u \to v$ and $u \to x_i$ by v_i, where $v_i = u$ is not ruled out. Without loss of generality assume that v_1 is closest to u among all v_i $(i = 1, 2, \ldots, k)$, and that $v_1 \neq v_k$ since the case of $v_1 = v_k$ can be similarly treated. Then modify t to $t' = [u_1, v]$. We see that $e \in C(t)$ if and only if

$$(e \in C(t') \cap C(t_k)) \vee (e \in C(t') - C(t_1)),$$

where $\vee$ stands for the logical OR. (The second condition is not necessary unless v_k lies on the interior of $u \to v$.) Namely, whenever the fact $e \in C(t)$ or $e \notin C(t)$ was used in T' to distinguish edges, we can instead use the above characterization. This shows that the resulting test set T'' also distinguishes E. □

Lemma 2. *There is a minimal test set T such that $n_T(v) = 1$ for all $v \in V_M$.*

Proof. For a given minimal test set T', we use an argument similar to that used in the proof of Lemma 1. Consider two cases.

Case A. A vertex $v \in V_M$ has two tests $t_1 = [v, u_1]$ and $t_2 = [v, u_2]$ such that $u_1 \to v \to u_2$ is a path in G (see Fig. 2(a)). Then modify t_1 and t_2 to

$$t_1' = [w, v] \quad \text{and} \quad t_2' = [u_1, u_2],$$

where w (possibly equal to u_1) is a leaf such that $w \to u_1 \to v$ is a path in G. As easily seen, $e \in C(t_1)$ if and only if $e \in C(t_1') \cap C(t_2')$, and $e \in C(t_2)$ if and only if $e \in C(t_2') - C(t_1')$. Therefore the new set also distinguishes E and is a minimal test set. This argument is equally valid for such degenerate cases as $w = u_1$.

Case B. A vertex $v \in V_M$ has two tests $t_1 = [v, u_1]$ and $t_2 = [v, u_2]$ such that $v \to u_1$ and $v \to u_2$ have a common edge (see Fig. 2(b)). Let the edge $e = (v', v)$ be not covered by t_1 or t_2. Since T' covers E, there must be a test $t_0 = [v_0, u_0] \in T'$ which covers e, where $v_0 \to v$ contains e. Assume without loss of generality that paths $v \to u_0$ and $q \to u_1$ do not have a common edge, where q is the last

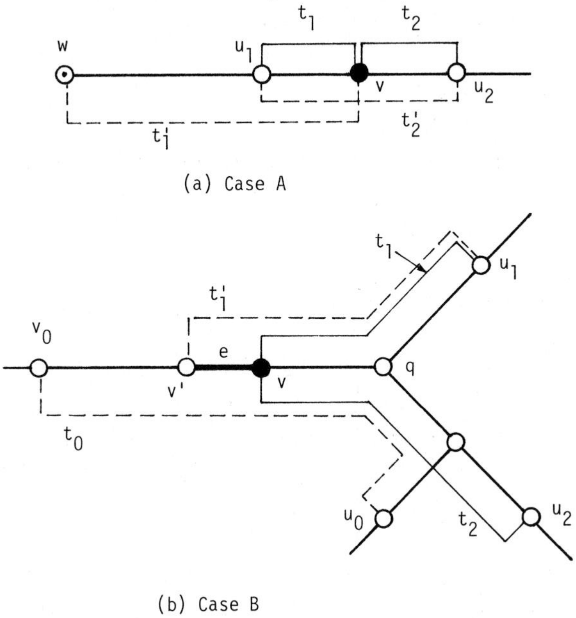

Fig. 2. Illustration for the proof of Lemma 2. (○ denotes a leaf and ● denotes an intermediate vertex. However, ⊙ denotes any type of vertex.)

vertex common to both $v \to u_1$ and $v \to u_2$ (as shown in Fig. 2(b)). Then modify t_1 to $t_1' = [v', u_1]$. Then $e \in C(t_1)$ is characterized by

$$(e \in C(t_1') - C(t_0)) \vee (e \in C(t_1') \cap C(t_2)).$$

Thus the new set is also a minimal test set.

Repeating these modifications, we eventually obtain a minimal test set satisfying the stated property. □

Lemma 3. *There is a minimal test set T such that*

$$n_T(v) = 0 \quad \textit{for all } v \in V_J,$$
$$n_T(v) = 1 \quad \textit{for all } v \in V_M,$$
$$n_T(v) \geq 1 \quad \textit{for all } v \in V_L.$$

Proof. Each $v \in V_M$ must satisfy $n_T(v) \geq 1$ because otherwise the two edges adjacent to v cannot be distinguished by T. Similarly for $v \in V_L$. The lemma then follows from Lemmas 1 and 2. □

In the remainder of this paper, we seek to find a minimal test set subject to the conditions of Lemma 3. Such a test set has

$$\frac{1}{2}\left(\sum_{v \in V_L} n_T(v)\right) + \tfrac{1}{2}|V_M| \tag{1}$$

tests, and hence a minimal test set is obtained by minimizing the first term of (1).

3. Minimal test set when $|V_L| = |V_M|$

Assuming that $|V_L| = |V_M|$ for a tree $G = (V, E)$, where $|V_L| \geq 2$ follows from the assumption $|V| \geq 3$, we construct a minimal test set with $\lceil \frac{1}{2}|V_L| + \frac{1}{2}|V_M| \rceil$ $(= |V_L|)$ tests. The minimality of this test set is obvious from (1). In the following algorithm, a test set is constructed while reconstructing G, starting from a single path and adding paths recursively. We maintain two graphs, G' and G''. G' is a partially reconstructed G and G'' is a homeomorphic image of G' such that a vertex in V_J is a vertex in G'' only if its degree is greater than two in G', and a vertex in V_M is a vertex in G'' if it is a test point of some already generated test. Otherwise, the two edges incident to such a vertex are considered to be just one edge. This is illustrated in Fig. 3. G'' will always have

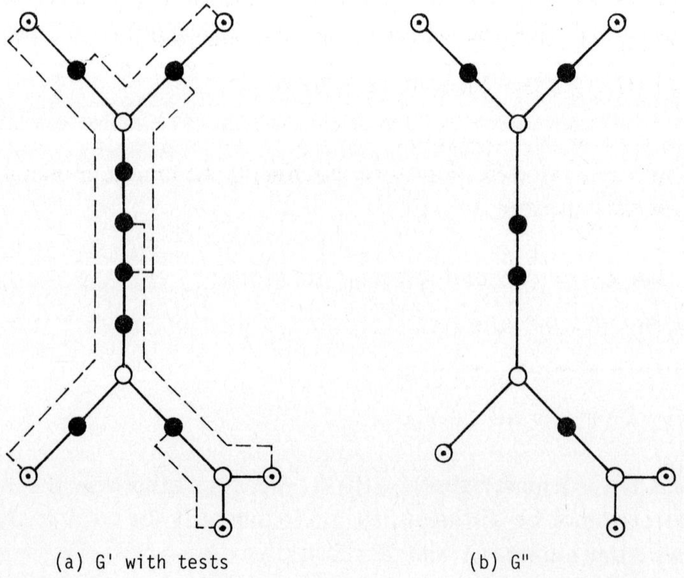

Fig. 3. Illustration of G'' constructed from G'. (⊙ denotes a leaf, ● an intermediate vertex, and ○ a junction vertex.)

the property that it is covered and distinguished by the tests generated so far. At each iteration, one new test is generated and some tests already generated are modified to preserve the covering and the distinguishability properties. Each test generated has a test point in V_L and the other test point in V_M.

For a tree, an edge is called *external* if one of its end vertices is a leaf; otherwise it is called *internal*. The graph G'' defined above will have the property that each external edge in it is covered by exactly one test and each internal edge is covered by at least two tests.

Algorithm A

Input. A tree $G=(V, E)$ satisfying $|V_L|=|V_M|\geq 2$.

Output. A minimal test set T for G.

Step 1. Find a path $u \to v$ in G such that $u, v \in V_L$ and $u \to v$ contains at least two intermediate vertices. (Such a path always exists by the assumption $|V_L|=|V_M|\geq 2$.) Let w_1 and w_2 be the intermediate vertices on $u \to v$ closest to u and v, respectively. Call the path $u \to v$ graph G'. Generate two tests $t_1=[u, w_2]$ and $t_2=[w_1, v]$ (see Fig. 4), and let $T'=\{t_1, t_2\}$.

Step 2. If there is a path $u \to v$ in G such that $u \in V_L$, v is a vertex in G', no part of $u \to v$ (except v) is contained in G' and $u \to v$ contains at least one intermediate vertex, then add $u \to v$ to G' and call the resulting graph G'. This is illustrated in Fig. 5(a) and (b). Let w be the intermediate vertex on $u \to v$ closest to u.

Case 2A. If there is a test $[x_1, x_2] \in T'$, $x_1 \in V_L$, which does not cover v (as illustrated in Fig. 5(a)), denote it by t.

Case 2B. Otherwise, take a test $t=[x_1, x_2] \in T'$, $x_1 \in V_L$, which covers v (as illustrated in Fig. 5(b)).

In either case, delete t from T', and add $t_1=[u, x_2]$ and $t_2=[w, x_1]$ to T'. Repeat Step 2 as many times as possible.

Step 3. If all $v \in V_L$ are in G', stop; $T=T'$ is a minimal test set for G. Otherwise take a vertex $u \in V_L$ not contained in G' and find the path $u \to v$ in G such that v is a vertex of G' but no part of $u \to v$ (except v) is contained in G'. Add $u \to v$ to G' and call the resulting graph G'. Let G'' be as defined above.

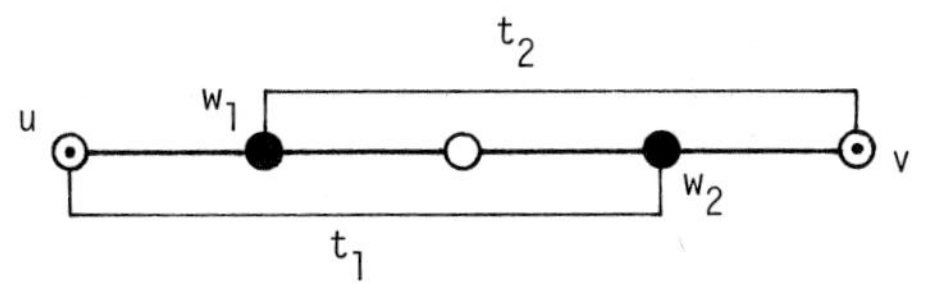

Fig. 4. Two tests t_1 and t_2 generated in Step 1 of Algorithm A. (⊙ denotes a leaf, ● an intermediate vertex, and ○ a vertex of any type.)

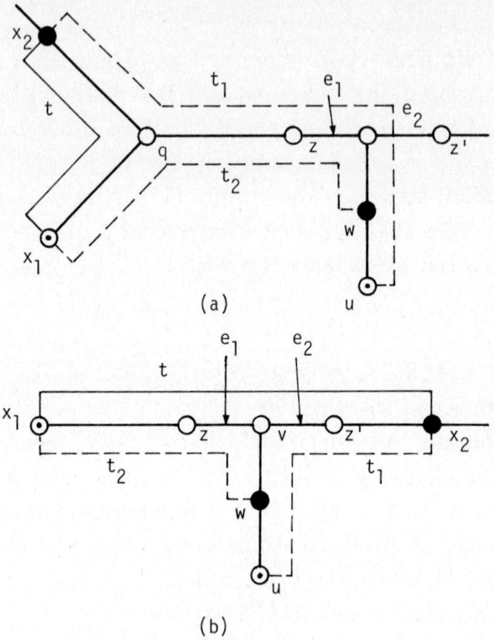

Fig. 5. Two tests t_1 and t_2 generated in Step 2 of Algorithm A.

Proceed as follows depending on the case applicable (use the first applicable case in the order given below).

Case 3A. If v has degree at least four in G'' (after v has been added), then choose any unused $w \in V_M$. Otherwise pick an unused $w \in V_M$, if any, such that the edge on which w lies is not incident to v in G'' (see Fig. 6). In either case generate test $t = [u, w]$ and add it to T'.

If Case 3A is not applicable, then all unused $w \in V_M$ lie on edges adjacent to v in G''. Select one of them arbitrarily as w.

Case 3B. There is a test $t = [x_1, x_2] \in T'$ with $x_1 \in V_L$ and $x_2 \in V_M$ forming a path $v \to w \to x_2 \to x_1$ in G' (see Fig. 7). Delete t from T' and add $t_1 = [u, x_2]$ and $t_2 = [w, x_1]$ to T'.

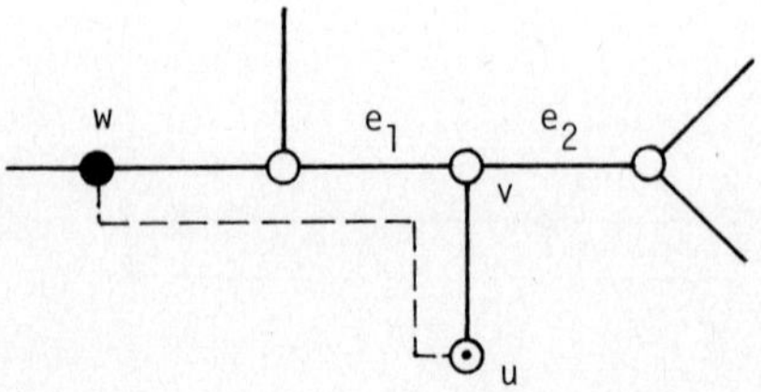

Fig. 6. Illustration for Case 3A in Step 3 of Algorithm A.

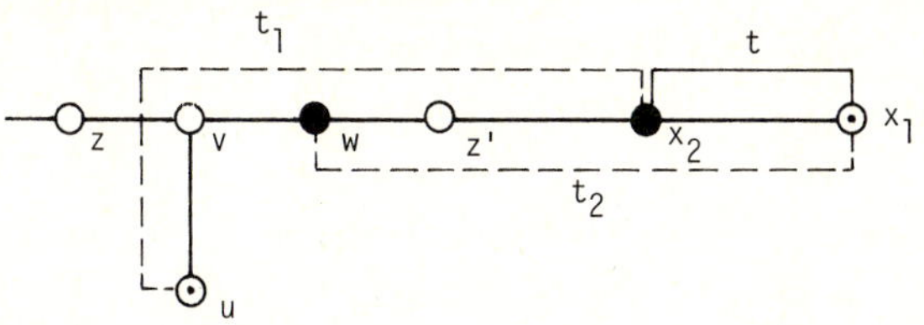

Fig. 7. Illustration for Case 3B of Algorithm A.

Case 3C. There is a test $t=[x_1, x_2]\in T'$ with $x_1\in V_L$ and $x_2\in V_M$ forming a path $x_1\to v\to w\to x_2$ in G':

(i) x_1 is not adjacent to v in G'' after adding $u\to v$ (see Fig. 8(a)). Delete t from T' and add $t_1=[u, x_2]$ and $t_2=[w, x_1]$ to T'.

(ii) x_1 is adjacent to v in G'' (see Fig. 8(b)). Find a test $t'=[y_1, y_2]$, where $y_1\in V_L$ and $y_2\in V_M$, covering x_2. (Since 3B is not the case, y_2 is not on $v\to y_1$.) Delete t' from T', and add $t_1=[u, y_2]$ and $t_2=[w, y_1]$ to T'.

Case 3D. None of Cases 3A, 3B and 3C are applicable. Then any test $t=[x_1, x_2]$ with $x_1\in V_L$ and $x_2\in V_M$ such that $v\to w\to x_2$ is a path in G' has the property that $v\to w\to x_1$ is a path in G' and x_2 is not on it (see Fig. 9(i)–(iii)). Select one of them arbitrarily as t, and let q be the last vertex common to $v\to x_1$ and $v\to x_2$. Now any test $t'=[y_1, y_2]$, with $y_1\in V_L$ and $y_2\in V_M$, covering v forms a path $y_2\to v\to w\to y_1$ in G' (otherwise Case 3C results). For such t', let p be the last vertex common to $v\to w\to y_1$ and $v\to w\to x_1$ (or x_2).

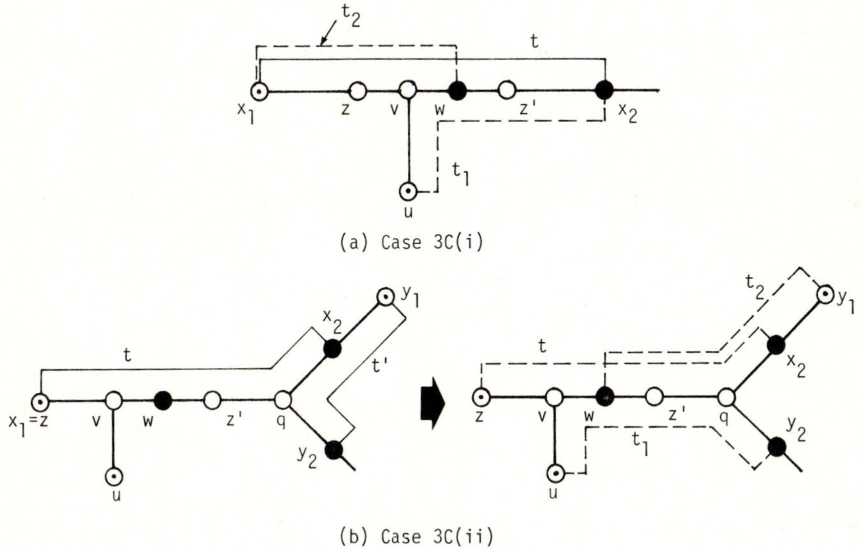

Fig. 8. Illustration for Case 3C of Algorithm A.

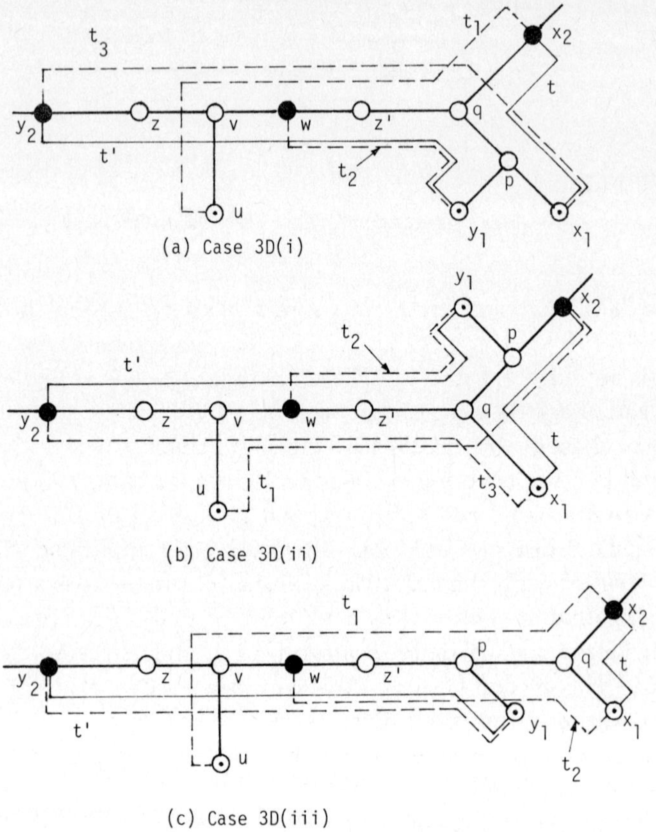

Fig. 9. Illustration for Case 3D of Algorithm A.

(i) There is a test $t' = [y_1, y_2] \in T'$ as defined above such that the vertex p lies on $q \to x_1$. In this case, select the test t' whose p is closest to x_1 (see Fig. 9(a)). Delete t and t', and add $t_1 = [u, x_2]$, $t_2 = [w, y_1]$ and $t_3 = [y_2, x_1]$ to T'.

(ii) Case (i) is not applicable, but there is a test t' such that vertex p lies on $q \to x_2$ (see Fig. 9(b)). This includes the degenerate case of $p = x_2$. Select one such test as t'. Delete t and t' from T', and add $t_1 = [u, x_2]$, $t_2 = [w, y_1]$ and $t_3 = [y_2, x_1]$ to T'.

(iii) Neither (i) nor (ii) is applicable, i.e., vertex p lies on $v \to q$ (see Fig. 9(c)). Delete t from T', and add $t_1 = [u, x_2]$ and $t_2 = [w, x_1]$ to T'.

Repeat Step 3 as many times as possible.

Before proving the validity of the above algorithm, we work on an example to illustrate how our algorithm works.

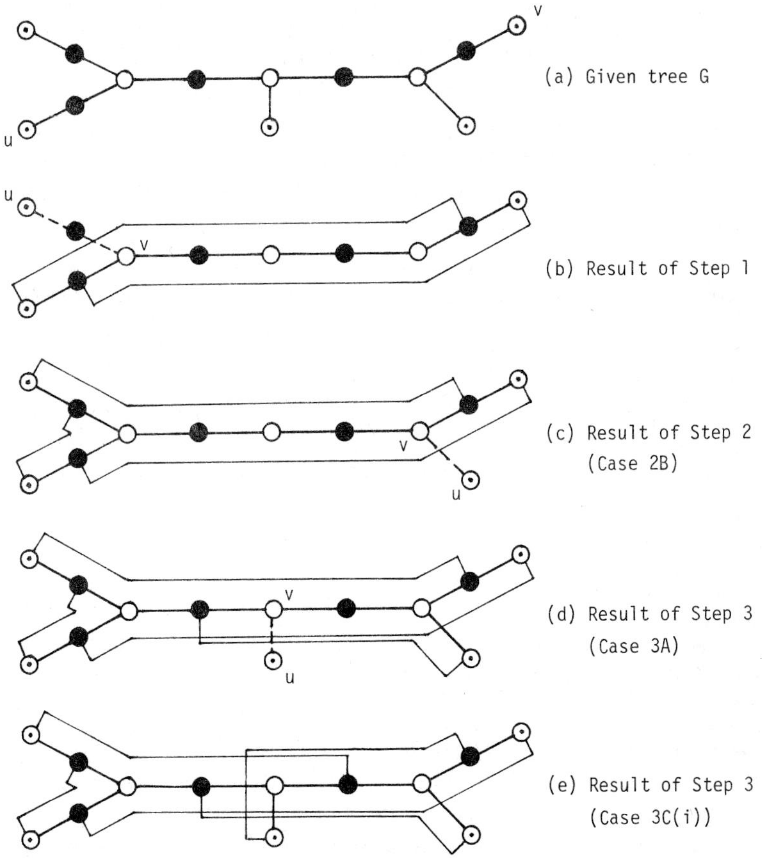

Fig. 10. Application of Algorithm A to a tree G with $|V_L| = |V_M|$. (⊙ denotes a leaf, ○ a junction vertex and ● an intermediate vertex.)

Example 1. Consider a tree given in Fig. 10(a). As a path $u \to v$ in Step 1, we choose the one indicated in Fig. 10(a), though other choices are also possible. Fig. 10(b) shows G' and the generated tests. The path $u \to v$ shown by broken lines in Fig. 10(b) is then added in Step 2. The computation then proceeds as illustrated in Fig. 10(b)–(e). The five tests given in Fig. 10(e) constitute a minimal test set T for G.

Lemma 4. *Algorithm* A *correctly generates a minimal test set* T *for a tree* $G(V, E)$ *with* $|V_L| = |V_M|$, *where* T *has* $\lceil \frac{1}{2}|V_L| + \frac{1}{2}|V_M| \rceil$ *tests.*

Proof. The result on the number of tests is obvious because each generated test has one leaf and one intermediate vertex as its test points. The minimality is then an obvious consequence of observation (1). Before proving the correctness, it is necessary to show that Algorithm A never stops unless all $v \in V_L$ are used as test points (Step 3). For this, we need to show that a used $x_2 \in V_M$ such that $v \to w \to x_2$ is a path in G' can always be found in Cases 3B, 3C and 3D of Step 3. This follows from the selection rule of an unused $w \in V_M$ in Steps 1 and 2, where the one closest to leaf u is always selected. It guarantees that, during computation, any unused $w \in V_M$ has at least two used $x, y \in V_M$ such that $x \to w \to y$ is a path in G', and hence the above x_2 can always be selected.

The current test set T' constructed during the application of Algorithm A satisfies

$$|T'(e)| = 1 \quad \text{if } e \text{ is an external edge in } G'',$$
$$\geqslant 2 \quad \text{if } e \text{ is an internal edge in } G''. \tag{2}$$

This is easily proved by induction, noting that, at each modification of T', each old edge is covered by at least the same number of tests compared with the previous test set and each newly created edge is covered by some of the modified or newly generated tests.

Let G'' be the tree defined as above corresponding to the graph G' maintained during the application of Algorithm A. Since (2) implies that all edges in G'' are covered by T', we need only to show by induction that T' distinguishes all edges.

Induction hypothesis: T' distinguishes all edges in G''.

For the G'' obtained after Step 1, this is trivial. We thus consider each case of Steps 2 and 3.

In the case of Fig. 5(a) of Step 2, notice that the edge (z, z') containing v in G'' (before the addition of $u \to v$) becomes two edges $e_1 = (z, v)$ and $e_2 = (v, z')$. Therefore we need to show that the set of edges e_1, e_2 and (u, w), (w, v) (these are new), as well as all the old edges of G'', are distinguished. By the argument used in the proofs of Lemmas 1 and 2, it is easy to see that we only need to show that $x_2 \to q$ and $w \to u$ (both are covered by t_1 but not by t_2) are mutually distinguished (more precisely, the set of edges on $x_2 \to q$ and the set of edges on $w \to u$ are mutually distinguished), where q is the last vertex common to both $v \to x_1$ and $v \to x_2$, and that $q \to v$ and $v \to w$ (both are covered by t_1 and t_2) are mutually distinguished. The first half is obvious because $x_2 \to q$ consists of internal edges and hence covered by some tests other than t_1 (by (2)); these tests mutually distinguish $x_2 \to q$ and $w \to u$. Also $q \to v$ consists of internal edges and covered by some tests not shown in the figure, which mutually distinguish $q \to v$ and $v \to w$. The case of Fig. 5(b) can be similarly treated. We just note that, since we assume that no test in T' satisfies the condition of Fig. 5(a), all edges in $v \to x_2$ are internal (i.e., v is not on an external edge of G' before adding $u \to v$) and hence covered by some other tests, which mutually distinguish $u \to w$ and $v \to x_2$.

Cases 3A to 3D of Step 3 can be similarly treated. we omit the details but only point out the following. (1) $v \to w$ of Fig. 7 is covered by some tests other than t_1, which mutually distinguish $u \to v$ and $v \to w$. (2) $w \to x_2$ of Fig. 8(a) is covered by some other tests (because the edges are internal), which mutually distinguish $w \to x_2$ and $u \to v$. (3) $q \to y_2$ of Fig. 8(b) is covered by some other tests, which mutually distinguish $q \to y_2$ and $u \to v$. (4) There are tests other than t covering $q \to x_2$ in Fig. 9(a), which mutually distinguish $q \to x_2$ and $u \to v$. (5) There are some tests other than t' covering $y_2 \to v$ in Fig. 9(a) and these tests do not cover any portion of $p \to x_1$ by definition of p. Thus $y_2 \to v$ and $p \to x_1$ are mutually distinguished. (6) $p \to x_2$ of Fig. 9(b) is covered by some other tests not shown, which mutually distinguish $p \to x_2$ and $u \to v$. (7) $y_2 \to v$ of Fig. 9(b) is covered by some other tests and these tests cover no portion of $q \to x_1$ (otherwise Case 3D(i) results). Thus $y_2 \to v$ and $q \to x_1$ are mutually distinguished. (8) $q \to x_2$ of Fig. 9(c) is covered by some other tests not shown, which mutually distinguish $u \to v$ and $q \to x_2$. By induction hypothesis, there are some tests distinguishing $y_2 \to z'$ and $p \to y_1$ in the old G'' (i.e., before adding $u \to v$). Such tests distinguish $y_2 \to v$ and $p \to y_1$ in the new G'' (i.e., after adding $u \to v$).

This completes the proof of the induction hypothesis after each execution of Step 2 or Step 3. When all $v \in V_L$ are used for constructing tests, graph G'' becomes identical to G and $T = T'$ is a test set covering and distinguishing all edges in E. □

Lemma 5. *Algorithm* A *can be carried out in* $O(|V|^2)$ time.

Proof. Considering that G has $|V|$ vertices and $|V|-1$ edges, it is not difficult to see that Step 1 and each iteration of Step 2 or Step 3 can be done in $O(|V|)$ time if appropriate data structures are used for storing G, G', G'', and T'. Since the total number of iterations of Steps 2 and 3 required is $|V_L|-2\,(<|V|)$, the total time is $O(|V|^2)$. □

4. Minimal test set when $|V_M|>|V_L|$

When a given tree $G=(V, E)$ satisfies $|V_M|>|V_L|$, let $m=|V_M|-|V_L|$. We first generate $\lceil \frac{1}{2}m \rceil$ tests in a certain way by pairing m intermediate vertices (when m is odd, one leaf is also used). Deleting these m intermediate vertices, we obtain the tree $\tilde{G}$ for which $|V_L|=|V_M|$ holds. Algorithm A is then applied to $\tilde{G}$. The union of the two test sets provides a minimal test set for G with $\lceil \frac{1}{2}|V_L|+\frac{1}{2}|V_M| \rceil$ tests.

We call a path $u \to v$ in G a *segment* if $u, v \in V_L \cup V_J$ and all vertices on $u \to v$ (other than u and v) are intermediate.

Algorithm B

Input. A tree $G=(V, E)$ with $|V_M|>|V_L|\geq 2$.

Output. A minimal test set T for G.

Step 1. Prepare two lists L and R, where L initially contains one leaf (chosen arbitrarily) if m is odd. Initially all segments in G are unscanned.

(a) Pick an unscanned segment and store all the intermediate vertices on it in L if the resulting L satisfies $|L|\leq\lceil\frac{1}{2}m\rceil$. The picked segment is now scanned. Otherwise call the segment S, and take $\lceil\frac{1}{2}m\rceil-|L|$ consecutive intermediate vertices from one end of S and store them in L. Repeat (a) until the latter case occurs.

(b) Construct R in a similar manner with L and R interchanged, by picking unscanned segments. If there remains no unscanned segment and still $|R|<\lceil\frac{1}{2}m\rceil$ holds, take the segment S and store in R $\lceil\frac{1}{2}m\rceil-|R|$ consecutive intermediate vertices chosen from the other end of S.

(c) Construct $\lceil\frac{1}{2}m\rceil$ tests by pairing a vertex in L and a vertex in R, where each vertex is used for exactly one test. Denote the resulting test set by T_1.

Step 2. Delete m intermediate vertices stored in L or R from G (in the same fashion as G'' was obtained from G'), and denote the resulting graph by $\tilde{G}$. Apply algorithm A to $\tilde{G}$ to obtain test set T_2.

Step 3. Let $T=T_1\cup T_2$. T is a minimal test set for G.

Lemma 6. *Algorithm* B *generates a minimal test set T for a tree $G=(V, E)$ with $|V_M|>|V_L|\geq 2$, where T has $\lceil\frac{1}{2}|V_L|+\frac{1}{2}|V_M|\rceil$ tests.*

Proof. The result on the number of tests and the minimality are obvious. We show that the obtained T covers and distinguishes E. By Lemma 4, the test set T_2 covers and distinguishes $\tilde{G}$. This also implies that T_2 covers E. Now let (u, v) of $\tilde{G}$ correspond to a series of edges of G, (u_0, u_1), $(u_1, u_2), \ldots, (u_{k-1}, u_k)$, where $u=u_0$ and $v=u_k$. By construction, if a test $t\in T_1$ has u_i $(1\leq i\leq k-1)$ as its test point, then the other test point belongs to a different edge of $\tilde{G}$ (or is a leaf). (A remark would be necessary when the segment S is simultaneously used to generate vertices in L and R in Step 1. In this case, by $|V_L|>0$, there remain some unchosen intermediate vertices on S between the vertices chosen for L and the vertices chosen for R. This guarantees that the test points of t belong to different edges of $\tilde{G}$.) This implies edges $(u_0, u_1), (u_1, u_2), \ldots, (u_{k-1}, u_k)$ are distinguished by T_1. Furthermore, any two edges (u_a, v_a) and (u_b, v_b) of G belonging to different edges of $\tilde{G}$ are distinguished by T_2. Combining these two properties, we see that all edges in G are distinguished. □

Lemma 7. *Algorithm* B *can be carried out in* $O(|V|^2)$ *time.*

Proof. It is not difficult to see that Step 1 of Algorithm B can be executed in $O(|V|)$ time. Step 2 is done in $O(|V|^2)$ time as shown by Lemma 5. Time required for Step 3 is a constant if T_1 and T_2 are in the form of linked lists. □

5. Minimal test set when $|V_M| = 0$

As a preliminary step to consider the general case $|V_L| > |V_M|$, we present a way of finding a minimal test set when $G = (V, E)$ has no intermediate vertex. The assumption that $|V| \geq 3$ then implies $|V_L| \geq 3$.

Algorithm C

Input. A tree $G = (V, E)$ with $|V_M| = 0$ and $|V_L| \geq 3$.

Output. A minimal test set T for G.

Step 1. Embed a given tree $G = (V, E)$ on the plane, in such a way that all leaves are located on a circle and no edges cross each other. Number the leaves clockwise from 0 to $k-1$, where $k = |V_L|$, and let v_i denote the ith leaf.

Step 2. Let $k = 3p + q$, where $0 \leq q \leq 2$, and generate the following tests.

$$[v_i, v_{i+2}] \quad \text{for } i = 3j \text{ and } i = 3j+1,\ j = 0, 1, \ldots, p-1,$$

$$[v_{k-1}, v_0] \quad \text{if } q = 1, \quad \text{and}$$

$$[v_{k-2}, v_{k-1}] \quad \text{and} \quad [v_{k-1}, v_0] \quad \text{if } q = 2.$$

(Fig. 11 shows the cases of $k = 9$, 10, 11.) Call the resulting test set T.

Lemma 8. *Algorithm* C *generates a minimal test set* T *for a tree* $G = (V, E)$ *with* $|V_M| = 0$, *where* T *has* $\lceil \frac{2}{3} |V_L| \rceil$ *tests.*

Proof. The result on the number of tests is obvious. Its minimality will be shown in Section 8. It is shown here only that T covers and distinguishes E. Note that T has the following properties.

(a) For any leaves v_i and v_{i+1} (indices are taken by modulo k), either at least one of them is a test point v_j of a test of the form $[v_j, v_{j+2}]$ or v_{i+1} is a test point of test $[v_{i+1}, v_{i+2}]$.

(b) For each test $[v_j, v_{j'}]$, either v_j or $v_{j'}$ (possibly both) is also a test point of another test.

To show that T covers E, assume that there is an edge $e \in E$ which is not covered. Obviously e is not an external edge of G. Let G_1 and G_2 be the two trees obtained from G by removing e. Since e is internal, both G_1 and G_2 have more than one leaf. Property (a) then guarantees that there is a test one of whose test points is in G_1 and the other is in G_2. Intuitively, this amounts to

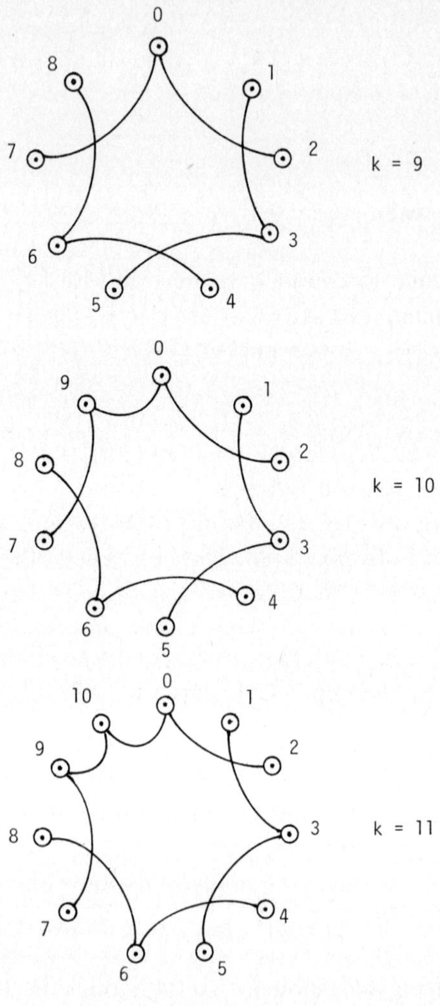

Fig. 11. Illustration for the tests generated by Algorithm C. (Each vertex represents a leaf).

saying that in a diagram as Fig. 11, one cannot draw a straight line through it without crossing some arcs connecting leaves. Thus T covers e, leading to a contradiction.

To show that T distinguishes E, assume that two edges e_1 and e_2 are not distinguished. Removing e_1 and e_2 from G, we obtain three trees G_1, G_2 and G_3 such that any path from a vertex in G_1 to a vertex in G_2 passes a vertex in G_3, as illustrated in Fig. 12. Classify the leaves in G_3 into two sets, V_1 and V_2, where V_1 is the set of those leaves which come after the leaves in G_1 but before the leaves in G_2 in the numbering used in Step 1, and V_2 contains the remaining leaves. If V_1 and V_2 are both empty, then G_3 consists only of

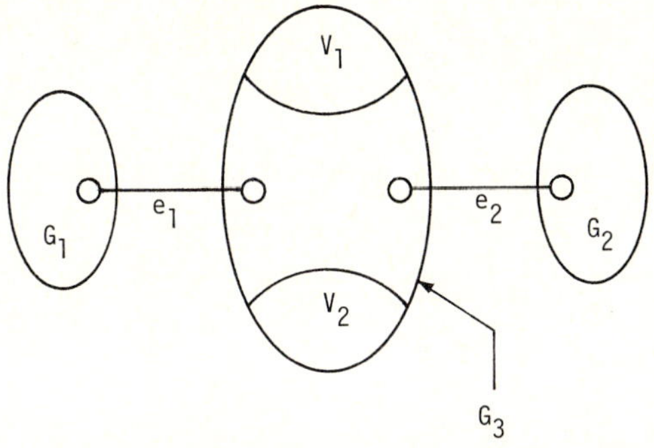

Fig. 12. Illustration for the proof of Lemma 8.

intermediate vertices, contradicting the assumption that $|V_M|=0$. If $|V_1|\geq 2$, it is obvious by property (a) that there is a test with one of its test points in V_1 and the other in G_1 or G_2, *distinguishing* e_1 and e_2. Similarly for V_2. Thus only possibility left is that $|V_1|=|V_2|=1$ holds and there is a test with its test points in V_1 and V_2. But at least one leaf in $V_1 \cup V_2$ is a test point of another test by property (b), and it distinguishes e_1 and e_2. □

The next lemma is easy and the proof is omitted. (The numbering of leaves can be carried out by the depth-first search in $O(|V|)$ time.)

Lemma 9. *Algorithm* C *can be carried out in* $O(|V|)$ *time.*

6. Minimal test set when $|V_L|>|V_M|$

When a given tree $G=(V, E)$ satisfies $|V_L|>|V_M|$, we decompose G into two (in general, overlapping) trees G_M and G_L such that the number of leaves of G_M is the same as that of the intermediate vertices, and G_L has no intermediate vertices. Algorithms A and C are then applied to G_M and G_L, respectively, and the two obtained test sets, T_M and T_L, are merged with certain modifications to give a minimal test set T for G.

We first assume here that $|V_M|\geq 2$; the case of $|V_M|=1$ will be discussed in the next section. In this case the resulting test set T will have

$$\lceil \tfrac{2}{3}|V_L|+\tfrac{1}{3}|V_M|\rceil \tag{3}$$

tests. Note that

$$|V_L|\geq |V_M|+3 \tag{4}$$

can be assumed without loss of generality. For, if $|V_L|=|V_M|+1$, we add an intermediate vertex to G arbitrarily, and obtain a test set for the resulting tree satisfying $|V_L|=|V_M|$ by means of Algorithm A. A test set for G is easily obtained by removing the added vertex and moving the test point which uses the added vertex to one of the vertices adjacent to it. If $|V_L|=|V_M|+2$, we add a leaf to G (together with a junction point to make the leaf connected to G). A test set for the resulting tree can be similarly modified to a test set for G. Notice that these modifications do not change the number in (3), and the obtained test set is minimal because, as we shall show in Section 8, at least $\lceil \frac{2}{3}|V_L|+\frac{1}{3}|V_M| \rceil$ tests are necessary.

We start with an algorithm to decompose G.

Algorithm DECOMP

Input. A tree $G=(V, E)$ with $|V_M| \geq 2$ and $|V_L| \geq |V_M|+3$.

Output. Two trees G_M and G_L with the properties stated above.

Step 1. Find a path $u \to v$ such that $u, v \in V_L$ and $u \to v$ contains at least two intermediate vertices. Denote this by G'.

Step 2. If there is a path $u \to v$ in G such that $u \in V_L$, v is a vertex in G', no part of $u \to v$ (except v) is contained in G', and $u \to v$ contains at least one intermediate vertex, then add $u \to v$ to G'. Call the resulting tree G'. Repeat Step 2 as many times as possible.

Step 3. Repeat the following until $|V_L'|=|V_M'|$ is reached, where V_L' and V_M' are the sets of leaves and intermediate vertices of G', respectively. Find $u \in V_L$ and $v \in V_J \cap V'$ such that no part of $u \to v$ (except v) is contained in G', and add $u \to v$ to G'. Call the resulting tree G'.

Step 4. Remove from G' all the junction vertices of G which have degree 2 (two edges adjacent to a removed vertex are redefined as an edge; recall the definition of graph G''). Call the resulting tree G_M.

Step 5. Let $\bar{V}_L$ be the set of leaves in G not contained in G_M. Find the minimal connected subtree of G that spans all leaves in $\bar{V}_L$. Remove all the intermediate vertices and all the junction vertices with degree 2 in the same manner as in Step 4, and call the resulting tree G_L.

Fig. 13 shows an example of the decomposition of G into G_M and G_L.

In reconstructing G from G_M and G_L, three cases are conceivable, depending on how G_M and G_L intersect. The intersection is, for example, indicated in Fig. 13 by thickened lines. Note that the intersection is always a connected subtree of G if it is nonempty.

(i) G_M and G_L do not intersect (see Fig. 14(a)). In this case there is an edge of G (indicated by double lines in Fig. 14(a)) which connects G_M and G_L.

(ii) The intersection of G_M and G_L is not empty but is properly contained in an edge of G_M or G_L ('properly' means that the end vertices of the edge do not belong to the intersection (see Fig. 14(b)).

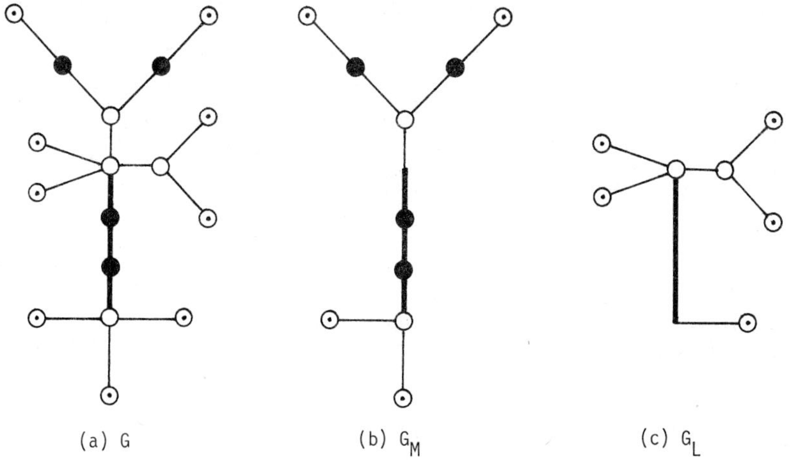

Fig. 13. Decomposition of G into G_M and G_L (the common portion of G_M and G_L is indicated by thick lines).

(iii) Neither (i) nor (ii) (see Fig. 14(c)).

We now proceed to an algorithm for generating a minimal test set for G.

Algorithm D

Input. A tree $G=(V, E)$ with $|V_M|\geqslant 2$ and $|V_L|\geqslant|V_M|+3$.

Output. A minimal test set T for G.

Step 1. Apply Algorithm DECOMP and obtain G_M and G_L.

Step 2. Apply Algorithm A to G_M to obtain T_M. Apply Algorithm C to G_L to obtain T_L.

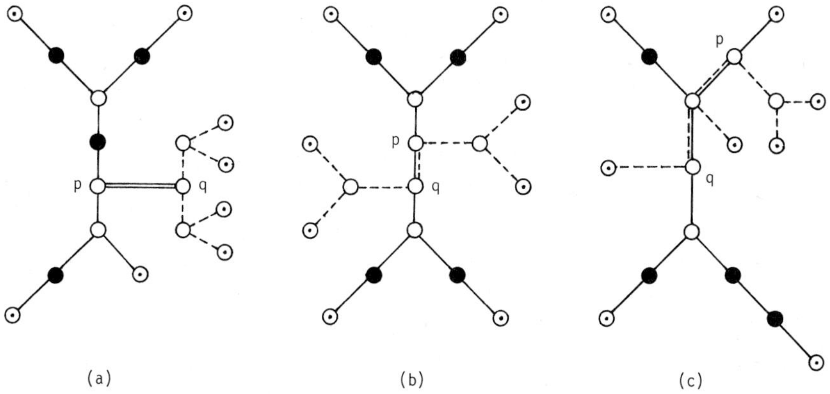

Fig. 14. Three cases of the intersection of G_M and G_L. (Solid lines indicate G_M and broken lines indicate G_L. The edge represented by double lines does not belong to G_M or G_L. Note that vertices p and q of G appear neither in G_M nor in G_L.)

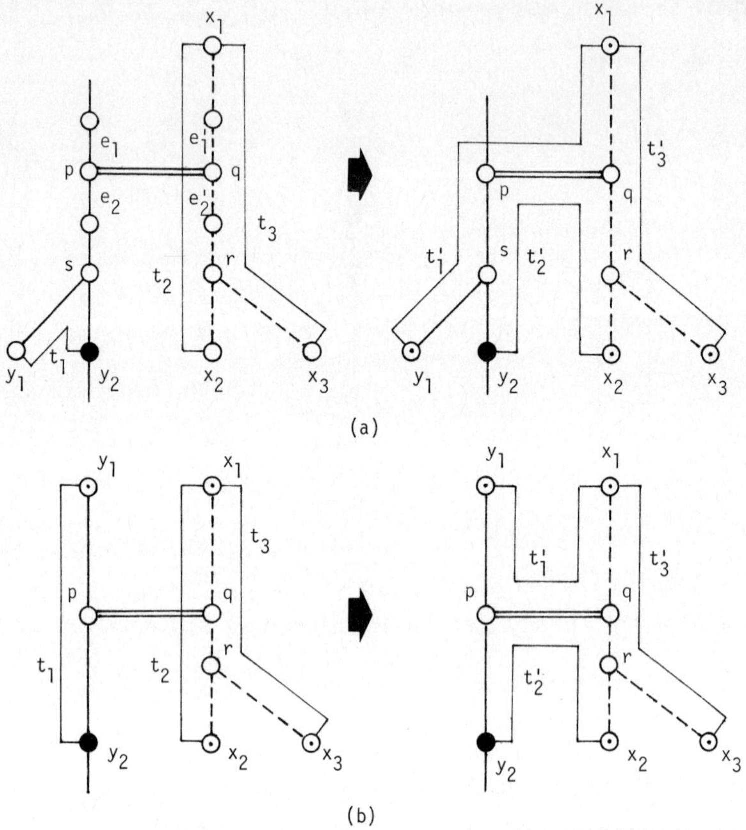

Fig. 15. Modification of tests in Case (i) of Step 3 of Algorithm D.

Step 3. Corresponding to cases (i), (ii) and (iii) above, T_M and T_L are modified as follows.

Case (i). Let (p, q) be the edge of G connecting G_M and G_L, where p is on an edge of G_M. If there is a test in T_M which does not cover p, take one and let it be $t_1 = [y_1, y_2]$, where $y_1 \in V_L$ and $y_2 \in V_M$ (see Fig. 15(a)). Otherwise take a test in T_M covering p and let it be $t_1 = [y_1, y_2]$, where $y_1 \in V_L$ and $y_2 \in V_M$ (see Fig. 15(b)). Then take tests $t_2 = [x_1, x_2]$ and $t_3 = [x_1, x_3]$ in T_L, such that t_2 covers q and t_3 shares a test point with t_2 (let x_1 be the common test point without loss of generality). Furthermore we assume the path $x_1 \to r$, where r is the last vertex common to $x_1 \to x_2$ and $x_1 \to x_3$, contains q; if not, we interchange the roles of x_1 and x_2 in (5) below. Modify t_1, t_2, and t_3 to

$$t'_1 = [x_1, y_1], \qquad t'_2 = [x_2, y_2], \qquad t'_3 = t_3 (= [x_1, x_3]). \tag{5}$$

Case (ii). Denote the two vertices in G showing the ends of the intersection of G_M and G_L by p and q. (See Fig. 16; although Fig. 16 shows the case in which an edge of G_M properly contains the intersection, the following procedure can be equally applied to the other cases.) If there is a test in T_M which does not cover p, take one and let it be $t_1=[y_1, y_2]$, where $y_1 \in V_L$ and $y_2 \in V_M$ (see Fig. 16(a); y_1 and y_2 can also be on the v side of G_M). Otherwise take a test in T_M covering p and let it be $t_1=[y_1, y_2]$, where $y_1 \in V_L$ and $y_2 \in V_M$ (see Fig. 16(b); the positions of y_1 and y_2 with respect to p can also be interchanged). Then take a test $t_2=[x_1, x_2] \in T_L$ covering p and $t_3=[x_1, x_3]$ sharing a test point with t_2 (let x_1 be the common test point without loss of generality). We assume that $x_1 \to p$ contains $x_1 \to r$, where r is the last vertex common to $x_1 \to x_2$ and $x_1 \to x_3$; if not, the roles of x_1 and x_2 are interchanged in (6) below. Modify t_1, t_2, and t_3 to

$$t_1'=[x_1, y_1], \qquad t_2'=t_2(=[x_1, x_2]), \qquad t_3'=[y_2, x_3]. \tag{6}$$

Case (iii). Take the union of T_M and T_L.

In each of the above cases, call the resulting test set T and stop.

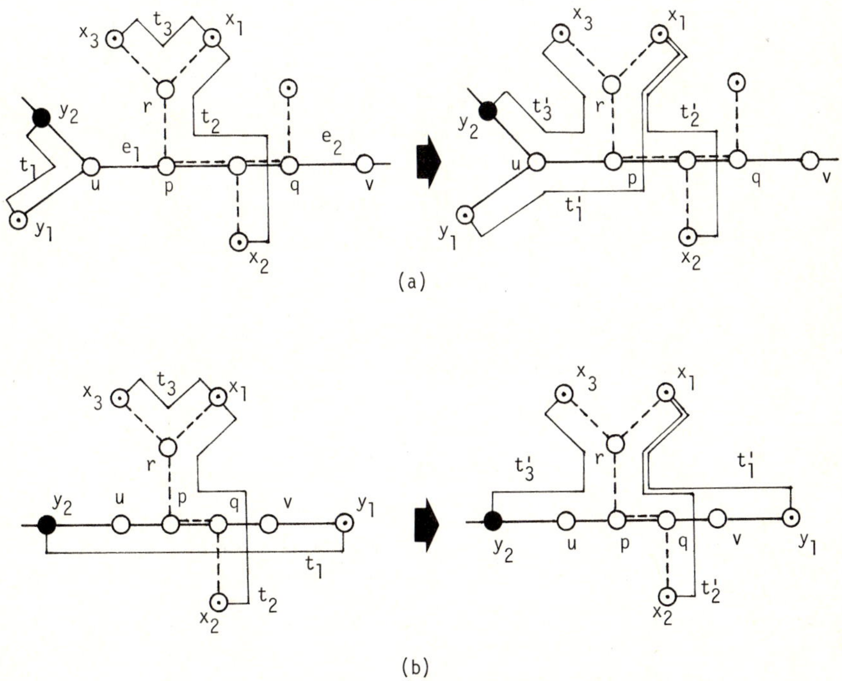

Fig. 16. Illustration for Case (ii) of Step 3 of Algorithm D. (Solid edges indicate G_M and broken edges indicate G_L.)

Lemma 10. *Algorithm* D *generates a minimal test set* T *for a tree* $G=(V,E)$ *with* $|V_M|\geq 2$ *and* $|V_L|\geq|V_M|+3$, *where* T *has* $\lceil\frac{2}{3}|V_L|+\frac{1}{3}|V_M|\rceil$ *tests.*

Proof. Since the modification of Step 5 does not change the number of tests, we have

$$\begin{aligned}|T|&=|T_M|+|T_L|\\&=|V_M|+\lceil\tfrac{2}{3}(|V_L|-|V_M|)\rceil\\&=\lceil\tfrac{2}{3}|V_L|+\tfrac{1}{3}|V_M|\rceil\end{aligned}$$

by Lemmas 4 and 8. The minimality will be proved in Sections 8. To show that T covers and distinguishes E, we consider cases (i), (ii) and (iii) of Step 3 separately.

Case (i). $T_M\cup T_L$ does not cover edge (p,q), does not distinguish e_1 and e_2, where e_1 and e_2 are two edges ($\neq(p,q)$) adjacent to p in G (in the degenerate case such that p is already a vertex in G_M, we may just ignore this consideration), and does not distinguish e_1' and e_2', where these are two edges ($\neq(p,q)$) adjacent to q in G (see Fig. 15(a)). These are remedied by modifying t_1, t_2, and t_3 to t_1', t_2', and t_3', respectively, as easily seen. By the argument similar to the ones used so far, we only need to show that $x_2\to r$ and $s\to y_2$ are distinguished in Fig. 15(a), where s is the last vertex common to $p\to y_1$ and $p\to y_2$, and that $x_2\to r$ and $p\to y_2$ are distinguished in Fig. 15(b). In Fig. 15(a), all edges on $y_2\to s$ are internal with respect to G_M and covered by some tests of T_M other than t_1. These tests distinguish $x_2\to r$ and $s\leftarrow y_2$. In the case of Fig. 15(b), the edges on $p\to y_2$ are all internal with respect to G_M (i.e., p is not on an external edge of G_M), since otherwise there must be a test not covering p, and the case of Fig. 15(a) must have resulted. These tests mutually distinguish $x_2\to r$ and $p\to y_2$.

Case (ii) can be similarly treated.

In *Case* (iii) (see Fig. 14(c) for example, no two edges e_1 and e_2 of G can both belong to one edge of G_M and to one edge of G_L simultaneously. This implies that e_1 and e_2 are distinguished by a test in $T_M\cup T_L$. □

The next result is easy to show and the proof is omitted.

Lemma 11. *Algorithm* D *can be carried out in* $O(|V|^2)$ *time.*

7. Minimal test set when $|V_M|=1$

When a tree $G=(V,E)$ satisfies $|V_M|=1$, we consider G_M of Section 6 as consisting of one edge, one end v_1 of which is a leaf and the other end v_2 is the

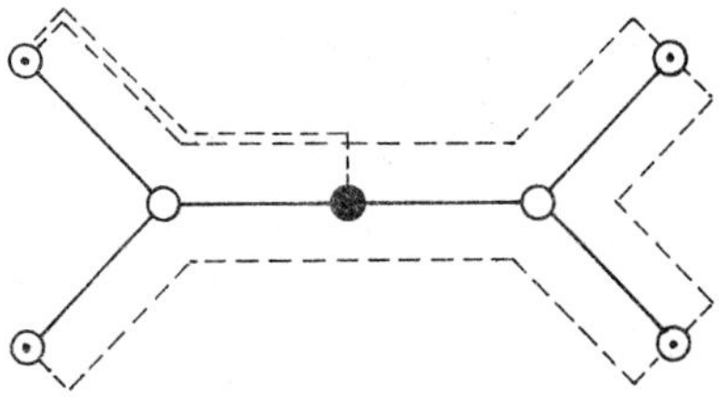

Fig. 17. Tree G which does not have a minimal test set T with $\lceil\frac{2}{3}|V_L|+\frac{1}{3}|V_M|\rceil$ (=3) tests (a minimal test set with four tests is indicated).

intermediate vertex. Deleting these vertices from G, G_L is defined as in Section 6. Let $T_M=\{[v_1, v_2]\}$ and construct T_L by applying Algorithm C to G_L. A minimal test set T for G is then constructed from T_M and T_L.

Lemma 12. *If a tree $G=(V, E)$ satisfies $|V_M|=1$, then there is a minimal test set T with*

$$|T| = \lceil\tfrac{2}{3}|V_L|+\tfrac{1}{3}|V_M|\rceil,$$

unless G is isomorphic to the graph shown in Fig. 17, *which has a minimal test set T with*

$$|T| = \lceil\tfrac{2}{3}|V_L|+\tfrac{1}{3}|V_M|\rceil+1=4.$$

Proof. G_L has $|V_L|-1$ leaves and T_L has $\lceil\frac{2}{3}(|V_L|-1)\rceil$ tests by Lemma 8. We show that, unless G is isomorphic to the graph of Fig. 15, a test set T with

$$|T|=|T_L|+1=\lceil\tfrac{2}{3}|V_L|+\tfrac{1}{3}|V_M|\rceil$$

covers and distinguishes E. The minimality will be shown in Section 8.

We can assume without loss of generality that $|V_L|\geq|V_M|+3\geq 4$, by the same argument as given in Section 6. Then reconstruct G by adding edge (v_1, v_2) of G_M back to G_L. Let u be the first vertex of G on $v_1\to v_2$ that is contained in an edge of G_L. If the intersection $u\to v_2$ of G_M and G_L is not *properly* contained in an edge of G_L (illustrated in Fig. 18), then the union $T=T_L\cup\{[v_1, v_2]\}$ covers and distinguishes E. This can be proved in a manner

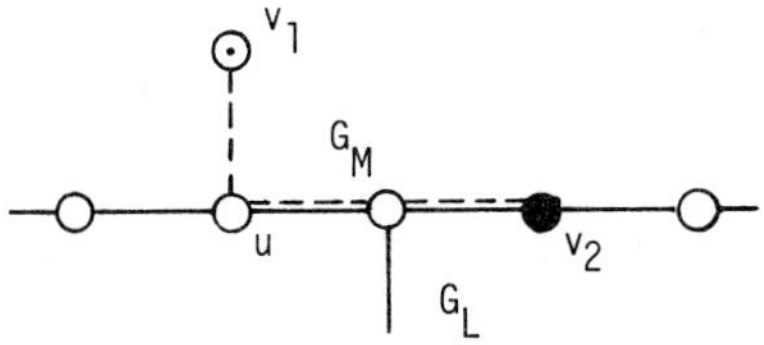

Fig. 18. Illustration of the case in which the intersection of G_M and G_L is not properly contained in an edge of G.

similar to Case (iii) in Step 3 of Algorithm D. Otherwise let G_1 and G_2 be the trees obtained from G by deleting the two edges adjacent to the intermediate vertex v_2. If G_1 or G_2 contains at least three leaves, it is easily seen that the pair v_1 and v_2 of G_M can be reselected from G so that the intersection of G_M and G_L is not properly contained in an edge of G_L. By the assumption that $|V_L| \geq 4$ mentioned above, G_1 and G_2 can both have at most two leaves only if G_1 and G_2 have exactly two leaves. Such a tree G is isomorphic to the one shown in Fig. 17. It is easy to show by exhausting all possible cases that testing such a tree requires four tests. □

The following result is easy to prove and is left for the reader.

Lemma 13. *The test set T of Lemma* 12 *can be constructed in* $O(|V|^2)$ *time.*

8. Minimality of test set T

The next lemma shows that the test sets constructed so far are minimal.

Lemma 14. *Any test set T covering and distinguishing a tree* $G=(V, E)$ *satisfies*

$$|T| \geq \lceil \tfrac{1}{2}|V_L| + \tfrac{1}{2}|V_M| \rceil \quad \text{if } |V_L| \leq |V_M|,$$
$$\geq \lceil \tfrac{2}{3}|V_L| + \tfrac{1}{3}|V_M| \rceil \quad \text{if } |V_L| > |V_M|.$$

Proof. The case of $|V_L| \leq |V_M|$ is obvious from (1). In order to prove the case of $|V_L| > |V_M|$, let

x_1 = the number of tests $t=[u, v]$ satisfying $u, v \in V_M$,
x_2 = the number of tests $t=[u, v]$ satisfying either $u \in V_L$ and $v \in V_M$ or $u \in V_M$ and $v \in V_L$,
x_3 = the number of tests $t=[u, v]$ satisfying $u, v \in V_L$.

Obviously, we have $2x_1 + x_2 \geq |V_M|$. Furthermore, at least $|V_L| - x_2$ leaves are the test points of tests $[u, v]$ of type u, $v \in V_L$ only. Denote the set of such leaves by V_L^*. To cover the external edges connected to V_L^*, each $v \in V_L^*$ must have at least one test connected to it. In addition, if a test $t=[u, v]$ satisfies u, $v \in V_L^*$, then at least one of u and v must be the test point of another test t'; otherwise, the two external edges incident to u and v would not be distinguished. Let

y_1 = the number of $v \in V_L^*$ which are the test points of exactly one test,
y_2 = the number of $v \in V_L^*$ which are the test points of more than one test.

By the definition of these variables,

$$y_1 \leqslant x_3,$$
$$y_1 + y_2 \geqslant |V_L| - x_2,$$
$$y_1 + 2y_2 \leqslant 2x_3.$$

To minimize the number of tests, we have the following integer programming problem:

$$\begin{aligned} \text{Minimize} \quad & z = x_1 + x_2 + x_3, \\ \text{subject to} \quad & 2x_1 + x_2 \geqslant |V_M|, \\ & x_3 - y_1 \geqslant 0, \\ & x_2 + y_1 + y_2 \geqslant |V_L|, \\ & 2x_3 - y_1 - 2y_2 \geqslant 0, \\ & x_1, x_2, x_3, y_1, y_2\text{: nonnegative integers.} \end{aligned} \tag{7}$$

To derive a lower bound on the objective function, remove the integrality constraint on variables, and consider the problem as a linear programming problem. It is easy to show that an optimal solution is

$$x_1 = 0, \qquad x_2 = |V_M|, \qquad x_3 = \tfrac{2}{3}(|V_L| - |V_M|),$$
$$y_1 = \tfrac{2}{3}(|V_L| - |V_M|), \qquad y_2 = \tfrac{1}{3}(|V_L| - |V_M|),$$

and the optimal objective value is

$$z = \tfrac{2}{3}|V_L| + \tfrac{1}{3}|V_M|. \tag{8}$$

(For example, consider the dual problem and let w_1, w_2, w_3, and w_4 be the dual variables corresponding to the four conditions of (7) from top to bottom. It is easily seen that a solution

$$w_1 = \tfrac{1}{3}, \qquad w_2 = \tfrac{1}{3}, \qquad w_3 = \tfrac{2}{3}, \qquad w_4 = \tfrac{1}{3}$$

is dual feasible and has the same objective value as (8), proving the optimality of both primal and dual solutions.)

Since z must be an integer in the original integer programming problem, we have the stated lower bound. □

Combining the results obtained so far, we have the following final result.

Theorem 1. *For a tree $G = (V, E)$ with $|V| \geqslant 3$, a minimal test set T satisfies*

$$\begin{aligned} |T| &= \lceil \tfrac{1}{2}|V_L| + \tfrac{1}{2}|V_M| \rceil \quad \textit{if } |V_M| \geqslant |V_L|, \\ &= \lceil \tfrac{2}{3}|V_L| + \tfrac{1}{3}|V_M| \rceil \quad \textit{if } |V_M| < |V_L|, \end{aligned}$$

except the tree isomorphic to the one shown in Fig. 17, *which has a minimal test set T with $|T| = 4$. Moreover, a minimal test set can be computed in* $O(|V|^2)$ *time.*

References

[1] M.R. Garey and D.S. Johnson, Computers and Intractability: A Guide to the Theory of NP-Completeness (Freeman, San Francisco, 1979).

[2] J.P. Hayes and A.D. Friedman, Test point placement to simplify fault detection, IEEE Trans. Computers C-23 (1974) 727–735.

[3] T. Ibaraki, T. Kameda, and S. Toida, Some NP-complete diagnosis problems on system graphs, Trans. IECE of Japan E62 (1979) 81–88.

[4] T. Ibaraki, T. Kameda, and S. Toida, On minimal test sets for locating single link failures in networks, IEEE Trans. Computers C-30 (1981) 182–190.

[5] M. Malek and S. Toida, Fault diagnosis in systems represented by acyclic digraphs, Technical Report 27-P-180577, Dept. of Systems Design, University of Waterloo, 1977.

[6] W. Mayeda and C. V. Ramamoorthy, Distinguishability criteria in oriented graphs and its application to computer diagnosis—I, IEEE Trans. Circuit Theory CT-16 (1969) 448–454.

[7] W. Mayeda, Graph Theory (Wiley-Interscience, New York, 1972).

[8] F.P. Preparata, G. Metze, and R.T. Chen, On the connection assignment problem of diagnosable systems, IEEE Trans. Elec. Computers EC-16 (1967) 849–854.

[9] C.V. Ramamoorthy and W. Mayeda, Computer diagnosis using the blocking gate approach, IEEE Trans. Computers C-20 (1971) 1294–1299.

[10] J.D. Russel and C.R. Kime, Structual factors in the fault diagnosis of combinatorial networks, IEEE Trans. Computers C-20 (1971) 1276–1285.

[11] J.D. Russel and C.R. Kime, System fault diagnosis: closure and diagnosability with repair, IEEE Trans. Computers C-24 (1975) 1078–1089.

P. Hansen, ed., Studies on Graphs and Discrete Programming

SUFFICIENT CONDITIONS FOR GRAPHS TO HAVE THRESHOLD NUMBER 2

T. IBARAKI

Department of Applied Mathematics and Physics, Kyoto University, Kyoto, Japan

U. N. PELED*

Department of Mathematics, University of Toronto, Toronto, Ontario M5S 1A1, Canada

The threshold number $t(G)$ of a graph $G=(V,E)$ is the smallest number of linear inequalities that separate the characteristic vectors of the independent sets of vertices from those of the dependent sets. The graph $G^*=(V^*,E^*)$ is such that $V^*=E$ and $(\alpha^*,\beta^*)\in E^*$ when $\alpha^*=(u,v)$, $\beta^*=(u',v')\in E$ with (u,u'), $(v,v')\notin E$. The chromatic number of G^* is denoted $\chi(G^*)$. Chvátal and Hammer showed that $t(G)=1$ if and only if $\chi(G^*)=1$, that $t(G)\geqslant\chi(G^*)$ and that computing $t(G)$ is NP-hard. We show that $t(G)=2$ if $\chi(G^*)=2$ and one of the following conditions holds: (i) V can be partitioned into a clique and an independent set; (ii) G^* has at most two non-singleton components. Under these conditions the corresponding two linear inequalities can be found in polynomial time.

1. Introduction

Let $G=(V,E)$ be an (undirected) graph with a set of vertices V and a set of edges $E\subset V^2$. V^2 denotes the set of unordered pairs (u,v) of distinct vertices. G has neither loops nor multiple edges. A graph $G'=(V',E')$ is a *threshold graph* when there exist numbers a_i, $i\in V'$ such that for $S\subset V'$, S contains an edge of G' iff $\sum_{i\in S}a_i>1$. It has been shown that G' is a threshold graph iff G' has no four distinct vertices u_1,u_2,v_1,v_2 such that (u_1,v_1), $(u_2,v_2)\in E'$, and (u_1,u_2), $(v_1,v_2)\notin E'$. These forbidden vertices are shown in Fig. 1, where a broken line indicates an absent edge. Some other equivalent characterizations of a threshold graph are also known [1].

For a graph $G=(V,E)$, a set of threshold graphs $G_i=(V,E_i)$, $i=1,2,\ldots,t$, with the property $E=E_1\cup E_2\cup\cdots\cup E_t$ is called a *threshold cover* of G. The *threshold number* $t(G)$ of G is the minimum t of threshold covers. The significance of $t(G)$ is, as proved in [1], that $t(G)$ is the smallest number of linear inequalities whose 0–1 solutions are precisely the incidence vectors of the independent subsets of V. For a graph $G=(V,E)$, define $G^*=(V^*,E^*)$ by

* Present address: Computer Science Department, Columbia University, New York, NY 10027, USA.

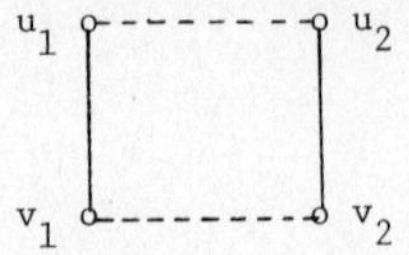

Fig. 1. Forbidden vertices of a threshold graph. (A broken line indicates an absent edge.)

$V^*=E$ and $(\alpha_1, \alpha_2)\in E^*$, where $\alpha_1=(u_1, v_1)\in E$ and $\alpha_2=(u_2, v_2)\in E$, iff u_1, u_2, v_1, v_2 are distinct and $(u_1, u_2), (v_1, v_2)\notin E$. Throughout this paper, edges of G (i.e., vertices of G^*) are denoted by α, β, γ etc., and vertices of G are denoted by u, v, etc., possibly with subscripts. By definition, no two edges $\alpha_1, \alpha_2\in E$ are in the same E_i of a threshold graph $G_i=(V, E_i)$ if $(\alpha_1, \alpha_2)\in E^*$. Thus

$$t(G)\geqslant\chi(G^*). \tag{1.1}$$

where $\chi(G^*)$ is the (vertex) chromatic number of G^*. Obviously $t(G)=\chi(G^*)$ when $\chi(G^*)\leqslant 1$, because $\chi(G^*)=0$ is equivalent to G consisting of isolated vertices so that $t(G)=0$, and $\chi(G^*)=1$ is equivalent to G being a threshold graph with some edges so that $t(G)=1$.

In [1], Chvátal and Hammer posed the following problem: '*Is there a graph G such that $t(G)>\chi(G^*)$?*'

Although we are still unable to give the final answer, it is shown in this paper that $t(G)=\chi(G^*)$ holds in the following two special cases.

(i) G^* is bipartite and G is a split graph.

(ii) G^* is bipartite and has at most two nonsingleton components.

$G=(V, E)$ is a *split graph* if V can be partitioned into K and I, i.e., $V=K\cup I$ and $K\cap I=\emptyset$, such that $(u, v)\in E$ for any $u\neq v\in K$ (K induces a clique), and $(u, v)\notin E$ for any $u\neq v\in I$ (I induces an independent set). Properties of split graphs are discussed in [2, 5].

Another class with the $t(G)=\chi(G^*)$ property is the class of matroidal graphs introduced and characterized in [6]; a graph G is matroidal if each connected component of G^* is a clique.

In concluding this section, we state the following conjecture as an extension of the results (i) and (ii).

'*$t(G)=\chi(G^*)$ holds if G^* is bipartite*'.

2. Review of previous results

For a given set of vertices V, $C=(V, E_1, E_2)$ is called a *configuration* if $E_1\cap E_2=\emptyset$ and $E_1, E_2\subset V^2$. Edges in E_1 and E_2 are called *present edges* and *absent edges* respectively. A graph $G=(V, E)$ is a *completion* of C if $E\supset E_1$ and $E_2\subset\bar{E}$ $(=V^2-E)$. It is known [3] that $G=(V, E)$ has a threshold cover of size t iff there exist t configurations $C_i = (V, E_i, \bar{E})$, $i = 1, 2, \ldots, t$, satisfying

$$E = E_1 \cup E_2 \cup \cdots \cup E_t, \qquad E_i \cap E_j = \emptyset \quad \text{for } i \neq j \tag{2.1}$$

and having threshold completions. Thus $t(G)$ is obtained by finding a minimal partition of E into E_i $(i = 1, 2, \ldots, t)$ such that each $C_i = (V, E_i, \bar{E})$ has a threshold completion.

Hereafter we assume that the associated graph $G^* = (V^*, E^*)$ is *bipartite*, i.e., $\chi(G^*) \leq 2$: the vertices of G^* (i.e., the edges of G) are 2-colourable. As remarked above, we may assume $\chi(G^*) = 2$. Let G^* have p components H_j^* $(j = 1, 2, \ldots, p)$. In any 2-colouring of G^*, the colour of one vertex α uniquely determines the colours of all other vertices in $H^*(\alpha)$, the component containing α. Thus each H_j^* has exactly two possible colourings, and G^* has 2^p 2-colourings. Denote the two colours by 0 and 1, and use binary variables x, y, z to represent these colours, with the notation $\bar{x} = 1 - x$. The colour of an edge α in G (i.e., a vertex in G^*) is denoted by $c(\alpha)$.

For a 2-colouring of G^*, let

$$F_x = \{\alpha \in V^* \mid c(\alpha) = x\}, \quad x = 0, 1. \tag{2.2}$$

(F_0, F_1) is a partition of E. It follows that G with $\chi(G^*) = 2$ has $t(G) = 2$ iff the configurations

$$C_x = (V, F_x, \bar{E}), \quad x = 0, 1 \tag{2.3}$$

induced by some 2-colouring of G^* have threshold completions. To check the last condition, the following theorem [3] is useful.

Theorem 2.1. *Let G^* of a graph $G = (V, E)$ be bipartite. Then a configuration C_x $(x = 0$ or $1)$ of* (2.3) *obtained from a 2-colouring of G^* has a threshold completion iff C_x contains neither configuration* AP_6 *nor configuration* AP_5, *defined below (see* Fig. 2).

$$AP_6 = (V, \{(u_1, u_2), (u_3, u_4), (u_5, u_6)\}, \{(u_2, u_3), (u_4, u_5), (u_6, u_1)\}),$$
$$AP_5 = (V, \{(u_1, u_2), (u_1, u_5), (u_3, u_4)\}, \{(u_2, u_5), (u_1, u_3), (u_1, u_4)\})$$

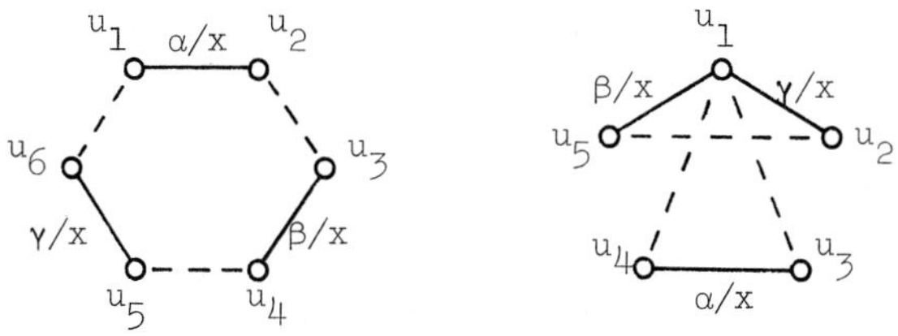

Fig. 2. Forbidden configurations AP_6 and AP_5 for a configuration to have threshold completions. (— shows a present edge and - - - shows an absent edge. α/x denotes that edge α has colour x $(=0$ or 1), etc.).

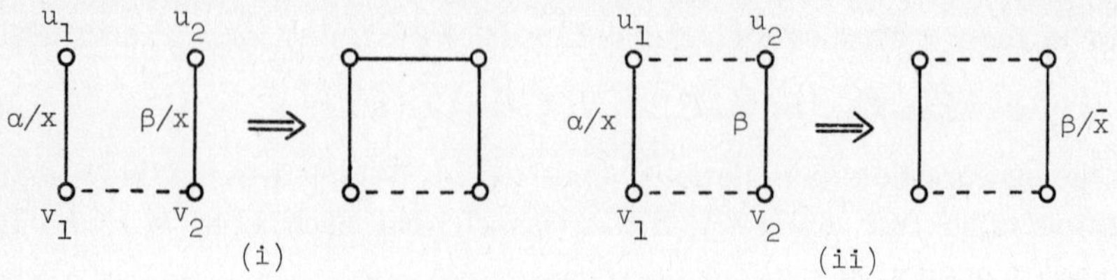

Fig. 3. Illustration of Lemma 3.1.

Consequently the subsequent discussion will endeavour to show that, under certain conditions, some 2-colouring of G^* gives C_0 and C_1 not containing AP_6 or AP_5.

3. Preliminaries 1: the ladder argument

Before proceeding to the main results, a series of lemmas will be proved in this section and the next one.

Lemma 3.1. *Let G^* of a graph $G=(V,E)$ satisfy $\chi(G^*)=2$, and consider the edge colours determined by a 2-colouring of G^*.*

(i) *Two edges $\alpha=(u_1,v_1)$ and $\beta=(u_2,v_2)$ with $c(\alpha)=c(\beta)=x$, $(v_1,v_2)\notin E$ and $u_1\neq u_2$ imply $\delta=(u_1,u_2)\in E$ (see* Fig. 3(i)).

(ii) *Two edges $\alpha=(u_1,v_1)$ and $\beta=(u_2,v_2)$ with (u_1,u_2), $(v_1,v_2)\notin E$ and $c(\alpha)=x$ must satisfy $c(\beta)=\bar{x}$ (see* Fig. 3(ii)).

Proof. (i) $\delta\notin E$ implies that α and β are adjacent in G^*, and $c(\alpha)\neq c(\beta)$ in any 2-colouring of G^*, a contradiction. (ii) can be similarly proved. □

Example 3.1. Consider configurations AP_6 and AP_5 given in Fig. 2(a) and (b). For each case we can deduce several other edges and their colours. The results are shown in Fig. 4(a) and (b). For example, that $(u_1,u_4)\in E$ in an AP_6 follows

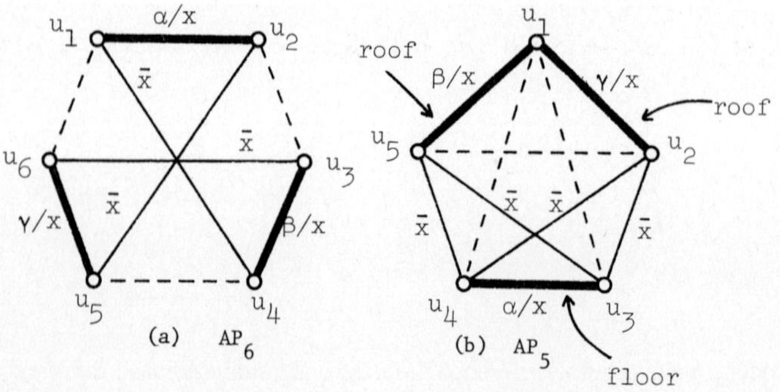

Fig. 4. Implied edges and their colours in the configurations AP_6 and AP_5. (Bold solid lines indicate edges $\in E$ in the original configurations of Fig. 2. Colours x and $\bar{x}$ are also indicated beside edges.)

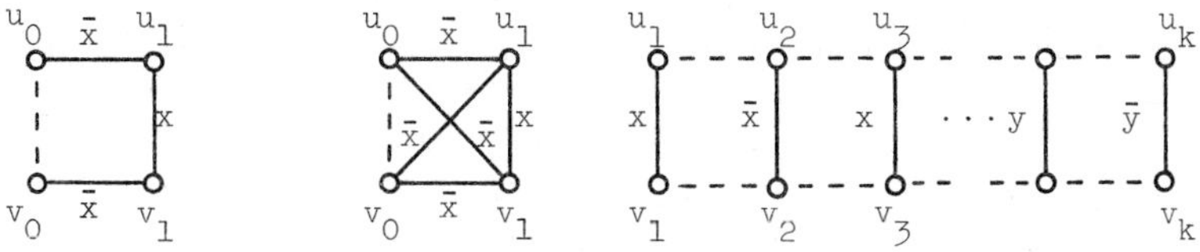

Fig. 5. Three special configurations with colours.

from Lemma 3.1(i) (consider $\alpha, \beta \in E$ and $(u_2, u_3) \notin E$), and its colour is determined from $c(\gamma) = x$ by Lemma 3.1(ii). In an AP_5, the edges β and γ are called the *roofs* and edge α is called the *floor.*

The following three special configurations with colours play an important role in the subsequent discussion.

(a) *Single starter* $S(u_0, v_0; u_1, v_1)$: configuration $(V, \{(u_1, v_1), (u_0, u_1), (v_0, v_1)\}, \{(u_0, v_0)\})$ with colours $c(u_1, v_1) = x$ and $c(u_0, u_1) = c(v_0, v_1) = \bar{x}$ (see Fig. 5(a)).

(b) *Double starter* $D(u_0, v_0; u_1, v_1)$: configuration $(V, \{(u_1, v_1), (u_0, u_1), (u_0, v_1), (v_0, u_1), (v_0, v_1)\}, \{(u_0, v_0)\})$ with $c(u_1, v_1) = x$, $c(u_0, u_1) = c(u_0, v_1) = c(v_0, u_1) = c(v_0, v_1) = \bar{x}$ (see Fig. 5(b)).

(c) *Ladder* L $(u_1, v_1; u_2, v_2; \ldots; u_k, v_k)$: configuration $(V, \{(u_1, v_1), (u_2, v_2), \ldots, (u_k, v_k)\}, \{(u_1, u_2), (u_2, u_3), \ldots, (u_{k-1}, u_k), (v_1, v_2), (v_2, v_3), \ldots, (v_{k-1}, v_k)\})$ with $c(u_i, v_i) = x$ for $i = 1, 3, \ldots, k'$ and $c(u_i, v_i) = \bar{x}$ for $i = 2, 4, \ldots, k''$ (see Fig. 5(c)), where $k \geq 1$ and

$$k' = 2\lfloor \tfrac{1}{2}(k-1) \rfloor + 1 \quad \text{and} \quad k'' = 2\lfloor \tfrac{1}{2}k \rfloor. \tag{3.1}$$

The order of vertices in each pair of vertices in a ladder L is essential. However, notations such as $L(\alpha; \ldots; \beta)$, where $\alpha = (u_1, v_1)$ and $\beta = (u_k, v_k)$ are used to denote either $L(u_1, v_1; \ldots, ; u_k, v_k)$ or $L(v_1, u_1; \ldots; v_k, u_k)$. Note that $L(u_1, v_1; \ldots; u_k, v_k)$ and $L(v_1, u_1; \ldots; v_k, u_k)$ denote the same ladder. It is important to see that $H^*(\alpha) = H^*(\beta)$ holds iff G contains a ladder $L(\alpha; \ldots; \beta)$.

Lemma 3.2. *Let G^* of a graph G satisfy $\chi(G^*) = 2$, and consider edge colours determined by a 2-colouring of G^*. Assume that G contains a single starter $S(u_0, v_0; u_1, v_1)$ attached to a ladder $L(u_1, v_1; \ldots; u_k, v_k)$. (This is illustrated in Fig. 6, where $y = x$ or $\bar{x}$ depending upon whether k is even or odd). In addition, assume the following.*

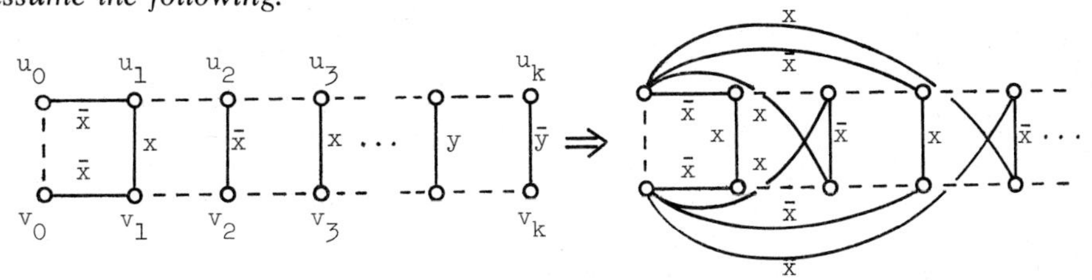

Fig. 6. The ladder argument from a single starter.

Nondegeneracy: $u_0 \neq u_1, v_2, u_3, v_4, u_5, \ldots, w_{k-1}$ ($w = u$ *if* k *is even, otherwise* $w = v$), *and* $v_0 \neq v_1, u_2, v_3, u_4, \ldots, w_{k-1}$ ($w = v$ *if* k *is even, otherwise* $w = u$). *(Note that the last edge is not included in this definition.)*

Then G *contains the following edges (see Fig. 6).*

$$(u_0, u_i),\quad i = 1, 3, 5, \ldots, k', \quad \textit{with colour } \bar{x} \textit{ unless } u_0 = u_i,$$

$$(v_0, v_i),\quad i = 1, 3, 5, \ldots, k', \quad \textit{with colour } \bar{x} \textit{ unless } v_0 = v_i,$$

$$(u_0, v_i),\quad i = 2, 4, 6, \ldots, k'', \quad \textit{with colour } x \textit{ unless } u_0 = v_i,$$

$$(v_0, u_i),\quad i = 2, 4, 6, \ldots, k'', \quad \textit{with colour } x \textit{ unless } v_0 = u_i.$$

Note that the conclusion of the Lemma says that there exist starters $S(u_0, v_0; v_2, u_2)$, $S(u_0, v_0; u_3, v_3)$ etc., i.e. the starter propagates along the ladder.

Proof. The proof is by induction. First edges (u_0, u_1) and (v_0, v_1) with colour $\bar{x}$ are contained in the starter. Assume edges (u_0, u_{2l-1}) and (v_0, v_{2l-1}) exist with colour $\bar{x}$. (The case of (u_0, v_{2l}) and (v_0, u_{2l}) with colour x is similar.) Then (u_0, u_{2l-1}), $(u_{2l}, v_{2l}) \in E$ with colour $\bar{x}$ and $(u_{2l-1}, u_{2l}) \notin E$ imply $(u_0, v_{2l}) \in E$ by Lemma 3.1(i). $c(u_0, v_{2l}) = x$ follows from $c(v_0, v_{2l-1}) = \bar{x}$ and (u_0, v_0), $(v_{2l-1}, v_{2l}) \notin E$ by Lemma 3.1(ii). A similar argument also applies to $(v_0, u_{2l}) \in E$ with $c(v_0, u_{2l}) = x$. $\square$

Next consider a double starter attached to a ladder. Since a double starter may be regarded as two single starters superimposed in different polarities, the above ladder argument can be applied twice through the ladder. Thus we have the next lemma.

Lemma 3.3. *Let* G *be a graph with colours as defined in Lemma 3.2. Let* G *contain a* $D(u_0, v_0; u_1, v_1)$ *attached to* $L(u_1, v_1; \ldots; u_k, v_k)$ $(k \geq 1)$, *as illustrated in Fig. 7. Furthermore assume the following.*

Nondegeneracy: u_0 *and* v_0 *are both distinct from* $u_1, u_2, \ldots, u_{k-1}$ *and* $v_1, v_2, \ldots, v_{k-1}$.

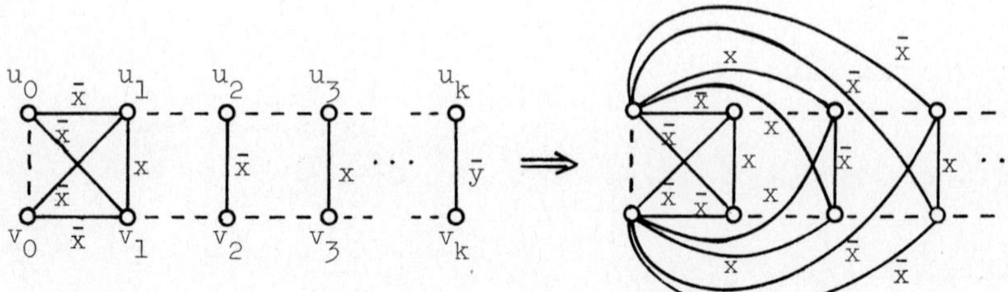

Fig. 7. The ladder argument from a double starter.

Then G contains the following edges (see Fig. 7).

$$(u_0, u_i), (u_0, v_i), (v_0, u_i), (v_0, v_i), \quad i = 1, 3, 5, \ldots, k', \quad \textit{with color } \bar{x}.$$

$$(u_0, u_i), (u_0, v_i), (v_0, u_i), (v_0, v_i), \quad i = 2, 4, 6, \ldots, k'', \quad \textit{with color } x.$$

unless $u_0 = u_i$ *etc.* □

In the subsequent discussion, we do not usually specify which of the above two nondegeneracy assumptions is meant, as it would be obvious from the context. Note that configurations AP_6 and AP_5 (see Fig. 2 and Fig. 4) contain starters; an AP_6 contains a single starter $S(u_4, u_5; u_1, u_2)$ and others, and an AP_5 contains a double starter $D(u_2, u_5; u_3, u_4)$.

If the nondegeneracy assumption does not hold, the ladder argument cannot proceed beyond the degenerate vertex. However, the following lemmas hold.

Lemma 3.4. *Let G be a graph with colours as defined in Lemma 3.2. Assume that G contains an* $S(u_0, v_0; u_1, v_1)$ *attached to* $L(u_1, v_1; \ldots; u_k, v_k)$, *as shown in Fig. 6. If the nondegeneracy assumption of Lemma 3.2, does not hold, G contains an* AP_5 *whose three edges* α, β, γ *are connected respectively to the three edges* (u_0, u_1), (u_1, v_1), (v_0, v_1) *of the starter* $S(u_0, v_0; u_1, v_1)$ *through ladders.*

Proof. Assume that u_0 does not coincide with $u_1, v_2, u_3, \ldots, v_{2l}$, and v_0 does not coincide with $v_1, u_2, v_3, \ldots, u_{2l}$, but u_0 coincides with u_{2l+1} (this implies $v_0 \neq v_{2l+1}$). Other cases are similarly treated. As illustrated in Fig. 8, edges (u_0, v_{2l}) and $(v_0, u_{2l}) \in E$ with colour x exist by the ladder argument of Lemma 3.2. Now $\alpha = (v_0, u_{2l})$, $\beta = (u_0, v_{2l})$, $\gamma = (u_0, v_{2l+1}) \in E$ with colour x and (v_{2l}, v_{2l+1}), (u_0, u_{2l}), $(u_0, v_0) \notin E$ form an AP_5. Finally it is easy to see that α, β, γ are respectively connected to (u_0, u_1), (v_0, v_1), (u_1, v_1) through ladders. □

Lemma 3.5. *Let G be a graph with colours as defined in Lemma 3.2. Assume that G contains a* $D(u_0, v_0; u_1, v_1)$ *attached to* $L(u_1, v_1; \ldots; u_k, v_k)$ *as shown in Fig. 7. Then this always satisfies the nondegeneracy assumption of Lemma 3.3.*

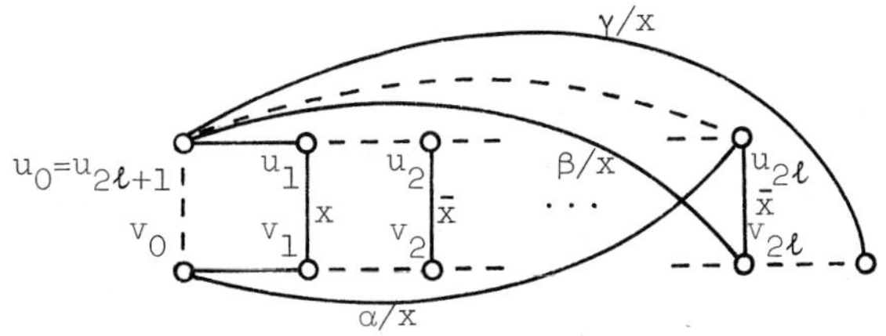

Fig. 8. Illustration of the proof of Lemma 3.4.

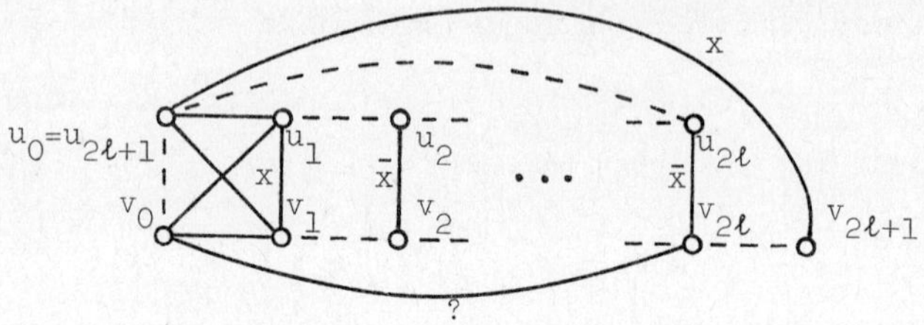

Fig. 9. Illustration of the proof of Lemma 3.5.

Proof. Assume that u_0 and v_0 do not coincide with $u_1, u_2, \ldots, u_{2l}$ or $v_1, v_2, \ldots, v_{2l}$, but u_0 coincides with u_{2l+1}. Other cases are similarly treated. Then, as shown in Fig. 9, $(v_0, v_{2l}) \in E$ and $c(v_0, v_{2l}) = x$ follows from the ladder argument. This is however a contradiction because $c(v_0, v_{2l}) = \bar{x}$ must hold, since (v_0, v_{2l}) is adjacent in G^* to (u_{2l+1}, v_{2l+1}) with colour x. □

Based on the above argument, we now find several configurations with colours that cannot be contained in G if G^* is bipartite.

Lemma 3.6. *Let G^* of a graph $G = (V, E)$ satisfy $\chi(G^*) = 2$, and let the edge colours of G be determined by a 2-colouring of G^*. Then G contains none of the following configurations with colours.*

(A) *$S(u_0, v_0; u_1, v_1)$ and two edges $\alpha = (u_0, w_1)$, $\beta = (v_0, w_2) \in E$, where* (1) *$c(\alpha) = c(\beta) = y$, and* (2) *α and β are reachable from (u_1, v_1) through nondegenerate ladders (see Fig.* 10(A)).

(B) *$S(u_0, v_0; u_1, v_1)$ and two edges $\alpha = (u_0, w)$, $\beta = (v_0, w) \in E$, where* (1) *$c(\alpha) = c(\beta) = y$, and* (2) *α is reachable from (u_1, v_1) through a nondegenerate ladder (see Fig.* 10(B)).

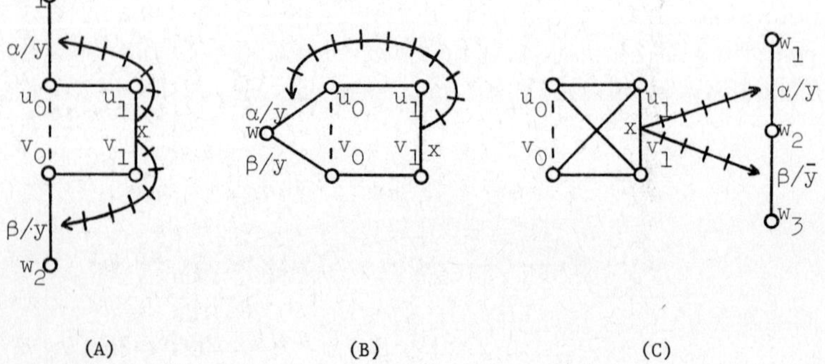

Fig. 10. Three configurations with colours not contained in G with bipartite G^*. (++ denotes a ladder.)

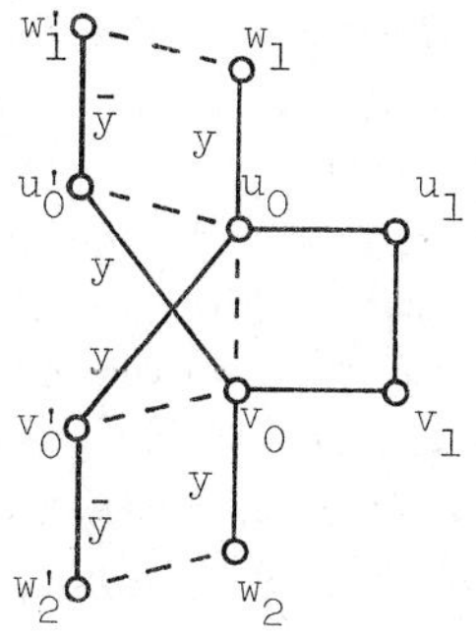

Fig. 11. Illustration of the proof of Lemma 3.6(A).

(C) *$D(u_0, v_0; u_1, v_1)$ and two edges $\alpha = (w_1, w_2)$, $\beta = (w_2, w_3) \in E$, where* (1) *$c(\alpha) \neq c(\beta)$, and* (2) *α and β are reachable from (u_1, v_1) through ladders (see Fig.* 10(C)).

Proof. (A) Let (w_1', u_0') and (v_0', w_2') (possibly identical with (u_1, v_1)) be the last edges (before α, β) in the ladders $L(u_1, v_1; \ldots; \alpha)$ and $L(u_1, v_1; \ldots; \beta)$ respectively. See Fig. 11. Their colours are both $\bar{y}$. By the ladder argument of Lemma 3.2, at least one of (u_0, w_1') and (u_0, u_0') (depending upon the polarity of the ladder) must be an edge in E. Since $(u_0, u_0') \notin E$ (by assumption), we have $(u_0, w_1') \in E$ and $c(u_0, w_1') = y$. The ladder argument then implies $(v_0, u_0') \in E$ and $c(v_0, u_0') = y$. Similarly, $(u_0, v_0') \in E$ and $c(u_0, v_0') = y$. In particular $u_0' \neq v_0'$, as $(u_0', v_0) \in E$, $(v_0', v_0) \notin E$. This is however a contradiction since (v_0, u_0') and (u_0, v_0') are adjacent in G^* and must have different colours.

(B) By the ladder argument, there must be an edge (v_0, w) or (v_0, u_0) in E with colour $\bar{y}$. Neither is possible, however, since $c(v_0, w) = y$ and $(v_0, u_0) \notin E$ by assumption.

(C) The ladders are nondegenerate by Lemma 3.5. By the ladder argument, we obtain $(u_0, w_2) \in E$ with $c(u_0, w_2) \neq c(\alpha)$ and $c(u_0, w_2) \neq c(\beta)$, a contradiction to $c(\alpha) \neq c(\beta)$. □

Corollary 3.7. *Let G be a graph with colours as defined in Lemma* 3.6. *Assume that G (with colours) contains an* AP_6 *or an* AP_5. *Then:*

(1) *The three edges α, β and γ of an* AP_6 *(see Fig.* 2 *and Fig.* 4) *cannot all belong to the same component H_j^* of G^*.*

(2) *The floor α of an* AP_5 *cannot belong to the same component H_j^* as the roof β or the roof γ.*

Proof. (1) If α, β and γ of an AP_6 belong to the same component H_j^*, α is connected by ladders to β and to γ. First assume that these ladders are nondegenerate. Then $S(u_4, u_5; u_1, u_2)$ and edges β, γ form a configuration (with colours) not permitted by Lemma 3.6(A). On the other hand, if one of these ladders is degenerate, G contains an AP_5 as shown by Lemma 3.4. Since the three edges (u_1, u_2), (u_1, u_4), $(u_2, u_5) \in E$ forming the starter belong to the same component H_i^* by assumption, the three edges of this AP_5 also belong to H_i^*. This however contradicts (2) below.

(2) If γ and α of an AP_5 are in the same component H_i^*, then a configuration (with colours) of Lemma 3.6(B) is easily identified. The ladder attached to a double starter $D(u_2, u_5; u_3, u_4)$ is always nondegenerate by Lemma 3.5. □

4. Preliminaries 2: the pseudo ladder argument

This section extends the ladder argument to more general cases. To motivate this extension, we first present a procedure trying to obtain a 2-colouring of G^* such that the induced configurations C_x $(x=0, 1)$ of (2.3) have threshold completions.

Procedure 4.1

Step 1. Given a graph G and the associated G^* with $\chi(G^*)=2$, choose a 2-colouring of G^*.

Step 2. If G with the current edge colours does not contain an AP_6 or an AP_5, terminate. C_x $(x=0, 1)$ have threshold completions (by Theorem 2.1). Otherwise go to Step 3.

Step 3. Select an AP_6 or an AP_5 contained in G, and let δ be one of the edges α, β and γ therein. Change the colours of all the edges in $H^*(\delta)$ (i.e., those reachable from δ through ladders), where $H^*(\delta)$ denotes the component of G^* containing δ. With the resulting 2-colouring of G^*, return to Step 2.

Note that Step 3 destroys the selected AP_6, since α, β and γ therein do not all belong to the same component of G^* by Corollary 3.7. However, Step 3 may create new AP_6 or AP_5's. Procedure 4.1 never terminates if $t(G)>2$.

Let us call a configuration with colours a *pseudo* AP_6 or a *pseudo* AP_5 if it becomes an AP_6 or an AP_5 by changing the colours of all edges in $H^*(\delta)$ for one edge δ in it. In other words, an AP_6 or an AP_5 newly generated in Step 3 was always a pseudo AP_6 or a pseudo AP_5 in the previous colouring. Two particular pseudo AP_i's are shown in Fig. 12. For example, the pseudo AP_6 of Fig. 12(a) becomes an AP_6 by changing the colour of α to x.

The above two pseudo AP_i's are particularly important, as we can show that the ladder argument can be applied in a somewhat weaker fashion even if the

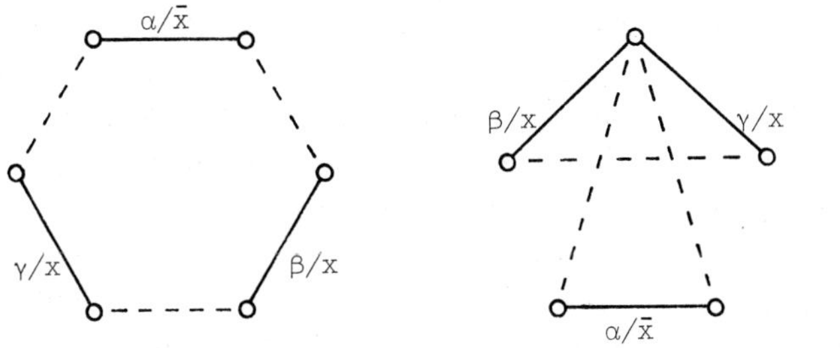

(a) A pseudo AP_6 ($H^*(\alpha)\neq H^*(\beta), H^*(\gamma)$)

(b) A pseudo AP_5 ($H^*(\alpha)\neq H^*(\beta), H^*(\gamma)$)

Fig. 12. Examples of a pseudo AP_6 and a pseudo AP_5.

ladders pass through such pseudo AP_j's. This will enable the ladder argument to proceed to other connected components of G^*.

Lemma 4.1. *Let G^* of a graph G satisfy $\chi(G^*)=2$, and consider the edge colours determined by a 2-colouring of G^*. Assume that $\alpha=(u_k, v_k)$ of a pseudo AP_6 of Fig.* 12(a) *(or a pseudo AP_5 of Fig.* 12(b)) *is reachable from $S(u_0, v_0; u_1, v_1)$ through a ladder $L=L(u_1, v_1; \ldots; u_k, v_k)$, as shown in Fig.* 13(a) *(or Fig.* 13(b)). *Let the ladders $L_1=L(u_{k+1}, v_{k+1}; \ldots)$ and $L_2=L(u'_{k+1}, v'_{k+1}; \ldots)$ start at β and γ respectively. Suppose further that L is*

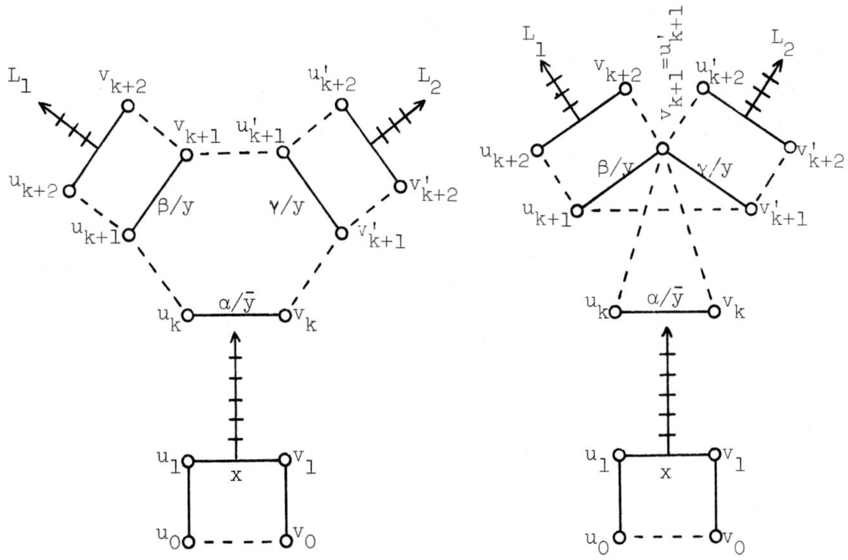

(a) A ladder through a pseudo AP_6.

(b) A ladder through a pseudo AP_5.

Fig. 13. Generalization of the ladder argument.

nondegenerate with respect to $S(u_0, v_0; u_1, v_1)$, *and that* u_0 *and* v_0 *are distinct from all the vertices of the* AP_6 *or* AP_5 *in question. Then there is a starter* $S(u_0, v_0; u_{k+1}, v_{k+1})$ *or* $S(u_0, v_0; u'_{k+1}, v'_{k+1})$ *or* $S(u_0, v_0; v_{k+1}, u_{k+1})$ *or* $S(u_0, v_0; v'_{k+1}, u'_{k+1})$, *which can start the ladder argument for* L_1 *or* L_2. *(We say in this case that the ladder argument proceeds to* L_1 *or* L_2, *respectively.)*

Proof. We consider the case of a pseudo AP_6 of Fig. 13(a) only. The case of a pseudo AP_5 can be similarly proved. In addition, assume for simplicity that k is even, i.e. $x = y$. This situation is shown in Fig. 14. By the ladder argument applied to $L(u_1, v_1; \ldots; u_k, v_k; u_{k+1}, v'_{k+1})$, we have (note that $(u_{k+1}, v'_{k+1}) \in E$ and has colour y as in Fig. 4(a))

$$(u_0, v_k), (v_0, u_k) \in E \quad \text{with colour } x, \text{ and}$$

$$(u_0, u_{k+1}), (v_0, v'_{k+1}) \in E \quad \text{with colour } \bar{x}.$$

Then $c(u_{k+1}, v_{k+1}) = c(u_k, v_0) = x$ and $(u_{k+1}, u_k) \notin E$ implies $(v_{k+1}, v_0) \in E$ by Lemma 3.1(i). Similarly $(u'_{k+1}, u_0) \in E$. These two edges satisfy $c(v_{k+1}, v_0) = \bar{y}$ and $c(u'_{k+1}, u_0) = y$ since they are adjacent in G^*. If $y = x$, we have two edges (u_0, u_{k+1}) and (v_0, v_{k+1}) with colour $\bar{x}$. Then there is an $S(u_0, v_0; u_{k+1}, v_{k+1})$ to start the ladder argument for L_1. On the other hand, if $y = \bar{x}$, there is an $S(u_0, v_0; u'_{k+1}, v'_{k+1})$. □

The result of Lemma 4.1, can easily be extended to the case in which ladders go through more than one pseudo AP_6 and/or pseudo AP_5 of the types shown in Fig. 12. A series of (ordered) pairs of vertices $u_1, v_1; u_2, v_2; \ldots; u_l, v_l$ is called a *pseudo ladder* with respect to a single starter $S(u_0, v_0; u_1, v_1)$ and denoted by $PL(u_1, v_1; \ldots; u_l, v_l)$ if

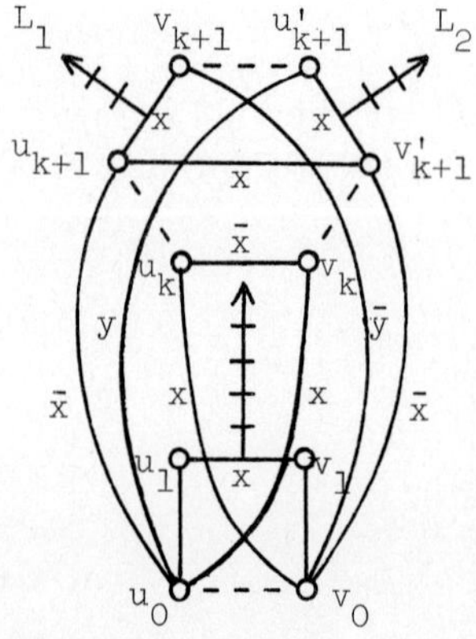

Fig. 14. Illustration of the proof of Lemma 4.1.

(1) for each i $(1 \leqslant i \leqslant l-1)$, either (a) $(u_i, v_i), (u_{i+1}, v_{i+1}) \in E$ and (u_i, u_{i+1}), $(v_i, v_{i+1}) \notin E$, or (b) (u_i, v_i), (u_{i+1}, v_{i+1}) are two edges in a pseudo AP_6 of Fig. 13(a) or in a pseudo AP_5 of Fig. 13(b) such that

$$(u_i, v_i) = (u_k, v_k),$$

$$(u_{i+1}, v_{i+1}) = (u_{k+1}, v_{k+1}) \quad \text{or} \quad (u'_{k+1}, v'_{k+1}),$$

(2) provided that $u_0 \neq u_1, v_2, u_3, v_4, \ldots$ and $v_0 \neq v_1, u_2, v_3, u_4, \ldots$ (nondegeneracy), the ladder argument proceeds through $u_1, v_1; \ldots; u_l, v_l$. This means that the choice for (u_{i+1}, v_{i+1}) between (u_{k+1}, v_{k+1}) and (u'_{k+1}, v'_{k+1}) in (b) of (1) and the choice which endpoint of this edge is u_{i+1} and which is v_{i+1} are such that a starter $S(u_0, v_0; u_{i+1}, v_{i+1})$ (i even) or $S(u_0, v_0; v_{i+1}, u_{i+1})$ (i odd) is contained in G.

Any ladder $L = (u_1, v_1; \ldots; u_l, v_l)$ is also a pseudo ladder if it is attached to $S(u_0, v_0; u_1, v_1)$. In Fig. 14, for example, we find $PL(u_1, v_1; \ldots; u_k, v_k; u_{k+1}, v_{k+1}; \ldots)$ if the ladder argument proceeds to L_1.

We do not consider here the generalization of the ladder argument starting from a double starter, since it is not required in the subsequent discussion.

Corollary 4.2. *Consider G and a configuration with colours as in Lemma* 4.1. *Assume in addition that* $H^*(\beta) = H^*(\gamma)$ *holds in Fig.* 13(a) (*or Fig.* 13(b)). *Then the ladder argument can proceed to both* L_1 *and* L_2.

Proof. If the ladder argument can proceed to L_1, for example, then follow the ladder $L(\beta; \ldots; \gamma)$ from β to γ. The ladder argument then proceeds to L_2 since L_2 starts with γ. □

By using the pseudo ladder argument instead of the ladder argument, some results of Lemma 3.6 can be easily generalized as follows.

Corollary 4.3. *Consider cases* (A) *and* (B) *of Lemma* 3.6 (*see Fig.* 10(A) *and* (B)) *except that the ladders therein are replaced by pseudo ladders. Then both* (A) *and* (B) *lead to contradictions.* □

5. Split graphs

Based on the previous results, we shall prove in this section the first main result that $t(G) = \chi(G^*)$ if G^* is bipartite and G is a split graph.

Lemma 5.1. *Let G be a split graph whose vertices are partitioned into a clique K and an independent set I. Then any four vertices* u_1, u_2, v_1, v_2 *with* $(u_1, v_1), (u_2, v_2) \in$

E and (u_1, u_2), $(v_1, v_2) \notin E$ *(see Fig.* 1) *satisfy either* $u_1, v_2 \in K$, $u_2, v_1 \in I$ *or* $u_1, v_2 \in I$, $u_2, v_1 \in K$.

Proof. If $u_1 \in K$, then $u_2 \in I$ since K induces a clique and $(u_1, u_2) \notin E$. Similarly, $u_2 \in I$ implies $v_2 \in K$, and then $v_2 \in K$ implies $v_1 \in I$. Thus $u_1, v_2 \in K$ and $u_2, v_1 \in I$. If $u_1 \in I$, on the other hand, we obtain $u_1, v_2 \in I$ and $u_2, v_1 \in K$. □

Lemma 5.2. *Let G be a split graph as in Lemma* 5.1.

(a) *If G contains an* AP_6 *or a pseudo* AP_6 *involving six vertices* $u_1, u_2, \ldots, u_6$ *(see Fig.* 2(a)), *then either* $u_1, u_3, u_5 \in K$, $u_2, u_4, u_6 \in I$ *or* $u_1, u_3, u_5 \in I$, $u_2, u_4, u_6 \in K$ *holds. See Fig.* 15.

(b) *G does not contain an* AP_5 *or a pseudo* AP_5 *(see Fig.* 2(b)). *(Properties* (a) *and* (b) *hold irrespectively of the bipartiteness of* G^* *or the edge colours of G.)*

Proof. (a) Apply the argument used in the proof of Lemma 5.1.

(b) Assume vertex u_1 of Fig. 2(b) is in K. Then $u_3, u_4 \in I$ holds by the argument of Lemma 5.1. This is a contradiction since $(u_3, u_4) \in E$. On the other hand, $u_1 \in I$ implies $u_2, u_5 \in K$, a contradiction to $(u_2, u_5) \notin E$. □

Lemma 5.3. *Let* G^* *of a split graph G satisfy* $\chi(G^*) = 2$ *and consider the edge colours of G determined by a 2-colouring of* G^*. *If* $\mathrm{PL}(u_1, v_1; u_2, v_2; \ldots; u_k, v_k)$ *is a pseudo ladder satisfying* $u_1 \in K$, $v_1 \in I$ *attached to* $S(u_0, v_0; u_1, v_1)$, *then* $u_i \in K$, $v_i \in I$ *for i odd and* $u_i \in I$, $v_i \in K$ *for i even (including* $i = 0$). *Consequently there is no degeneracy. A similar result holds with K and I exchanged.*

Proof. Since $v_1 \in I$ and $(v_0, v_1) \in E$, $v_0 \in K$. Since $(u_0, v_0) \notin E$, $u_0 \in I$. We may now take i positive and proceed by induction. By definition of a pseudo ladder, either (a) (u_i, u_{i+1}), $(v_i, v_{i+1}) \notin E$ and Lemma 5.1 applies, or else (b) there is a pseudo AP_6 containing $v_i, u_i, u_{i+1}, v_{i+1}$ or $u_i, v_i, v_{i+1}, u_{i+1}$ as consecutive vertices

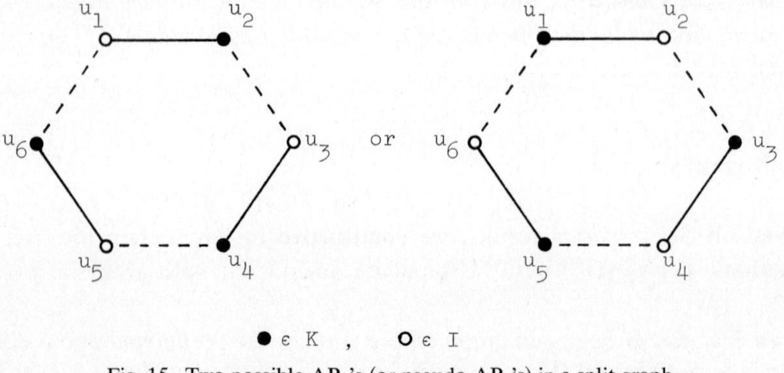

Fig. 15. Two possible AP_6's (or pseudo AP_6's) in a split graph.

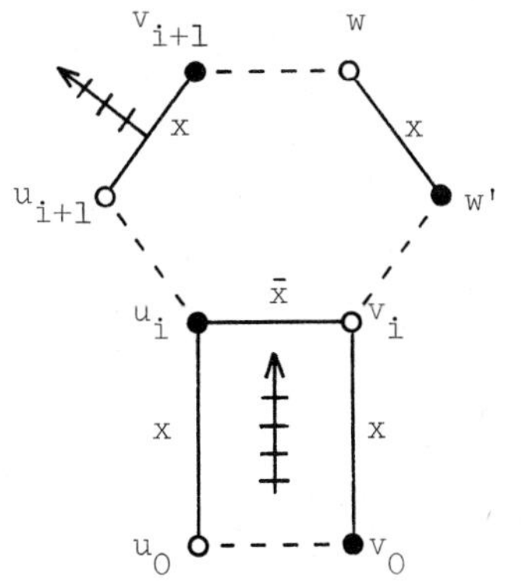

Fig. 16. Illustrating the proof of Lemma 5.3, case (b).

(Fig. 16 illustrates the first possibility for i odd) and Lemma 5.2(a) applies. [Note that in case (b), u_0 and v_0 are distinct from all the vertices of the pseudo AP_6 as required by Lemma 4.1. For example, in the case illustrated in Fig. 16, $u_0 \neq u_{i+1}$ because $(u_0, u_i) \in E$ by the pseudo ladder argument and non-degeneracy and $(u_{i+1}, u_i) \notin E$; similarly $u_0 \neq v_i$; $u_0 \neq v_{i+1}, w'$ because $u_0 \in I$, $v_{i+1}, w' \in K$; $u_0 \neq w$ because $(w, v_{i+1}) \notin E$ and $(u_0, v_{i+1}) \in E$.] □

Lemma 5.4. *Let G^* of a split graph G satisfy $\chi(G^*) = 2$ and consider edge colours of G determined by a 2-colouring of G^*. If α and β are two distinct edges of an AP_6, then there is no pseudo ladder (including the case of a ladder) $PL(\alpha; \ldots; \beta)$.*

Proof. Let the consecutive vertices of the AP_6 be $w_1, w_2, \ldots, w_6$ with $\alpha = (w_1, w_2)$, $\beta = (w_3, w_4)$. Suppose there is a pseudo ladder $PL(u_1, v_1; \ldots; u_k, v_k)$ where $u_1 = w_1$, $v_1 = w_2$. There is a starter $S(u_0, v_0; u_1, v_1)$ where $u_0 = w_4$, $v_0 = w_5$. By Lemma 5.2(a) we may assume $w_i \in K$ for $i = 1, 3, 5$ and $w_i \in I$ for $i = 2, 4, 6$ without losing generality (Fig. 17).

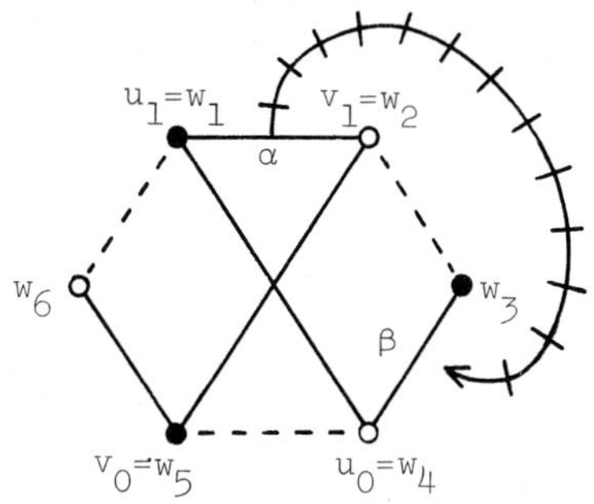

Fig. 17. A pseudo ladder connecting two edges of an AP_6.

Since $c(\alpha)=c(\beta)$, k is odd. By Lemma 5.3 $v_k \in I$. But then v_k cannot be w_3 since $w_3 \in K$, and v_k cannot be $w_4 = u_0$ since that means $(v_0, u_0) \in E$ by Lemma 4.1 contradicting the assumption. □

We now give the main result of this section.

Theorem 5.5. *Let G^* of a split graph G be bipartite. Then*

$$t(G)=\chi(G^*).$$

Proof. We may assume $\chi(G^*)=2$ as discussed in the Introduction. If $t(G)>2$, then Procedure 4.1 never terminates, no matter which AP_6 P is selected in Step 3 and which edge δ of P changes colour. In particular we may execute Procedure 4.1 so that, after the first execution of Step 3, P is among the AP_6's that were generated in the immediately preceding Step 3. Moreover, if the edges of P are α, β, γ and P has just become an AP_6 because $c(\alpha)$ has changed, then we may choose edge δ (to next change colour) from β and γ so that the ladder argument leading to α could proceed to δ. Define an abstract graph H whose vertices are the AP_6's and pseudo AP_6's of G (note that what is an AP_6 or a pseudo AP_6 in one colouring is either an AP_6 or a pseudo AP_6 in any other colouring, i.e. the set of vertices of H is independent of the colouring). Two vertices P_1 and P_2 of H are joined when G contains a ladder joining distinct edges $\alpha \in P_1$, $\beta \in P_2$. H has no loops (P, P) by Lemma 5.4 (applied to the colours such that P is an AP_6). The non-terminating computation proceeds along a path in H without backtracking an edge that has just been traversed. Since H is finite, the computation progresses along elementary cycles of H. Let $M=(P_0, P_1, \ldots, P_{n-1})$ be such an elementary cycle of minimum length n traversed by any computation as described above. Let P_0 with edges $\alpha_0, \beta_0, \gamma_0$ be an AP_6 in the colouring C_0. The computation switches to colouring C_1 by changing $c(\beta_0)$, which also changes $c(\alpha_1)$ and makes P_1 an AP_6 with edges $\alpha_1, \beta_1, \gamma_1$. The computation switches to colouring C_2 by changing $c(\beta_1)$, and so on. Finally the computation returns to P_0 by changing $c(\beta_{n-1})$, which also changes $c(\alpha_0)$. See Fig. 18. We may rule out the possibility that changing $c(\beta_{n-1})$ changes $c(\beta_0)$ rather than $c(\alpha_0)$, because in that case the computation could bypass P_0 and traverse the shorter cycle $(P_1, P_2, \ldots, P_{n-1})$. We claim that M has no diagonals, i.e. that P_i and P_j are not adjacent in H unless $i-j=\pm 1 \pmod{n}$. Suppose P_i and P_j are adjacent in H, with $i-j \neq \pm 1 \pmod{n}$. Without losing generality we may assume that β_i is connected by a ladder to α_j or β_j or γ_j. If there is a ladder $L=L(\beta_i; \ldots; \alpha_j)$, then the computation could develop along the shorter cycle $(P_i, \ldots, P_j)$ by proceeding from β_i to α_j along M and returning to β_i via L. Similarly there cannot be a ladder $L(\beta_i; \ldots; \beta_j)$. If there is a ladder $L=L(\beta_i; \ldots; \gamma_j)$, we can apply the

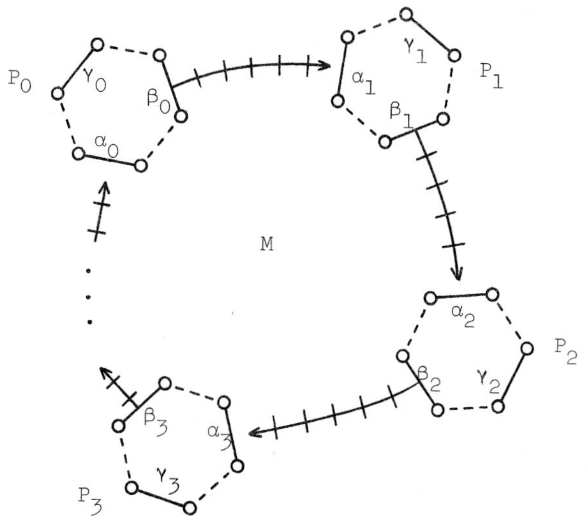

Fig. 18. A cycle in H traversed by Procedure 4.1.

ladder argument (in colouring C_i) from the starter $S(u_0, v_0; \beta_i)$ contained in P_i along L to γ_j, and therefore u_0 is adjacent to one end of γ_j and v_0 to the other end. But then there is a computation traversing the shorter cycle $(P_i, \ldots, P_j)$ by proceeding from β_i to α_j along M, continuing to γ_j (rather than to β_j) and returning to P_i via L. This proves the claim that M has no diagonals. From this it follows that in colouring C_0, P_0 is AP_6 and all the other P_i are pseudo AP_6 satisfying $c(\alpha_i) \neq c(\beta_i) = c(\gamma_i)$. Indeed P_i satisfies this condition in colouring C_{i-1}, but then it must have satisfied it already in colouring C_0, since switching C_0 to C_1 does not affect P_i as there is no ladder from β_0 to P_i, switching C_1 to C_2 does not affect P_i as there is no ladder from β_1 to P_i, and so on. Therefore in colouring C_0 there is a pseudo ladder $PL(\beta_0; \ldots; \alpha_1; \beta_1; \ldots; \beta_{n-1}; \alpha_0)$ joining distinct edges of the AP_6 P_0, contradicting Lemma 5.4. □

6. G^* with at most two nonsingleton components

In this section, we place a restriction that components H_j^*, $j = 3, 4, \ldots, p$, of G^* are all singletons, i.e., $|H_j^*| = 1$. We start with a special case in which only H_1^* can be nonsingleton.

Theorem 6.1. *Let G^* of a graph G be bipartite, and furthermore let $|H_j^*| = 1$ for $j = 2, 3, \ldots, p$, where H_j^* $(j = 1, 2, \ldots, p)$ are the components of G^* and $|H^*|$ denotes the number of vertices in H^*. Then $t(G) = \chi(G^*)$ holds.*

Proof. We may assume that $\chi(G^*)=2$ as remarked in the Introduction. Consider the edge colours determined by a 2-colouring of G^*. We first show that G does not contain an AP_6. The three edges α, β, γ of an AP_6 (see Fig. 4(a)) must all belong to H_1^* because α is adjacent in G^* to (u_3, u_6), etc. This is a contradiction to Corollary 3.7(1).

If G contains an AP_5 (see Fig. 4(b)), $\beta, \gamma \in H_1^*$ and $\alpha \in H_j^*$ $(j \geq 2)$ must hold because the roofs β, γ are not isolated in G^*, and $\alpha \in H_1^*$ is not allowed by Corollary 3.7(2). When the first AP_5 is selected in Step 3 of Procedure 4.1, we can select the floor α to change its colour. If $t(G)>2$, this must generate another AP_5. Since α is isolated in G^*, α itself belongs to the second AP_5 and is in fact its floor. This situation is illustrated in Fig. 19, where bold lines indicate the original edges of these two AP_5's. Obviously, (u_1, u_2), (u_1, u_5), (u_1', u_2'), $(u_1', u_5') \in H_1^*$ and $(u_3, u_4)(=(u_3', u_4')) \in H_j^*$ $(j \geq 2)$. We derive a contradiction in each of the following two cases.

(i) $u_1 \neq u_1'$. Then $c(u_1', u_2') = c(u_4, u_2) = \bar{x}$ and $(u_4, u_1') \notin E$ imply $(u_2, u_2') \in E$ by Lemma 3.1(i) ($u_2 \neq u_2'$ because $c(u_2, u_3) \neq c(u_2', u_3)$). Similarly $(u_5, u_5') \in E$. $(\bar{y}=)$ $c(u_2, u_2') \neq c(u_5, u_5')(=y)$ holds by Lemma 3.1(ii). Let $x=y$ without loss of generality. Then $c(u_2, u_2') = c(u_5', u_1') = \bar{x}$ and $(u_2', u_5') \notin E$ imply $(u_2, u_1') \in E$ by Lemma 3.1(i) ($u_2 \neq u_1'$ because $(u_2, u_4) \in E$, $(u_1', u_4) \notin E$). Next note that $(u_1, u_1') \notin E$ since otherwise (u_3, u_4) is adjacent in G^* to (u_1, u_1'), a contradiction to $(u_3, u_4) \in H_j^*$ $(j \geq 2)$. Thus $c(u_2, u_1') \neq c(u_1, u_5) = x$ and $c(u_2, u_1') \neq c(u_5, u_4) = \bar{x}$ by Lemma 3.1(ii), a contradiction.

(ii) $u_1 = u_1'$. This case is shown in Fig. 20. $(u_5, u_5') \in E$ holds because $c(u_5, u_4) = c(u_5', u_1) = \bar{x}$ and $(u_1, u_4) \notin E$ (it is easy to see that $u_5 \neq u_5'$). Similarly $(u_2, u_2') \in E$. Since these are adjacent in G^*, let $c(u_5, u_5') = y$ and $c(u_2, u_2') = \bar{y}$. Similarly (u_5', u_2), $(u_2', u_5) \in E$ and $c(u_5', u_2) = z$, $c(u_2', u_5) = \bar{z}$. Assume $y=x$ without loss of generality, and let $y=z$ for simplicity (the case of $y=\bar{z}$ can be

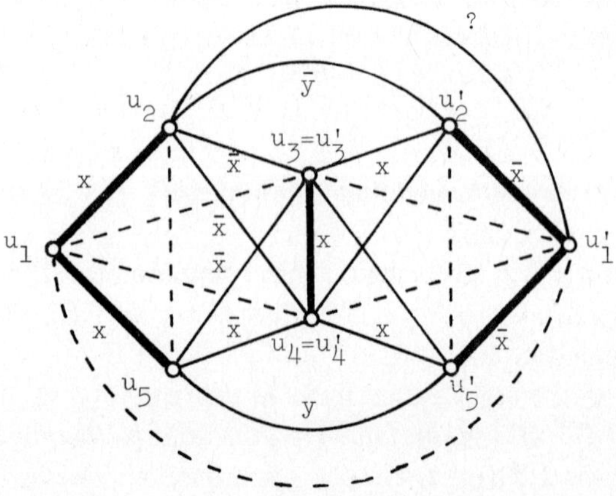

Fig. 19. Two AP_5's with a common floor. (Bold lines indicate original edges in the two AP_5's.)

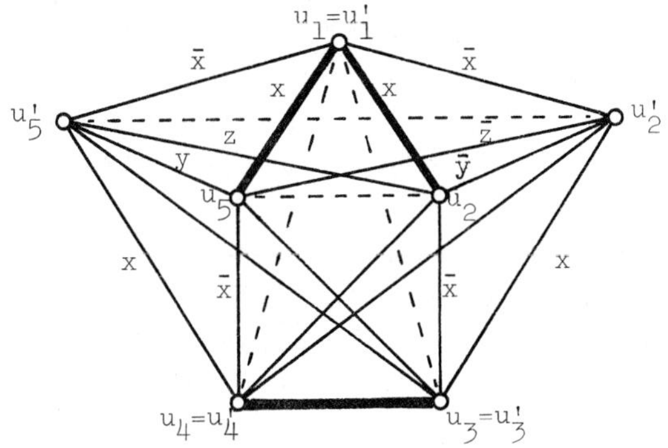

Fig. 20. Two AP_5's with common floor and top. (Bold lines indicate the original edges in the two AP_5's.)

similarly treated). Now there is a double starter $D(u_2, u_5; u_1, u_5')$ in G. This leads to a contradiction by Lemma 3.6(C) since the edges (u_2, u_3) and (u_3, u_2') (for example) can be considered as the edges (w_1, w_2) and (w_2, w_3) of Lemma 3.6(C). (Note here that all edges in Fig. 20 except (u_3, u_4) belong to H_1^* since they are not isolated in G^*. Thus any edge $(\neq (u_3, u_4))$ is reachable from any edge $(\neq (u_3, u_4))$ through a ladder.) □

Theorem 6.2. *Let G^* of a graph G be bipartite, and let $|H_j^*| = 1$, $j = 3, 4, \ldots, p$, where H_j^* $(j = 1, 2, \ldots, p)$ are the components of G^*. Then $t(G) = \chi(G^*)$.*

Proof. We may assume $\chi(G^*) = 2$. Assume $t(G) > 2$ and apply Procedure 4.1. Then G with the initial edge colours contains an AP_6 or AP_5, P, and there is a computation path in which Step 3 generates another AP_6 or AP_5, P', in the next edge colours. From these P and P' we shall derive a contradiction. Before enumerating all possible cases, note that, if P (or P') is an AP_6 with edges α, β, γ, then each of α, β, γ belongs to either H_1^* or H_2^*. Since α, β, γ do not all belong to the same component by Corollary 3.7(1), one or two edges out of α, β and γ belong to H_1^*, and the rest belong to H_2^*. On the other hand, if P (or P') is an AP_5, the floor α can belong to any component H_j^* but the roofs β, γ belong to either H_1^* or H_2^*. In addition, floor and roof do not belong to the same component by Corollary 3.7(2).

We first assume that all ladders and pseudo ladders appearing in the argument are *nondegenerate*.

(A) *P and P' are both AP_6's.* Let P and P' have edges α, β, γ and α', β', γ', respectively. Without loss of generality, we assume

(a) $\alpha, \beta \in H_2^*$, $\gamma \in H_1^*$,
(b) $\alpha' \in H_1^*$, $\gamma' \in H_2^*$,
(c) $c(\alpha) = c(\beta) = c(\gamma) = x$, $c(\alpha') = y$, $c(\gamma') = \bar{y}$ in the initial edge colours,

i.e., P is an AP_6 and P' is a pseudo AP_6. P' becomes an AP_6 by switching all colours in H_1^*. The following two subcases exhaust all possible combinations of the two AP_6's.

(A1) $\beta' \in H_2^*$. See Fig. 21(1), where straight lines indicate edges in H_1^*, and wavy lines indicate edges in H_2^*. (Here $c(\beta') = \bar{y}$). P contains a single starter $S(u_1, u_6; u_4, u_3)$ (see Fig. 4(a)). The ladder argument proceeds through $\mathrm{PL}(\gamma; \ldots; \alpha'; \beta'; \ldots; \alpha)$ and further through $L(\alpha; \ldots; \beta)$ (see Lemma 4.1 and Corollary 4.2). This leads to a contradiction by Corollary 4.3(A).

(A2) $\beta' \in H_1^*$. See Fig. 21(2). (Here $c(\beta') = y$). Consider the second edge colours obtained by changing colours of all edges in H_1^*. Then P' is an AP_6 and contains $S(u_1', u_2'; u_4', u_5')$. Apply the ladder argument through $L(\alpha'; \ldots; \beta')$ and through $\mathrm{PL}(\alpha'; \ldots; \gamma; \beta; \ldots; \gamma')$. This leads to a contradiction by Corollary 4.3(A).

(B) *One of the P and P' is an* AP_6 *and the other is an* AP_5. Without loss of generality, let P be an AP_6 and P' be an AP_5 (otherwise start Procedure 4.1 with the second colouring, and obtain the first colouring after executing Step 3). Let the three edges of P be α, β, γ and let the floor and the roofs of P' be α' and β', γ', respectively. The following assumptions do not lose generality.

(a) $\alpha' \in H_1^*$, $\beta', \gamma' \in H_2^*$,

(b) $\alpha \in H_2^*$, $\gamma \in H_1^*$,

(c) $c(\alpha) = c(\beta) = c(\gamma) = x$, $c(\alpha') = y$, $c(\beta') = c(\gamma') = \bar{y}$.

Assumption (a) is justified as follows. α' does not belong to $H^*(\beta')$ or $H^*(\gamma')$ by Corollary 3.7(2). If $\alpha' \in H_j^*$ $(j \geq 3)$, then change $c(\alpha')$ by selecting α' as δ of Step 3 when the AP_5 P' is generated. This may create another AP_5, but does not create an AP_6 since none of the edges in an AP_6 belongs to H_j^* $(j \geq 3)$. If no AP_5 is created, we simply resume Procedure 4.1. If an AP_5 is created, we

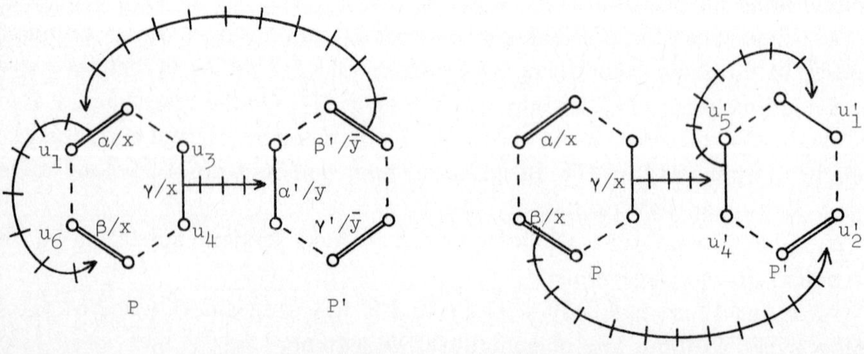

(1) $\beta' \in H_2^*$ (2) $\beta' \in H_1^*$

Fig. 21. Two AP_6's indicating a contradiction. (— shows an edge in H_1^*, and = an edge in H_2^*. ++> stands for a ladder. All edge colours shown are original.)

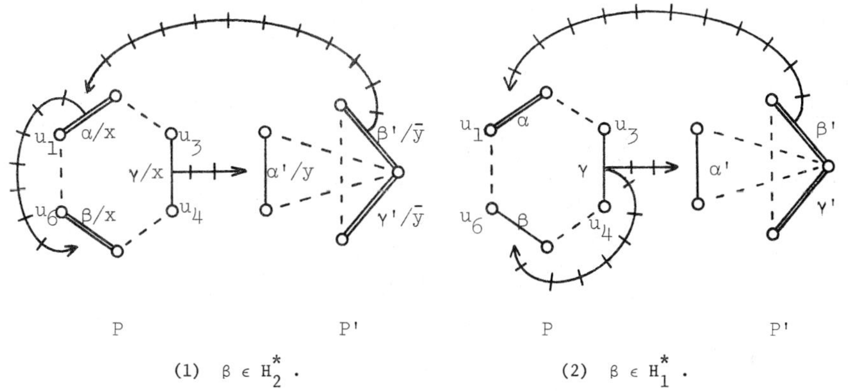

Fig. 22. AP_6 and AP_5 indicating contradictions.

have two AP_5's and this case will be treated in (C) below. Consequently, we may assume $\alpha' \in H_1^*$ or H_2^*, and take $\alpha' \in H_1^*$ without loss of generality. The edge colours for P' in (c) above follow from (a) and the fact that P' becomes an AP_5 after changing all colours in H_1^* or H_2^*.

The following exhaust all possible cases.

(B1) $\beta \in H_2^*$. See Fig. 22(1). Starting from $S(u_1, u_6; u_4, u_3)$, the ladder argument proceeds through $PL(\gamma; \ldots; \alpha'; \beta'; \ldots; \alpha)$ and further through $L(\alpha; \ldots; \beta)$. This leads to a contradiction by Corollary 4.3(A).

(B2) $\beta \in H_1^*$. See Fig. 22(2). The ladder argument starts from $S(u_1, u_6; u_4, u_3)$ and proceeds through $L(\gamma; \ldots; \beta)$ as well as through $PL(\gamma; \ldots; \alpha'; \beta'; \ldots; \alpha)$. This also leads to a contradiction by Corollary 4.3(A).

(C) *P and P' are both* AP_5*'s.* Let α and α' be the floors of P and P', respectively, and let β, γ and β', γ' be the roofs of P and P', respectively. If $\alpha \in H_j^*$ $(j \geq 3)$, then we need consider only the case in which changing $c(\alpha)$ creates another AP_5, P', as discussed in case (B) above. This occurs only if $H^*(\alpha) = H^*(\alpha')$, i.e., $\alpha = \alpha'$. If $\alpha \notin H_j^*$ $(j \geq 3)$, we assume $\alpha \in H_1^*$ without loss of generality. Then $\beta, \gamma \in H_2^*$ by Corollary 3.7(2). P' satisfies either $\alpha' \in H_1^*$, β', $\gamma' \in H_2^*$ or $\alpha' \in H_2^*$, β', $\gamma' \in H_1^*$. In all three cases, we consider edge colours such that $c(\alpha) = c(\beta) = c(\gamma) = x$ and $c(\alpha') = y$, $c(\beta') = c(\gamma') = \bar{y}$, i.e., P is an AP_5 and P' is a pseudo AP_5.

(C1) $\alpha \in H_j^*$ $(j \geq 3)$, i.e., $\alpha = \alpha'$. This case was already treated in the proof of Theorem 6.1. It differs only in that β, γ, β', γ' here may belong to either H_1^* or H_2^*, while these belong only to H_1^* in Theorem 6.1. Case (i) of the proof of Theorem 6.1 can be carried over to the present case without change. Case (ii) requires some modification since (u_2, u_3) and (u_3, u_2') in Fig. 20 may belong to components different from $H^*(u_1, u_5')$ (i.e., the component containing edge

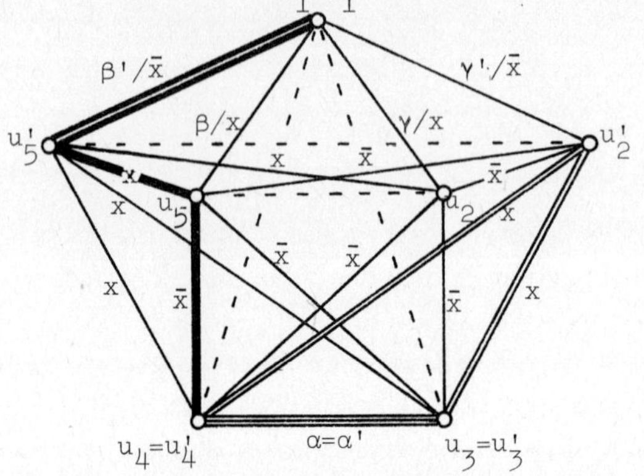

Fig. 23. Two AP_5's treated in (C1) of the proofs of Theorem 6.2. (— denotes an edge in H_1^*, = an edge in H_2^* and ≡ an edge in H_j^* ($j \geq 3$). Bold lines indicate a single starter.)

$(u_1, u_5'))$. In most cases arising from possible assignments of $\beta, \gamma, \beta', \gamma'$ to H_1^* and H_2^*, however, we can find two edges δ and δ' with a common vertex such that $H^*(\delta) = H^*(\delta') = H^*(u_1, u_5')$ and $c(\delta) \neq c(\delta')$, where (u_1, u_5') is an edge of the double starter in Fig. 20. Thus we get contradictions by Lemma 3.6(C). The only exception (under the assumption $x = y = z$ in Fig. 20; other cases can be similarly treated) is the case shown in Fig. 23. In this case, however, there are a single starter $S(u_1', u_4'; u_5', u_5)$ and edges (u_1', u_2'), (u_4', u_2) with colour $\bar{x}$. Since these three edges (u_5, u_5'), (u_1', u_2') and (u_4', u_2) belong to H_1^* by assumption (i.e., connected by ladders), we get a contradiction by Lemma 3.6(A).

(C2) $\alpha, \alpha' \in H_1^*$. See Fig. 24(1). Then the ladder argument starting from $S(u_2, u_5; u_3, u_4)$ proceeds through $PL(\alpha; \ldots; \alpha'; \beta'; \ldots; \beta)$. This leads to a contradiction by Corollary 4.3(B).

(C3) $\alpha \in H_1^*$ and $\alpha' \in H_2^*$. See Fig. 24(2). As is obvious from Fig. 4(b), $(v_4, v_5) \in E$ and $c(v_4, v_5) = y$ holds. Thus a double starter $D(u_2, u_5; u_3, u_4)$ is

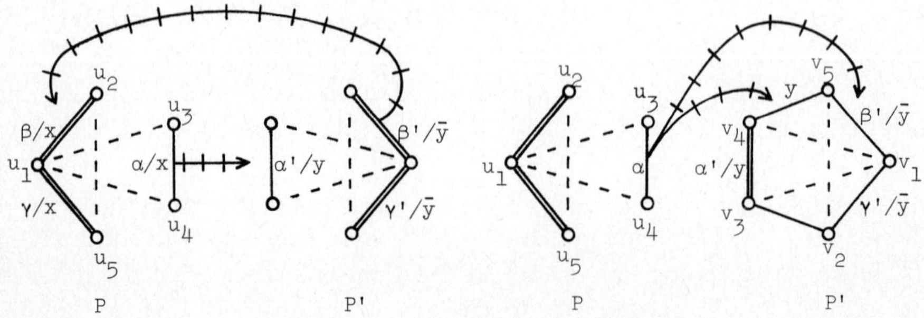

(1) $\alpha, \alpha' \in H_1^*$ (2) $\alpha \in H_1^*$ and $\alpha' \in H_2^*$

Fig. 24. Two AP_5's indicating contradictions.

connected to (v_1, v_5) and (v_5, v_4) through ladders (since (u_3, u_4), (v_1, v_5), $(v_5, v_4) \in H_1^*$), and $c(v_1, v_5) \neq c(v_5, v_4)$. This is a contradiction by Lemma 3.6(C).

This completes the proof for the nondegenerate case. We now turn to the *degenerate case.*

Recall Lemmas 3.4 and 3.5. Lemma 3.5 states that a ladder following a double starter is never degenerate, while Lemma 3.4 states that the degeneracy of a ladder following a single starter implies the existence of an AP_5. The result of Lemma 3.4 can be extended to a pseudo ladder $PL(u_1, v_1; u_2, v_2; \ldots; u_k, v_k)$ attached to $S(u_0, v_0; u_1, v_1)$ as will be shown later in Lemma 6.3.

Now consider cases (A) and (B) above. We have used the ladder argument through a pseudo ladder starting from a single starter. Therefore, if the degeneracy is possible, it means that G contains an AP_5 different from P. Thus case (A) is reduced to case (B), and case (B) is reduced to case (C) (strictly speaking, to case (C2) or case (C3), because it is easy to see in the proofs of Lemmas 3.4 and 6.3 that the three edges of the AP_5 belong to H_1^* or H_2^*). Consequently it suffices to consider case (C) only.

(C1)′ $\alpha \in H_j^*$ $(j \geq 3)$. Following the proof of case (C1) above (i.e., the proof of Theorem 6.1 and the proof for Fig. 23), one sees easily that there is no room for degeneracy except the case of Fig. 23. In that case, if Lemma 3.6(A) cannot be applied due to degeneracy, then by Lemma 3.4 G contains an AP_5 whose three edges belong to H_1^* or H_2^* (since the three edges of the single starter in Fig. 23 belong to H_1^* or H_2^*). Therefore, this case is reduced to case (B), case (C2) or case (C3), and case (B) in turn reduces to case (C2) or case (C3).

(C2)′ $\alpha, \alpha' \in H_1^*$. See Fig. 24(1). Since P contains a double starter $D(u_2, u_5; u_3, u_4)$, $L(\alpha; \ldots; \alpha')$ is nondegenerate by Lemma 3.5. As will be shown in the proof of Lemma 6.3, the edge β' of $L(\beta'; \ldots; \beta)$ is not degenerate. Now assume without loss of generality that P and P' and the original 2-colouring of G^* are chosen so that $L(\beta'; \ldots; \beta)$ is as short as possible. Let $\delta = (w_{l+1}, w'_{l+1})$ with $c(\delta) = z$ be the first degenerate edge between β' and β in $L(\beta'; \ldots; \beta)$ (see Fig. 25). Assume for simplicity that $u_5 = w'_{l+1}$. (The case of $u_2 = w_{l+1}$ is similar.) Then (u_5, w_l), $(u_2, w'_l) \in E$ and $c(u_5, w_l) = c(u_2, w'_l) = z$ follows from the ladder argument. An AP_5 consisting of the roofs $\delta = (u_5, w_{l+1})$, (u_5, w_l) and the floor (u_2, w'_l) is easily identified. $\delta \in H_2^*$ holds since δ is reachable from β' through a ladder. By Corollary 3.7(2), since the roof δ is in H_2^*, the floor (u_2, w'_l) must be in H_1^*, and then the other roof (u_5, w_l) must be in H_2^*, as in Fig. 25. This new AP_5 can be considered as P of Fig. 24(1) (with respect to the same P'). This is however a contradiction because $L(\beta'; \ldots; \delta)$ connecting the roofs of P' and the new P is shorter than $L(\beta'; \ldots; \beta)$. This is because $\delta \neq \beta$, since δ contains the vertex u_5 and β does not.

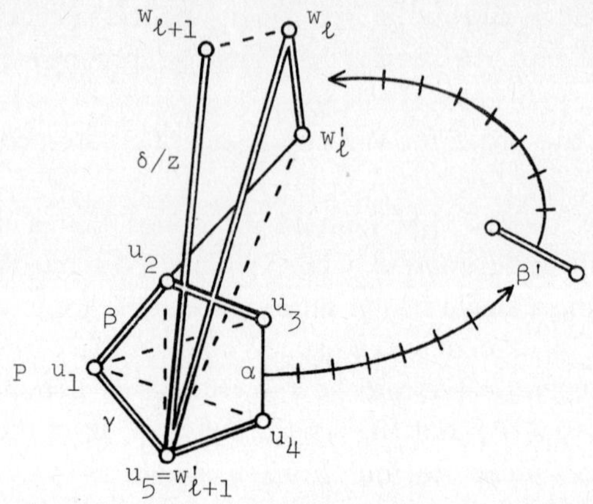

Fig. 25. Degeneracy of edge δ in $L(\beta'; \ldots; \beta)$. (— denotes an edge in H_1^*, and = denotes an edge in H_2^*.)

(C3)′ $\alpha \in H_1^*$ and $\alpha' \in H_2^*$. See Fig. 24(2). In this case the ladders $L(\alpha; \ldots; \beta')$ and $L(\alpha; \ldots; v_4, v_5)$ are not degenerate by Lemma 3.5, and this case cannot arise.

This exhausts all possible cases of P and P'. □

Lemma 6.3. *Let G^* of a graph G be bipartite, and let $|H_j^*| = 1$ for $j = 3, 4, \ldots, p$, where H_j^* $(j = 1, 2, \ldots, p)$ are the components of G^*. Let* PL = PL$(u_1, v_1; u_2, v_2; \ldots; u_k, v_k)$ *be a pseudo ladder in G starting from a single starter $S(u_0, v_0; u_1, v_1)$. Assume that* PL *passes through exactly one pseudo* AP_6 *or pseudo* AP_5 *from floor to roof. If* PL *is degenerate, then G contains an* AP_5.

Proof. Without loss of generality, let (u_{2l+1}, v_{2l+1}) be the first degenerate edge in PL, i.e., $u_0 = u_{2l+1}$ or $v_0 = v_{2l+1}$. Furthermore assume that (u_{2l}, v_{2l}) and (u_{2l+1}, v_{2l+1}) are edges of a pseudo AP_6 or a pseudo AP_5, since otherwise the proof of Lemma 3.4 can be directly applied. First let (u_{2l}, v_{2l}) and (u_{2l+1}, v_{2l+1}) be edges of a pseudo AP_6 P''. There are two cases.

(1) $u_0 = u_{2l+1}$. Let the consecutive vertices of P'' be $v_{2l+1}, u_0 = u_{2l+1}, u_{2l}, v_{2l}, w, w'$. The situation is shown in Fig. 26(1). We have $(v_0, u_{2l}), (u_0, v_{2l}) \in E$ and $c(v_0, v_{2l}) = c(u_0, v_{2l}) = x$ by the ladder argument. Furthermore, $w \neq u_0$ and $(w, u_0) \in E$ (consider the edge colours that make P'' an AP_6, and see Fig. 4(a)). $c(w, u_0) = x$ is also obvious. Now (u_{2l}, v_0) (floor) and (u_0, w), (u_0, v_{2l}) (roofs) form an AP_5.

(2) $v_0 = v_{2l+1}$. Let the consecutive vertices of P'' be $v_0 = v_{2l+1}, u_{2l+1}, u_{2l}, v_{2l}, w, w'$. See Fig. 26(2). In a manner similar to case (1), we obtain (u_0, v_{2l}),

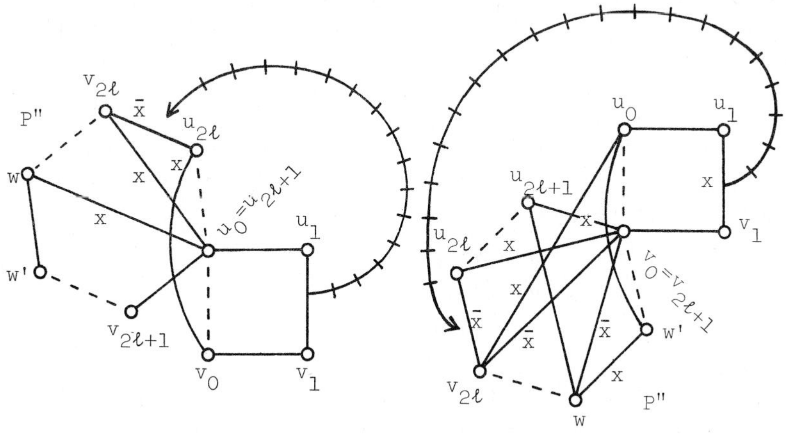

(1) $u_0 = u_{2\ell+1}$ (2) $v_0 = v_{2\ell+1}$

Fig. 26. A pseudo AP_6 degenerates with $S(u_0, v_0; u_1, v_1)$.

(v_0, u_{2l}), (v_0, v_{2l}), $(v_0, w) \in E$ and $c(v_0, u_{2l}) = c(v_0, u_{2l+1}) = c(u_0, v_{2l}) = c(w, w') = x$, $c(v_0, v_{2l}) = c(v_0, w) = \bar{x}$. ($c(v_0, w) = \bar{x}$ is obtained by the ladder argument applied to (u_{2l+1}, w).) First assume $u_0 \neq w'$. Then $(u_0, w') \in E$ by $c(u_0, v_{2l}) = c(w, w') = x$ and $(v_{2l}, w) \notin E$ (see Lemma 3.1(i)). If $c(u_0, w') = x$, we have an AP_5 consisting of (u_0, w') (floor) and (v_0, u_{2l}), (v_0, u_{2l+1}) (roofs). On the other hand, if $c(u_0, w') = \bar{x}$, an AP_5 consisting of (u_0, w') (floor) and (v_0, v_{2l}), (v_0, w) (roofs) is identified.

Next assume $u_0 = w'$. This case is shown in Fig. 27. (w, u_{2l+1}), (u_0, u_{2l}), $(v_0, v_{2l}) \in E$ and their colours (shown in Fig. 27) are determined in the manner similar to Fig. 4(a). (u_0, v_{2l}), (v_0, u_{2l}), (u_0, u_{2l+1}), $(v_0, w) \in E$ and their colours

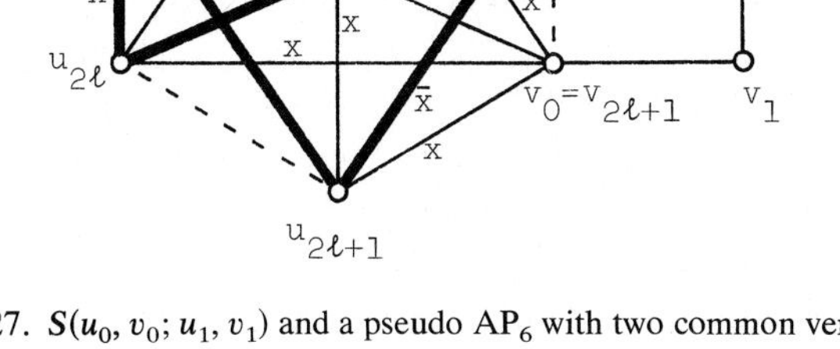

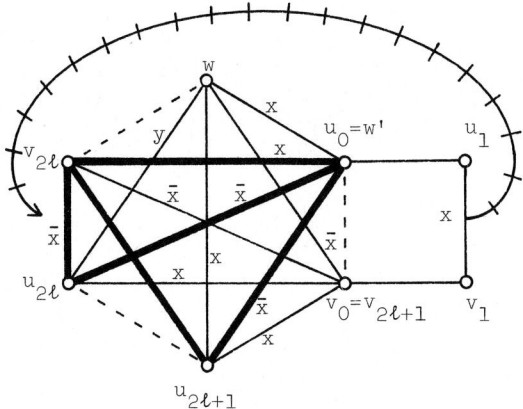

Fig. 27. $S(u_0, v_0; u_1, v_1)$ and a pseudo AP_6 with two common vertices.

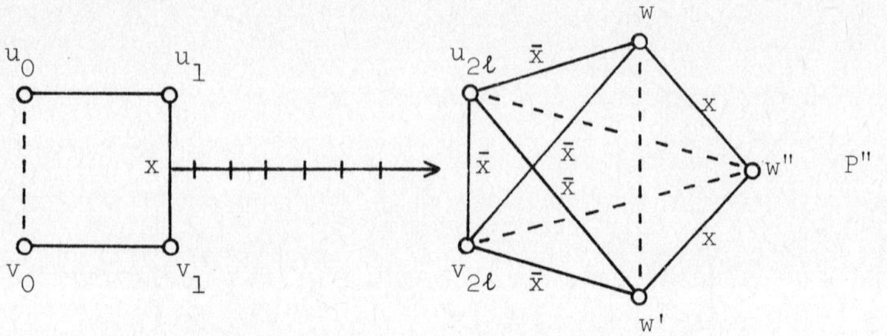

Fig. 28. A pseudo AP_5, P'', and a single starter $S(u_0, v_0; u_1, v_1)$.

follow from the ladder argument. $(w, u_{2l}) \in E$ follows from $c(w, u_0) = c(v_0, u_{2l}) = x$ and $(u_0, v_0) \notin E$ by Lemma 3.1(i). Similarly $(v_{2l}, u_{2l+1}) \in E$. Now let $c(w, u_{2l}) = y$ and $c(v_{2l}, u_{2l+1}) = \bar{y}$ because these are adjacent in G^*. Assume $y = x$ (the case of $y = \bar{x}$ is similarly treated). G contains a double starter $D(u_{2l}, u_{2l+1}; v_{2l}, u_0)$ as indicated by bold lines in Fig. 27. Note that (v_{2l}, u_0) belongs to H_1^* or H_2^*, and one or two edges out of the edges of (u_{2l}, v_{2l}), (u_{2l+1}, v_{2l+1}), (u_0, w) belong to H_1^* while the rest belong to H_2^*. In each case, we can find edges δ and δ' that share a common vertex, and satisfy $c(\delta) \neq c(\delta')$ and $H^*(\delta) = H^*(\delta') = H^*(v_{2l}, u_0)$, a contradiction by Lemma 3.6(C). Such δ and δ' exist because otherwise

$H^*(w, u_0) \neq H^*(w, v_0) = H^*(u_0, v_{2l})$ follows by considering vertex w,

$H^*(u_0, v_{2l}) \neq H^*(u_0, u_{2l})$ follows by considering vertex u_0,

$H^*(u_0, u_{2l}) \neq H^*(w, u_0)$ follows by considering vertex u_0,

together leading to a contradiction. The degeneracy never occurs in applying Lemma 3.6(C) because it starts from a double starter (see Lemma 3.5).

Finally we consider the case in which (u_{2l}, v_{2l}) and (u_{2l+1}, v_{2l+1}) are in a pseudo AP_5, P''. By the assumption in the lemma statement, (u_{2l}, v_{2l}) is the floor and (u_{2l+1}, v_{2l+1}) is a roof of P''. The situation is shown in Fig. 28. Then u_0 cannot be identical with w, w' or w'' because (u_0, v_{2l}) is an edge with $c(u_0, v_{2l}) = x$ by the ladder argument but (w, v_{2l}), (w', v_{2l}), (w'', v_{2l}) are not so. Similarly, v_0 cannot be identical with w, w' or w''. □

7. Conclusion

The results proved for two special cases strongly suggest that $t(G) = \chi(G^*)$ generally holds if G^* is bipartite. The methods developed in this paper seem to be useful for proving such general result. Probably, the case of G^* with at most three nonsingleton components could be proved without much difficulty. It seems, however, that some new ideas are necessary in order to reach the final goal. If G^* is not bipartite, we have at present no tool with which to settle the question whether $t(G)=\chi(G^*)$.

The interpretation of our results from the algorithmic point of view may also be worthwhile. Since our two conditions, i.e., (1) G is a split graph, and (2) G^* has at most two nonsingleton components, can be recognized in polynomial time, $t(G)$ can be computed in polynomial time if G^* is bipartite in these cases (note that the bipartiteness of G^* is tested in polynomial time). Moreover, under these conditions one can actually find in polynomial time a threshold cover of size 2 of G and the corresponding two linear inequalities. Indeed, under condition (1), if Procedure 4.1 is carried out as in the proof of Theorem 5.5, then it progresses along a simple path of H and hence terminates after at most $|H|$ iterations, where $|H|$ is the number of alternating hexagons of G and is polynomially bounded. Each iteration requires only polynomial time. Under condition (2) G^* has at most two 2-colourings up to complementation and the colours of the singleton components, and one of these 2-colourings has the required property. Thus under (1) or (2) one can find in polynomial time a 2-colouring of the edges of G such that each colour has a threshold completion in G. The threshold completions can be found in polynomial time by the methods of [3], and the corresponding linear inequalities by the methods of [1]. This contrasts with the known result [1] that the computation of $t(G)$ for an arbitrary graph G is NP-hard, and hence it is unlikely to have a polynomial-time algorithm.

Note added in proof

Recently M. Yannakakis has shown in his paper 'The Complexity of the Partial Order Dimension Problem' (submitted for publication) that it is NP-complete to determine whether a given bipartite graph can be covered by three chain subgraphs. As M. Golumbic has remarked (private communication), it is straightforward to use this result for showing that it is NP-complete to determine whether a given split graph has threshold number at most 3. Yannakakis also mentioned (without proof) in his paper that it is possible to determine in polynomial time whether a given bipartite graph can be covered by two chain subgraphs (i.e., whether a given split graph has threshold number ≤ 2). This appears to be the same conclusion as obtained in Section 7 from Theorem 5.5.

Acknowledgements

We wish to thank Professor P.L. Hammer for illuminating discussions and suggestions on the subject of this work. T.I.'s stay at the University of

Waterloo was supported by a Canada Council Grant for exchange of Scientists between Japan and Canada. U.P.'s work was supported by an NRC Postdoctoral Fellowship.

References

[1] V. Chvátal and P.L. Hammer, Aggregation of inequalities in integer programming, Annals Discrete Math. 1 (1977) 145–162.

[2] S. Földes and P.L. Hammer, Split graphs, in: F. Hoffman et al., eds., Proceedings of the Eighth Southeastern Conference on Combinatorics, Graph Theory and Computing, Louisiana State University, Baton Rouge (1977) 311–315.

[3] P.L. Hammer, T. Ibaraki and U. Peled, Threshold numbers and threshold completions, in: P. Hansen, ed., Studies on Graphs and Discrete Programming (North-Holland, Amsterdam, 1981), in this Volume.

[4] P.L. Hammer, T. Ibaraki and B. Simeone, Threshold sequences, SIAM J. Algebraic and Discrete Methods 2 (1981) 39–49.

[5] P.L. Hammer and B. Simeone, The splittance of a graph, Department of Combinatorics and Optimization, University of Waterloo, CORR 77-39, 1976; forthcoming in Canad. J. Math. under the title 'The blur of a graph'.

[6] U.N. Peled, Matroidal graphs, Discrete Math. 20 (1977) 263–286.

P. Hansen, ed., Studies on Graphs and Discrete Programming

OPTIMUM SYNTHESIS OF A NETWORK WITH NON-SIMULTANEOUS MULTICOMMODITY FLOW REQUIREMENTS*

M. MINOUX

Centre National d'Etudes des Télécommunications, Paris, France

The problem of designing a minimum cost network meeting multicommodity requirements *under security constraints* is a fundamental problem in the area of telecommunication network management and planning. It may be viewed as a generalization of the well-known optimum single-commodity network synthesis problem formulated and solved by Gomory and Hu.

Based on a generalization of the max-flow min-cut theorem, an algorithm to solve the multicommodity network synthesis problem is described. Convergence is proved and problems of implementation are discussed. Preliminary computational results are reported.

1. Introduction

The following well-known optimum network flow synthesis problem has been solved in 1962 by Gomory and Hu [4]: given a N-nodes M-edges connected graph $G=[X, U]$ and a collection of p single commodity flows $\psi^1, \psi^2, \ldots, \psi^p$ of known prescribed values $v^1, v^2, \ldots, v^p$ determine an M-vector $Y=(Y_i)_{i\in U}$ of capacities associated with the edges, meeting any one flow requirement, and minimizing the cost function $\sum_{u\in U} \gamma_u \cdot Y_u$, where γ_u is the unit capacity cost on arc $u \in U$.

However, such a model is not well-suited to many practical situations.

Our paper actually deals with a key-problem in telecommunication network design where p non-simultaneous *multicommodity flows* are to be considered, instead of single commodity flows. The standard node-arc formulation of the problem [11] commonly leads to very large scale linear programs of thousands of variables and constraints.

The purpose of this paper is to describe the basic elements of an algorithm for the practical solution of this problem, exploiting its structure (a two-level block angular structure in the terminology of Ladson [7]) and combining some theoretical results concerning multicommodity flows with the generalized linear

* Invited paper, presented at the workshop 'Applications of graph theory and combinatorics to management' European Institute for Advanced Studies in Management, Brussels March 20–21, 1979.

programming technique [2]. This algorithm may be viewed as a two-level decomposition procedure, closely related to the Benders partitioning methodology [1].

The algorithm has been implemented and tested on a number of real network optimization problems, and is now currently in use as the core of a general software tool for network planning applications. A number of typical computational results, illustrating the efficiency of the method, are presented at the end of the paper, and problems of implementation are discussed.

2. Problem statement and formulation

We consider a N nodes M edges unoriented connected graph $G=[X, U]$.

The problem is to determine an M-vector of capacities $Y=[Y_u]$ assigned to the various edges $u \in U$ such that:

(1) graph G together with the capacities Y_u $(u \in U)$ is feasible for any one of p multicommodity flow requirements $\psi^1, \ldots, \psi^p$;

(2) total cost $\sum_{u \in U} \gamma_u \cdot Y_u$ is minimized (γ_u is the unit capacity cost on edge u).

For each $r=1, 2, \ldots, p$, the multicommodity flow ψ^r is defined by K_r single commodity flows: $\psi^{r,1}, \psi^{r,2}, \ldots, \psi^{r,K_r}$. Every $\psi^{r,k}$ $(k=1, \ldots, K_r)$ satisfies the conservation equations:

$$A \cdot \psi^{r,k} = b^{r,k}$$

where:

(1) A is the node-arc incidence matrix of the directed graph G' obtained from G by replacing each edge $u=(i, j)$ by two arcs $u^+=(i, j)$ and $u^-=(j, i)$;

(2) $b^{r,k}$ is a N-vector the ith component of which is either the amount of flow entering i, with a $+$ sign (if i is a source), or the amount of flow leaving i, with a $-$ sign (if i is a sink);

(3) all the components of $\psi^{r,k}$ are non-negative.

Note that the above model could easily be extended to the situation where each $\psi^{r,k}$ is supported by a subgraph or a partial graph $G^{r,k}$ instead of G itself. The only change consists in replacing A by $A^{r,k}$, the node-arc incidence matrix of the directed graph deduced from $G^{r,k}$.

Multicommodity flow ψ^r is then defined (in node-arc formulation) as an M-vector $(\psi^r_u)_{u \in U}$ satisfying the following set of equations:

$$\text{(I)} \qquad \begin{aligned} & A \cdot \psi^{r,k} = b^{r,k} \quad (k=1, 2, \ldots, K_r), \\ & \psi^r_u = \sum_{k=1}^{K_r} \psi^{r,k}_{u+} + \psi^{r,k}_{u-} \quad (u=1, 2, \ldots, M), \\ & \psi^{r,k} \geq 0 \quad (k=1, 2, \ldots, K_r). \end{aligned}$$

It follows from (I) that the set of all possible ψ^r is a *convex* (generally unbounded) *polytope* of $\mathbb{R}^{M+}$, which will be noted $\mathcal{D}^r$.

In other words, given an M-vector $Y=(Y_u)$ of capacities assigned to the edges $u \in U$ of G, there exists a feasible multicommodity flow ψ^r if and only if $Y \in \mathcal{D}^r$.

It is worth while noting that the number K_r of single commodity flows forming ψ^r may be always be made less or equal to N (for each $i \in Y$, all the flows $Q^{r,k}$ having the same source i may be bunched into a unique flow originating at i). Thus, the linear system defining each ψ^r consists in about $2\,M \cdot N$ variables and N^2 constraints.

Now, the problem to be solved is the following: determine $Y=(Y_u)$ such that:

$$\text{(P)} \qquad \min \quad \gamma \cdot Y = \sum_{u \in U} \gamma_u \cdot Y_u,$$

$$\text{s.t.} \quad Y \in \mathcal{D}^r \quad (r=1,2,\ldots,p).$$

This is a very general model which arises whenever a network, meeting multicommodity requirements (a communication network for instance), is to be designed at *minimum cost* under *security constraints.*

The various constitutive elements of the network (nodes or links) being subject to failures, the network has to be dimensioned in order to be able to route all (or a minimum imposed fraction of) the requirements in each of the most likely failure cases. Very often, the probability of simultaneous failures of two or more elements can be reasonably neglected, and it is sufficient to take into account the 'elementary' failures (those affecting a single element, node or link). The model, however, is very general, and readily allows to take into account non-elementary failures. For a more detailed exposition of the problem and its mathematical modelling, see [8, 9, 10].

Problem (P) may be viewed as a generalization, to the multicommodity case, of a problem formulated and solved by Gomory and Hu [4] for single-commodity flows: the optimal synthesis problem of a network with non simultaneous single commodity flow requirements. The solution method proposed by Gomory and Hu was based on generalized linear programming and made extensive use of the max-flow min-cut theorem.

In the multicommodity case, which is considered here, the difficulty is twofold:

(1) the max-flow min-cut theorem is no longer applicable;

(2) the size of the problem grows very rapidly: even moderate sized examples lead to very large scale linear programs (64 000 variables and 16 000 constraints in node-arc formulation for $N=20$ nodes, $M=40$ edges, $p=40$); these cannot be treated by standard linear programming techniques.

However, we note that problem (P) displays a quite special structure, namely a *two-level block angular structure* (in the terminology of Lasdon (7]): at the higher level, Y are the coupling variables, and the constraints decompose into p blocks each block r corresponding to the equations defining multicommodity flow r; at the lower level, each multicommodity block r itself decomposes into subblocks corresponding to the various constitutive single commodity flows, and a number of (coupling) capacity constraints. Not surprisingly, the solution method suggested below leads to a *two-level decomposition algorithm:* at the higher level, the coupling variables Y are dealt with in the same way as in Benders' partitioning methodology [1] via a constraint generation scheme, allowing separate solving of the r multicommodity flow problems; at the lower level, each multicommodity feasible flow problem, in turn, is solved via decomposition.

3. Necessary and sufficient condition for a feasible multicommodity flow. Another formulation of the problem

Since the constraints of problem (P) express feasibility conditions for each ψ^r $(r=1,\ldots,p)$ with respect to capacities Y_u $(u\in U)$ it is natural to look for a simple expression of $Y\in\mathcal{D}^r$.

As the following applies to any one of the ψ^r, the r superscript will be temporarily omitted for the sake of simplicity.

We assume that multicommodity flow ψ is composed of Q single commodity-flows, and that, for $q=1,2,\ldots,Q$, $s(q)\in X$ is the source, $t(q)\in X$, the sink, and d^q is the required flow value.

We now recall the edge-chain formulation of the feasible multicommodity flow problem [3, 11]. For each $q=1,\ldots,Q$, let us consider the set of all elementary chains between $s(q)$ and $t(q)$ in G: $L_1^q, L_2^q,\ldots,L_{\gamma(q)}^q$ (where $Y(q)$ is the total number of such chains). Let x_j^q denote the fraction of flow q on the L_j^q chain; and P_j^q, the M-vector, the uth component of which is 1 if $u\in L_j^q$, otherwise 0. Then, a feasible multicommodity flow with respect to capacities $Y=(Y_u)$ exists if and only if the following linear system has a solution:

$$\text{(II)}\qquad \begin{aligned} &\sum_{q=1}^{Q}\sum_{j=1}^{\nu(q)} P_j^q x_j^q \leqslant Y,\\ &\sum_{j=1}^{\nu(q)} x_j^q = d^q \quad (q=1,\ldots,Q),\\ &x_j^q \geqslant 0.\end{aligned}$$

Equivalently, the problem may be transformed into the following linear optimization problem:

$$
\text{(III)}\quad
\begin{aligned}
&\text{Min} \quad \eta,\\
&\text{s.t.} \quad \sum_{q=1}^{Q}\sum_{j=1}^{\nu(q)} P_j^q \cdot x_j^q - \eta \cdot \mathbb{1} \leqslant Y, && (1)\\
&\qquad \sum_{j=1}^{\nu(q)} x_j^q = d^q, && (2)\\
&\qquad x_j^q \geqslant 0,
\end{aligned}
$$

where $\mathbb{1}$ is the M-vector with all components equal to 1, and η a real number.

Problem (III) may be interpreted as follows: if η is added to all capacities, and η is large enough, then a feasible multicommodity flow will always exist.

In (III) the minimum value η^* of η is looked for. If $\eta^* \leqslant 0$, then (II) has a solution and the problem is feasible. If $\eta^* > 0$, there is no solution.

Now, associate with the constraints (1) of (III) a row-vector $\pi = (\pi_u)_{u \in U}$ and with constraints (2) a row-vector $\mu = (\mu^q)_{q=1,\ldots,Q}$ of *dual variables*. Let $d = (d^1, d^2, \ldots, d^Q)^T$. The problem, dual of (III) is:

$$
\text{(D)}\quad
\begin{aligned}
&\text{Max} \quad -\pi \cdot Y + \mu \cdot d,\\
&\text{s.t.} \quad -\pi \cdot P_j^q + \mu^q \leqslant 0 \quad (q = 1, \ldots, Q;\; j = 1, \ldots, \nu(q)), && (3)\\
&\qquad \pi \cdot \mathbb{1} = 1, \qquad \pi \geqslant 0.
\end{aligned}
$$

Since $d^q \geqslant 0$, it is easily seen that, at the optimum (π, μ), μ^q must take the largest possible value:

$$\mu^q = \underset{j=1,\ldots,\nu(q)}{\text{Min}} \{\pi \cdot P_j^q\}$$

which is nothing but the length of the shortest chain between $s(q)$ and $t(q)$ with respect to the lengths π_u on the edges $u \in U$. It follows that a feasible multicommodity flow will exist if and only if, for every $\pi \in \mathbb{R}^{M+}$ (in particular, for the optimum of (D)):

$$-\pi \cdot Y + \mu \cdot d \leqslant 0$$

with μ^q the length of the shortest chain between $s(q)$ and $t(q)$ for the π_u; and $\sum_{u \in U} \pi_u = 1$, $\pi \geqslant 0$.

Observe that this condition may be considered as a generalization of the well-known max-flow–min-cut theorem (to see this, consider the special case of a single commodity flow and take as π vectors the 0–1 characteristic vectors of cuts separating sources and sinks). It is important to note that, in order to test this condition, it is sufficient to restrict to those π vectors corresponding to extreme points of the dual polyhedron (D), the number of which is *finite* (though usually very large).

Making use of this condition, we are now in a position to give another simple representation of the $\mathcal{D}$ polytope, associated with multicommodity flow ψ.

Suppose that we enumerate all the extreme points of the constraint polyhedron of the dual (D). These correspond to vectors: $\pi^1, \pi^2, \ldots, \pi^\alpha$. For each π^i, let

$$\theta^i = \mu \cdot d = \sum_{q=1}^{Q} \mu^q d^q$$

(with μ^q the shortest chain between $s(q)$ and $t(q)$ with respect to lengths π^i_u).

We conclude that $Y \in \mathcal{D}$ if and only if Y satisfies the set of inequalities:

$$\text{(IV)} \qquad \begin{aligned} &\pi^1 \cdot Y \geq \theta^1, \\ &\pi^2 \cdot Y \geq \theta^2, \\ &\ldots \\ &\pi^\alpha \cdot Y \geq \theta^\alpha, \\ &Y \geq 0. \end{aligned}$$

This suggests another formulation of the (P) problem of Section 2 where p multicommodity flows were considered:

$$\begin{aligned} &\text{Min} \quad \gamma \cdot Y, \\ &\text{s.t.} \quad Y \in \mathcal{D}^r \quad (r = 1, \ldots, p). \end{aligned}$$

To express the condition $Y \in \mathcal{D}^r$, we write down system (IV) relative to multicommodity flow ψ^r:

$$\begin{aligned} &\pi^{r,1} \cdot Y \geq \theta^{r,1}, \\ &\ldots \\ &\pi^{r,\alpha_r} \cdot Y \geq \theta^{r,\alpha_r}, \\ &Y \geq 0. \end{aligned}$$

Problem (P) is thus equivalent to:

$$\text{(P1)} \qquad \begin{aligned} &\text{Minimize} \quad \gamma \cdot Y, \\ &\text{subject to} \quad \pi^{r,j} \cdot Y \geq \theta^{r,j} \quad (r = 1, \ldots, p; \;\; j = 1, \ldots, \alpha_r), \\ &\qquad\qquad\quad Y \geq 0. \end{aligned}$$

4. An algorithm using a constraint generation technique

Except for very small sized problems, (P1) cannot be written explicitly, due to the enormous number of constraints.

4.1. *Description of the algorithm*

A natural way of circumventing the difficulty, consists in using an iterative procedure, in which at each step k:

(1) a restricted problem (PR_k) consisting of a subset of constraints of (P1) is solved;

(2) depending on the solutions obtained, a number of constraints are added to (PR_k) thus providing a new restricted problem (PR_{k+1}).

More precisely, let us see how the new constraints are generated, at step k, when Y^k, the solution of (PR_k), has been computed. A possible method consists in successively testing feasibility of each multicommodity flow ψ^r with respect to capacities Y^k. This is done by solving the corresponding problem (III).

Optimization of (III) can be performed either exactly (using linear programming with a column generation scheme, see [3] for instance), or approximately with a subgradient algorithm (cf. [4, Chapter 6]).

Let $(\bar{\pi}, \bar{\mu})$ be the optimal (or suboptimal) dual variables obtained. If the necessary and sufficient condition for a feasible multicommodity flow:

$$\bar{\pi} \cdot Y \geqslant \bar{\mu} \cdot d$$

is satisfied, we proceed to examine the next multicommodity flow ψ^{r+1}. If $\bar{\pi} \cdot Y < \bar{\mu} \cdot d$, then the constraint $\bar{\pi} \cdot Y \geqslant \bar{\mu} \cdot d$ is added to the current restricted problem. Afterwards, we turn to testing feasibility of ψ^{r+1}.

Once, the p multicommodity flows have been examined, and if new constraints have been added, step $k+1$ is started by solving the new restricted problem (PR_{k+1}).

If no constraint has been generated, this means that the current solution Y^k satisfies all the constraints of (P1) and clearly an optimal solution is obtained.

4.2. *Convergence*

In order to prove finite convergence, it is necessary to assume that the feasibility problem (III) is solved exactly (using linear programming and column generation for instance). Under this assumption, at each step and for each multicommodity ψ^r, the generated constraint $\bar{\pi} \cdot Y \geqslant \bar{\mu} \cdot d$ corresponds to an extreme point of the constraint polytope of the dual (D); moreover this extreme point is necessarily different from those previously generated. Finite convergence is thus obtained.

In practice, it should be noted that the convergence will seldom be achieved, either because of round-off errors, or because the time limits will be reached before. Nevertheless, when the algorithm stops (as step k, say) the cost $\gamma \cdot Y^k$ of the current solution gives a lower bound of the optimal cost $\gamma \cdot Y^*$.

Moreover, if $\bar{\eta}$ in the largest of the objectives obtained when solving the p feasibility problems (III) at step k, a solution of (P) is readily obtained by adding $\bar{\eta}$ to all capacities Y_u^k $(u=1,\ldots,M)$. This provides an upper bound $\gamma\cdot(Y^k+\bar{\eta}\cdot\mathbb{1})$ of the optimal cost, and a double sided inequality:

$$\gamma\cdot Y^k\leq\gamma\cdot Y^*\leq\gamma\cdot(Y^k+\bar{\eta}\cdot\mathbb{1}).$$

If $\bar{\eta}$ is small enough, it is seen that good approximate solutions are obtained.

4.3. *Implementation and results*

For an efficient implementation of the algorithm, it is better to solve the dual (DR_k) of the restricted problem (PR_k) instead of (PR_k) itself (since the latter may have much more constraints than variables). The algorithm then turns out to be a *column generation technique* for the dual of (P1).

A crucial point is the initialization of the algorithm, in other words, the determination of the starting restricted program (PR_0).

One method suggested in [8] consists in using cut conditions (the π vectors being 0–1 characteristic vectors of cuts in G). For example, the elementary cocycles $\omega(i)$, consisting in all the edges incident to $i\in X$, can be used. Another similar idea is to consider, for each multicommodity flow, the set of elementary cocycles associated with the minimum spanning tree of the graph of point-to-point requirements. Computational experience, however, leads to the conclusion that these heuristic rules usually don't provide good starting solutions, and thus result in slow overall convergence of the algorithm.

Recent work has shown that a better approach is to initialize the restricted problem with the p constraints (one for each ψ^r) obtained by Lagrangian relaxation of the coupling constraints in (P) and optimization of the dual problem with a subgradient algorithm. This procedure has two main advantages:

(1) it allows decomposition of problem (P) into p subproblems—one for each multicommodity flow—hence its efficiency in solving the dual problem;

(2) it generally provides tight lower bounds and starting solutions very close to the optimum (for more details about the initializing procedure, please refer to Minoux–Serrault [10].

Presently, the algorithm, based on the above principles, can be used to solve average-sized problems ($N=20$, $M=40$, $p=40$ say) in about 10 minutes of CPU with an accuracy of about 5–10%. As an illustration, Table 1 shows typical results obtained on some 12 nodes 25 edges multicommodity network synthesis problems with $p=25$ nonsimultaneous multicommodity requirements. It may be observed that the number of necessary main iterations is usually very small—5 to 10—and that the total number of constraints in the restricted problems (PR) at the end of the algorithm is quite moderate (since

Table 1
Typical results obtained on various 12 nodes 25 edges and 25 multicommodity flows test problems

	Number of iterations	Total number of constraints in (PR)	Lower bound	Upper bound	Max. relat. difference with optimum value
Problem 1	6	58	35.2	36.9	5%
Problem 2	4	83	31.5	31.9	13%
Problem 3	4	72	44.9	49.6	9%
Problem 4	4	76	47.5	55.8	17%
Problem 5	4	69	37.3	41.7	12%
Problem 6	4	71	34.6	37.6	9%

$p = 25$, the number of constraints in the starting restricted problems is always 25). Moreover, it can be seen that the accuracy of the solutions (the maximum relative difference with the true—unknown—optimum value) derived from the lower and upper bounds obtained, is generally quite satisfactory for the needs of most practical applications.

The algorithm has also been tested on a number of real network optimization problems, and is now currently in use as the core of a general software tool for network planning purposes.

References

[1] J.F. Benders, Partitioning procedures for solving mixed variables programming problems, Numer. Math. 4 (1962) 238–252.
[2] G.B. Dantzig, Linear Programming and Extensions (Princeton University Press, Princeton, NJ, 1963).
[3] L.R. Ford and D.R. Fulkerson, A suggested computation for maximal multicommodity network flows, Management Sci. 5 (1) (1958) 97–101.
[4] R.E. Gomory and T.C. Hu, An application of generalized linear programming to network flows, SIAM J. Appl. Math. 10 (2) (1962) 260–283.
[5] M. Gondran and M. Minoux, Graphes et Algorithmes (Eyrolles, France, 1979).
[6] J.L. Kennington, (1978) A survey of linear cost multicommodity network flows, Oper. Res. 26 (2) (1978) 209–236.
[7] L.S. Lasdon, Optimization theory for large systems, Macmillan Series for Operations Research (1970).
[8] M. Minoux, La synthèse des réseaux de télécommunication avec contraintes de sécurité: une approche théorique, Note technique ITD/CES/68 du Centre National d'Etudes des Télécommunications, France (1972).
[9] M. Minoux, Flots équilibrés et flots avec sécurité, Bulletin Dir. Et. Recherches E.D.F. no. 1, (1976) 5–16.
[10] M. Minoux and J.Y. Serreault, Synthèse optimale d'un réseau de télécommunications avec contraintes de sécurité, Annales de Télécommunications 36 (3, 4) (1981).
[11] J.A. Tomlin, Minimum cost multicommodity network flows, Oper. Res. 14 (1966) 45–51.

P. Hansen, ed., Studies on Graphs and Discrete Programming
© North-Holland Publishing Company (1981) 279–301

MAXIMIZING SUBMODULAR SET FUNCTIONS: FORMULATIONS AND ANALYSIS OF ALGORITHMS*

G.L. NEMHAUSER
School of Operations Research and Industrial Engineering, Cornell University, Ithaca, NY, USA

L.A. WOLSEY
CORE, Université Catholique de Louvain, Louvain-la-Neuve, Belgium

We consider integer programming formulations of problems that involve the maximization of submodular functions. A location problem and a 0–1 quadratic program are well-known special cases. We give a constraint generation algorithm and a branch-and-bound algorithm that uses linear programming relaxations. These algorithms are familiar ones except for their particular selections of starting constraints, subproblems and partitioning rules. The algorithms use greedy heuristics to produce feasible solutions, which, in turn, are used to generate upper bounds. The novel features of the algorithms are the performance guarantees they provide on the ratio of lower to upper bounds on the optimal value.

1. Introduction

The following four problems are fundamental to the practical solution of integer linear programming problems.

(a) *Formulation of the model.* There are, in many instances, a wide variety of different linear constraints that can, together with integrality conditions, be used to represent the same set of points of a (mixed) integer linear program. Generally these equivalent integer programming formulations give different linear programs when the integrality constraints are suppressed (see [18] for several examples). Solving these linear programs is an essential part of the fundamental algorithms (branch-and-bound and cutting plane) of integer programming. Both the size of the linear programs and the quality of the bounds they produce determine the success of an integer programming algorithm. Thus it would be highly desirable to have systematic ways of producing different formulations and to have criteria for comparing them.

(b) *Selection of an initial set of constraints.* There are integer programs whose formulations require a very large number of linear constraints. A particular example is the well-known formulation of the traveling salesman

* This work has been supported by a NATO Systems Science special research and U.S. National Science Foundation grant to Cornell University

problem, which requires an exponential number of constraints to eliminate subtours [6]; a general class of these problems is the formulation obtained by applying Benders' decomposition [2] to mixed integer linear programs. In these instances a general algorithmic approach is to begin with a relaxation that contains only a relatively small number of linear equalities and to generate others only if they are found to be violated in the course of the algorithm. One supposedly tries to choose the initial constraints on the basis of their importance or significance in defining the feasible region in a neighborhood of an optimal solution. It would be desirable to begin with a set of constraints having the property that the optimal value to the relaxed problem (at some stage) was guaranteed to be no larger than a fixed multiple of the optimal value of the original problem.

(c) *Decisions in a branch-and-bound algorithm.* It has been widely observed that the criterion for choosing the order in which subproblems will be solved can have a substantial effect on the running time of a branch-and-bound algorithm. In fact, most commercial codes use a combination of rules in order to balance the objectives of obtaining feasible solutions quickly and minimizing the amount of enumeration [12]. Another important decision in a branch-and-bound algorithm is the type of partition and the rule for choosing a variable (or set of variables) at each node of the tree to form the partition.

It would be desirable to have criteria for choosing subproblems and partitions that would, with a given amount of computation, guarantee a feasible solution whose value is a specified fraction of the upper bound.

(d) *Finding good feasible solutions.* Since an integer programming algorithm may have to be terminated before it completes the solution of a problem, an important feature of an algorithm is its capability of producing good feasible solutions. Although feasible solutions can be generated in a distinct heuristic phase, a highly desirable feature is an optimizing algorithm that yields feasible solutions (hopefully of increasing value) as intermediate results.

There is almost no mathematical theory that addresses these problems for general integer programs. It seems that some structure must be imposed even to obtain partial answers. In the few instances where there is enough structure to formulate the integer program directly with a set of linear constraints whose extreme points are the feasible integer solutions (e.g., network flows and matching) the formulation question is solved. In the case of network flows this formulation is compact and there are efficient algorithms so that questions (b), (c) and (d) are irrelevant. In the case of matching the number of linear constraints required is exponential in the number of variables. But questions (c) and (d) are still irrelevant since there are efficient algorithms that begin with a small number of constraints and then generate only a small number of the remaining ones.

For most integer programs all four of the problems are relevant and current practice is to cope with them largely by intuition and experience. Our intention is to provide some theoretical and possibly computationally useful answers to these problems for a very limited but nontrivial class of combinatorial optimization problems. This class contains some problems of practical interest. Furthermore our results suggest potentially useful algorithms for a much larger class of problems.

Our approach will be to design algorithms that with a specified amount of computation give both a lower bound (from a feasible solution) and an upper bound on the optimal value. These bounds have the property that the lower bound is at least a certain fraction of the upper bound. They will be obtained by giving appropriate initial sets of constraints and by specifying decision rules for a branch-and-bound algorithm. We will also show how the structure produces alternate integer linear programming formulations.

The structure that we require is submodularity. We will introduce this structure in terms of a practical and well-known integer programming problem. This problem involves the location of K facilities to maximize the profit from supplying a commodity to m clients. The set of available locations is $N=\{1,\ldots,n\}$ and $K<n$. There is no limit on the number of clients that can be supplied from a given facility; $c_{ij}\geqslant 0$ is the profit obtained by supplying client i from a facility at location j. Thus a particular problem is specified by an $m\times n$ nonnegative matrix $C=\{c_{ij}\}$ and a positive integer K.

A standard mixed integer linear programming formulation of this problem is

$$
\begin{aligned}
V=\max \quad & \sum_{i=1}^{m}\sum_{j=1}^{n} c_{ij}x_{ij},\\
& \sum_{j=1}^{n} x_{ij}=1, \quad i=1,\ldots,m,\\
& x_{ij}-y_j\leqslant 0, \quad i=1,\ldots,m, \quad j=1,\ldots,n, \qquad\qquad (1.1)\\
& \sum_{j=1}^{n} y_j=K,\\
& x_{ij}\geqslant 0, \quad i=1,\ldots,m, \quad j=1,\ldots,n\\
& y_j\in\{0,1\}, \quad j=1,\ldots,n
\end{aligned}
$$

where $y_j=1$ means that a facility is placed at location j.

The y_j's are the strategic variables since given an $S\subseteq N$ and its characteristic vector y^S ($y_j^S=1$, $j\in S$ and $y_j^S=0$, otherwise) an optimal set of x_{ij}'s is given by

$$
x_{ij}^S=\begin{cases} 1 & \text{for some } j \text{ such that } c_{ij}=\max_{k\in S} c_{ik},\\ 0 & \text{otherwise,}\end{cases} \qquad i=1,\ldots,m.
$$

The value of a solution (x^S, y^S) is

$$v(S)=\sum_{i=1}^{m}\sum_{j=1}^{n} c_{ij}x_{ij}^S=\sum_{i=1}^{m}\max_{j\in S} c_{ij}.$$

We can therefore restate the problem as

$$V=\max(v(S)\colon S\subseteq N, |S|=K). \tag{1.2}$$

With $v(\emptyset)=0$, it is well-known and easy to see that the set function $v(S)$ satisfies for all $S, T\subseteq N$

$$v(S)+v(T)\geqslant v(S\cup T)+v(S\cap T). \tag{1.3}$$

Set functions that satisfy (1.3) are called *submodular*. In addition, $v(S)$ satisfies for all $S\subset N$ and $j\notin S$

$$v(S\cup\{j\})-v(S)\geqslant 0. \tag{1.4}$$

Set functions that satisfy (1.4) are called *nondecreasing*.

Thus a natural generalization of this K-location problem is problem $\mathcal{P}_K$ given by

$$Z=\max(z(S)\colon S\subseteq N, |S|=K, \quad z \text{ submodular and nondecreasing}, z(\emptyset)=0). \tag{1.5}$$

Problem family $\mathcal{P}_K$ is the class of problems for which we obtain theoretical results.

In [3] it was shown that a simple greedy heuristic for problem (1.1) produces a value V^G that satisfies for all nonnegative $m\times n$ matrices C and all m and n

$$\frac{V^G}{V^{LP}}\geqslant 1-\left(\frac{K-1}{K}\right)^K\geqslant\frac{e-1}{e}\cong 0.63 \tag{1.6}$$

where V^{LP} is the value of the linear programming relaxation of (1.1) obtained by replacing $y_j\in\{0,1\}$ by $y_j\geqslant 0$ and e is the base of the natural logarithm. Since $V^G\leqslant V\leqslant V^{LP}$, (1.6) implies

$$\frac{V^G}{V}\geqslant 1-\left(\frac{K-1}{K}\right)^K. \tag{1.7}$$

It was shown in [17] that (1.7) generalized to the problem $\mathcal{P}_K$, in other words to all nondecreasing submodular functions. A slightly more general result of [17] is that partial enumeration of all solutions of cardinality q together with a greedy choice of the remaining $K-q$ elements yields for problem (1.5)

$$\frac{Z^{G(q)}}{Z}\geqslant 1-\left(\frac{K-q}{K}\right)\left(\frac{K-q-1}{K-q}\right)^{K-q}, \tag{1.8}$$

where $Z^{G(q)}$ is the value of the q-enumeration plus greedy solution. Note that putting $q=0$ (the greedy algorithm) in (1.8) yields (1.7).

Our interest in the q-enumeration plus greedy family of algorithms is that in a certain sense they are an optimal family of heuristics for $\mathcal{P}_K$. For fixed K, the q-enumeration plus greedy algorithm requires $O(n^{q+1})$ values[1] of the function z. In [16] we showed that any other algorithm for $\mathcal{P}_K$ that could give a performance guarantee (bound) larger than the right-hand side of (1.8) would require at least $O(n^{q+2})$ values of the function z.

In contrast with this prior work on heuristics, here we will be concerned with exact algorithms for solving $\mathcal{P}_K$ and particular cases of it. Nevertheless, the results on heuristics will be relevant since we will adopt a point of view motivated by the questions raised at the beginning of this section.

In Section 2 we give a linearization of nondecreasing, submodular functions the leads immediately to an integer linear programming formulation of $\mathcal{P}_K$ containing a large number of constraints. In Section 3 we propose a constraint generation algorithm for this formulation, which is similar to Benders' algorithm [2]. Section 4 gives a branch-and-bound algorithm that uses linear programming relaxations. These algorithms are familiar ones except for their particular selections of starting constraints, subproblems, and partitioning rules. The algorithms use heuristics to produce feasible solutions, which, in turn, are used to generate upper bounds. The novel features of the algorithms are the performance guarantee they provide on the ratio of lower to upper bounds on the optimal value.

In Section 5 we consider the maximization of a general submodular function subject to linear constraints. These are the problems for which the earlier results may have computational significance as the algorithms extend rather naturally to this larger class. Section 6 addresses the question of simplified linearizations for cases in which the submodular function has additional structure. We apply these ideas to the K-location problem and the minimization of some quadratic 0–1 programs.

2. Linearization of nondecreasing submodular functions and an integer programming formulation

Let $\rho_j(S)=z(S\cup\{j\})-z(S)$, $\forall S\subset N$ and $j\in N$.

Proposition 1 [17]. *A real-valued function z on the subsets of N is submodular*

[1] We use $O(n^r)$ for the family of functions that is bounded below by c_1n^r and above by c_2n^r for some $0<c_1<c_2$.

and nondecreasing if and only if

(a) $\rho_j(S) \geq \rho_j(T) \geq 0$, $\forall S \subset T \subset N$ *and* $j \in N-T$, *or*

(b) $z(T) \leq z(S) + \sum_{j\in T-S} \rho_j(S)$, $\forall S, T \subseteq N$.

Recall that y^U is the characteristic vector of $U \subseteq N$. Consider the set

$$X = \left\{(\eta, y) \colon \eta \leq z(S) + \sum_{j\in N-S} \rho_j(S) y_j, \forall S \subseteq N, y_j \in \{0, 1\}, j \in N\right\}.$$

Lemma 1. *If z is submodular and nondecreasing, then* $(\xi, y^U) \in X$ *if and only if* $\xi \leq z(U)$.

Proof. Suppose $\xi \leq z(U)$. Then for all $S \subseteq N$

$$z(S) + \sum_{j\in N-S} \rho_j(S) y_j^U = z(S) + \sum_{j\in U-S} \rho_j(S) \geq z(U) \geq \xi,$$

where the first inequality follows from Proposition 1(b). Conversely $(\xi, y^U) \in X$ implies in particular that

$$\xi \leq z(U) + \sum_{j\in N-U} \rho_j(U) y_j^U = z(U).$$

Consider problem (1.5) and the linear integer program

$$\begin{aligned} \max \quad & \eta \\ & \eta \leq z(S) + \sum_{j\in N-S} \rho_j(S) y_j, \quad \forall S \subseteq N, \\ & \sum_{j\in N} y_j = K, \\ & y_j \in \{0, 1\}, \quad j \in N. \end{aligned} \tag{2.1}$$

Problem (2.1) is a reformulation of (1.5) since Lemma 1 implies:

Theorem 1. $(\eta, y) = (z(U), y^U)$ *is an optimal solution to* (2.1) *if and only if U is an optimal solution to* (1.5).

Note that although the inequalities of (2.1) are valid for all $S \subseteq N$, they are needed only for $|S| = K$. However, as we shall see later, there may be computational advantages in using some of the inequalities corresponding to sets of cardinality smaller than K.

When the particular class of submodular functions v obtained from the K-location problem are used in (2.1) we obtain an integer linear programming formulation of the location problem. Although we will not give the details here, a surprising observation is that this integer linear program is precisely the integer program obtained by applying Benders' decomposition (with natural choice of dual extreme points) to the mixed integer linear programming formulation (1.1), see [10, 15]. The results of the next section will suggest how we might choose an initial set of constraints if we are going to solve the K-location problem by Benders' algorithm.

3. A constraint generation algorithm

First we describe an algorithm for problem (2.1), and then we analyze its behavior for a particular initialization. An example is given in an appendix.

Algorithm 1

Initialization. Let $Q^t = \{R^0, \ldots, R^{t-1}\}$ be a nonempty set of distinct subsets of N. Set $p = t$.

Iteration p

Step 1. Solve the problem

$$\begin{aligned} \eta^p = \max \quad & \eta, \\ & \eta \leqslant z(S) + \sum_{j \in N-S} \rho_j(S) y_j, \quad S \in Q^p, \\ & \sum_{j \in N} y_j = K, \\ & y_j \in \{0, 1\}, \quad j \in N. \end{aligned}$$

Let (η^p, y^{R^p}) be an optimal solution.

Step 2. (a) If $\eta^p = z(R^p)$, terminate. R^p is optimal.

(b) If $\eta^p > z(R^p)$, set $Q^{p+1} = Q^p \cup \{R^p\}$, $p \leftarrow p+1$, and return to Step 1.

The optimality of R^p when Step 2(a) occurs is implied by Lemma 1. To prove finiteness we note that if Step 2(b) occurs, $\eta^p > z(R^p)$ implies $R^p \notin Q^p$. Also, since R^p is feasible the algorithm must terminate after at most $\binom{n}{K}$ iterations of Step 2.

We note that $\eta^t \geqslant \cdots \geqslant \eta^p \geqslant \eta^{p+1} \geqslant \cdots \geqslant Z$ and $Z \geqslant z(S)\ \forall S \in Q^p$, so that the nonincreasing upper bound η^p and the nondecreasing lower bound $\zeta^p = \max_{S \in Q^p} z(S)$ give some measure of the progress of the algorithm.

With an appropriate choice of $Q = Q^t$, a performance guarantee can be obtained relating η^t and ζ^t. The value of the guarantee depends on t. For fixed K, we will give $O(n^q)$ initial constraints for $q = 0, \ldots, K-1$. The set of constraints comes from the q-enumeration plus greedy algorithm.

The q-enumeration plus greedy algorithm $G(q)$

The algorithm has two parts.

(i) *q-enumeration.* Produce a list of all subsets S_h^q of N of cardinality q, $h=1,\ldots,\binom{n}{q}$.

(ii) *greedy.* Do for all sets of the list

Initialization. Let S_h^q be the set chosen from the list, $N_h^q = N - S_h^q$ and $t=q+1$.

Iteration t. Let $\rho_j(S) = z(S\cup\{j\}) - z(S)$. Select $i(t)\in N_h^{t-1}$ for which

$$\rho_{i(t)}(S_h^{t-1}) = \max_{i\in N_h^{t-1}} \rho_i(S_h^{t-1})$$

with ties settled arbitrarily. Set $\rho_{t-1} = \rho_{i(t)}(S_h^{t-1})$. Set $S_h^t = S_h^{t-1}\cup\{i(t)\}$ and $N_h^t = N_h^{t-1} - \{i(t)\}$. If $t=K$ stop with the set S_h^K; otherwise set $t\leftarrow t+1$ and continue.

Solution. Output the best solution found in the $\binom{n}{q}$ passes i.e.,

$$Z^{G(q)} = \max_{h=1,\ldots,\binom{n}{q}} z(S_h^K).$$

Let $F(q)=\{S_h^t\colon t=q,\ldots,K-1;\ h=1,\ldots,\binom{n}{q}\}$ and consider the relaxation of (2.1) given by

$$\begin{aligned} \eta^{G(q)} = \max \quad & \eta, \\ & \eta \leqslant z(S) + \sum_{j\in N-S} \rho_j(S) y_j, \quad S\in F(q), \\ & \sum_{j\in N} y_j = K, \\ & y_j \in \{0,1\}, \quad j\in N. \end{aligned} \tag{3.1}$$

Note that for fixed K problem (3.1) has $O(n^q)$ constraints, each of which is constructed from $O(n)$ values of the function z. Thus the formulation of (3.1) requires $O(n^{q+1})$ values of z.

Theorem 2. *For any solution of the q-enumeration plus greedy algorithm to* $\mathcal{P}_K$,

$$\frac{Z^{G(q)}}{\eta^{G(q)}} \geqslant 1 - \left(\frac{K-q}{K}\right)\left(\frac{K-q-1}{K-q}\right)^{K-q}.$$

Proof. Let $(\eta^{G(q)}, y^T)$ be an optimal solution to (3.1). Applying q steps of the greedy algorithm to the set T, we obtain a set $S_{h^*}^q = \{i_1,\ldots,i_q\}$ for which the submodularity of z implies

$$\rho_j(S_{h^*}^q) \leqslant z(S_{h^*}^q)/q, \quad \forall j\in T - S_{h^*}^q. \tag{3.2}$$

Now as $S^q_{h^*}\in F(q)$ and $(\eta^{G(q)}, y^T)$ is feasible to (3.1), it follows that

$$\begin{aligned}\eta^{G(q)} &\leqslant z(S^q_{h^*})+\sum_{j\in N-S^q_{h^*}}\rho_j(S^q_{h^*})y_j^T\\ &\leqslant z(S^q_{h^*})+(K-q)\max_{j\in T-S^q_{h^*}}\rho_j(S^q_{h^*})\\ &\leqslant z(S^q_{h^*})+\frac{K-q}{q}z(S^q_{h^*})\\ &=\frac{K}{q}z(S^q_{h^*})\end{aligned}\tag{3.3}$$

where the middle inequality follows from $|T-S^q_{h^*}|=K-q$ and the last one from (3.2).

Consider the $K-q$ constraints of (3.1)

$$\eta\leqslant z(S^t_{h^*})+\sum_{j\in N-S^t_{h^*}}\rho_j(S^t_{h^*})y_j,\quad t=q,\dots,K-1.$$

Let $\rho_{q+i}=z(S^{q+i+1}_{h^*})-z(S^{q+i}_{h^*})$, $i=0,\dots,K-q-1$. Since $(\eta^{G(q)}, y^T)$ is a feasible solution to (3.1) we obtain

$$\begin{aligned}\eta^{G(q)} &\leqslant z(S^t_{h^*})+\sum_{j\in N-S^t_{h^*}}\rho_j(S^t_{h^*})y_j^T\\ &= z(S^q_{h^*})+\sum_{i=0}^{t-q-1}\rho_{q+i}+\sum_{j\in T-S^t_{h^*}}\rho_j(S^t_{h^*})\\ &\leqslant z(S^q_{h^*})+\sum_{i=0}^{t-q-1}\rho_{q+i}+(K-q)\rho_t,\quad t=q,\dots,K-1\end{aligned}$$

where the last inequality follows from $\rho_t\geqslant\rho_j(S^t_{h^*})$ by the greedy algorithm and $|T-S^t_{h^*}|\leqslant K-q$.

With $\eta^{G(q)}-z(S^q_{h^*})$ normalized to one, the problem of minimizing

$$\frac{z(S^K_{h^*})-z(S^q_{h^*})}{\eta^{G(q)}-z(S^q_{h^*})}=\sum_{i=0}^{K-1-q}\rho_{q+i}$$

subject to

$$\eta^{G(q)}\leqslant z(S^q_{h^*})+\sum_{i=0}^{s-1}\rho_{q+i}+(K-q)\rho_{q+s},\quad s=0,\dots,K-q-1$$

is a linear program with variables $(\rho_q,\dots,\rho_{K-1})$. Using the weak duality theorem of linear programming we obtain a lower bound on its value given by [17, Theorem 4.1]

$$\frac{z(S^K_{h^*})-z(S^q_{h^*})}{\eta^{G(q)}-z(S^q_{h^*})}\geqslant 1-\left(\frac{K-q-1}{K-q}\right)^{K-q}.\tag{3.4}$$

Finally using $Z^{G(q)} \geqslant z(S^K_{h^*})$ and $\eta^{G(q)} \leqslant (K/q) z(S^q_{h^*})$ from (3.3) in (3.4), we obtain

$$\eta^{G(q)} - Z^{G(q)} \leqslant \left(\frac{K-q-1}{K-q}\right)^{K-q} \left(\frac{K-q}{K}\right) \eta^{G(q)}.$$

An important observation about the set $F(q)$ is that it contains subsets of cardinality less than K. Since $\rho_j(S) \geqslant 0$, the algorithm will always produce subsets of size K, even if we replace $\sum_{j \in N} y_j = K$ by $\sum_{j \in N} y_j \leqslant K$. However the small subsets are needed to obtain the bound of Theorem 2.

Let Z be the value of an optimal solution to $\mathcal{P}_K$. Since $Z^{G(q)} \leqslant Z \leqslant \eta^{G(q)}$ we obtain

Corollary 1. *For any solution of the q-enumeration plus greedy algorithm to* $\mathcal{P}_K$,

$$\frac{Z^{G(q)}}{Z} \geqslant 1 - \left(\frac{K-q}{K}\right)\left(\frac{K-q-1}{K-q}\right)^{K-q}.$$

Corollary 2

$$\frac{Z}{\eta^{G(q)}} \geqslant 1 - \left(\frac{K-q}{K}\right)\left(\frac{K-q-1}{K-q}\right)^{K-q}.$$

Note that Corollary 1 is the result (1.8) given in the introduction. Our result on best algorithms [16], which states that the bound of (1.8) cannot be greater for any heuristic requiring $O(n^{q+1})$ values of z, also applies to the result of Theorem 2. In particular, suppose we initialize Algorithm 1 by a collection of constraints generated from $O(n^r)$ values of z. Then if we obtain the bound

$$\frac{\max_{S \in Q^t} z(S) \colon |S| = K}{\eta^t} > 1 - \left(\frac{K-q}{K}\right)\left(\frac{K-q-1}{K-q}\right)^{K-q}$$

for all nondecreasing submodular functions, it follows that $r \geqslant q+2$.

For the K-location problem Algorithm 1 is precisely Benders' algorithm and thus Theorem 2 suggests how we might initialize Benders' algorithm. Although our result on best algorithms for $(\mathcal{P}_k)$ does not apply to the subclass of K-location problems, we do not know of any better initialization in terms of a guarantee on the ratio of the lower to the upper bound generated by Benders' algorithm. Furthermore, the results in [4] imply that, in the worst case, Benders' algorithm will require $O(n^K)$ values of the function v to verify optimality.

To solve the integer program (3.1), we might first solve its linear programming relaxation. For $q=0$ we obtain the linear program

$$Z^{\mathrm{LP(G)}}=\max \quad \eta,$$
$$\eta \leqslant z(S^t)+\sum_{j\in N-S^t} \rho_j(S^t)y_j, \quad t=0,\ldots,K-1,$$
$$\sum_{j\in N} y_j = K,$$
$$0\leqslant y_j \leqslant 1, \quad j\in N$$

where $\{S^t\}_{t=0}^{K-1}$ are the sets generated by the greedy algorithm. Since $Z^{\mathrm{LP(G)}}\geqslant \eta^{G(0)}$ a stronger version of Theorem 2 for $q=0$ is:

Theorem 3. *For any solution of the greedy algorithm to* $\mathcal{P}_K$,

$$\frac{Z^{G(0)}}{Z^{\mathrm{LP(G)}}}\geqslant 1-\left(\frac{K-1}{K}\right)^K.$$

Proof. Since $\rho_t = z(S^t)-z(S^{t-1})\geqslant \rho_j(S^t)\geqslant 0$ for all $j\in N-S^t$, we have that $\sum_{j\in N-S^t}\rho_j(S^t)y_j\leqslant K\rho_t$ for any (real) y satisfying $\sum_{j\in N} y_j = K$. Now from

$$Z^{\mathrm{LP(G)}}\leqslant z(S^t)+\sum_{j\in N-S^t}\rho_j(S^t)y_j,$$

we obtain $Z^{\mathrm{LP(G)}}\leqslant\sum_{i=0}^{t-1}\rho_i+K\rho_t$, $t=0,\ldots,K-1$. The result then follows from the linear programming arguments sketched in the proof of Theorem 2.

For some very special cases, including the K-location problem, a result similar to Theorem 3 has been given in Section 6 of [17].

4. A branch-and-bound algorithm for $\mathcal{P}_K$

Algorithm 1 requires the solution of a sequence of integer programs. Here we propose a branch-and-bound algorithm that uses linear programming relaxations. The basic idea is to enumerate implicitly all subsets of cardinality K. For a given subset of cardinality $k<K$, we use the greedy algorithm to augment it by $K-k$ elements and thus determine a lower bound on the value of all solutions containing the given k-element subset. We also use the $K-k$ elements generated by the greedy algorithm to construct an integer programming relaxation of the formulation of $\mathcal{P}_K$ given in (2.1). We get an upper bound on the value of all solutions containing the k-element subset by solving the linear program obtained by dropping the integrality conditions in the integer program.

A node at level k of the enumeration tree corresponds to a partial solution in which a subset S of cardinality k has been selected, $N^S \subseteq N - S$ remains and $N - S - N^S$ has been discarded. We denote such a partial solution (or node) by the pair (S, N^S).

Algorithm 2

Initialization. The list conists only of the node (ϕ, N). The upper bound for this node is $\bar{Z}^{\phi} = \infty$. The problem lower bound is $\underset{\sim}{Z} = 0$. The incumbent is $I = \phi$.

General Iteration

Step 1. If the list is empty, stop; the incumbent is optimal. Otherwise remove a node (S, N^S) from the top of the list.

Step 2. If $\bar{Z}^S \leq \underset{\sim}{Z}$ return to Step 1. Otherwise let $k = |S|$. Compute a greedy solution consisting of S plus $K - k$ elements from N^S. Let $\underset{\sim}{Z}^S$ denote its value. Suppose the greedy algorithm generates the sets $(S = S^k, \ldots, S^K)$. Compute the linear programming upper bound

$$\begin{aligned} \max \quad & \eta, \\ & \eta \leq z(S^t) + \sum_{j \in N^S - S^t} \rho_j(S^t) y_j, \quad t = k, \ldots, K-1, \\ & \sum_{j \in N^S} y_j = K - k, \\ & 0 \leq y_j \leq 1, \quad j \in N^S. \end{aligned} \tag{4.1}$$

Set $\bar{Z}^S = \max \eta$.

Step 3. If $\underset{\sim}{Z}^S > \underset{\sim}{Z}$, then $\underset{\sim}{Z} \leftarrow \underset{\sim}{Z}^S$ and $I \leftarrow S^K$. If $\underset{\sim}{Z}^S = \bar{Z}^S$ or if $\bar{Z}^S \leq \underset{\sim}{Z}$ go to Step 1, otherwise go to Step 4.

Step 4. Suppose $N^S = \{j_1, \ldots, j_r\}$ and $\rho_{j_1}(S) \geq \cdots \geq \rho_{j_r}(S)$. Add the nodes $(S \cup \{j_t\}, N^S - \{j_1, \ldots, j_t\})$ to the bottom of the list unless

$$|(S \cup \{j_t\}) \cup (N^S - \{j_1, \ldots, j_t\})| < K, \quad t = 1, \ldots, r.$$

Assign each of these nodes an upper bound of $\bar{Z}^S$. Go to Step 1.

An example of Algorithm 2 is given in the Appendix.

Note that after any iteration of the algorithm all nodes on the list are such that $|S| = k$ or $k+1$ for some $k < K$ and that all nodes with $|S| = k$ are removed before any nodes with $|S| = k+1$.

We now consider the situation in which we have completed Step 3 for the last node corresponding to $|S| = q$. At this point we have, for fixed K, evaluated the function z $O(n^{q+1})$ times. Also $\underset{\sim}{Z} = \max_{|S|=q} \underset{\sim}{Z}^S = Z^{G(q)}$, from the implicit application of the q-enumeration plus greedy algorithm. Define $\bar{Z}^{LP(q)} = \max_{|S|=q} \bar{Z}^S$, where $\bar{Z}^S$ is the linear programming bound computed from (4.1).

Theorem 4

$$\frac{Z^{G(q)}}{\bar{Z}^{LP(q)}} \geqslant 1 - \left(\frac{K-q}{K}\right)\left(\frac{K-q-1}{K-q}\right)^{K-q}.$$

Proof. Theorem 3 implies that

$$\frac{\underline{Z}^S - z(S)}{\bar{Z}^S - z(S)} \geqslant 1 - \left(\frac{K-q-1}{K-q}\right)^{K-q}$$

for each node (S, N^S) with $|S| = q$. Alternatively,

$$\bar{Z}^S - \underline{Z}^S \leqslant \left(\frac{K-q-1}{k-q}\right)^{K-q} (\bar{Z}^S - z(S)). \tag{4.2}$$

From the first inequality of (4.1),

$$\eta \leqslant z(S) + \sum_{j \in N^S} \rho_j(S) y_j \quad \text{and} \quad \sum_{j \in N^S} y_j = K - q,$$

we have that

$$\bar{Z}^S \leqslant z(S) + (K-q) \max_{j \in N^S} \rho_j(S).$$

From the construction of nodes in Step 4 of the algorithm we have that $\max_{j \in N^S} \rho_j(S) \leqslant z(S)/q$. Hence

$$\bar{Z}^S \leqslant z(S) + \frac{K-q}{q} z(S) = \frac{K}{q} z(S). \tag{4.3}$$

Using (4.3) and $Z^{G(q)} \geqslant \underline{Z}^S$ in (4.2) yields

$$\frac{Z^{G(q)}}{\bar{Z}^S} \geqslant 1 - \left(\frac{K-q}{K}\right)\left(\frac{K-q-1}{K-q}\right)^{K-q}. \tag{4.4}$$

The theorem follows since (4.4) holds for every node (S, N^S) with $|S| = q$.

Theorem 4 shows that Algorithm 2 has properties similar to those of Algorithm 1. For fixed K and with $O(n^{q+1})$ function evaluations, Algorithm 2 can be run down to level q, and we can guarantee

$$\frac{\text{value of best feasible solution}}{\text{linear programming upper bound}} \geqslant 1 - \left(\frac{K-q}{K}\right)\left(\frac{K-q-1}{K-q}\right)^{K-q}.$$

As before, our result of [16] implies that no better performance guarantee can be obtained with $O(n^{q+1})$ values of z.

Theorem 4 is based on implicit enumeration of the sets S with $|S| \leqslant q$, which is the reason for the particular subproblem selection rule of the algorithm. Its proof requires $\max_{j\in N^S}\rho_j(S)\leq z(S)/q$, which is the reason for the branching rule of Step 4. Without these subproblem selection and branching rules, the bound cannot be guaranteed.

We note that the linear programming problem (4.1) at node (S, N^S) does not necessarily contain any of the inequalities from previously solved subproblems. Using these additional inequalities can improve the bounds for particular problems but not the worst case bound of Theorem 4.

It is possible to do a similar analysis for the more conventional binary tree in which the two sons of a node (S, N^S) correspond to selecting and discarding a $j^*\in N^S$ such that $\rho_{j^*}(S)=\max_{j\in N^S}\rho_j(S)$. It is still, however, necessary to choose nodes with S of minimum cardinality first to guarantee the bound of Theorem 4. In the binary tree, this approach requires $O(n^{q+2})$ function values in the worst case.

When Algorithm 2 is applied to the K-location problem, at node (S, N^S) we could solve the linear program

$$
\begin{aligned}
\bar{V}^S = \max \quad & \sum_{i=1}^{m}\sum_{j=1}^{n} c_{ij}x_{ij}, \\
& \sum_{j=1}^{n} x_{ij}=1, \quad i=1,\ldots,m, \\
& \sum_{j\in N^S} y_j = K-|S|, \\
& x_{ij}-y_j\leq 0, \quad i=1,\ldots,m, \quad j\in N^S, \\
& x_{ij}=0, \quad i=1,\ldots,m, \quad j\in N-N^S-S, \\
& x_{ij}\geq 0, \quad i=1,\ldots,m, \quad j\in S\cup N^S, \\
& y_j\geq 0, \quad j\in N^S.
\end{aligned}
\tag{4.5}
$$

For every node (S, N^S), (4.1) is a relaxation of (4.5). This follows from the fact that (4.1) is obtained from a Benders' decomposition of (4.5) using only those constraints generated from a subset of the dual extreme points of (4.5). Therefore

$$\bar{V}^S\leq \bar{Z}^S \quad \text{and} \quad \bar{V}^{LP(q)}=\max_{|S|=q}\bar{V}^S\leq \bar{Z}^{LP(q)}$$

so that Theorem 4 yields:

Theorem 5. *For the K-location problem* (1.2),

$$\frac{V^{G(q)}}{\bar{V}^{LP(q)}}\geq 1-\left(\frac{K-q}{K}\right)\left(\frac{K-q-1}{K-q}\right)^{K-q}.$$

5. Maximizing submodular functions subject to linear constraints

Many submodular functions that arise in practice are not nondecreasing and many problems have more complicated feasibility conditions than the cardinality constraint that we have supposed so far. Here we indicate how the formulations and algorithms presented previously can be adapted to the maximization of general submodular functions subject to linear constraints.

Proposition 2 [17]. *A real-valued function z on the subsets of N is submodular if and only if*

$$\text{(a)} \quad z(T) \leq z(S) + \sum_{j \in T-S} \rho_j(S) - \sum_{j \in S-T} \rho_j(S \cup T - \{j\}) \qquad \forall S, T \subseteq N$$

or

$$\text{(b)} \quad z(T) \leq z(S) + \sum_{j \in T-S} \rho_j(S \cap T) - \sum_{j \in S-T} \rho_j(S - \{j\}) \qquad \forall S, T \subseteq N.$$

Consider the problem

$$\max_{S \in \mathcal{S}} (z(S)\text{: } z \text{ submodular, } \mathcal{S} = \{S\text{: } Ay^S = b,\ y_j \in \{0, 1\},\ j \in N\}). \tag{5.1}$$

Let U be an optimal solution to (5.1) and let $g_S + h_S y$ represent one of the linear forms

$$\text{(a)} \quad z(S) + \sum_{j \in N-S} \rho_j(S) y_j - \sum_{j \in S} \alpha_j (1 - y_j)$$

where for $j \in S$, $\alpha_j \leq \rho_j(S \cup U - \{j\})$, e.g., $\alpha_j = \rho_j(N - \{j\})$, or

$$\text{(b)} \quad z(S) + \sum_{j \in N-S} \gamma_j y_j - \sum_{j \in S} \rho_j(S - \{j\})(1 - y_j)$$

where for $j \in N - S$, $\gamma_j \geq \rho_j(S \cap U - \{j\})$, e.g., $\gamma_j = \rho_j(\emptyset)$.

Now consider the integer linear program

$$\begin{aligned} \max \quad & \eta, \\ & \eta \leq g_S + h_S y, \quad \forall S \subseteq N, \\ & Ay = b, \\ & y_j \in \{0, 1\},\ j \in N, \end{aligned} \tag{5.2}$$

where $g_S + h_S y$ is one of the linear forms (a) or (b).

Precisely as in Section 2, we see that problem (5.2) is a reformulation of problem (5.1) since:

Theorem 6. $(z(U), y^U)$ *is an optimal solution to* (5.2) *if and only if* U *is an optimal solution to* (5.1).

Algorithms 1 and 2 can be adapted to problem (5.2). This is straightforward for Algorithm 1 and Theorem 2 suggests the use of the greedy algorithm in selecting the intial constraints. When $Ay = b$ is the simple cardinality constraint $\sum_{j\in N} y_i = K$, the greedy algorithm can be applied directly and, in fact, a performance guarantee that is related to the one in Theorem 2 (but weaker) can be obtained [17]. For a general set of linear constraints, the greedy algorithm would have to be modified so that maximum improving elements were selected sequentially subject to $Ay = b$. In this case no performance guarantee is known.

We would propose a similar adaptation for Algorithm 2. The constraints for the linear program of Step 2 would be obtained from the sets generated by the modified greedy algorithm just mentioned. The branching rule of Step 4 would be based on decreasing ρ_j's subject to feasibility. This type of algorithm resembles the implementation given in [14] of Benders' decomposition of mixed integer programs since it synthesizes constraint generation, linear programming relaxation and implicit enumeration.

6. Structure, Simplifications and applications

Here we show how simplified linearizations can be obtained when the submodular functions have structure. We apply these ideas to the location problem and the minimization of some quadratic 0–1 programs. (Unless stated otherwise we use constraints of type (a) in (5.2).)

A (nondecreasing) submodular function z on the set N is said to be *separable* if $z \equiv \sum_{i\in I} v^i$, where I is a finite set with $|I|>1$, and each function v^i is (nondecreasing) submodular on N. Let $\rho_j^i(S) = v^i(S\cup\{j\}) - v^i(S)$.

Consider the formulation

$$\begin{aligned} \max\ & \sum_{i\in I} \eta^i, \\ & \eta^i \leqslant v^i(S) + \sum_{j\in N-S} \rho_j^i(S) y_j - \sum \rho_j^i(N-\{j\})(1-y_j), \quad \forall i \in I, \quad \forall S \subseteq N, \qquad (6.1) \\ & Ay = b, \\ & y_j \in \{0, 1\}, \quad j \in N. \end{aligned}$$

This is a new formulation of (5.1) as:

Theorem 7. $(z(U), y^U)$ *is an optimal solution to* (6.1) *if and only if* U *is an optimal solution of* (5.1).

Proof. As in Lemma 1, (ξ^i, y^U) is feasible in (6.1) if and only if y^U is feasible and $\xi^i \leqslant v^i(U)$, $\forall i \in I$.

The value of separability evidently depends to some extent on whether the resulting functions v^i have a simpler structure than z. We now examine two cases, where the majority of the inequalities in formulation (5.2) are redundant.

Let $\{c_j\}_{j\in N}$, $N = \{1, \ldots, n\}$ be a set of nonnegative numbers and consider the submodular function $z^*(S) = \max_{j\in S} c_j$, $\forall S \subseteq N$. Note that $v(S)$, the function associated with the K-location problem, is a sum of such functions. Define $c_0 \equiv 0$ and suppose, without loss of generality, that $c_n \geqslant c_{n-1} \geqslant \cdots \geqslant c_1 \geqslant c_0$. Let $\chi^+ = \max(0, \chi)$.

Theorem 8. *For the function z^*, problem* (5.2) *can be formulated as*

$$\begin{aligned} \max \quad & \eta, \\ & \eta \leqslant c_r + \sum_{j\in N} (c_j - c_r)^+ y_j, \quad r = 0, \ldots, n-1, \\ & Ay = b, \\ & y_j \in \{0, 1\}, \quad j \in N. \end{aligned} \tag{6.2}$$

Proof. Note first that the above formulation is a relaxation of (5.2), as the inequalities in (6.2) can be obtained from the sets $S_0 = \{\emptyset\}$, and $S_i = \{i\}$, $i = 1, \ldots, n-1$ with $\alpha_j = 0$. Hence it suffices to show that (ξ, y^U) is infeasible if $\xi > z^*(U)$. Let $c_r = \max_{j\in U} \{c_j\}$. From the rth inequality we obtain

$$\xi \leqslant c_{r-1} + \sum_{j\in N} (c_j - c_{r-1})^+ y_j^U = c_{r-1} + (c_r - c_{r-1}) = c_r = z^*(U).$$

As an immediate consequence of Theorem 8, letting $c_{i_n} \geqslant c_{i_{n-1}} \geqslant \cdots \geqslant c_{i_1} \geqslant c_{i_0} \equiv 0$ be an ordering of $\{c_{ij}\}$, $j \in N$, we obtain the formulation of the K-location problem (1.2) given by

$$\begin{aligned} \max \quad & \sum_{i=1}^{m} \eta^i, \\ & \eta^i \leqslant c_{i_r} + \sum_{j\in N} (c_{ij} - c_{i_r})^+ y_j, \quad i = 1, \ldots, m, \quad r = 0, \ldots, n-1, \\ & \sum_{j\in N} y_j = K, \\ & y_j \in \{0, 1\}, \quad j \in N. \end{aligned}$$

This formulation requires only mn linear inequalities to represent $v(S)$. Again we note that an appropriate application of Benders' procedure also leads to this formulation. See [5] for alternative reformulations.

A different simplification may occur by expressing the set function as a polynomial in 0–1 variables. Define $g_\emptyset = z(\emptyset)$ and recursively $g_S = z(S) - \sum_{T \subset S} g_T$. Let $f(y) = \sum_{T \subseteq N} g_T \prod_{i \in T} y_i$ so that $z(S) = f(y^S)$.

A representation of f in the form:

$$f(y) = \sum_{T \subseteq T^*} \prod_{i \in T} y_i \Big(d_0^T + \sum_{j \in N} d_j^T y_j \Big)$$

will be called a *polynomial linearization* of z. Clearly every set function has a trivial polynomial linearization with $T^* = N$, $d_0^T = g_T$, $d_j^T = 0$, $\forall j \in N$. We are interested in cases where $|T^*| < n$ since:

Theorem 9. *If z is submodular and has a polynomial linearization, the following integer program is a reformulation of* (5.1):

$$\begin{aligned} \max \quad & \eta, \\ & \eta \leqslant z(S) + \sum_{j \in N-S} \rho_j(S) y_j - \sum_{j \in S} \alpha_j (1 - y_j), \quad \forall S \subseteq T^*, \\ & Ay = b, \\ & y_j \in \{0, 1\}, \quad j \in N. \end{aligned} \tag{6.3}$$

Proof. From Lemma 1 it suffices to show that (ξ, y^U) is infeasible in (6.3) if $\xi > z(U)$. Let $S = U \cap T^*$, and consider the corresponding inequality. We obtain

$$\begin{aligned} \xi & \leqslant z(S) + \sum_{j \in U-S} \rho_j(S) \\ & = \sum_{T \subseteq S} \Big(d_0^T + \sum_{j \in S} d_j^T \Big) + \sum_{j \in U-S} \sum_{T \subseteq S} d_j^T \qquad (\text{as } (U-S) \cap T^* = \emptyset) \\ & = \sum_{T \subseteq S} \Big(d_0^T + \sum_{j \in S} d_j^T + \sum_{j \in U-S} d_j^T \Big) \\ & = \sum_{T \subseteq U \cap T^*} \Big(d_0^T + \sum_{j \in U} d_j^T \Big) = z(U). \end{aligned}$$

As an example we consider the quadratic 0–1 programming problem

$$\min(cy + y'Qy \colon Ay = b,\ y_j \in \{0, 1\},\ j \in N) \tag{6.4}$$

where $q_{ii} = 0$, $q_{ij} = q_{ji} \geqslant 0$, $i \neq j$. This example combines the use of separability, polynomial linearization, and the choice of alternative inequalities of type (b). Let

$$m_j = \min\Big\{ \sum_{i \colon i \neq j} q_{ij} y_i \colon Ay = b,\ y_i \in \{0, 1\},\ i \in N \Big\},$$

and

$$M_j = \max\Big\{ \sum_{i:\, i\neq j} q_{ij} y_i : \ Ay = b,\ y_i \in \{0,1\},\ i \in N \Big\}.$$

Now consider the problem

$$\begin{aligned} \min \quad & \sum_{j\in N} [(c_j + m_j) y_j + \lambda_j], \\ & \lambda_j \geq -M_j + \sum_{i:\, i\neq j} q_{ij} y_i + (M_j - m_j) y_j, \quad j \in N, \\ & Ay = b, \\ & y_j \in \{0,1\}, \quad \lambda_j \geq 0, \quad j \in N. \end{aligned} \tag{6.5}$$

Theorem 10. *y^U is an optimal solution to problem (6.4) if and only if (y^U, λ^U) is an optimal solution to (6.5), where*

$$\lambda_j^U = \Big(-M_j + \sum_{i:\, i\neq j} q_{ij} y_i^U + (M_j - m_j) y_j^U \Big)^+.$$

Proof. Consider the objective $\max(-cy - y'Qy)$. Let $f(y) = -y'Qy$. With the given conditions on the $\{q_{ij}\}$, f is submodular, [17]. Taking $f(y) = \sum_{j=1}^n f^j(y)$ with $f^j(y) = y_j(-\sum_{i\neq j} q_{ij} y_i)$, we see that f is separable. Note that f^j has a polynomial linearization with $T^* = \{j\}$. As $|T^*| = 1$ Theorem 9 implies that precisely two inequalities are required to linearize f^j. Taking $S = \emptyset$ we obtain the inequality $\eta^j \leq 0$. Taking $S = \{j\}$ we obtain $\eta^j \leq -\sum_{i\neq j} q_{ij} y_i - \alpha_j(1 - y_j)$ where $\alpha_j \leq \rho_j(U - \{j\})$ and U is an optimal solution. Noting that $\rho_j(U - \{j\}) = -\sum_{i\neq j} q_{ij} y_i^U$, we can take $\alpha_j = -M_j$, and the second inequality becomes

$$\eta^j \leq -\sum_{i\neq j} q_{ij} y_i + M_j(1 - y_j).$$

Now note that inequality 5.2(b) applied to $f^j(y)$ with $S = N - \{j\}$ yields $\eta^j \leq \gamma_j y_j$, where $\gamma_j \geq \rho_j(U - \{j\})$. Since $-m_j \geq \rho_j(U - \{j\})$ and $m_j \geq 0$, the inequality $\eta^j \leq -m_j y_j$ dominates $\eta^j \leq 0$ and can be used in its place.

Hence we obtain the reformulation

$$\begin{aligned} \max \quad & -cy + \sum_{j\in N} \eta^j, \\ & \eta^j \leq -m_j y_j, \quad j \in N, \\ & \eta^j \leq -\sum_{i:\, i\neq j} q_{ij} y_i + M_j(1 - y_j), \quad j \in N, \\ & Ay = b, \\ & y_j \in \{0,1\}, \quad j \in N. \end{aligned}$$

Now setting $\lambda_j = -m_j y_j - \eta^j$, $j \in N$, the result follows.

This formulation can also be obtained by the methods in [9].

7. Conclusion

We have indicated how the structure of submodularity arises in combinatorial optimization problems, how it may be used in the design of algorithms (both approximate and exact) and in the generation of problem reformulations. Our point of view has been to use results on the behavior of heuristics to design exact algorithms. From intermediate stages of these algorithms we obtain performance guarantees on the optimal solution.

Other well-known problems to which these models are applicable include a capacitated location problem [7], multiproduct, and even multi-level distribution systems, see for example [7, 8] where Benders' algorithm was successfully used, and the quadratic assignment problem [1, 11, 13], a special case of the quadratic 0–1 program. Furthermore the formulations that we have given here do not exhaust the possibilities since other linearizations of submodular functions can be constructed as well.

Appendix

We consider the K-location problem (1.1) with $K=3$ and

$$C=\begin{bmatrix} 2 & 2 & 0 & 0 & 0 & 1 \\ 2 & 0 & 2 & 0 & 0 & 1 \\ 2 & 0 & 0 & 2 & 0 & 1 \\ 0 & 2 & 2 & 0 & 0 & 1 \\ 0 & 2 & 0 & 2 & 0 & 1 \\ 0 & 0 & 2 & 2 & 0 & 1 \\ 0 & 0 & 0 & 0 & 4 & 2 \end{bmatrix}.$$

Applying the greedy algorithm to this problem, with ties broken by choosing the lowest index element, yields the subsets $S^0=\emptyset$, $S^1=\{6\}$, $S^2=\{6,1\}$, $S^3=\{6,1,2\}$ and $Z^{G(0)}=13$.

Algorithm 1

Initialization. Choose the sets S^0, S^1, and S^2 given by the greedy algorithm to obtain the relaxed problem (see 3.1):

$$\begin{aligned} \eta^{G(0)} = \max \quad & \eta \\ & \eta \leqslant 0+6y_1+6y_2+6y_3+6y_4+4y_5+8y_6, \\ & \eta \leqslant 8+3y_1+3y_2+3y_3+3y_4+2y_5, \\ & \eta \leqslant 11 \qquad +2y_2+2y_3+2y_4+2y_5, \\ & \sum_{j=1}^{6} y_j = 3, \\ & y_j \in \{0,1\}, \quad j=1,\ldots,6. \end{aligned} \tag{A.1}$$

The solution is $y_2 = y_3 = y_4 = 1$, $y_j = 0$, otherwise and $\eta^{G(0)} = 17$. Note that

$$\frac{Z^{G(0)}}{\eta^{G(0)}} = \frac{13}{17} \geq 1 - \left(\frac{2}{3}\right)^3 = \frac{19}{27},$$

which is the bound given by Theorem 2.

Iterative Phase. $R^3 = \{2, 3, 4\}$, $z(R^3) = 12$. Add the constraint

$$\eta \leq 12 + 4y_5 + 2y_6$$

to (A.1) and re-solve. There are three optimal solutions, $y_2 = y_3 = y_5 = 1$, $y_j = 0$ otherwise, $y_2 = y_4 = y_5 = 1$, $y_j = 0$ otherwise, $y_3 = y_4 = y_5 = 1$, $y_j = 0$ otherwise, and $\eta^4 = 16$. Also $z(R^4) = 14$. The constraints generated by these solutions are respectively

$$\eta \leq 14 + 2y_1 + 2y_4 + y_6,$$
$$\eta \leq 14 + 2y_1 + 2y_3 + y_6;$$
$$\eta \leq 14 + 2y_1 + 2y_2 + y_6.$$

These constraints are added successively in the next 3 iterations. We then obtain the optimal solutions $y_1 = y_2 = y_5 = 1$, $y_j = 0$ otherwise, $y_1 = y_3 = y_5 = 1$, $y_j = 0$ otherwise, $y_1 = y_4 = y_5 = 1$, $y_j = 0$ otherwise and $\eta^7 = 15$. Also $z(R^7) = 14$. The constraints generated by these solutions are respectively

$$\eta \leq 14 + 2y_3 + 2y_4 + y_6,$$
$$\eta \leq 14 + 2y_2 + 2y_4 + y_6,$$
$$\eta \leq 14 + 2y_2 + 2y_3 + y_6.$$

These constraints are added successively in the next three iterations and we finally obtain $\eta^{10} = 14$, which verifies the optimality of selecting any two of the first four columns and the fifth; i.e., the solutions produced in the last six iterations of the algorithm.

As we noted in Section 3, Benders' algorithm gives results identical to the ones we have just given.

Algorithm 2

Initialization. The list consists only of $(\emptyset, N)$. $\underline{Z} = 0$, $I = \emptyset$.

Iterative Phase. We select $(\emptyset, N)$ from the list, apply the greedy algorithm and obtain $\underline{Z}^{\emptyset} = 13$. Next the linear programming relaxation of (A.1) is solved and we obtain $\bar{Z}^{\emptyset} = 17$. Thus $\underline{Z} = 13$ and $I = \{1, 2, 6\}$.

We have $\rho_6(\emptyset) > \rho_1(\emptyset) = \rho_2(\emptyset) = \rho_3(\emptyset) = \rho_4(\emptyset) > \rho_5(\emptyset)$. The list consists of $(\{6\}, N - \{6\})$, $(\{1\}, N - \{1, 6\})$, $(\{2\}, \{3, 4, 5\})$, $(\{3\}, \{4, 5\})$. Each of these nodes has $\bar{Z}^S = 17$.

Select $(\{6\}, N - \{6\})$ from the list. Set $k = 1$ and compute the greedy solution

$\{6, 1, 2\}$ of value $\underline{Z}^{\{6\}} = 13$. Solve the linear program (see 4.1):

$$\begin{aligned} \max \quad & \eta, \\ & \eta \leq 8 + 3y_1 + 3y_2 + 3y_3 + 3y_4 + 2y_5, \\ & \eta \leq 11 + \quad 2y_2 + 2y_3 + 2y_4 + 2y_5, \\ & \sum_{j=1}^{5} y_j = 2, \\ & 0 \leq y_j \leq 1, \quad j = 1, \ldots, 5. \end{aligned}$$

An optimal solution is $\eta = 14$, $y_2 = y_3 = 1$, $y_j = 0$ otherwise so that $\bar{Z}^{\{6\}} = 14$. $\rho_1(\{6\}) = \rho_2(\{6\}) = \rho_3(\{6\}) = \rho_4(\{6\}) > \rho_5(\{6\})$. We add $(\{1, 6\}, N - \{1, 6\})$, $(\{2, 6\}, \{3, 4, 5\})$, $(\{3, 6\}, \{4, 5\})$ $(\{4, 6\}, \{5\})$ to the list. Each of these nodes has $\bar{Z}^S = 14$.

Select $(\{1\}, N - \{1, 6\})$ from the list. Set $k = 1$ and compute the greedy solution $\{1, 2, 5\}$ of value $\underline{Z}^{\{1\}} = 14$. Solve the linear program

$$\begin{aligned} \max \quad & \eta, \\ & \eta \leq 6 + 4y_2 + 4y_3 + 4y_4 + 4y_5, \\ & \eta \leq 10 + \quad + 2y_3 + 2y_4 + 4y_5, \\ & \sum_{j=2}^{5} y_j = 2, \\ & 0 \leq y_j \leq 1, \quad j = 2, \ldots, 5. \end{aligned}$$

An optimal solution is $\eta = 14$, $y_3 = y_4 = 1$, $y_j = 0$ otherwise so that $\underline{Z}^{\{1\}} = 14$. We update the incumbent to $I = \{1, 2, 5\}$ and $\underline{Z} = 14$.

The problems generated from the nodes $(\{2\}, \{3, 4, 5\})$ and $(\{3\}, \{4, 5\})$ also give lower and upper bounds of 14. The remaining problems on the list have upper bounds of 14 so the problem is solved.

Since our problem is a K-location problem, we could have solved (4.5) instead of (4.1) for each node of the list. For the node $(\emptyset, N)$ we obtain $y_1 = y_2 = \frac{1}{2}$, $y_3 = y_5 = 1$, $y_4 = y_6 = 0$ and $\bar{V}^{\emptyset} = 16$, which is better than the bound obtained from (4.1). However, for this example the same amount of enumeration is required.

References

[1] E.M.L. Beale, Integer Programming, in: T.B. Boffey, ed., Proceedings of CP 77 (University of Liverpool, 1977) 269–279.

[2] J.F. Benders, Partitioning procedures for solving mixed-variables programming problems, Numer. Math. 4 (1962) 238–252.

[3] G. Cornuéjols, M.L. Fisher and G.L. Nemhauser, Location of bank accounts to optimize float: An analytic study of exact and approximate algorithms, Management Sci. 23 (1977) 789–810.

[4] G. Cornuéjols, G.L. Nemhauser and L.A. Wolsey, Worst-case and probabilistic analysis of algorithms for a location problem, Oper. Res. 28 (1980) 847–858.

[5] G. Cornuéjols, G.L. Nemhauser and L.A. Wolsey, A canonical representation of simple plant locations problems and its applications, SIAM J. Algebraic and Discrete Methods 1 (1980) 261–272.

[6] G. Dantzig, D.R. Fulkerson and S. Johnson, Solution of a large-scale travelling salesman problem, Oper. Res. 2 (1954) 393–410.

[7] A.M. Geoffrion, How can specialized discrete and convex optimization methods be married, in: P.L. Hammer, E.L. Johnson, B.H. Korte and G.L. Nemhauser, eds., Studies in Integer Programming (North-Holland, Amsterdam, 1977) 205–220.

[8] A.M. Geoffrion and G.W. Graves, Multicommodity distribution system design by Benders' decomposition, Management Sci. 20 (1974) 822–844.

[9] F. Glover, Improved linear integer programming formulations of nonlinear integer problems, Management Sci. 22 (1975) 455–460.

[10] M. Guignard and K. Spielberg, Algorithms for exploiting the structure of the simple plant location problem, in: P.L. Hammer, E.L. Johnson, B.H. Korte and G.L. Nemhauser, eds., Studies in Integer programming (North-Holland, Amsterdam, 1977) 247–272.

[11] L. Kaufman and F. Broeckx, An algorithm for the quadratic assignment problem using Benders' decomposition, European J. Oper. Res. 2 (1978) 207–211.

[12] A Land and S. Powell, Computer codes for problems of integer programming, Ann. Discrete Math. 5 (1979) 221–269.

[13] E.L. Lawler, The quadratic assignment problem, Management Sci. 9 (1963) 586–599.

[14] C.E. Lemke and K. Spielberg, Direct search zero–one and mixed integer programming, Oper. Res. 15 (1967) 892–914.

[15] T.L. Magnanti and R.T. Wong, Accelerating Benders' decomposition: Algorithmic enhancements and model selection criteria, CORE Discussion Paper 8003, Louvain-la-Neuve, Belgium (January 1980).

[16] G.L. Nemhauser and L.A. Wolsey, Best algorithms for approximating the maximum of a submodular set function, Math. Oper. Res. 3 (1978) 177–188.

[17] G.L. Nemhauser, L.A. Wolsey and M.L. Fisher, An analysis of approximations for maximizing a submodular set function—I, Math. Programming 14 (1978) 265–294.

[18] H.P. Williams, Experiments in the formulation of integer programming problems, Math. Programming Studies 2 (1974) 180–197.

P. Hansen, ed., Studies on Graphs and Discrete Programming

A SOLVABLE MACHINE MAINTENANCE MODEL WITH APPLICATIONS

A.H.G. RINNOOY KAN and J. TELGEN
Erasmus University, P.O. Box 1738, Rotterdam, The Netherlands

At the end of each time period maintenance work can be performed on a machine at constant cost, reducing operating cost in the next period to zero. If this option is not exercised, the operating cost in the next period increases by a constant. We derive a closed form expression for the maintenance policy that minimizes total cost. Various applications of this model are discussed, one of which is connected with the optimal reinversion policy for linear programming.

1. Introduction

Consider the following *machine maintenance problem.* At the end of each time period, the possibility exists of performing maintenance work on a machine at given constant cost. If this option is not exercised, the operating cost for the next period increases by a constant; if it is, this cost decreases to zero. At some given point in the future, the entire machine will be scrapped. What is the maintenance policy that minimizes total cost?

More explicitly, consider a system with denumerable state space $\{0, 1, 2, 3, \ldots\}$ over a time interval consisting of T periods. At the end of period $t \in \{1, \ldots, T\}$, we can reduce the state of the system to 0 at cost a or increase it from $i-1$ to i at cost bi, where a and b are given nonnegative constants. Initially, at time 0, the system is in state i_0. The problem is to determine a *policy* that minimizes total cost. As will be shown in Section 2, an integer programming formulation of this problem can be solved to yield a closed form solution, characterizing the optimal policy as a function of a, b, i_0 and T. If the value of the latter parameter is not known in advance, one can resort to various devices, depending on the information that is available. For example, the probability distribution of T may be known, we may be able to estimate T or a bound on T from the past behaviour of the system or—in a Bayesian framework—we may know the type of probability distribution of T, together with some prior information on the parameters of the distribution.

In Section 3 we turn to applications of the model. The most interesting one occurs in the context of the *product form of the inverse* variant of the simplex method for linear programming. Here, the state of the system corresponds to the number of elementary matrices that are currently stored in a file, the product of which determines the basis inverse. After each simplex iteration, we can either perform a reinversion or add another elementary matrix to the file. The results of the above model can be immediately applied to this case. As will be illustrated in the same section, the model readily lends itself to various other interpretations as well.

The final section contains concluding remarks and some topics for further research.

2. A closed form solution

In this section we assume that the four parameters of the model, a, b, i_0 and T, are fixed and known in advance.

A maintenance policy corresponds to a partition of the time horizon in *maintenance intervals*. Let $t_1 < t_2 < \cdots < t_{k-1}$ be the points in time at which maintenance occurs, at total cost $(k-1)a$. The operating cost is

$$\sum_{i=i_0}^{t_0+t_1-1} ib + \sum_{j=2}^{k-1} \sum_{i=0}^{t_j-t_{j-1}-1} ib + \sum_{i=0}^{T-t_{k-1}-1} ib.$$

We define the *state* in which the maintenance is done as follows:

$$x_1 = t_1 + i_0,$$
$$x_j = t_j - t_{j-1}, \quad (j = 2, \ldots, k-1)$$
$$x_k = T - t_{k-1},$$

and rewrite the combined maintenance and operating cost C as

$$C = (k-1)a + b\left(\sum_{i=i_0}^{x_1-1} i + \sum_{j=2}^{k} \sum_{i=1}^{x_j-1} i\right)$$
$$= (k-1)a + \tfrac{1}{2}b\left(\sum_{j=1}^{k} x_j^2 - \sum_{j=1}^{k} x_j - i_0^2 + i_0\right)$$
$$= \tfrac{1}{2}b(-i_0^2 - T) + (k-1)a + \tfrac{1}{2}b \sum_{j=1}^{k} x_j^2.$$

Ignoring constant terms, we arrive at the following mathematical programming problem:

$$\text{minimize} \quad \alpha k + \sum_{j=1}^{k} x_j^2,$$
$$\text{subject to} \quad \sum_{j=1}^{k} x_j = \tau,$$
$$k \in \{1, \ldots, T+1\},$$
$$x_j \geq 0 \text{ and integer} \quad (j = 1, \ldots, k)$$

with

$$\alpha = 2a/b, \qquad \tau = T + i_0.$$

First of all, let us note that the problem obtained by relaxation of the integrality constraints on x_j ($j = 1, \ldots, k$) can be solved easily. For given k, we calculate

$$g(k) = \min\left\{\sum_{j=1}^{k} x_j^2 \;\middle|\; \sum_{j=1}^{k} x_j = \tau,\ x_j \geqslant 0\ (j = 1, \ldots, k)\right\}$$

by forming the Lagrangean function

$$\bar{g}(k; x, \lambda) = \sum_{j=1}^{k} x_j^2 - \lambda\left(\sum_{j=1}^{k} x_j - \tau\right)$$

and setting the derivatives equal to 0:

$$\frac{\partial \bar{g}}{\partial x_j} = 2x_j - \lambda = 0 \quad (j = 1, \ldots, k),$$

$$\frac{\partial \bar{g}}{\partial \lambda} = \sum_{j=1}^{k} x_j - \tau = 0.$$

We obtain the well-known result:

$$x_1^* = x_2^* = \cdots = x_k^* = \tau/k,$$

with $x_j^* \geqslant 0$ ($j = 1, \ldots, k$) as required. These values are easily verified to define a minimum with value τ^2/k, so that the relaxed problem becomes:

$$\begin{aligned} &\text{minimize} && G(k) = \alpha k + \tau^2/k, \\ &\text{subject to} && k \geqslant 0, \\ & && k \in \{1, \ldots, T+1\}. \end{aligned}$$

The objective function G is unimodal and convex so that the global minimum G^* will be attained at the integer round-down or round-up of the continuous solution $k = \tau/\sqrt{\alpha}$:

$$k^* \in \{\lfloor \tau/\sqrt{\alpha} \rfloor, \lceil \tau/\sqrt{\alpha} \rceil\}$$

under the reasonable assumption that at least one of these values is in the interval $[1, T+1]$.

If the integrality constraints on x_j ($j = 1, \ldots, k$) are not relaxed, the problem can be approached in a similar manner. We start by calculating

$$f(k) = \min\left\{\sum_{j=1}^{k} x_j^2 \;\middle|\; \sum_{j=1}^{k} x_j = \tau,\ x_j \geqslant 0 \text{ and integer } (j = 1, \ldots, k)\right\}.$$

We first verify that $(x_1, \ldots, x_k)$ minimizes $\sum_{j=1}^{k} x_j^2$ subject to the above constraints if and only if $|x_i - x_j| \leqslant 1$ for all (i, j). In fact, we simply consider a pair of feasible solutions $(x_1, \ldots, x_i, \ldots, x_j, \ldots, x_k)$ and $(x_1, \ldots, x_i - 1, \ldots, x_j + 1, \ldots, x_k)$ and note that the latter one will be superior to the former one if

$$(x_i - 1)^2 + (x_j + 1)^2 < x_i^2 + x_j^2,$$

i.e., if $x_i - x_j > 1$.

Hence, we obtain an optimal solution $x_1^*, \ldots, x_k^*$ determining $f(k)$ by defining

$$\tau(k) \equiv \tau \pmod{k}$$

and setting

$$x_1^* = \cdots = x_{\tau(k)}^* = \frac{\tau - \tau(k)}{k} + 1,$$

$$x_{\tau(k)+1}^* = \cdots = x_k^* = \frac{\tau - \tau(k)}{k}.$$

It follows that

$$f(k) = \tau(k)\left(\frac{\tau - \tau(k)}{k} + 1\right)^2 + (k - \tau(k))\left(\frac{\tau - \tau(k)}{k}\right)^2,$$

and the original problem can be rewritten as follows:

$$\begin{aligned} &\text{minimize} \quad F(k) = \alpha k + \frac{\tau^2}{k} + \frac{1}{k}\,\tau(k)(k - \tau(k)), \\ &\text{subject to} \quad k \in \{1, \ldots, T+1\}. \end{aligned}$$

To solve this problem, note that $\tau(k)$ is a discontinuous, piecewise linear function of k: on each interval

$$I_i = \left\{ k \,\middle|\, \frac{\tau}{i+1} < k \leqslant \frac{\tau}{i},\ k \text{ integer} \right\} \quad (i = 1, \ldots, \tau)$$

it is the case that

$$(i+1)k > \tau, \qquad ik \leqslant \tau$$

and hence on I_i

$$\tau(k) \equiv \tau \pmod{k} = \tau - ik.$$

Of course, one or more of the intervals I_i may be empty. This, however, can easily be verified not to affect the subsequent arguments.

It follows that on I_i the function $F(k)$ can be written as

$$F(k) = \alpha k + \frac{\tau^2}{k} + \frac{1}{k}(\tau - ik)(k - \tau + ik)$$
$$= (\alpha - i(i+1))k + (2i+1)\tau.$$

Thus, on each interval I_i the function F is linear in k, and its minimum F_i^* is assumed at one of the endpoints, depending on the sign of $\alpha - i(i+1)$. The only positive root of $\alpha - i(i+1) = 0$ is equal to

$$i^* = -\tfrac{1}{2} + \tfrac{1}{2}\sqrt{1+4\alpha}.$$

It follows that:

(i) if $i \geqslant i^*$, the minimum value F_i^* is attained at $k = \lfloor \tau/i \rfloor$;

(ii) if $i \leqslant i^*$, the minimum value F_i^* is attained at $k = \lfloor \tau/(i+1) \rfloor + 1$.

We shall now show that in case (i) $F_i^* \leqslant F_{i+1}^*$, and that in case (ii) $F_i^* \leqslant F_{i-1}^*$.

(i) $$F_i^* - F_{i+1}^* = (\alpha - i(i+1))\left\lfloor \frac{\tau}{i} \right\rfloor - (\alpha - (i+1)(i+2))\left\lfloor \frac{\tau}{i+1} \right\rfloor - 2\tau$$
$$= (\alpha - i(i+1))\left(\left\lfloor \frac{\tau}{i} \right\rfloor - \left\lfloor \frac{\tau}{i+1} \right\rfloor\right) + \left((2i+2)\left\lfloor \frac{\tau}{i+1} \right\rfloor - 2\tau\right).$$

The first term is nonpositive because $i \geqslant i^*$ implies that $\alpha \leqslant i(i+1)$, the last term is nonpositive because

$$\left\lfloor \frac{\tau}{i+1} \right\rfloor \leqslant \frac{\tau}{i+1} = \frac{2\tau}{2i+2}.$$

(ii) $$F_i^* - F_{i-1}^* = (\alpha - i(i+1))\left(\left\lfloor \frac{\tau}{i+1} \right\rfloor - \left\lfloor \frac{\tau}{i} \right\rfloor\right)$$
$$+ \left((i(i-1) - i(i+1))\left(\left\lfloor \frac{\tau}{i} \right\rfloor + 1\right) + 2\tau\right).$$

This expression can be proved to be nonpositive by means of an argument similar to the one above.

It follows that the values F_i^* form a nonincreasing sequence for $i \leqslant i^*$ and a nondecreasing sequence for $i \geqslant i^*$. The global minimum F^* will either be realized by the right endpoint of $I_{\lceil i^* \rceil}$ for

$$k = \lfloor \tau / \lceil i^* \rceil \rfloor$$

or by the left endpoint of $I_{\lfloor i^* \rfloor}$ for

$$k = \left\lfloor \frac{\tau}{\lfloor i^* \rfloor + 1} \right\rfloor + 1 = \left\lceil \frac{\tau}{\lceil i^* \rceil} \right\rceil,$$

where the last equality holds only if i^* and $\tau/(\lfloor i^* \rfloor + 1)$ are not integers.

Assuming that this is the case, we have that k^* is the integer round-down or round-up of $\tau/\lceil i^* \rceil$:

$$k^* \in \{\lfloor \tau/\lceil i^* \rceil \rfloor, \lceil \tau/\lceil i^* \rceil \rceil\}.$$

It is instructive to compare this result to the outcome of the continuous approximation

$$k^* \in \{\lfloor \tau/\sqrt{\alpha} \rfloor, \lceil \tau/\sqrt{\alpha} \rceil\}.$$

The two denominators, $\lceil i^* \rceil = \lceil -\frac{1}{2} + \frac{1}{2}\sqrt{1+4\alpha} \rceil$ and $\sqrt{\alpha}$ are close to each other. In particular,

$$\sqrt{\alpha} - \tfrac{1}{2} < -\tfrac{1}{2} + \tfrac{1}{2}\sqrt{1+4\alpha} \leq \sqrt{\alpha}$$

so that

$$|\sqrt{\alpha} - \lceil -\tfrac{1}{2} + \tfrac{1}{2}\sqrt{1+4\alpha} \rceil| \leq 1.$$

Of course, depending on the value of τ the continuous approximation can be arbitrarily far away from the real optimum. In particular, if the cost of maintenance and operation in the first period is less than the operating cost in the second period (i.e. $a + b < 2b$ or $a < b$ and hence $\lceil i^* \rceil = 1$), maintenance will be performed in every time period. But in the continuous approximation maintenance will be performed in every time period only if $2a \leq b$.

Our conclusion is that we have obtained a closed form solution to an integer programming problem. It can be calculated just as simply as the continuous approximation and, depending on the particular values of the parameters a, b, i_0 and T it can be of significantly better quality.

What happens if some of these parameters are not known *a priori*? The obvious candidate to consider for this role is T, since it is reasonable to suspect that the life time of the machine may not be perfectly known in advance. In a few simple cases, the previous results are directly applicable. For instance, if the expected value μ_T of T is known, then the expected value of the objective function over the interval I_i is given by

$$(\alpha - i(i+1))k + (2i+1)(\mu_T + i_0)$$

and the same method can be applied to minimize expected cost. If upper and lower bounds on T are known, we can postulate a distribution for T over the interval between these values and use the corresponding expected value in the above calculations.

It seems more realistic to assume that information about T becomes gradually available as time goes on. This information should then be used to update the current estimate of T in a Bayesian fashion and to adjust the maintenance policy accordingly. The application dealt with in the next section provides an example of such a situation.

It is simple to extend the analysis to the case of an infinite time horizon ($T \to \infty$). Clearly, in this case the maintenance points will occur at regular intervals of length $x+1$. The cost per period is given by

$$\frac{1}{x}\left(a+\sum_{i=0}^{x} bi\right)=\frac{a}{x}+\frac{1}{2}bx+\frac{1}{2}b$$

which is minimized by choosing x equal to the integer round-up or round-down of $\sqrt{2a/b}=\sqrt{\alpha}$. These can be viewed as limiting cases of the continuous approximation values derived above.

A less trivial extension, that seems worthy of further exploration, is to describe the state transitions as a stochastic rather than as a deterministic process and to invoke techniques from Markov programming. We shall not investigate this possibility any further in this paper.

3. Applications

The model described and solved in the preceding sections can be applied in the context of linear programming routines. In the *product form of the inverse* variation of the simplex method, the inverse of the current basis is not explicitly available, but only implicitly as the product of a sequence of elementary matrices stored in the so called *ETA-file* [8]. In each simplex iteration, a new elementary matrix describing the corresponding pivot transformation is added to the file. At certain moments a complete basis *reinversion* is performed, as a result of which the file length is reduced to a constant. The decision when to reinvert can be based on considerations of numerical accuracy, but also on computational arguments: the amount of work in each iteration is proportional to the current file length and at some point a time-consuming reinversion will become preferable.

The problem of choosing the proper reinversion point on other than numerical arguments has been recognized since the early days of linear programming. Hoffman wrote in 1955: 'Dantzig has informed us that the use of the product form leads to exciting moments for the operator. If a large number of vectors have accumulated in the product, one is tempted to clean up in the manner previously described. This takes substantial time, however, and one is also tempted to hang on a few iterations longer in the hope that the problem will be solved'. [6]

Nowadays, all major linear programming codes [3; 7] contain *triggering mechanisms* that decide upon the moment of reinversion (*e.g.* [2, p. 225]). A few examples of some triggering criteria are the following:

(i) a fixed number of iterations since the last reinversion, e.g.: reinversion after every 100 or 150 iterations;

(ii) a problem dependent number of iterations, e.g.: reinversion after every $\frac{1}{2}m$ iterations, where m is the number of rows;

(iii) an absolute amount of time, e.g.: reinversion after every 10 seconds of CPU time;

(iv) a problem dependent amount of time, e.g.: reinversion after every $\frac{1}{25}m$ seconds of CPU time;

(v) reinversion whenever the total time required for the iterations since the last reinversion plus reinversion time, divided by the number of these iterations, starts to increase [1, p. 15].

Quite often, one finds a mixture of these triggers; for instance, in the IBM MPSX/370 package, reinversion takes place after min{max{20, $m/10$}, 150} iterations [7, p. 91].

Our maintenance model from the previous section can be used to determine an *optimal reinversion point.* The parameter a represents the cost of reinversion, b represents the cost increase due to the appearance of an extra elementary matrix in the file, i_0 can be taken equal to 0 or to the current file length and T is the number of simplex iterations required to arrive at the final solution. Interestingly enough, the outcome of our infinite time horizon analysis

$$x_1^* = \cdots = x_{k^*}^* \in \{\lfloor\sqrt{2a/b}\rfloor, \lceil\sqrt{2a/b}\rceil\}$$

supports the triggering mechanism described under (v) above. This mechanism calls for a reinversion whenever

$$\frac{1}{x}\left(a + \sum_{i=0}^{x} bi\right) = \frac{a}{x} + \frac{1}{2}bx + \frac{1}{2}b$$

starts to increase, and this is easily seen to correspond to the above rule.

However, as stated before, this approximation can be arbitrarily far away from the optimal solution, indicating that the theoretical quality of triggering mechanisms used in practice is relatively poor. This is caused mainly by the fact that the decision when to reinvert is based on information from the past only. Thus a reinversion might be called for even in situations in which the process will terminate after one more iteration.

Of course, in linear programming the number of iterations T will generally not be known in advance. This provides a typical example of a situation in which more and more reliable estimates of its value can be made as the objective function approaches its optimum value. Some empirical studies suggest that this happens according to the regular pattern illustrated in Fig. 1, taken from [5].

In that case, an estimate of T could be based on appropriately fitting an exponential or linear curve to the available data. However, some preliminary

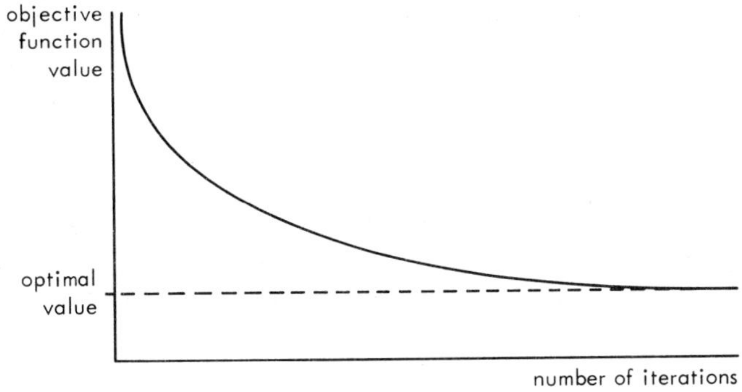

Fig. 1. A typical pattern of objective function value against number of iterations.

experiments that we have carried out indicate that this pattern need not always occur, and additional work is required to investigate the applicability of a more sophisticated reinversion criterion than those currently in use.

The basic ingredient of the model, *i.e.* a confrontation with the increasing cost of postponing a costly decision, can be recognized in many other situations as well. Consider for instance the *transportation of natural gas* through a *pipe line.* As the gas flows through the pipe, its *pressure* decreases linearly with the distance travelled, and at certain points it becomes attractive to install a *repressurization point* where the original pressure is restored. Apart from a certain fixed factor, the costs associated with such a point are quadratically proportional to the pressure loss. It is easy to verify that our closed form formula can be used to yield the optimal number of repressurization points.

We leave it to the reader to concoct other similar situations. (Consider, for instance, the problem of the organization which has to decide when to switch to strong and expensive arguments in order to extract a regular payment from its steady customers.) We do note that if, in addition to the fixed maintenance costs, maintenance charges are incurred that are linearly proportional to the length of the maintenance interval, the model can still be applied since this only adds a constant term to the objective function. For example, by way of one more algorithmic application, the model can be used to determine with which frequency an increasingly large set of linear constraints (*e.g.*, cutting planes) should be checked for possible redundancies [9].

4. Concluding remarks

We have found a closed form solution to the maintenance problem described in the Introduction by means of a linearization technique that may be applicable to other periodic optimization problems. An obvious question is under what extension the model will lose its property of *polynomial-time solvability*. The model involves only four parameters, but in spite of that it should be possible to identify certain generalizations that can be proved to be *NP-complete* [4] and thus are unlikely to admit of a polynomial-bounded algorithm.

Certain stochastic extensions mentioned at the end of Section 2 are worthy of further investigation and should facilitate the application of the model within linear programming routines or in other practical situations.

Acknowledgements

We are grateful to J.K. Lenstra and R.S. Garfinkel for many useful comments. The research of the second author was partially supported by a NATO Science Fellowship awarded by the Netherlands Organization for the Advancement of Pure Research (Z.W.O.).

References

[1] E.M.L. Beale, Sparseness in linear programming, in: J.K. Reid, ed., Large Sparse Sets of Linear Equations (Academic Press, New York, 1971).

[2] S.P. Bradley, A.C. Hax and T.L. Magnanti, Applied Mathematical Programming (Addison-Wesley, Reading, MA, 1977).

[3] C.D.C. APEX III, Reference Manual (1977).

[4] M.R. Garey and D.S. Johnson, Computers and Intractability: A Guide to the Theory of NP-completeness (Freeman, San Francisco, 1979).

[5] P.M.J. Harris, Pivot selection methods of the Devex LP code, Mathematical Programming (1) 1–28 (1973).

[6] A.J. Hoffman, How to solve a linear programming problem, in: H.A. Antosiewicz, ed., Proceedings of the Second Symposium on Linear Programming, Washington D.C. (1955).

[7] I.B.M. MPSX/370, Program Reference Manual (1976).

[8] W. Orchard-Hays, Advanced Linear Programming Computer Techniques (McGraw-Hill, New York, 1968).

[9] J. Telgen, Redundancy and Linear Programs, Ph.D. Thesis, Erasmus University Rotterdam (1979).

P. Hansen, ed., Studies on Graphs and Discrete Programming

CONVERGENCE OF DECOMPOSITION ALGORITHMS FOR THE TRAFFIC ASSIGNMENT PROBLEM

Pablo SERRA and Andrés WEINTRAUB

Departamento de Industrias, Universidad de Chile, Av. Bernardo O'Higgins 1058, Casilla 10–D, Santiago, Chile

One approach that has been proposed for solving the traffic assignment problem is through decomposition by products or origin-flows. These algorithms try to obtain improved solutions working on one product at a time, while flows corresponding to all other products remain fixed. It is relatively simple to prove that for convex differentiable cost functions, if no improvements can be obtained sequencially in all products, the present solution is optimal.

This paper presents a proof that for the typical cost function used in the traffic assignment problem, decomposition algorithms converge to an optimal solution.

1. The traffic assignment problem

Consider the problem of finding an equilibrium flow in a traffic network among drivers who are trying each to minimize the total trip 'cost'. This problem, called the traffic assignment can be solved through a cost function transformation by solving an optimization problem [2].

Let E be a directed network of m nodes and n arcs, where (ij) represents the arc joining nodes i and j. In some nodes a flow f_k, $k = 1, \ldots, K$ is originated. Each such flow goes to a set of destination nodes D_k, such that

$$f_k = \sum_{i \in D_k} f_{ki},$$

f_{ki} being the flow from node k attracted by node i.

Product k is defined by the flow originated at node k. We call X_{ijk} the flow on arc (ij) corresponding to product k.

The traffic assignment problem can be expressed as:

$$\text{Min} \quad Z = \sum_{(i,j)} C_{ij}\left(\sum_{k=1}^{K} X_{ijk} \right),$$

$$\sum_{j \in A(i)} X_{ijk} - \sum_{j \in B(i)} X_{jik} = \begin{cases} f_k, & \text{if } i = k, \\ 0, & \text{if } i \neq k, \quad i \notin D_k, \\ -f_{ki}, & \text{if } i \neq k, \quad i \in D_k, \end{cases}$$

$$X_{ijk} \geq 0, \quad \forall (i,j) \in E, \quad k = 1, \ldots, K$$

where $A(i)$ is the set of arcs leaving node i to some other node in E, $B(i)$ is the set of arcs entering node i and C_{ij} is the cost function on arc (ij).

Typical cost functions are of the form $C_{ij}(x) = a_{ij}x + b_{ij}x^s$ ($s = 5$ being a common value) [4].

2. Algorithms based on product-decomposition

An algorithm based on product-decomposition tries to improve the objective function considering that all flows, except those corresponding to product k are fixed. First, a feasible solution is obtained. Then, each product is analyzed separately, reassigning flow of that product to improve the cost associated to it while considering the flows of all other products as fixed. If in succession no improvement can be obtained in any product, the present solution is optimal. In practice since these methods can only converge in the limit, approximate solutions are obtained based on predefined tolerance parameters [5, 3].

The steps of a product decomposition algorithm are the following, where m indicates the iteration number and p the number of successive products for which no improvement in the objective function has been obtained.

(1) Determine a feasible solution $m = 0$, $p = 0$.

(2) $k = 0$, $m = m + 1$.

(3) $k = k + 1$, $p = p + 1$.

(4) $d_{ij} = \sum_{h=1}^{k-1} X_{ijh}^{m+1} + \sum_{h=k+1}^{K} X_{ijh}^{m}$.

(5) Solve the problem:

$$\text{Min} \quad Z = \sum_{(i,j)} C_{ij}(X_{ijk}^{m+1} + d_{ij}),$$

$$\text{s.t.} \quad X_{ijk}^{m+1} \geqslant 0.$$

(6) If an improved solution is obtained, $X_{ijk}^{m+1} \neq X_{ijk}^{m}$ for some (ij), then $p = 0$.

(7) If $p = K$ stop.

(8) If $k = K$ go to (2).

(9) Go to (3).

If the algorithm stops in Step 7, a critical point has been reached, that is, a solution in which it is impossible to improve the objective value by considering each product independently.

The algorithms proposed following the above scheme differ in the form of carrying out Step 5 to determine improved solutions. Nguyen [5] uses the convex simplex specialized to networks to determine a direction of flow variation, followed by a unidimensional search to determine the best step size. Weintraub–González [6] solve an assignment problem to determine a set of disjoint circuits along which flow is reassigned, followed by a unidimensional search.

3. Convergence of the algorithms

Since the cost function is convex and differentiable, it is relatively simple to show that any critical point will be an optimal solution [1, 6]. Hence if the described algorithms stop in Step 7, that solution is optimal.

We first show, that if cost functions satisfy the property of being strongly convex, then convergence can be proved. (This is based on a theorem by Auslander [1].)

Then, we show that the typical cost functions used in traffic assignment belong to a family that satisfies the property of being strongly convex.

Theorem 1. *If the real functions C_{ij} are positive and differentiable in R^+ and if two strictly positive real numbers λ and θ exist such that for each pair (ij) the relation:*

$$C_{ij}(z) \geqslant C_{ij}(y) + C'_{ij}(y)(z-y) + \lambda \, \|z-y\|^{2\theta}$$

is satisfied for any positive real numbers z and y, then the described algorithms converge to a critical point.

Proof. Let Z^m be defined as

$$Z^m = \sum_{i,j} C_{ij}\left(\sum_{k=1}^{K} X^m_{ijk}\right).$$

By construction

$$\sum_{i,j} C_{ij}\left(\sum_{h=1}^{k} X^{m+1}_{ijh} + \sum_{h=k+1}^{K} X^m_{ijh}\right) \leqslant \sum_{i,j} C_{ij}\left(\sum_{h=1}^{k-1} X^m_{ijh} + \sum_{h=k}^{K} X^m_{ijh}\right),$$

hence $Z^{m+1} \leqslant Z^m$. Thus, the sequence $\{Z^m\}$ contained in the compact $[0, Z^0]$ is decreasing and must converge to a real number Z^*, $Z^m \to Z^*$. By hypothesis

$$\begin{aligned}
&\sum_{i,j} C_{ij}\left(\sum_{h=1}^{k-1} X^{m+1}_{ijh} + \sum_{h=k}^{K} X^m_{ijh}\right)\\
&\quad \geqslant \sum_{i,j} C_{ij}\left(\sum_{h=1}^{k} X^{m+1}_{ijh} + \sum_{h=k+1}^{K} X^m_{ijh}\right)\\
&\quad + \sum_{i,j} C'_{ij}\left(\sum_{h=1}^{k} X^{m+1}_{ijh} + \sum_{h=k+1}^{K} X^m_{ijh}\right)(X^m_{ijk} - X^{m+1}_{ijk})\\
&\quad + \lambda \sum_{i,j} \|X^m_{ijk} - X^{m+1}_{ijk}\|^{2\theta}.
\end{aligned}$$

By X^{m+1}_{ijk} being the optimal solution of the problem in Step 5 of the algorithm and the definition of the functions C_{ij} we have

$$\sum_{i,j} C'_{ij}\left(\sum_{h=1}^{k} X_{ijh}^{m+1} + \sum_{h=k+1}^{K} X_{ijh}^{m}\right)(X_{ijk}^{m} - X_{ijk}^{m+1}) \geqslant 0,$$

hence

$$\sum_{i,j} C_{ij}\left(\sum_{h=1}^{k-1} X_{ijh}^{m+1} + \sum_{h=k}^{K} X_{ijh}^{m}\right) - \sum_{i,j} C_{ij}\left(\sum_{h=1}^{k} X_{ijh}^{m+1} + \sum_{h=k+1}^{K} X_{ijh}^{m}\right)$$

$$\geqslant \sum_{i,j} \|X_{ijk}^{m} - X_{ijk}^{m+1}\|^{2\theta}.$$

$$\therefore \quad Z^{m} - Z^{m+1} \geqslant \sum_{i,j} \|X_{ijk}^{m} - X_{ijk}^{m+1}\|^{2\theta},$$

then

$$\lim_{m\to\infty} \|X_{ijk}^{m} - X_{ijk}^{m+1}\|^{2\theta} = 0. \tag{1}$$

The sequence X^m is contained in the compact defined by the feasible space (which is compact), and a subsequence must exist which converges to a vector, $X^{m_s} \to X^*$.

By definition of X^{m_s+1} we have

$$\sum_{i,j} C_{ij}\left(\sum_{h=1}^{k} X_{ijh}^{m_s+1} + \sum_{h=k+1}^{K} X_{ijh}^{m_s}\right)$$

$$\leqslant \sum_{i,j} C_{ij}\left(\sum_{h=1}^{k-1} X_{ijh}^{m_s+1} + Y_{ij} + \sum_{h=k+1}^{K} X_{ijh}^{m_s}\right) \quad \forall Y_{ij} \in \mathbb{R}^+$$

Considering (1) and the fact that the objective function is continuous, in the limit we will have

$$\sum_{i,j} C_{ij}\left(\sum_{h=1}^{K} X_{ijh}^{*}\right) \leqslant \sum_{i,j} C_{ij}\left(\sum_{\substack{h=1\\ h\neq k}}^{K} X_{ijh}^{*} + Y_{ij}\right) \quad \forall Y_{ij} \in \mathbb{R}^+$$

which implies X^* is a critical point. By differentiability and convexity of the objective function, the critical point is a global optimum.

If the sequence X^m does not converge to X^*, then there must exist another subsequence X^{m_t} converging to a point $\hat{X}$. Then, $\hat{X}$ will be a critical point and global optimum. By hypothesis,

$$\sum_{i,j} C_{ij}\left(\sum_{k=1}^{K} X_{ijk}^{*}\right)$$

$$\geqslant \sum_{i,j} C_{ij}\left(\sum_{k=1}^{K} \hat{X}_{ijk}\right)$$

$$+ \sum_{i,j} C'_{ij}\left(\sum_{k=1}^{K} \hat{X}_{ijk}\right)\left(\sum_{k=1}^{K} X_{ijk}^{*} - \sum_{k=1}^{K} X_{ijk}\right)$$

$$+ \sum_{i,j} \lambda \left\| \sum_{k=1}^{K} X_{ijk}^{*} - \sum_{k=1}^{K} \hat{X}_{ijk} \right\|^{2\theta},$$

but since both X^* and $\hat{X}$ are global optimum:

$$\sum_{i,j} C'_{ij}\left(\sum_{k=1}^{K} \hat{X}_{ijk}\right)\left(\sum_{k=1}^{K} X^*_{ijk} - \sum_{k=1}^{K} \hat{X}_{ijk}\right) \leqslant 0$$

which implies that at the optimal point there exists a directional derivative in a feasible direction which is negative. This contradicts the condition of optimality and hence the sequence converges to X^*.

4. Case of typical cost functions in traffic assignment

We consider the family of polynomial cost functions, which includes the ones used in traffic assignment, and show that for that case, the hypothesis of the Theorem 1 is satisfied.

Let

$$C_{ij}(z) = a_{0,ij} + \sum_{l=1}^{L,ij} a_{l,ij} z^{b_{l,ij}}$$

where the numbers $a_{l,ij}$ and $b_{l,ij}$ are positive with $b_{l,ij} \geqslant 1$ and $b_{L,ij} \geqslant 2$.

For the purpose of simplicity we will no longer use the subindices i, j. We first show that there exist two real positive numbers λ and θ such that the relation.

$$C_{ij}(z) \geqslant C_{ij}(y) + C'_{ij}(y)(z-y) + \lambda \, \|z-y\|^{2\theta}$$

is satisfied for any real positive numbers z and y. Replacing the C_{ij} functions we must prove

$$a_0 + \sum_{l=1}^{L} a_l z^{b_l} \geqslant a_0 + \sum_{l=1}^{L} a_l y^{b_l} + \sum_{l=1}^{L} a_l b_l y^{b_l - 1}(z-y) + \lambda \, \|z-y\|^{2\theta}.$$

or equivalently:

$$\sum_{l=1}^{L} a_l\{z^{b_l} - y^{b_l} - b_l y^{b_l-1}(z-y)\} - \lambda \, \|z-y\|^{2\theta} \geqslant 0. \tag{2}$$

To prove (2) we will take $\lambda = a_L$ and $\theta = b_L/2$. Defining $\varepsilon = z - y$ we can rewrite (2) as

$$\sum_{l=1}^{L} a_l\{(y+\varepsilon)^{b_l} - y^{b_l} - b_l y^{b_l-1}\varepsilon\} - a_L \, |\varepsilon|^{b_L} \geqslant 0.$$

We define the functions

$$h_l(\varepsilon)=(y+\varepsilon)^{b_l}-y^{b_l}-b_l y^{b_l-1}\varepsilon, \quad l=1,\dots,L-1, \tag{3}$$

$$g(\varepsilon)=(y+\varepsilon)^{b_L}-y^{b_L}-b_L y^{b_L-1}\varepsilon-|\varepsilon|^{b_L} \tag{4}$$

with domain of definition $[-y,\infty]$. It will be sufficient to show that (3) and (4) are positive in their domain of definition to prove (2).

$$h_l(0)=0, \qquad h_l'(\varepsilon)=b_l\{(y+\varepsilon)^{b_l-1}-y^{b_l-1}\}.$$

The function h is decreasing in the interval $[-y,0]$ and increasing in $(0,\infty)$. Hence, it is a positive function in its domain of definition.

$$g(0)=0,$$

$$g'(\varepsilon)=\begin{cases} b_L\{(y+\varepsilon)^{b_L-1}-y^{b_L-1}-\varepsilon^{b_L-1}\}, & \text{if } \varepsilon\geqslant 0,\\ b_L\{(y+\varepsilon)^{b_L-1}-y^{b_L-1}+(-\varepsilon)^{b_L-1}\}, & \text{if } -y\leqslant\varepsilon\leqslant 0.\end{cases}$$

By considering that $(a+b)^x\geqslant a^x+b^x$ if a, b are positive real numbers and $x\geqslant 1$, we can see that the function g is decreasing in the interval $[-y,0]$ and increasing in $(0,\infty)$. Since $g(0)=0$, g must be positive in its domain of definition.

Taking then $\lambda=\mathrm{Min}_{ij}\, a_{L,ij}$ and $\theta=\frac{1}{2}\mathrm{Min}_{ij}\, b_{L,ij}$, the inequality (2) is proved.

References

[1] A. Auslander, Optimisation (Masson, 1976).
[2] S.C. Dafermos and F.T. Sparrow, The traffic assignment problem for a general network, J. Res. Nat. Bur. Standards B 37 (2) (1969).
[3] L.J. LeBlanc, E.K. Morlock and W.P. Pierskalla, An efficient approach to solving the road networks equilibrium traffic assignment problem, Transportation Res. 9 (1975) 309–318.
[4] R.M. Michaels, Attitudes of driver determine choice between alternate highways, Public Roads 34 (1965) 225–236.
[5] S. Nguyen, An algorithm for the traffic assignment, Transportation Sci. 8 (1974) 203–216.
[6] A. Weintraub and J. González, An efficient algorithm for the traffic assignment problem, Networks 10 (1980) 197–209.

P. Hansen, ed., Studies on Graphs and Discrete Programming

A RECURSIVE METHOD FOR SOLVING ASSIGNMENT PROBLEMS*

Gerald L. THOMPSON

Graduate School of Industrial Administration, Carnegie-Mellon University, Pittsburgh, PA 15213, USA

The recursive algorithm is a polynomially bounded nonsimplex method for solving assignment problems. It begins by finding the optimum solution for a problem defined from the first row, then finding the optimum for a problem defined from rows one and two, etc., continuing until it solves the problem consisting of all the rows. It is thus a dimension expanding rather than an improvement method such as the simplex. During the method the row duals are non-increasing and the column duals non-decreasing.

Best and worst case behavior is analyzed. It is shown that some problems can be solved in one pass through the data, while others may require many passes. The number of zero shifts (comparable to degenerate pivots in the primal method) is shown to be at most $\frac{1}{2}n^2$.

Extensive computational experience on the DEC-20 computer shows the method to be competitive for at least some kinds of assignment problems. Further tests on other computers are planned.

1. Introduction

The recursive algorithm is a nonsimplex method for solving assignment problems. It begins by finding the optimum solution for a problem defined from the first row, then finding the optimum for a problem defined from rows one and two, etc., continuing until it obtains the optimum for the problem containing all the rows. It is thus a dimension-expanding rather than an improvement method such as the simplex. Throughout the steps of the method the dual variables vary monotonically, with the row duals being non-increasing and column duals non-decreasing.

In this paper it is shown that in the best case the recursive method is a one data pass algorithm for some special kinds of assignment problems. That is, the method can solve such problems after looking each piece of data exactly once. (This should be compared with simplex type methods which require several data passes to get feasible starting solutions, additional passes to choose pivots, and a final complete data pass to verify optimality.) Although the classes of problems for which these one data pass solutions are possible are not

* This report was prepared as part of the activities of the Management Sciences Research Group, Carnegie-Mellon University, under contracts N00014-76-C-0932 and N00014-75-C-0621 NR 047-048 with the Office of Naval Research.

very interesting, computation experience is presented to show that more interesting problems can be solved by the recursive method with only slightly more than one data pass. Because only a relatively slow computer was available for testing, the number of data passes as well as computational times were recorded and presented here. It appears from this computational experience that the recursive method is competitive with other methods on most problems and unbeatable by other methods for the one pass and near one pass problems.

Worst case analysis is also presented which shows that the worst case bound for the recursive method is a polynomial which is less than half of the bound for other polynomially bounded assignment algorithms such as Ford–Fulkerson [5], Balinski–Gomory [1], and Srinivasan–Thompson [10]. Examples which exhibit both best and worst case bounds are presented.

This paper came out of earlier work by the author on auctions and market games [11, 12] and is based on previous work in these areas by Shapley and Shubik [7] and Barr and Shaftel [3]. The idea of solving bottleneck assignment problems by adding rows one by one has been previously used by Derigs and Zimmermann [4]. The backshift operation used in this paper is similar to analogous operations in the Derigs–Zimmermann paper and also to the one in the relaxation method of Hung and Rom [6]. Finally, the alternating path basis, which is shown in this paper to correspond to Orden's perturbation method, and also to the dutch auction procedure of Barr and Shaftel, was previously employed by Barr, Glover, and Klingman [2]. In [13] Tomizawa describes an inductive method for the assignment problem which begins by solving a 2×2 subproblem, then a 3×3 subproblem containing the previous one, etc. His solution technique involves solving shortest route subproblems.

The recursive method has been extended to transportation problems, and to bottleneck assignment and transportation problems. Those extensions will be presented elsewhere.

2. Notation and preliminary results

We consider the assignment problem to be a market game with the rows representing sellers and the columns buyers. The index sets of the sellers and buyers are

$$I' = \{1, \ldots, n\}, \tag{1}$$

$$J' = \{1, \ldots, n\}. \tag{2}$$

We assume each seller has one unit to sell and each buyer wants to buy one unit. Let

$$c_{ij} \geqslant 0 \tag{3}$$

be the bid of buyer j for seller i's good. For technical reasons we add a dummy seller and a dummy buyer so that the index sets (1) and (2) become

$$I = I' \cup \{n+1\} = \{1, \ldots, n, n+1\}, \tag{4}$$

$$J = J' \cup \{n+1\} = \{1, \ldots, n, n+1\}. \tag{5}$$

The bids for these dummy players are

$$c_{n+1,j} = 0 \quad \text{for } j \in J, \quad \text{and} \quad c_{i,n+1} = 0 \quad \text{for } i \in I. \tag{6}$$

The assignment problem can now be stated as

$$\text{maximize} \quad Z = \sum_{i \in I} \sum_{j \in J} c_{ij} x_{ij}, \tag{7}$$

$$\text{subject to} \quad \sum_{j \in J} x_{ij} = 1 \quad \text{for } i \in I', \tag{8}$$

$$\sum_{i \in I} x_{ij} = 1 \quad \text{for } j \in J', \tag{9}$$

$$\sum_{j \in J} x_{n+1,j} = 0, \tag{10}$$

$$\sum_{i \in I} x_{i,n+1} = 0, \tag{11}$$

$$x_{ij} \geqslant 0. \tag{12}$$

It is well known that the assignment problem is massively (primal) degenerate. One way of handling this problem is to perturb the right hand sides of (8), (10) and (1)) as follows:

$$\text{(P1)} \quad \begin{cases} \sum_{j \in J} x_{ij} = 1 + \varepsilon \quad \text{for } i \in I', & (8') \\ \sum_{j \in J} x_{n+1,j} = \varepsilon, & (10') \\ \sum_{i \in I} x_{i,n+1} = (n+1)\varepsilon, & (11') \end{cases}$$

where $0 \leqslant \varepsilon < \frac{1}{2}(n+1)$. We call this perturbation (P1). (Another perturbation (P2) which adds ε to the right hand sides of (9) and (11) and $(n+1)\varepsilon$ to the right hand side of (10) leads to an equivalent algorithm which we do not discuss here.)

The following facts are well known. For a small positive ε, a *basic* feasible solution to (8′), (9), (10′) and (11′) has exactly $2n+1$ cells (i, j) with $x_{ij}>0$; all other $x_{ij}=0$. If B is the set of basis cells then the graph $G=(I \cup J, B)$ is a *tree*. Let $X(\varepsilon)$ be a basic solution, let x_{ij} be the optimal shipping amount for basis cell (i, j), and let $R(x_{ij})$ be the scientifically rounded value of x_{ij}; if $R(x_{ij})=1$ then we say cell (i, j) ships 1, and if $R(x_{ij})=0$ then we say cell (i, j) ships 0.

Lemma 1. *Given perturbation* (P1) *a feasible solution is basic if and only if each row* (*except* $n+1$) *has exactly* 2 *basis cells, one shipping* 1 *and the other shipping* 0.

Proof. Each row $i \in I'$ has a supply of $1+\varepsilon$ and row $n+1$ has a supply of ε, while each column $j \in J'$ has a demand of 1 and column $n+1$ has demand of $(n+1)\varepsilon$. Hence each row $i \in I'$ must have at least two basis cells to use up its supply, and row $n+1$ must have at least one basis cell. Since this adds up to $2n+1$ basis cells, each row $i \in I'$ has exactly 2 basis cells and row $n+1$ has 1. To show the rest of the statement, recall that every tree has at least one pendant node, i.e., there is at least one cell (i, j) which is the only basis cell in column j for some j. Since the solution is assumed feasible, $x_{ij}=1$ i.e., (i, j) ships 1. The other basis cell in row must therefore ship 0. Now cross out row i and column j and repeat the same argument for the rest of the rows.

We now introduce some graph terminology. Assume the tree $G=(I \cup J, B)$ is drawn with row $n+1$ as the root, see [8], which appears as the node at the top of the tree, and which downward pointing arcs going to other nodes. Let a downward arc be the son (or successor) relation and the reverse relation be the father (or predecessor) relation. All nodes, except row $n+1$, have fathers.

Definition. A basis tree has the *unique row-son property* if every row node (except possibly the root node) has one and only one son.

Lemma 2. *Given perturbation* (P1).

(a) *The graph of a feasible solution has the unique row-son property.*

(b) *The graph of a feasible solution is an alternating path basis*[1] *in the sense of Barr, Glover, and Klingman* [2].

Part (a) can be proved by using the crossing out routine given in the proof of Lemma 1. The unique row-son property is simply another way of stating the definition of an alternating path basis, see [2].

[1] An alternating path basis is a tree (graph) satisfying the following: the root of the tree is row $n+1$; each arc of the tree connecting a column node to its father ships 1; each arc connecting a row node (except row $n+1$) to its father ships 0.

The dual problem to the perturbed problem (7), (8′), (9), (10′), (11′), (12) is

$$\text{minimize} \quad \left\{\sum_{i\in I'} u_i(1+\varepsilon)+u_{n+1}\varepsilon+\sum_{j\in J'} v_j+v_{n+1}(n+1)\varepsilon\right\}, \tag{13}$$

$$\text{subject to} \quad u_i+v_j \geqslant c_{ij} \quad \text{for } i\in I,\ j\in J. \tag{14}$$

Because of assumption (3), it can be shown (see [11]) that we can also impose nonnegativity constraints

$$u_{n+1}\geqslant 0 \quad \text{and} \quad v_{n+1}\geqslant 0. \tag{15}$$

This, together with (6), implies $u_{n+1}=v_{n+1}=0$ so that (13) becomes

$$\text{minimize} \quad \left\{\left[\sum_{i\in I'} u_i+\sum_{j\in J'} v_j\right]+\varepsilon\sum_{i\in I'} u_i\right\}. \tag{16}$$

Definition. We call u_i the *price* for seller i, and v_j the *surplus* of buyer j.

Lemma 3. *The solution of the problem with perturbation* (P1) *simultaneously*
(a) *minimizes the sum of the sellers prices;*
(b) *maximizes the sum of the buyers surpluses.*

Proof. An optimal solution to (7), (8′), (9), (10′), (11′) and (12) for $\varepsilon>0$ remains optimal when ε is replaced by 0, see [10]. Let Z be the value of the optimal solution when $\varepsilon=0$. By the duality theorem if u_i and v_j are optimal solutions to (14), (15), and (16) when $\varepsilon=0$

$$Z=\sum_{i\in I'} u_i+\sum_{j\in J'} v_j. \tag{17}$$

Substituting (17) in to (16) gives for $\varepsilon>0$

$$\text{minimize} \quad \left\{Z+\varepsilon\sum_{i\in I'} u_i\right\} \tag{18}$$

which, since Z is constant, proves (a). Solving (17) for $\sum u_i$ and substituting into (18) gives

$$\text{minimize} \quad \left\{Z(1+\varepsilon)-\varepsilon\sum_{j\in J'} v_j\right\} \tag{19}$$

which proves (b).

The set of all solutions to (14), (15), and (16) with $\varepsilon=0$ is called the *core* of the game by Shapley and Shubik [7]. In [11], the author gives an algorithm for finding all basic solutions in the core, starting with the maximum buyer surplus solution of Lemma 3.

One last bit of terminology is appropriate. A *dutch auction* is a concealed bid auction in which the price (indicated by a 'clock') starts high and steadily drops; when the highest bidder makes his (concealed) bid he is certain to get the object being sold; however, the price continues to drop until the second highest bidder makes his (concealed) bid which 'stops the clock'; the price the highest bidder pays is therefore determined by the bid of the second highest bidder. The solution procedure to be proposed is formally like a dutch auction. As we saw above, the perturbation (P1), the solution has two basis cells in each row, one shipping 1, corresponding to the highest bidder who gets the good, and the other shipping 0, corresponding to the second highest bidder who determines the price. In the description of the algorithm of the next section, economic as well as mathematical terminology will frequently be used because it provides intuitive reasons as to why the procedures are carried out as they are.

3. Outline of the recursive method

As noted earlier, the recursive method is not a simplex method even though it uses some of the terminology and concepts of that method. In order to emphasize this difference we state two characteristics of the recursive method which are distinctly different from those of the simplex method.

(a) In the recursive method every feasible primal or dual solution found is also optimal.

(b) In the recursive method a primal solution is not proved to be optimal by means of the dual; in fact, it is possible to find an optimal primal solution to an assignment problem without ever finding its optimal dual solution.

Definition. Problem P_k consists of the first k sellers, the dummy seller $n+1$ who is given $n-k$ dummy units to sell, and all the buyers including the dummy buyer. In other words, problem P_k can be obtained from (7)–(12) by replacing I by $I_k=\{1, \ldots, k, n+1\}$, and the right hand side of (10) by $n-k$.

A brief outline of the recursive method is: solve P_0; from its solution find the solution to P_1; from P_1 solve P_2; etc.; continue until the solution to P_n is produced from the solution to P_{n-1}.

A more detailed outline is given in Fig. 1. As indicated we first find the optimal primal and dual solutions to P_0 which is the trivial one row problem conisting of the slack row only. From them we find the optimal primal solution to P_1, which consists of the first row of the assignment problem, and the slack row. Then we find the optimal dual solution to P_1. Then we find the optimal primal solution to P_2 followed by its optimal dual solution. Etc. Finally, the optimal primal solution to P_n is constructed from the optimal solution to P_{n-1}. As noted in Fig. 1, the actual finding of the optimal dual solution to P_n is

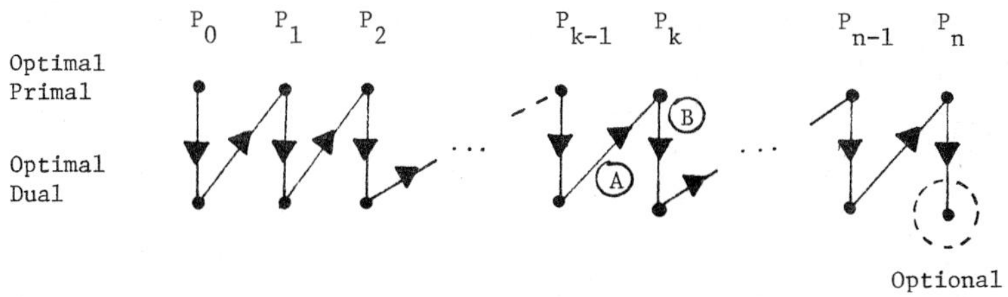

Fig. 1. Order of solution of problems P_k in the recursive method.

optional, and need not be carried out if only the optimal primal solution is wanted. As noted in the figure there are two recursive Steps A and B, where Step A constructs the optimal primal solution to P_k from the optimal solution to P_{k-1}, and Step B constructs the optimal dual solution to P_k from its optimal primal. These two steps will be discussed in detail in the next two sections. In the remainder of this section we discuss the solutions of P_0 and P_1.

The initial basis graph for P_0 is shown in Fig. 2. It corresponds to the solution

$$x_{n+1,j}=1 \quad \text{for } j\in J', \qquad x_{n+1,n+1}=0,$$

$$u_{n+1}=0 \qquad \text{and} \qquad v_j=0 \quad \text{for } j\in J.$$

In order to solve P_1 we find the two largest reduced bids in row 1. Let C_l and C_s be columns in which the largest and second largest reduced bids are, i.e., since all $v_j=0$,

$$c_{1l}=\operatorname*{Max}_{j\in J} c_{1j}, \qquad c_{1s}=\operatorname*{Max}_{j\in J-\{l\}} c_{1j}.$$

We then perform an *easy backshift* to get the optimal primal and dual solutions for P_1 as illustrated in Fig. 3. First node C_l is cut out of the P_0 basis graph; then node R_1 is made the son of C_s, node C_l the son of R_1, and finally we set $u_1=c_{1s}$ and $v_l=c_{1l}-c_{1s}$. As is evident, only a few instructions are needed to perform the easy backshift. Also it is easy to check that the indicated solution is optimal for problem P_1.

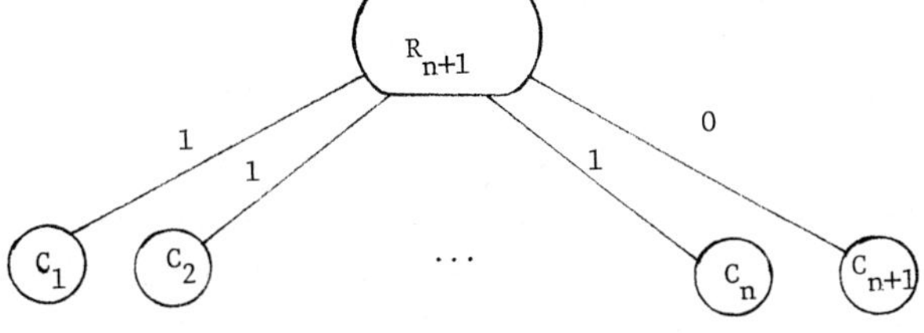

Fig. 2. Optimal basis graph for P_0.

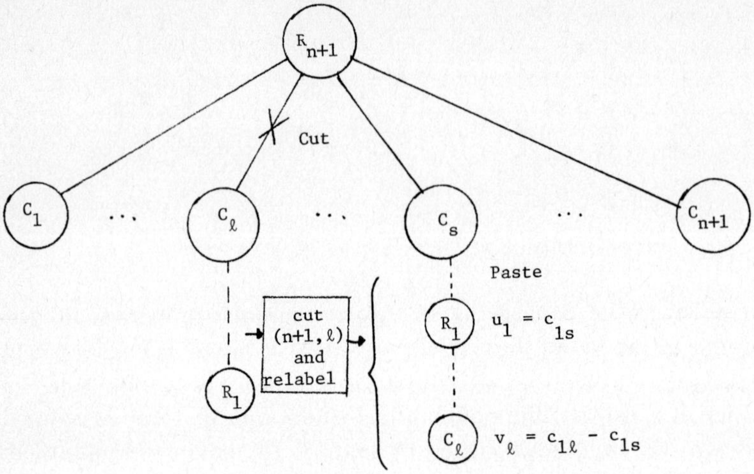

Fig. 3. An easy backshift. It is always possible to solve P_1 by an easy backshift.

Surprisingly, it turns out that the same easy backshift operation suffices to solve many of the subproblems. As will be seen in Section 8 on computational experience, for most problems from 40 to 95 percent or more of the subproblems P_k can be solved in this extremely simple fashion.

4. Recursive Step A

The purpose of recursive Step A is: from the known basic optimal primal and dual solutions to problem P_{k-1} find an optimal primal solution to P_k. This will be done by carrying out a backshift step, to be defined next. Recall that the basis tree T_{k-1} for P_{k-1} is always stored with node R_{n+1} as the root.

The (Ordinary) Backshift Step

1. In one pass through row k find the two largest reduced cost entries; i.e.

$$c_{kl} - v_l = \underset{j \in J}{\text{Maximum}}(c_{kj} - v_j),$$

$$c_{kt} - v_t = \underset{j \in J-\{l\}}{\text{Maximum}}(c_{kj} - v_j).$$

2. Set $u_k = c_{kt} - v_t$.
3. Add node R_k and arc (k, l) to T_{k-1}.
4. Suppose the backward path (found by applying the father relation) from R_k to R_{n+1} is $R_k, C_l, \ldots, C_p, R_{n+1}$. Interchange fathers and sons along this backward path from R_k to C_p.
5. Cut (remove) arc $(p, n+1)$ from T_{k-1}.

A picture of the ordinary backshift step is shown in Fig. 4. Note that the unique row son property is preserved in the relabelling of Step 4. Notice also that at Step 5 of the backshift step the tree is broken into two subtrees: tree T_k^* which has root R_k, and tree T_{n+1} which has root R_{n+1}. Also, as marked in the figure, because no dual variables are changed during the backshift step, all arcs are dual feasible except (possibly) for the new arc (k, l).

The comparision of the backshift operation with a primal pivot step is instructive. The search for the incoming cell in Step 1 is confined to a single row; there is no cycle finding; there is no search for the outgoing cell, since it is known in advance to be $(n+1, p)$; finally no dual changes are made during the backshift operation. Hence the backshift step requires less than half the work of a primal pivot.

Counting easy backshifts, there are always exactly n backshift steps made in the course of solving an $n \times n$ assignment problem. As will be seen empirically in Section 8, from 40–95 per cent of the backshifts are easy and the rest are ordinary. A final remark is that the backshift step is the only operation of the recursive method in which the father-son relationships at a node are interchanged.

Lemma 4. *The backshift operation produces a primal feasible solution whose objective function satisfies*

$$Z_k = Z_{k-1} + (c_{kl} - v_l) \tag{20}$$

where Z_h is the objective function value for problem P_h and v_l is the dual variable for column l in problem P_{k-1}.

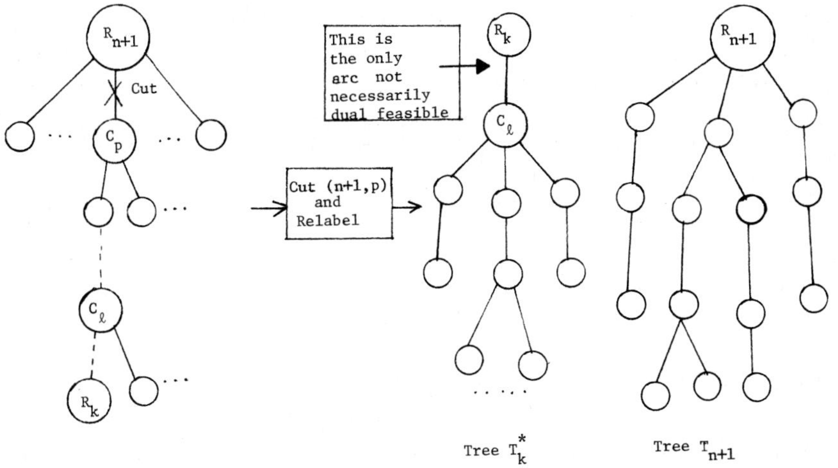

Fig. 4. An ordinary backshift step.

Proof. The fact that the backshift step preserves the unique row-son property and hence produces a primal feasible solution was noted above. Let s and s' be the son functions for the solutions of P_{k-1} and P_k, respectively, let f be the father functon for P_{k-1}, let $I_h = \{1, \ldots, h\}$, and let B_{k-1} be the set of rows on the backward path in T_{k-1} from node C_l to node C_p. Then we have

$$\begin{aligned} Z_k - Z_{k-1} &= \sum_{h \in I_k} c_{h,s'(h)} - \sum_{h \in I_{k-1}} c_{h,s(h)} \\ &= c_{k,l} + \sum_{h \in I_{k-1}} [c_{h,s'(h)} - c_{h,s(h)}] \\ &= c_{k,l} - \sum_{h \in B_{k-1}} [c_{h,s(h)} - c_{h,f(h)}] \\ &= c_{k,l} - v_l. \end{aligned}$$

The third step follows from the fact that $s'(h) = s(h)$ for $h \notin B_{k-1}$ and $s'(h) = f(h)$ for $h \in B_{k-1}$. The last step follows from the facts that $u_{n+1} = v_p = 0$ in the solution to P_{k-1}, and v_l can be computed by working down the tree from node C_p to node C_l.

Theorem 1. *The backshift step produces a primal optimal solution to Problem P_k.*

Proof. We present two proofs of this theorem. The first rests on an economic argument which some readers may not find completely convincing. The second proof, however, rests on standard linear programming reasoning.

Economic Optimality Proof. Think of the change from P_{k-1} to P_k as that of an auction in which one more seller is added. The buyers positions cannot be worsened by this change, i.e., buyer surpluses v_j will stay the same or increase. Consequently buyer j's bid in P_k for seller k's good is $c_{kj} - v_j$, i.e., it is his original bid c_{kj} less his current surplus v_j. The maximum change $\Delta Z_{k-1} = Z_k - Z_{k-1}$ that can be made in the objective function is obtained by 'selling to the highest bidder', buyer l. By Lemma 4, this sale actually produces the maximum possible change $c_{kl} - v_l$ in the objective function, hence the theorem.

Duality Proof. As will be shown in the next section (see Theorem 2), recursive Step B produces in a finite number (at most k) of steps a feasible dual solution to P_k corresponding to the primal solution obtained from the backshift step. From the duality theorem of linear programming both the primal and dual solutions to P_k are therefore optimal.

5. Recursive Step B

The purpose of recursive Step B is: from the primal solution to P_k found at the end of recursive Step A, construct a basic (maximal buyer surplus) dual feasible (hence optimal) solution to P_k without changing the primal solution. As can be seen in Fig. 4, this step has two objectives: (a) to make arc (k, l) dual feasible, and (b) to join together the trees T_k^* and T_{n+1}.

To help intuition concerning recursive Step B is is best to revert to economic terminology. In Fig. 4, the row and column nodes in tree T_{n+1} represent sellers and buyers whose prices and surpluses, respectively, are correct for problem P_k; however, the sellers (row nodes) in tree T_k^* have prices which are (possibly) to high while the buyers (column nodes) in T_k^* have surpluses which are (possibly) too low.

In the first part of recursive Step B, called Step B1, we simultaneously reduce all prices for sellers in T_k^* and increase all surpluses for buyers in T_k^*. As prices drop, a buyer in T_{n+1} may become a second highest bidder for the good of a seller in T_k^*. Then, as illustrated in Fig. 5, part of tree T_k^* is removed and attached to T_{n+1}. This is called a *zero shift* operation. Recursive Step B1 continues with zero shifts being made as needed until it is seller k whose good has a second highest bidder in tree T_{n+1}. At this point, no zero shift is made,

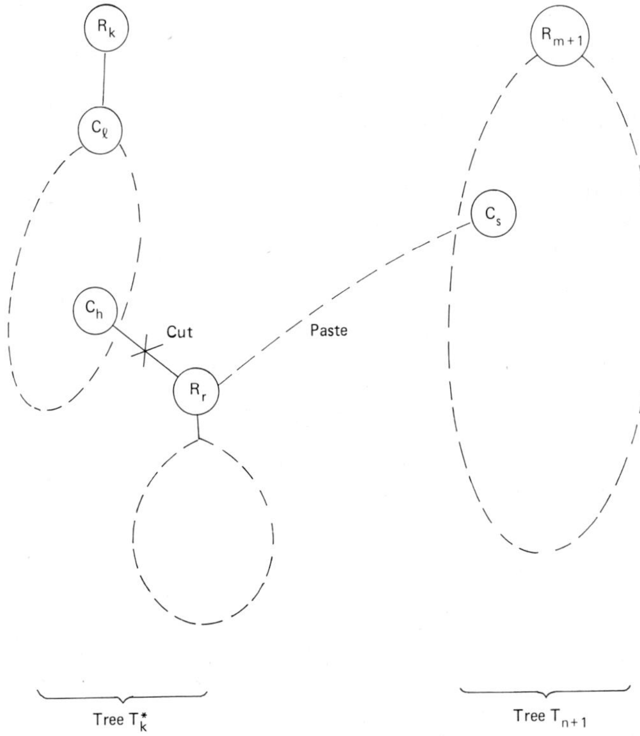

Fig. 5. A zero shift step.

and recursive Step B2 is initiated. Note that the infeasibility of arc (k, l) has not changed during Step B1. What has been accomplished during this step is that the correct selling price u_k for seller k has been determined.

During recursive Step B2 the selling price u_k is held fixed while all other sellers' prices in T_k^* are reduced, and all buyer surpluses v_j for j in T_k^* are increased simultaneously, with zero shifts being made as needed. Since the reduced cost of arc (k, l) is $c_{kl} - u_k - v_j$ we see that arc (k, l) becomes steadily less infeasible as v_j increases. When v_j has become large enough so that arc (k, l) is just feasible, changes in seller prices and buyer surpluses stop, tree T_k^* is pasted to tree T_{n+1} and recursive Step B2 ends with dual and primal feasible (hence optimal) solutions to problem P_k.

In order to describe these steps precisely we define two subroutines which depend on a subset of rows, called ROWS, of T_k^* and a subset of columns, called COLS, of T_{n+1}. The first subroutine, SEARCH, is defined as follows:

SEARCH. Calculate

$$\lambda = \underset{\substack{i \in \text{ROWS} \\ j \in \text{COLS}}}{\text{Minimum}} (u_i + v_j - c_{ij}).$$

Let (r, s) be a cell on which the minimum is taken on.
Let h be the father of r in T_k^*.
Let T_r be the subtree of T_k^* which is below r.

The second subroutine, DUAL, is defined as:

DUAL. Let

$$u_i \to u_i - \lambda \quad \text{for } i \in \text{ROWS},$$

$$v_j \to v_j + \lambda \quad \text{for } j \notin \text{COLS}.$$

Note that application of the dual operation, which is the only step in which dual variables are changed can only decrease seller prices and increase buyer surpluses.

Given these subroutines we can now define Recursive Step B1.

Step B1

(1) Let ROWS be the set of rows in T_k^*; let COLS be the set of columns in T_{n+1}.

(2) Use SEARCH to find λ, r, s, h, T_r.

(3) Change dual variables by applying DUAL.

(4) If $r = k$ go to recursive Step B2. Otherwise go to 5.

(5) Cut arc (r, h) and paste arc (r, s).

(6) Go to (1).

We can define Recursive Step B2 similarly.

Step B2

(1) Let ROWS be the set of rows in T_k^* except for row k; let COLS be the set of columns in T_{n+1}.

(2) Let $\lambda^* = c_{kl} - u_k - v_l$, where l was found as the largest reduced bid in row k.

(3) Use SEARCH to find λ, r, s, h, T_r.

(4) If $\lambda \geq \lambda^*$ go to (7); otherwise go to (5).

(5) Change dual variables by applying DUAL.

(6) Cut (r, h); paste (r, s). Go to (1).

(7) Let $\lambda \to \lambda^*$; apply DUAL.

(8) Paste (k, s).

(9) END

Each time the cut and paste operation is performed in either Step B1 or B2 we will say a *zero shift* has been made. Sometimes during the course of application of the SEARCH subroutine it finds $\lambda = 0$; when this happens it is possible to do the cut and paste operation without making a dual change and continue the SEARCH routine without finding ROWS and COLS again; such a zero shift is called an *easy zero shift.* Use of easy zero shifts has greatly speeded up the performance of the code, see Section 8.

The maximum number of elements in the set ROWS for either Step B1 or B2 is k. Each time the cut and paste step is made at least one row (and at least one column) is transferred from the tree T_k^* to tree T_{n+1} and ROWS is correspondingly made smaller. Therefore k is the maximum number of zero shifts that can be made by recursive Step B at step k.

Note that when T_k^* consists of just row k and its son, the set ROWS in (1) of Step B2 is empty, so that λ found in (3) is $+\infty$. In this case the algorithm is certain to terminate since $\lambda = \infty > \lambda^*$ in (4). Also when (8) of Step B2 is made, a basic feasible dual solution has been found.

We summarize the above in a theorem.

Theorem 2. *Given the primal feasible solution to P_k produced by recursive Step* A, *the application of recursive Steps* B1 *and* B2 *will produce a basic feasible dual solution to P_k after making at most k zero shifts. During the application of recursive Step* B *the seller prices are nonincreasing and the buyer surpluses are nondecreasing.*

6. Solution of an example

Consider the 3×3 example shown in Fig. 6(a) in which a slack row and column have also been added. Fig. 6(b) shows the solution to problem P_0,

-31-

15	14	17	0	1
19	22	20	0	1
17	21	14	0	1
0	0	0	0	0
1	1	1	0	

(a) Original Problem.

	0	0	0	0	
0	$\textcircled{0}^{1}$	$\textcircled{0}^{1}$	$\textcircled{0}^{1}$	$\textcircled{0}^{0}$	3
	1	1	1	0	

(b) Optimal Solution to P_0.

	0	0	2	0
15	$\textcircled{15}^{0}$	14	$\textcircled{17}^{1}$	0
0	$\textcircled{0}^{1}$	$\textcircled{0}^{1}$	0	$\textcircled{0}^{0}$

(c) Optimal Solution to P_1.

	0	$0+\lambda$	2	
15	$\textcircled{15}^{0}$	14	$\textcircled{17}^{1}$	0
19	19	$\textcircled{22}^{1}$	20	0
0	$\textcircled{0}^{1}$	0	0	$\textcircled{0}^{0}$

(d) End of Backshift Step for P_2.

	0	3	2	0
15	$\textcircled{15}^{0}$	14	$\textcircled{17}^{1}$	0
19	$\textcircled{19}^{0}$	$\textcircled{22}^{1}$	20	0
0	$\textcircled{0}^{1}$	0	0	$\textcircled{0}^{0}$

(e) Optimal Solution to P_2.

	$0+\lambda$	$3+\lambda$	$2+\lambda$	0
$15-\lambda$	$\textcircled{15}^{0}$	14	$\textcircled{17}^{1}$	0
$19-\lambda$	$\textcircled{19}^{1}$	$\textcircled{22}^{0}$	20	0
$17-\lambda$	17	$\textcircled{21}^{1}$	14	0
	0	0	0	$\textcircled{0}^{0}$

(f) End of Backshift Step for P_3. Note that the optimal assignments are now determined.

	$15+\lambda$	$18+\lambda$	17	0
0	15	14	$\textcircled{17}^{1}$	$\textcircled{0}^{0}$
$4-\lambda$	$\textcircled{19}^{1}$	$\textcircled{22}^{0}$	20	0
$2-\lambda$	17	$\textcircled{21}^{1}$	14	0
0	0	0	0	$\textcircled{0}^{0}$

(g) End of first zero shift.

	16	$19+\lambda$	17	0
0	15	14	$\textcircled{17}^{1}$	$\textcircled{0}^{0}$
3	$\textcircled{19}^{1}$	22	$\textcircled{20}^{0}$	0
$1-\lambda$	17	$\textcircled{21}^{1}$	14	0
	0	0	0	$\textcircled{0}^{0}$

(h) End of second zero shift.

	16	20	17	0
0	15	14	$\textcircled{17}^{1}$	$\textcircled{0}^{0}$
3	$\textcircled{19}^{1}$	22	$\textcircled{20}^{0}$	0
1	$\textcircled{17}^{0}$	$\textcircled{21}^{1}$	14	0
0	0	0	0	$\textcircled{0}^{0}$

(i) End of third zero shift. Optimal Solution to P_3.

Fig. 6. Solution of an example.

where the basis cells are circled, and the shipping amounts for each basis cell are marked above the circle. Dual variables are marked to the left and the top of the figure. These conventions hold for the rest of the figures as well.

Since 17 and 15 are the largest and second largest bids in the first row of the problem, the easy backshift operation applied to P_1 yields the optimal solution in Fig. 6(c); the optimal basis tree for P_1 is shown in Fig. 7(a).

As can be seen in Fig. 6(d) the two highest bids in row 2 of P_2 are 22 and 19, because the reduced cost in column 3 is 18. The situation at the end of the first backpivot is shown in Fig. 6(d). The corresponding to parts of the basis tree are shown in Fig. 7(b). Because T_k^* has only row k in it recursive Step B1 is empty, and Step B2 can be carried out by pasting R2 to C1 by means of arc (2, 1); also the λ in Fig. 6(d) is set equal to 3 as indicated in (7) of B2. The resulting tableau is shown in Fig. 6(e) and the optimal basis tree appears in Fig. 7(c).

When the third row of the original problem is added, it can be seen that the highest reduced bid of 18 occurs in column 2 and the second highest bid of 17

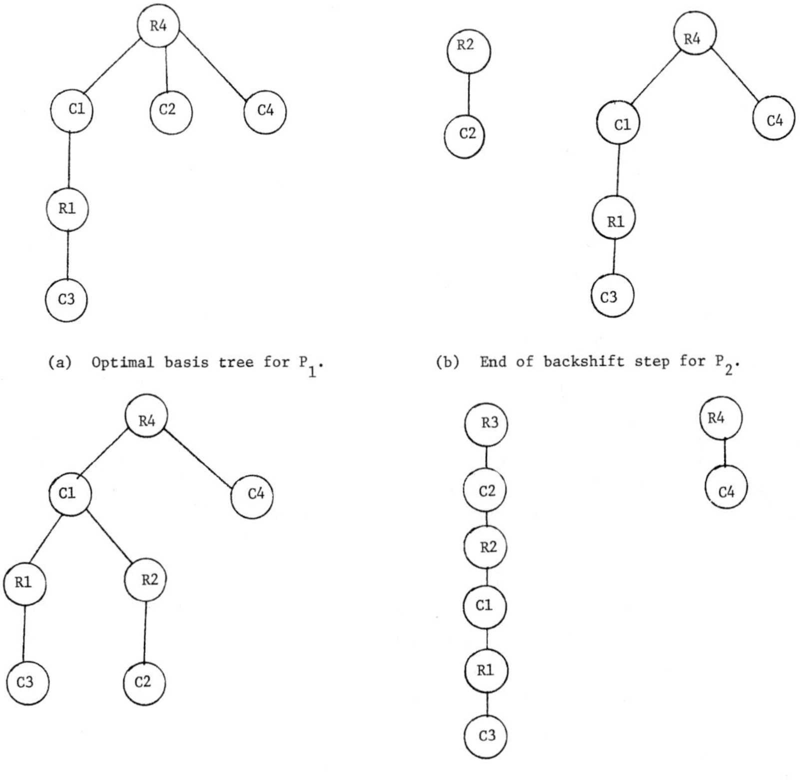

(a) Optimal basis tree for P_1.

(b) End of backshift step for P_2.

(c) Optimal basis tree for P_2.

(d) End of backshift step for P_3.

Fig. 7.

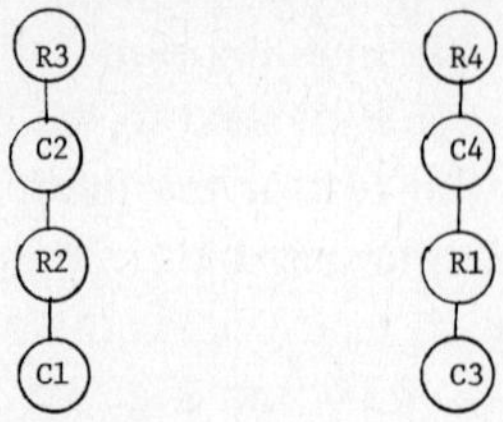

(e) End of first zero shift.

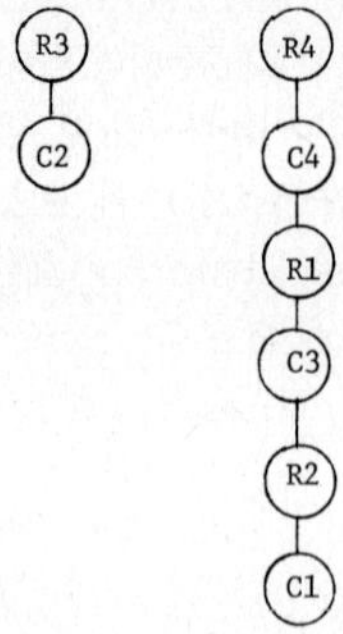

(f) End of second zero shift.

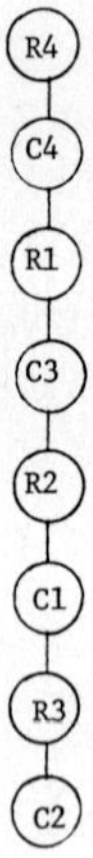

(g) End of third zero shift.
Optimal basis tree for P_3.

Fig. 7 (cont.).

is in column 1. The backpivot step involves adding R3 as a son of C2 in Fig. 7(c) relabelling the tree back to C1, and cutting arc (4, 1). The resulting two parts of the basis tree appear in Fig. 7(d). The tableau at the end of the backpivot step appears in Fig. 6(f). Note that the shipping amounts have changed as follows: x_{22} from 1 to 0, x_{21} from 0 to 1, and x_{41} from 1 to 0. The actual transfers of goods from sellers to buyers are indicated by the basis cells which ship 1 in Fig. 6(f). These transfers are, in fact, the optimum assignment as proved in Theorem 1. Of course, the dual solution in Fig. 6(f) is neither feasible nor basic and we go to recursive Steps B1 and B2 to achieve that. The direction of price and buyer surplus changes for Step B1 are indicated by the λ's marked in Fig. 6(f). Here ROWS$=\{1, 2, 3\}$ and COLS$=\{4\}$: hence $\lambda = 15$, $r=1$, $h=1$, and $s=4$. Hence we change duals by 15, cut arc (1,1), paste arc (1, 4) and arrive after one zero shift at the situation indicated in Figs. 6(g) and 7(e). We repeat Step B1 with ROWS$=\{2, 3\}$, COLS$=\{3, 4\}$ and λ changes as marked in Fig. 6(g). Now $\lambda = 1$, $r=2$, $h=2$, and $s=3$. Hence we change duals by 1, cut arc (2, 2), paste arc (2, 3) and arrive at the situation depicted in Figs.

6(h) and 7(f). Now ROWS = {3}, COLS = {1, 3, 4}, $\lambda = 0$, $r = 3$, h is not defined, and $s = 1$. We are now in Step B2 with $\lambda^* = 1$ and ROWS and COLS as before. Now $\lambda = \infty$ so we paste arc (3, 1) increase v_2 by 1 and come to the optimal tableau in Fig. 6(i) and its corresponding optimal basis tree in Fig. 7(g).

7. Best case and worst case analysis

In contrast to many other mathematical programming algorithms, it is easy to determine the best and worst case behavior for the recursive algorithm, and it is also easy to find examples which have either kinds of behavior. In fact, a single example will be given which exhibits both kinds of behavior, depending on the order of introduction of the sellers (rows)! A second example which shows not quite such extreme behavior is also presented.

The numbers of easy backshifts, ordinary backshifts, and zero shifts for the best and worst cases are shown in Table 1. These numbers for the best case are easy to determine. The maximum number of easy backshifts is $n-1$, since in the most favorable case all the second highest bidders can be chosen to be in the same column, and that column is therefore not available for an easy backshift. The minimum number of zero shifts is 1 (at least if the optimum dual to P_n is desired) since at step n the tree T_{n+1} has just two elements R_{n+1} and C_{n+1}, and at least one zero shift is needed to hook T_n^* and T_{n+1} together.

For the worst case analysis, recall that the solution to P_1 involves an easy backshift, and that there are always exactly n backshifts. Hence the worst case involves 1 easy backshift and $n-1$ ordinary backshifts. To determine the maximum number of zero shifts recall that no zero shift is needed for P_1; for problem P_k with $k \geq 2$ the maximum number of rows in T_k^* after the backshift is k, and at each zero shift at least one row is shifted from T_k^* to T_{n+1}; therefore the maximum total number of zero shifts is

$$2+3+\cdots+(n-1)+n=\tfrac{1}{2}(n+2)(n-1) \tag{21}$$

as indicated in Table 1.

Table 1
Number of operations for the best and worst cases

	Number of easy backshifts	Number of ordinary backshifts	Number of zero shifts
Best Case	$n-1$	1	1
Worst Case	1	$n-1$	$\frac{1}{2}(n+2)(n-1)$

Table 2
Best and worst case examples

Problem defined in eq.	Easy backshifts	Zero shifts	Easy zero shifts	Area search factor	Time (DEC-20 secs.)
(22)	99	1	0	1.00	.26
(23)	1	5049	4949	1.51	2.99
(24)	1	99	0	34.48	8.22
(25)	1	5049	0	859.48	166.20

To exhibit the various best and worst cases consider the following four examples with $m = n = 100$,

$$c_{ij} = \begin{cases} 1, & \text{if } j \neq 1, \\ 101 - i, & \text{if } j = 1, \end{cases} \tag{22}$$

$$c_{ij} = \begin{cases} 1, & \text{if } j \neq 1, \\ i, & \text{if } j = 1, \end{cases} \tag{23}$$

$$c_{ij} = (101 - i)(101 - j), \tag{24}$$

$$c_{ij} = i \cdot j. \tag{25}$$

Note that problems (22) and (23) are related in that the order of listing the rows is reversed. Problems (24) and (25) are similarly related.

The computer solution statistics for these four problems are given in Table 2. Note that problem (22) is a one-pass problem, while problem (23), which is the same except for order of rows, has the maximum number (5049) of zero shifts which is given by (21) when $n = 100$. However, the area search factor (that is, the ratio of the number of data calls to the number of data elements) which indicates the number of times the data is searched to solve (23), is only 50 percent more than one data pass and more than 98 percent of the zero shifts are easy zero shifts; also the time to solve (23) is more than 10 times that of (22).

Problem (25) is known to be of maximum difficulty for the Hungarian algorithm, and is also of maximum difficulty for the recursive method; as indicated in Table 2, its solution requires 5049 zero shifts none of which is easy. The time to solve (25) is nearly 640 times as long as the time to solve (22), which is somewhat less than the ratio of the areas searched. Problem (24), which is the same as (25) except for order of rows, is much easier in that it requires 99 zero shifts none of which is easy. The time to solve (25) is 20 times as long as (24), which somewhat less than the ratio of the areas searched. Note that the costs in problems (24) and (25) range from 1 to 10 000.

Table 3
Worst case bounds for various polynomially bounded algorithms

Ford–Fulkerson dual method [5]	$n^2 - 1$ non-breakthroughs
Balinski–Gomory primal method [1]	n^2 non-breakthroughs
Srinivasan–Thompson cost operator method [10]	$n(2n+1)$ primal pivots
Recursive method in this paper	$\frac{1}{2}(n+2)(n-1)$ zero shifts

Another way to evaluate the recursive method is to compare its worst case behavior with worst case behaviors for three other methods due to Ford–Fulkerson [5], Balinski–Gomory [1], and Srinivasan–Thompson [10], which are known to be polynomial bounded, see Table 3. The amount of work needed for primal or dual non-breakthrough steps is approximately the same as a primal pivot step in the cost operator method (see [10]) and is somewhat more than a zero shift step in the recursive method. However the recursive method's worst case behavior is less than half that of other methods.

Comparisons of worst case behavior are not usually good indicators of average case behavior. In the next section some computational results on randomly generated problems is discussed.

8. Computational experience

The recursive method has been programmed in FORTRAN and extensively tested. The computer used was a DEC-20 which is (a) relatively slow, (b) has a limited memory, and (c) is always operated in a multi-programming environment so that timing results can vary by as much as 10–50 percent on different runs of the same problem.

Table 4 presents computational experience for 150 randomly generated problems having $n = 100$, 200, or 300 so that the number of nodes ($2n$) is 200, 400, or 600, and approximately 4000 arcs with arc costs unformly distributed between 0 and the indicated maximum cost. Each row of the table presents average results for 5 randomly generated problems. The maximum cost ranged from 1 to 1000. By observing one of the problem sizes for various maximum costs the so-called 'minimum cost effect' (which we will call the 'maximum cost effect' since we have a maximizing rather than a minimizing problem) first observed by Srinivasan and Thompson in [9] will be seen; that is, for a fixed number of nodes and arcs, as the maximum cost increases from 1 to 1000 the time and area search factor increase while the percentages of easy backshifts and zero shifts decrease. In other words, for a fixed problem size, the larger the cost range, the harder the problem.

Table 4
Computational experience for problems having approximately 4000 arcs. Each row presents average results for five randomly generated test problems

Maximum cost	n	Average number of arcs	Percentage of easy backshifts	Number of zero shifts	Percentage of easy zero shifts	Area search factor	Time (DEC-20 secs.)
1	100	3980	90.4	81.8	85.46	1.08	.14
	200	3968	82.5	263.6	85.5	1.25	.27
	300	4267	78.1	630.6	89.36	1.60	.54
2	100	4002	87.0	100.8	81.93	1.14	.16
	200	4017	76.3	396.2	86.94	1.56	.34
	300	4278	64.9	858.6	86.75	2.42	.75
5	100	3973	73.8	144.4	80.91	1.39	.20
	200	3981	59.0	650.0	86.75	3.83	.74
	300	4256	54.3	1304.4	88.84	8.41	1.34
10	100	3994	60.6	233.4	86.68	2.79	.42
	200	4016	48.7	658.8	82.19	7.38	.93
	300	4298	47.6	1104.4	82.84	11.57	1.88
20	100	4003	53.2	275.2	80.52	5.74	.61
	200	4016	45.7	721.0	79.00	11.66	1.43
	300	4321	44.9	1078.6	77.36	15.24	2.18
50	100	3999	45.6	281.4	69.59	10.04	.97
	200	4023	44.9	586.2	64.73	16.24	1.92
	300	4336	43.7	889.0	60.21	20.70	2.88
100	100	4025	44.0	255.4	57.15	12.38	1.12
	200	4000	43.3	595.4	49.02	24.23	2.74
	300	4263	43.2	925.6	47.98	31.45	4.30
200	100	3997	43.8	275.4	46.54	17.79	1.63
	200	4019	42.0	525.8	35.48	28.65	3.09
	300	4308	42.7	882.4	35.16	43.06	5.74
500	100	3992	46.6	234.4	23.97	19.91	1.89
	200	3997	43.4	535.8	20.29	37.7	4.05
	300	4314	43.5	864.8	20.18	52.8	6.47
1000	100	4016	42.6	234.4	16.87	21.97	1.91
	200	4003	42.5	571.2	11.42	44.74	4.82
	300	4330	43.0	842.6	11.44	58.17	7.60

As indicated in the previous section the area search factor is the ratio of the number of times arc cost data is called by the code to the total number of arcs. For instance for $n = 100$ the area search factor in Table 4 is 1.08 (meaning 8 percent of the data is called twice, the rest once) when the maximum cost is 1 and 21.97 when it is 1000. The corresponding computation times are 0.14 and 1.91. The ratio of the area factors (20.3) is greater than the ratio of the times (13.6) for this example. The same result is true for other cases in Table 4, but it is clear that the two ratios are highly correlated. Since the area factors are independent of the computer being used, they give a measure of problem difficulty which is independent of the computer being used.

In Table 5 computational experience with 90 problems each having approximately 18 000 arcs and nodes ranging from 400 to 2000 ($n = 200$ to 1000). For each problem with a given set of nodes and arcs, two sets of costs are generated, one with maximum cost 1 and the other with maximum cost 100. The results are shown on successive lines, and indicate that the second type of problem is about 10 times as hard as the first.

Another result that was evident when step by step solution data was printed, was that the easy backshifts tend to occur early in the solution process, i.e., the time needed to solve problems $P_1, \ldots, P_{n/2}$ was much less than the time needed to solve problems $P_{n/2}, \ldots, P_n$. In fact the last step of finding the optimal dual solution to P_n always involves at least one and usually many zero shifts. An example of this is shown in Table 6, where cumulative data for the solutions of P_{92}–P_{100} are shown. Note that Problems P_1–P_{92} are solved by easy backshifts, without entering the main part of the program; P_{93} and P_{94} require 1 ordinary backshift and 1 zero shift each; P_{95} is solved by an easy backshift; P_{96} requires 1 ordinary backshift and 1 zero shift; P_{97} and P_{98} are solved by easy backshifts; P_{99} requires an ordinary backshift and 20 zero shifts, 19 of which are easy; finally, P_{100} requires 1 ordinary backshift and 2 zero shifts. The total area searched to solve this problem was only 4 percent more than a 1 pass solution

Table 5
Computational experience for large problems having about 18 000 arcs. Each row presents average results for five randomly generated test problems

n	Average number of arcs	Maximum cost	Percentage of easy backshifts	Number of zero shifts	Percentage of easy zero shifts	Area search factor	Time (DEC-20 secs.)
200	17948	1	96.8	106.4	74.11	1.03	.47
		100	44.9	679.8	77.32	12.17	4.72
300	18009	1	94.27	230.8	89.68	1.05	.63
		100	46.47	1000.0	74.14	16.24	6.61
400	17917	1	93.3	362.8	90.56	1.07	.82
		100	45.25	1457.0	72.44	22.43	9.99
500	18063	1	90.8	628.0	92.32	1.12	1.14
		100	44.28	1893.4	71.45	25.46	12.36
600	18048	1	88.9	1000.6	92.68	1.18	1.49
		100	43.67	2095.0	67.37	27.78	14.73
700	18704	1	86.2	1304.6	92.32	1.22	1.96
		100	43.14	2690.2	67.70	35.27	19.64
800	18627	1	83.68	1745.0	92.07	1.29	2.42
		100	43.50	2952.8	66.04	36.07	20.87
900	18292	1	81.82	2191.8	92.26	1.36	3.00
		100	42.33	3402.0	64.70	39.84	24.21
1000	17976	1	79.66	2958.4	92.93	1.49	3.66
		100	42.82	3877.8	64.00	46.34	31.17

Table 6
Cumulative data from the solution of a randomly generated 100×100 problem with 5028 arcs and maximum cost 1. The area search factor was 1.04

Step	Easy back shifts	Ordinary back shifts	Zero shifts	Easy zero shifts
92	92	0	0	0
93	92	1	1	0
94	92	2	2	0
95	93	2	2	0
96	93	3	3	0
97	94	3	3	0
98	95	3	3	0
99	95	4	23	19
100	95	5	25	19

would require even though the problem is only 50 per cent dense. The ease of its solution is due to the fact that the maximum cost was 1.

Problems which are larger, have larger maximum costs, or which are less dense than the one shown in Table 6, have more complicated solutions than the one illustrated there. Nevertheless their solution behavior is similar with the first few subproblems being very easy to solve and later ones becoming progressively harder.

The author is currently trying to test the recursive algorithm code on faster computers so that meaningful timing comparisons with other algorithms can be made.

9. Practical coding considerations

The computational results of the previous section indicate that the recursive method is a competitive algorithm for solving assignment problems. Several considerations should be taken into account in the design of a practical code for this purpose.

First, objective function (7) is maximizing rather than minimizing which is more common. If we change a minimizing objective to a maximizing one by multiplying each cost by -1 then the nonnegativity assumption (3) is violated. It is easy to get around that difficulty by adding 1 minus the most negative cost to each cost, and suitably adjusting the optimal objective function when found. Of course, finding the most negative cost requires one data pass, unless it is supplied by the user.

Another good idea would be to find the largest entry in each row and introduce the rows in decreasing order of these largest entries. This again would involve one pass through the data (which could be the same as the data pass to find the most negative cost). The evidence from examples (22)–(25) indicates that the value of having a good order in which to introduce rows can be considerable.

The memory requirements for the recursive method can be stated in terms of N the number of nodes and A the number of arcs. For the version of the code which solves completely dense problems (i.e., $A = n^2$) the memory requirements are $4N+A$; and for the sparse version of the code, the memory requirements are $4N+2A$. These compare favorably with similar requirements for the alternating path algorithm [2]. The number of FORTRAN instructions required for the recursive method is less than half of those needed for the primal code in [9].

Because of the relatively small memory requirements and relatively small running times it is likely that the recursive method will be a good choice for installation in minicomputers.

10. Comparisons with the simplex method

Although the recursive method is not a simplex method, it is instructive to make comparisons between the steps of the two methods. A detailed comparison between a primal pivot step and the ordinary backshift and zero shift operations is presented in Table 7. Note that aside from the search step which is extremely difficult to compare since it is problem dependent as well as algorithm dependent, both the backshift and zero shift operations involve less work than the primal pivot (roughly one-half as much). This comparison is even more pronounced in the case of easy backshifts and easy zero shifts.

Table 7
Comparison of steps in a primal pivot with those in a backshift or zero shift

	Primal pivot	Backshift	Zero shift
Find incoming arc	Area search	Row search	Area search
Find cycle	Yes	No	No
Find outgoing arc	Yes	No	No
Cut and paste	Yes	Yes	Yes
Relabel	Yes	Yes	No
Change duals	Yes	No	Yes (No for an easy zero shift)

In Table 1(B) of [9] it was reported that the solution by a primal algorithm of a 100×100 assignment problem having about 4000 arcs and maximum cost of 100 required 651 primal pivots and 2.187 UNIVAC-1108 seconds. In Table 4 it can be noted that the solution of a similar problem by the recursive method requires 66 ordinary and 44 easy backshifts, and 255.4 zero shifts of which more than 57 percent were easy, and took 1.12 DEC-20 seconds. Comparison of these times on two different machines is very difficult, but it is a common belief that the UNIVAC-1108 is considerably faster than the DEC-20.

The author is undertaking more extensive testing on a single computer of these and other methods. Others have also volunteered to help in this testing.

It should also be noted that the recursive method requires neither a starting solution nor a phase I part of the code. Also it does not require artificial arcs, even for sparse problems. It does not need degeneracy prevention techniques since as noted in Section 2, the dutch auction solution is, in fact, equivalent to the usual well-known perturbation technique.

Another feature of the recursive method is that it is able to take advantage of alternate optimal solutions when selecting the two highest bids in a row. When several choices are possible, it is a good idea to select, when possible, these two highest bids so that an easy backshift can be performed. Note in Table 4 that the number of easy backshifts goes down from over 90 percent to the low 40's as the maximum cost increases from 1 to 1000. The primary reason for this is the decreasing number of alternate optimal solutions as the maximum cost increases. This gives another explanation for the maximum (minimum) cost effect.

11. Conclusions

The recursive algorithm has been described and tested, and shown to be superior for at least problems which it is capable of solving in one pass or only slightly more than one pass. It's full comparison with other methods awaits further computational testing by the author and others. However, it is clearly a competitive algorithm for at least some kinds of assignment problems.

The author will report on further tests of the method elsewhere. Also extensions of these ideas to the solution of sum and bottleneck transportation problems will be discussed at a later time.

References

[1] M.L. Balinski and R.E. Gomory, A primal method for the assignment and transportation problem, Management Sci. 10 (1964) 578–593.

[2] R.S. Barr, F. Glover and D. Klingman, The alternating path algorithm for assignment problems, Math. Programming 13 (1977) 1–13.
[3] J.L. Barr and T.L. Shaftel, Solution properties of deterministic auctions, J. Financial and Quantitative Analysis (1976) 287–311.
[4] U. Derigs and V. Zimmerman, An augmenting path method for solving linear bottleneck transportation problems, Comput. 22 (1979) 1–15.
[5] L.R. Ford and D.R. Fulkerson, Flows in Networks (Princeton Univ. Press, Princeton, NJ, 1962).
[6] M.S. Hung and W.O. Rom, Solving the assignment problem by relaxation, Working Paper 78-5, 1978, College of Business Administration, Cleveland State University.
[7] L.S. Shapley and M. Shubik, The assignment game I: the core, Internat. J. Game Theory 1 (1972) 111–130.
[8] V. Srinivasan and G.L. Thompson, Accelerated algorithms for labelling and relabelling of trees, with applications to distribution problems, J. Assoc. Comput. Mach. 9 (1972) 712–726.
[9] V. Srinivasan and G.L. Thompson, Benefit-cost analysis of coding techniques for the primal transportation algorithm, J. Assoc. Comput. Mach. 20 (1973) 194–213.
[10] V. Srinivasan and G.L. Thompson, Cost operator algorithms for the transportation problem, Math. Programming 12 (1977) 372–391.
[11] G.L. Thompson, Computing the core of a market game, in: A. Fiacco and K. Kortanek, eds., Extremal Methods and Systems Analysis, Lecture Notes in Economics and Mathematical Systems (Springer-Verlag, Berlin, 1980) 312–334.
[12] G.L. Thompson, Pareto optimal deterministic models for bid and offer auctions, in: Methods of Operations Research, Vol. 35 (Athenaum/Hain/Scriptor/Hanstein, 1979), 517–530.
[13] N. Tomizawa, On some techniques useful for solution of transportation network problems, Networks 1 (1971) 173–194.

P. Hansen, ed., Studies on Graphs and Discrete Programming
© North-Holland Publishing Company (1981) 345–358

ASYMPTOTICAL ESTIMATIONS FOR THE NUMBER OF CLIQUES OF UNIFORM HYPERGRAPHS

Ioan TOMESCU
Department of Mathematics, University of Bucarest, Bucarest, Roumanie

In this paper it is shown that, when $n \to \infty$, for almost all $(h+1)$-hypergraphs with n vertices every clique contains at most $\{((h+1)!\log n)^{1/h} + \frac{1}{2}h\}$ vertices.

By using the method developed by V.V. Glagolev for some asymptotical estimations of disjunctive normal forms of Boolean functions, method applied to the study of graphs (the case $h=1$) by Phan Dinh Diêu, it is proved that when $n \to \infty$ for almost all $(h+1)$-hypergraphs of order n $(h \geqslant 2)$, the number of cliques (or maximal cliques) is of the order of $n^{(h/(h+1))(h!\log n)^{1/h}}$.

This fact implies that every enumerative algorithm for determining the maximum cardinality of a clique of a uniform hypergraph cannot work in polynomial time whenever it finds all cliques of the hypergraph.

1. Definitions and notations

An $(h+1)$-uniform hypergraph H is a pair $H=(X, \mathscr{E})$, where X is a set of vertices and $\mathscr{E}$ is a family of $(h+1)$-sets of X, i.e., $\mathscr{E}=(E_1, \ldots, E_m)$ so that $|E_i| = h+1$ for all i and $h \geqslant 1$. The sets E_i are called the edges of H. The order of the hypergraph H is equal to $|X|$.

A clique C of H is a set $C \subset X$ such that any $(h+1)$-set of C is an edge of H [1]. If $|C| = k$ the clique C will be called a k-clique of H. A clique C is called maximal, if it is not contained in any other clique of H.

An independent set of H is a subset of vertices $S \subset X$ which does not contain any edge of H.

In the sequel we shall consider the case $h \geqslant 2$ only (for the case of graphs, when $h=1$, see [3]).

We denote by $\alpha(H)$ the maximum cardinality of an independent set of H, or the independence number of H, and by $\omega(H)$ the maximum cardinality of a clique of H, or the cliquomatic number of H. It is obvious that $\omega(H) = \alpha(\bar{H})$, where $\bar{H} = (X, \bar{\mathscr{E}})$ and $\bar{\mathscr{E}} = \mathscr{P}_{h+1}(X) \setminus \mathscr{E}$.

Also, we denote by $C_k(H)$ the number of k-cliques of H, by $C(H)$ the number of all cliques of H, by $\mathrm{CM}_k(H)$ the number of maximal k-cliques of H, and by $\mathrm{CM}(H)$ the number of maximal cliques of H. The minimum number of classes of a partition of the vertex-set X, whose classes are independent sets is called the chromatic number of H, and it will be denoted by $\chi(H)$.

We shall use the following notations:

$\bar{C}(n)$—the mean value of $C(H)$ taken for all $(h+1)$-hypergraphs H of order n;

$\bar{C}_k(n)$—the mean value of $C_k(H)$ taken for all $(h+1)$-hypergraphs with n vertices;

$N_{n,k,h}(m)$—the number of $(h+1)$-hypergraphs H of order n having exactly m k-cliques;

$\xi_{n,k,h}$—the stochastic variable taking the value m with the probability

$$N_{n,k,h}(m)/2^{\binom{n}{h+1}};$$

$\mathbf{M}\xi_{n,k,h}$—the mathematical expectation of $\xi_{n,k,h}$;

$\mathbf{D}\xi_{n,k,h}$—the dispersion of $\xi_{n,k,h}$.

It is easy to see that the number of $(h+1)$-hypergraphs with the vertex-set $X=\{x_1,\ldots,x_n\}$ is equal to $2^{\binom{n}{h+1}}$ and the number of k-subsets of X is equal to $q=\binom{n}{k}$.

For any $m>q$ we have $N_{n,k,h}(m)=0$ and we derive also that $\mathbf{M}\xi_{n,k,h}=\bar{C}_k(n)$, because

$$N_{n,k,h}(m)/2^{\binom{n}{h+1}}$$

is the probability of the event that an $(h+1)$-hypergraph with n vertices has m k-cliques.

For any real x we denote by $[x]$ the integer part of x and by $\{x\}=-[-x]$, i.e., the smallest integer not less than x.

All the logarithms used here are in the base 2.

We denote $(x)_h=x(x-1)\cdots(x-h+1)$ for any real x and natural h.

2. Asymptotical estimation for the size of the cliques

In order to estimate either the size or the number of the cliques of uniform hypergraphs, we shall derive a useful result.

Proposition 1. *The positive root of the equation:*

$$(x)_h = a\log n \tag{1}$$

where $a>0$, is the following:

$$x=(a\log n)^{1/h}+\tfrac{1}{2}(h-1)+O((\log n)^{-1/h}) \tag{2}$$

as $n\to\infty$.

Proof. For $x>h-1$ the function $(x)_h$ is strictly increasing, so that (1) has a unique positive root when n is large enough.

For even h we deduce $(x-1)(x-h+2) \geq x(x-h+1)$, $(x-2)(x-h+3) > x(x-h+1), \ldots, (x-\frac{1}{2}h+1)(x-\frac{1}{2}h) > x(x-h+1)$, hence

$$(x)_h > (x(x-h+1))^{h/2}. \tag{3}$$

For odd h we can write

$$(x-1)(x-h+2) > x(x-h+1), \ldots,$$
$$(x-\tfrac{1}{2}(h-1)+1)(x-\tfrac{1}{2}(h-1)-1) \geq x(x-h+1)$$

and $x-\frac{1}{2}(h-1) > \sqrt{x(x-h+1)}$, hence

$$(x)_h > (x(x-h+1))^{(h-1)/2+\frac{1}{2}} = (x(x-h+1))^{h/2}$$

and (3) is also true for odd h.

The inequality between the arithmetical and geometrical mean implies also

$$((x)_h)^{1/h} < \frac{x+(x-1)+\cdots+(x-h+1)}{h} = x - \tfrac{1}{2}(h-1). \tag{4}$$

If $x>0$ is the root of (1), then (3) gives

$$x^2 - x(h-1) - (a \log n)^{2/h} < 0$$

or

$$x < \tfrac{1}{2}(h-1) + (\tfrac{1}{4}(h-1)^2 + (a \log n)^{2/h})^{\frac{1}{2}}.$$

Now (4) implies

$$0 < x - \tfrac{1}{2}(h-1) - (a \log n)^{1/h} < \varepsilon(n)$$

where

$$\lim_{n\to\infty} \varepsilon(n)(a \log n)^{1/h} = \tfrac{1}{8}(h-1)^2.$$

Proposition 2. *When $n \to \infty$ for almost all $(h+1)$-hypergraphs H with n vertices, every clique contains at most*

$$\{((h+1)! \log n)^{1/h} + \tfrac{1}{2}h\}$$

vertices.

Proof. Let $\alpha(n, h)$ be the positive root of the equation $(x)_h = (h+1)! \log n$ as $n \to \infty$ and $q = \{\alpha(n, h)\}$. The number of hypergraphs having k-cliques, where $k \geq q+1$ is less than or equal to

$$\binom{n}{q+1} 2^{\binom{n}{h+1} - \binom{q+1}{h+1}}.$$

Indeed, any $(q+1)$-clique can be chosen in $\binom{n}{q+1}$ ways and the remaining edges can be chosen in $2^{\binom{n}{h+1}-\binom{q+1}{h+1}}$ different ways. All hypergraphs obtained in this way have cliques containing at least $q+1$ vertices and any hypergraph verifying this property can be obtained in this way. Therefore we derive

$$\delta(n, h)=\frac{\binom{n}{q+1}2^{\binom{n}{h+1}-\binom{q+1}{h+1}}}{2^{\binom{n}{h+1}}}=\frac{\binom{n}{q+1}}{2^{\binom{q+1}{h+1}}}$$

$$<\frac{n^{q+1}}{(q+1)!2^{(q+1)q\cdots(q-h+1)/(h+1)!}}.$$

Since

$$\frac{q(q-1)\cdots(q-h+1)}{(h+1)!}>\frac{(\alpha(n, h))_h}{(h+1)!}=\log n,$$

we find

$$\delta(n, h)<\frac{n^{q+1}}{(q+1)!2^{(q+1)\log n}}=\frac{1}{(q+1)!}\to 0 \quad \text{as } n\to\infty,$$

because Proposition 1 implies that

$$\alpha(n, h)=((h+1)!\log n)^{1/h}+\tfrac{1}{2}(h-1)+O((\log n)^{-1/h}). \tag{5}$$

Hence when $n\to\infty$ almost all hypergraphs of order n contain only k-cliques, where $k\leq q$. But

$$q=\{((h+1)!\log n)^{1/h}+\tfrac{1}{2}(h-1)+O((\log n)^{-1/h})\}$$

following (5) and when $n\to\infty$ it follows $O((\log n)^{-1/h})<\frac{1}{2}$, therefore

$$q\leq\{((h+1)!\log n)^{1/h}+\tfrac{1}{2}h\}.$$

An independent set of H being a clique of $\bar{H}$, we derive that for almost all $(h+1)$-hypergraphs H of order n any independent set has no more than $\{((h+1)!\log n)^{1/h}+\frac{1}{2}h\}$ vertices when $n\to\infty$.

Corollary 1. *When $n\to\infty$ for almost all $(h+1)$-hypergraphs H of order n the cliquomatic number $\omega(H)$ and the independence number $\alpha(H)$ are less than or equal to $\{((h+1)!\log n)^{1/h}+\frac{1}{2}h\}$.*

Corollary 2. *When $n\to\infty$ for almost all $(h+1)$-hypergraphs H of order n, the chromatic number $\chi(G)$ satisfies the estimation*

$$\chi(H)\geq\frac{n}{\{((h+1)!\log n)^{1/h}+\frac{1}{2}h\}}.$$

Proof. This follows from Corollary 1 and the following inequality

$$\chi(H)\alpha(H)\geq n.$$

3. Asymptotical estimations of the number of cliques

In this section we shall give an asymptotical estimation of the number of cliques of an $(h+1)$-hypergraph. For this purpose we first calculate the mean $\mathbf{M}\xi_{n,k,h}$ and the dispersion $\mathbf{D}\xi_{n,k,h}$ in the same manner as Phan Dinh Diêu [3] for the case $h=1$.

In the sequel we shall denote

$$p=2^{\binom{n}{h+1}} \quad \text{and} \quad q=\binom{n}{k} \quad \text{for any } 0\leqslant k\leqslant n.$$

Let $H_1, \ldots, H_p$ be all $(h+1)$-hypergraphs with vertex-set $X=\{x_1, \ldots, x_n\}$ and $E_1, \ldots, E_q$ be all k-subsets of X.

Proposition 3. *The following equality holds*

$$\mathbf{M}\xi_{n,k,h}=\binom{n}{k}2^{-\binom{k}{h+1}}. \tag{6}$$

Proof. For any set $E_j\subset X$ we denote by $h_1(E_j)$ the number of hypergraphs $H=(X, \mathscr{E})$ such that E_j is a k-clique of H. It is obvious that

$$\sum_{i=1}^{p} C_k(H_i)=\sum_{j=1}^{q} h_1(E_j).$$

We now compute the right side of this equation. If E_j is a k-clique of H, it contains $\binom{k}{h+1}$ edges. It follows that

$$h_1(E_j)=2^{\binom{n}{h+1}-\binom{k}{h+1}}.$$

Hence we can write

$$\sum_{j=1}^{q} h_1(E_j)=q2^{\binom{n}{h+1}-\binom{k}{h+1}}=\binom{n}{k}2^{\binom{n}{h+1}-\binom{k}{h+1}},$$

which implies

$$\mathbf{M}\xi_{n,k,h}=\bar{C}_k(n)=\frac{1}{p}\sum_{i=1}^{p} C_k(H_i)=\binom{n}{k}2^{-\binom{k}{h+1}}.$$

Proposition 4. *The following equality holds*

$$\mathbf{D}\xi_{n,k,h}=\binom{n}{k}\Big/2^{2\binom{k}{h+1}}\sum_{j=h+1}^{k}\binom{k}{j}\binom{n-k}{k-j}(2^{\binom{j}{h+1}}-1). \tag{7}$$

Proof. We consider the set P of all ordered pairs (E_r, E_s), where $1\leqslant r, s\leqslant q$. For each hypergraph H_i $(i=1, \ldots, p)$ we denote by $p(H_i)$ the number of pairs

(E_r, E_s) such that both E_r and E_s are k-cliques of H_i. On the other hand, for each pair $(E_r, E_s) \in P$ we denote by $h_2(E_r, E_s)$ the number of hypergraphs H_i such that E_r and E_s are cliques of H_i. It is clear that

$$\sum_{i=1}^{p} p(H_i) = \sum_{(E_r, E_s)\in P} h_2(E_r, E_s). \tag{8}$$

Note that if the hypergraph H_i has m k-cliques, then $p(H_i) = m^2$, and the number of such hypergraphs H_i is equal to $N_{n,k,h}(m)$. Therefore we have

$$\sum_{i=1}^{p} p(H_i) = \sum_{m=0}^{q} m^2 N_{n,k,h}(m)$$

where $q = \binom{n}{k}$. Hence we obtain

$$\mathbf{M}\xi^2_{n,k,h} = \sum_{m=0}^{q} m^2 \frac{N_{n,k,h}(m)}{2^{\binom{n}{h+1}}} = \frac{1}{2^{\binom{n}{h+1}}} \sum_{i=1}^{p} p(H_i). \tag{9}$$

We now calculate the right side of (8). Let $(E_r, E_s) \in P$ such that $|E_r \cap E_s| = j$. The number of edges contained into $E_r \cup E_s$ is equal to $2\binom{k}{h+1} - \binom{j}{h+1}$ hence

$$h_2(E_r, E_s) = 2^{\binom{n}{h+1} - 2\binom{k}{h+1} + \binom{j}{h+1}},$$

where $\binom{i}{h+1} = 0$ for $i \leqslant h$. For any j $(0 \leqslant j \leqslant k)$ there exist $\binom{n}{j}\binom{n-j}{k-j}\binom{n-k}{k-j}$ pairs $(E_r, E_s) \in P$ such that $|E_r \cap E_s| = j$ and $|P| = q^2 = \binom{n}{k}^2$. Therefore we have

$$\begin{aligned}\sum_{(E_r, E_s)\in P} h_2(E_r, E_s) &= \left[\binom{n}{k}^2 - \sum_{j=h+1}^{k} \binom{n}{j}\binom{n-j}{k-j}\binom{n-k}{k-j}\right] 2^{\binom{n}{h+1} - 2\binom{k}{h+1}} \\ &\quad + \sum_{j=h+1}^{k} \binom{n}{j}\binom{n-j}{k-j}\binom{n-k}{k-j} 2^{\binom{n}{h+1} - 2\binom{k}{h+1} + \binom{j}{h+1}} \\ &= \sum_{j=h+1}^{k} \binom{n}{k}\binom{k}{j}\binom{n-k}{k-j} 2^{\binom{n}{h+1} - 2\binom{k}{h+1}} \left(2^{\binom{j}{h+1}} - 1\right) \\ &\quad + \binom{n}{k}^2 2^{\binom{n}{h+1} - 2\binom{k}{h+1}}.\end{aligned} \tag{10}$$

From (6), (8), (9) and (10) we obtain

$$\mathbf{M}\xi^2_{n,k,h} = \frac{\binom{n}{k}}{2^{2\binom{k}{h+1}}} \sum_{j=h+1}^{k} \binom{k}{j}\binom{n-k}{k-j}\left(2^{\binom{j}{h+1}} - 1\right) + (\mathbf{M}\xi_{n,k,h})^2,$$

hence $\mathbf{D}\xi_{n,k,h} = \mathbf{M}\xi^2_{n,k,h} - (\mathbf{M}\xi_{n,k,h})^2$ is given by (7). Let now $\beta(n, h)$ be the positive root of the equation $(x)_h = h! \log n$ as $n \to \infty$, therefore

$$\beta(n, h) = (h! \log n)^{1/h} + \tfrac{1}{2}(h-1) + O((\log n)^{-1/h}).$$

Proposition 5. *There is a natural number n_0 such that for any $n \geqslant n_0$ and for* *any* $k \leqslant [\beta(n,h)]$ *we have*

$$\mathbf{D}\xi_{n,k,h} < \frac{k^{2h+3}}{n^{h+1}}(\mathbf{M}\xi_{n,k,h})^2.$$

Proof. In order to estimate $\mathbf{D}\xi_{n,k,h}$ we put for every j $(h+1 \leqslant j \leqslant k)$

$$a_j = \binom{k}{j}\binom{n-k}{k-j}(2^{\binom{j}{h+1}}-1).$$

For any $j < k$ we deduce

$$\frac{a_{j+1}}{a_j} = \frac{(k-j)^2}{(j+1)(n-2k+j+1)} \cdot \frac{2^{\binom{j+1}{h+1}}-1}{2^{\binom{j}{h+1}}-1}.$$

But

$$\frac{2^{\binom{j+1}{h+1}}-1}{2^{\binom{j}{h+1}}-1} < 2^{\binom{j}{h}+1},$$

since

$$2^{\binom{j}{h+1}}2^{\binom{j}{h}} - 1 < 2 \cdot 2^{\binom{j}{h+1}}2^{\binom{j}{h}} - 2 \cdot 2^{\binom{j}{h}},$$

which is equivalent to

$$2^{\binom{j}{h+1}}2^{\binom{j}{h}} + 1 > 2 \cdot 2^{\binom{j}{h}}.$$

This inequality holds because $j \geqslant h+1$.

For $h+1 \leqslant j \leqslant k$ we obtain $4(k-j)^2 \leqslant 5 \cdot 2^{k-j} \leqslant 5 \cdot 2^{\binom{k}{h}-\binom{j}{h}}$, hence

$$\frac{a_{j+1}}{a_j} < \frac{2(k-j)^2 2^{\binom{j}{h}}}{(h+2)(n-2k)} \leqslant \frac{5}{2(h+2)} \cdot \frac{2^{\binom{k}{h}}}{n-2k} \leqslant \frac{5n}{2(h+2)(n-2k)}$$

since

$$2^{\binom{k}{h}} \leqslant 2^{(\beta)_h/h!} = n.$$

The inequality $5n/((2h+4)(n-2k)) \leqslant 1$ is equivalent to $n \geqslant 4(h+2)k/(2h-1)$. There exists an index $n_1 \in N$ such that this last inequality to be valid for any $n \geqslant n_1$, because we have $k = O((\log n)^{1/h})$. Therefore, it follows that

$$\max_{h+1 \leqslant j \leqslant k} a_j = a_{h+1} = \binom{k}{h+1}\binom{n-k}{k-(h+1)},$$

hence

$$\mathbf{D}\xi_{n,k,h} < \frac{\binom{n}{k}}{2^{2\binom{k}{h+1}}} k\binom{k}{h+1}\binom{n-k}{k-(h+1)}$$

$$= \frac{k\binom{k}{h+1}\binom{n-k}{k-(h+1)}}{\binom{n}{k}}(\mathbf{M}\xi_{n,k,h})^2,$$

where

$$\frac{k\binom{k}{h+1}\binom{n-k}{k-(h+1)}}{\binom{n}{k}}$$

$$=\frac{k^3(k-1)^2(k-2)\cdots(k-h)^2}{(h+1)!n(n-1)\cdots(n-h)}\cdot\frac{(n-k)(n-k-1)\cdots(n-2k+h+2)}{(n-h-1)(n-h-2)\cdots(n-k+1)}$$

$$\leqslant\frac{k^{2h+3}}{(h+1)!(n)_{h+1}},\quad \text{because } k\geqslant h+1.$$

There is an index $n_2\in N$ such that $(h+1)!(n)_{h+1}>n^{h+1}$ for any $n\geqslant n_2$, which implies

$$\mathbf{D}\xi_{n,k,h}<\frac{k^{2h+3}}{n^{h+1}}(\mathbf{M}\xi_{n,k,h})^2 \quad \text{for any } n\geqslant n_0,$$

where $n_0=\max(n_1, n_2)$.

Proposition 6. *Let $k=[\beta(n, h)]$. When $n\to\infty$ for almost all $(h+1)$-hypergraphs H with n vertices, the number $C_k(H)$ of k-cliques of H satisfies the following estimation*

$$\frac{\binom{n}{k}}{2^{\binom{k}{h+1}}}\left(1-\frac{k^{h+2}}{n^{(h+1)/2}}\right)<C_k(H)<\frac{\binom{n}{k}}{2^{\binom{k}{h+1}}}\left(1+\frac{k^{h+2}}{n^{(h+1)/2}}\right).$$

Proof. By using the Tchebychev's inequality in the probability theory we have for any $t>0$

$$\mathbf{P}(|\xi_{n,k,h}-\mathbf{M}\xi_{n,k,h}|\geqslant t)<\frac{\mathbf{D}\xi_{n,k,h}}{t^2}$$

where $\mathbf{P}(A)$ is the probability of the event A. Taking

$$t=\frac{k^{h+2}}{n^{(h+1)/2}}\mathbf{M}\xi_{n,k,h}=\frac{k^{h+2}}{n^{(h+1)/2}}\cdot\frac{\binom{n}{k}}{2^{\binom{k}{h+1}}},$$

we have

$$\frac{\mathbf{D}\xi_{n,k,h}}{t^2}<\frac{k^{2h+3}(\mathbf{M}\xi_{n,k,h})^2}{n^{h+1}}\cdot\frac{n^{h+1}}{k^{2h+4}(\mathbf{M}\xi_{n,k,h})^2}=\frac{1}{k}\to 0$$

as $n\to\infty$, since $k\geqslant(h!\log n)^{1/h}$ when n is large enough. The stochastic variable $\xi_{n,k,h}$ takes the value m with the probability $N_{n,k,h}(m)/2^{\binom{n}{h+1}}$, hence the variable $\xi_{n,k,h}$ takes the value $C_k(H)$ for each given hypergraph H with the same probability, equal to $2^{-\binom{n}{h+1}}$. It follows that when $n \to \infty$, for almost all hypergraphs H of order n we have

$$\left| C_k(H) - \frac{\binom{n}{k}}{2^{\binom{k}{h+1}}} \right| < \frac{k^{h+2}}{n^{(h+1)/2}} \cdot \frac{\binom{n}{k}}{2^{\binom{k}{h+1}}}.$$

Theorem 1. *When $n \to \infty$ for almost all $(h+1)$-hypergraphs H with n vertices $(h \geq 2)$, the number $C(H)$ of all cliques satisfies the estimation*

$$n^{(h/(h+1))(h!\log n)^{1/h}+(h-3)/2} < C(H) < n^{(h/(h+1))(h!\log n)^{1/h}+h(\frac{1}{2}+(h+1)^{1/h})}.$$

Proof. The lower bound. Let $k = [\beta(n, h)]$. When n is large enough we have $\binom{n}{k} > (n/k)^k$ because $(n-k+i)/i \geq n/k$ for $i \leq k$, with equality for $i = k$ only. Hence $\binom{n}{k} > (n/k)^k \geq (n/\beta)^{\beta-1}$ where $\beta = \beta(n, h)$ and

$$2^{\binom{k}{h+1}} = 2^{((k-h)/(h+1))\binom{k}{h}} < 2^{((k-h)/(h+1))\cdot(\beta)_h/h!} = n^{(k-h)/(h+1)} \leq n^{(\beta-h)/(h+1)}$$

because $(\beta)_h = h! \log n$. Therefore, from Proposition 6 we deduce that as $n \to \infty$ for almost all hypergraphs of order n we have

$$C_k(H) > \frac{\binom{n}{k}}{2^{\binom{k}{h+1}}}\left(1 - \frac{k^{h+2}}{n^{(h+1)/2}}\right) > \frac{(n/\beta)^{\beta-1}}{n^{\beta/(h+1)}}\, n^{h/(h+1)}\left(1 - \frac{k^{h+2}}{n^{(h+1)/2}}\right).$$

But $k = O((\log n)^{1/h})$, which implies

$$\lim_{n\to\infty} \frac{k^{h+2}}{n^{(h+1)/2}} = 0.$$

Hence when n is large enough we derive

$$n^{h/(h+1)}\left(1 - \frac{k^{h+2}}{n^{(h+1)/2}}\right) > n^{\frac{1}{2}+1/3(h+1)}$$

since the difference

$$\frac{h}{h+1} - \left(\frac{1}{2} + \frac{1}{3(h+1)}\right) = \frac{3h-5}{6(h+1)} > 0 \quad \text{for } h \geq 2.$$

It results that

$$C_k(H) > \frac{n^{\beta h/(h+1) - \frac{1}{2} + 1/3(h+1)}}{\beta^{\beta-1}}.$$

Also we obtain

$$\beta^{\beta-1} = n^{O((\log n)^{1/h-1}\log\log n)}$$

and Proposition 1 implies $\beta>(h!\log n)^{1/h}+\frac{1}{2}(h-1)$, hence

$$\frac{\beta h}{h+1}-\frac{1}{2}+\frac{1}{3(h+1)}>\frac{h}{h+1}(h!\log n)^{1/h}+\frac{h(h-1)}{2(h+1)}-\frac{1}{2}+\frac{1}{3(h+1)}$$
$$>\frac{h}{h+1}(h!\log n)^{1/h}+\tfrac{1}{2}(h-3)+\frac{1}{3(h+1)}.$$

If $h\geqslant 2$ and n is large enough we have

$$\frac{1}{3(h+1)}-O((\log n)^{1/h-1}\log\log n)>0.$$

Consequently

$$C(H)>C_k(H)>n^{(h/(h+1))(h!\log n)^{1/h}+(h-3)/2}$$

for almost all hypergraphs H with n vertices as $n\to\infty$.

(2) The upper bound. By Proposition 3 we have

$$\bar{C}_r(n)=\binom{n}{r}2^{-\binom{r}{h+1}}.$$

Therefore, for any $r<n$

$$\frac{\bar{C}_{r+1}(n)}{\bar{C}_r(n)}=\frac{n-r}{(r+1)2^{\binom{r}{h}}}.$$

If $r<k=[\beta(n,h)]$ we deduce $r\leqslant k-1\leqslant\beta-1$ and

$$(\beta-1)_h=(\beta-1)(\beta-2)\cdots(\beta-h)=(\beta)_h\frac{\beta-h}{\beta}=h!\frac{\beta-h}{\beta}\log n.$$

Consequently, for $r\leqslant k-1$ we have

$$\frac{\bar{C}_{r+1}(n)}{\bar{C}_r(n)}\geqslant\frac{n-\beta+1}{\beta 2^{(\beta-1)_h/h!}}=\frac{n-\beta+1}{\beta n^{1-h/\beta}}=\frac{n^{h/\beta}(n-\beta+1)}{n\beta}>\frac{1}{\beta}$$

when n is large enough.

If $r\geqslant k$ we deduce $r>\beta-1$, hence

$$\frac{\bar{C}_{r+1}(n)}{\bar{C}_r(n)}<\frac{n-\beta+1}{\beta n^{1-h/\beta}}<\frac{n^{h/\beta}}{\beta}.$$

Therefore, we obtain $\bar{C}_r(n)<\beta^{k-r}\bar{C}_k(n)$ for $r<k$ and

$$\bar{C}_r(n)<\left(\frac{n^{h/\beta}}{\beta}\right)^{r-k}\bar{C}_k(n)\quad\text{for } r\geqslant k.$$

Let us denote by $\bar{C}(n)$ the mean value of the number of cliques $C(H)$ taken for all hypergraphs H of order n. Then, when n is large enough, we have

$$\bar{C}(n)=\frac{1}{2^{\binom{n}{h+1}}}\sum_{i=1}^{p}\bar{C}(H_i)=\sum_{r=1}^{n}\bar{C}_r(n)$$

$$<\sum_{r=1}^{k-1}\bar{C}_k(n)\beta^{k-r}+\sum_{r=k}^{n}\bar{C}_r(n). \tag{11}$$

Now remember that $\alpha=\alpha(n,h)$ is the positive root of the equation $(x)_h=(h+1)!\log n$ and $q=\{\alpha\}$. If $p\geq q+1$ we have

$$\bar{C}_p(n)=\binom{n}{p}2^{-\binom{p}{h+1}}<\frac{n^p}{p!}\cdot\frac{1}{2^{p(p-1)\cdots(p-h)/(h+1)!}}$$

$$\leq\frac{n^p}{p!}\cdot\frac{1}{2^{p(\alpha)_h/(h+1)!}}=\frac{1}{p!},$$

hence

$$\sum_{r=q+1}^{n}\bar{C}_r(n)<\sum_{r=q+1}^{n}\frac{1}{r!}\to 0 \quad \text{as } n\to\infty, \tag{12}$$

because

$$\lim_{n\to\infty}q=\infty \quad \text{and} \quad \sum_{r=0}^{\infty}\frac{1}{r!}=\mathrm{e}.$$

Now (11) and (12) imply

$$\bar{C}(n)<\sum_{r=1}^{k-1}\bar{C}_k(n)\beta^{k-r}+\sum_{r=k}^{q}\bar{C}_k(n)\left(\frac{n^{h/\beta}}{\beta}\right)^{r-k}+\varepsilon_1(n) \tag{13}$$

where $\lim_{n\to\infty}\varepsilon_1(n)=0$, or

$$\bar{C}(n)<\bar{C}_k(n)\left[\frac{\beta^k-\beta}{\beta-1}+\left(\left(\frac{n^{h/\beta}}{\beta}\right)^{q-k+1}-1\right)\frac{1}{\frac{n^{h/\beta}}{\beta}-1}\right]+\varepsilon_1(n). \tag{14}$$

We have $\lim_{n\to\infty}(n^{h/\beta}/\beta)=\infty$ for any $h\geq 2$, hence the following inequality holds as $n\to\infty$:

$$\bar{C}(n)<\bar{C}_k(n)\left(\beta^\beta+\left(\frac{n^{h/\beta}}{\beta}\right)^{q-k+1}\right)+\varepsilon_1(n). \tag{15}$$

We shall estimate the values of $\bar{C}_k(n)$.

$$\bar{C}_k(n)=\binom{n}{k}2^{-\binom{k}{h+1}}$$

where $k=[\beta]$ and β satisfies $(\beta)_h=h!\log n$. We have $\binom{n}{k}<n^k/k!$ and when n is large enough we have

$$k!>\left(\frac{k+1}{\mathrm{e}}\right)^k>\left(\frac{\beta}{\mathrm{e}}\right)^{\beta-1}$$

by applying Stirling's formula. Hence we deduce

$$\binom{n}{k} < \frac{n^{\beta}}{(\beta/e)^{\beta-1}}.$$

On the other hand we have

$$2^{-\binom{k}{h+1}} < 2^{-(\beta-1)(\beta-2)\cdots(\beta-h-1)/(h+1)!} = n^{-(\beta-h)(\beta-h-1)/\beta(h+1)}.$$

But

$$\frac{(\beta-h)(\beta-h-1)}{h+1} > \beta\left(\frac{1}{h+1}\beta - h\right), \quad \text{hence } 2^{-\binom{k}{h+1}} < n^{h-(\beta/(h+1))},$$

which implies

$$\bar{C}_k(n) < \frac{n^{(h/(h+1))\beta+h}}{(\beta/e)^{\beta-1}}. \tag{16}$$

For the inequality (15) the exponent

$$\begin{aligned} q-k+1 &= ((h+1)!\log n)^{1/h} + \tfrac{1}{2}(h-1) + O((\log n)^{-1/h}) \\ &\quad - (h!\log n)^{1/h} - \tfrac{1}{2}(h-1) - O((\log n)^{-1/h}) \\ &= ((h+1)^{1/h}-1)(h!\log n)^{1/h} + O((\log n)^{-1/h}), \end{aligned}$$

hence for $h \geq 2$ we have

$$\lim_{n\to\infty} \beta^{\beta}\left(\frac{\beta}{n^{h/\beta}}\right)^{q-k+1} = 0.$$

From (15) we derive now

$$\bar{C}(n) < \bar{C}_k(n)\left(\frac{n^{h/\beta}}{\beta}\right)^{q-k+1}(1+\varepsilon_2(n)) < 2\bar{C}_k(n)\left(\frac{n^{h/\beta}}{\beta}\right)^{q-k+1}, \tag{17}$$

where $\lim_{n\to\infty} \varepsilon_2(n) = 0$. From (16) and (17) we obtain

$$\begin{aligned} \bar{C}(n) &< 2\frac{n^{(h/(h+1))\beta+h}}{(\beta/e)^{\beta-1}} \cdot \frac{n^{h(q-k+1)/\beta}}{\beta^{q-k+1}} \\ &= 2n^{(h/(h+1))(h!\log n)^{1/h}+h(h-1)/2(h+1)+h(h+1)^{1/h}+\varepsilon_3(n)}, \end{aligned}$$

because $h(q-k+1)/\beta = h((h+1)^{1/h}-1) + O((\log n)^{-1/h})$ and

$$(\beta/e)^{\beta-1}\beta^{q-k+1} = n^{O((\log n)^{1/h-1}\log\log n)}.$$

For any $h \geq 2$ we have $\lim_{n\to\infty} \varepsilon_3(n) = 0$. This implies that, when n is large enough we have

$$h(h-1)/2(h+1) + \varepsilon_3(n) < \tfrac{1}{2}(h-1),$$

hence

$$\bar{C}(n) < 2n^{(h/(h+1))(h!\log n)^{1/h}+h(\frac{1}{2}+(h+1)^{1/h})-\frac{1}{2}}. \tag{18}$$

Let $a(n)$ denote the number of $(h+1)$-hypergraphs H such that

$$C(H) > \tfrac{1}{2}\bar{C}(n)\sqrt{n}.$$

Then we have

$$\bar{C}(n) = \frac{1}{p}\sum_{i=1}^{p} C(H_i) = \frac{1}{p}\Bigg(\sum_{C(H_i)>(\sqrt{n}/2)\bar{C}(n)} C(H_i) + \sum_{C(H_i)\leqslant(\sqrt{n}/2)\bar{C}(n)} C(H_i)\Bigg)$$

$$> a(n)\sqrt{n}\bar{C}(n)/2p.$$

Hence $a(n)/p < 2/\sqrt{n}$, which implies $\lim_{n\to\infty} a(n)/p = 0$. Thus, when $n \to \infty$ for almost all hypergraphs H of order n we have

$$C(H) \leqslant \tfrac{1}{2}\bar{C}(n)\sqrt{n} < n^{(h/(h+1))(h!\log n)^{1/h} + h(\frac{1}{2}+(h+1)^{1/h})}.$$

Note that for any $h \geqslant 2$ we obtain $(h+1)^{1/h} \leqslant \sqrt{3} \cong 1.732$ and $\lim_{h\to\infty}(h+1)^{1/h} = 1$.

Theorem 2. *When $n \to \infty$ for almost all $(h+1)$-hypergraphs H of order n, the number of maximal cliques satisfies the estimation*

$$n^{(h/(h+1))(h!\log n)^{1/h} + (h-3)/2} < \mathrm{CM}(H) < n^{(h/(h+1))(h!\log n)^{1/h} + h(\frac{1}{2}+(h+1)^{1/h})}.$$

Proof. Because we have $\mathrm{CM}(H) \leqslant C(H)$, the upper bound follows immediately from Theorem 1. By Proposition 2 as $n \to \infty$ every clique of H does not contain more than $q = \{\alpha\}$ vertices for almost all hypergraphs H with n vertices.

Every maximal clique having not more than q vertices contains at most $\binom{q}{k}$ k-cliques, and on the other hand, every k-clique must be contained in some maximal clique.

Therefore, as $n \to \infty$ for almost all hypergraphs H of order n, the number of maximal cliques is not less than $1/\binom{q}{k}$ time of the number of k-cliques, i.e.,

$$\mathrm{CM}(H) \geqslant C_k(H) \Big/ \binom{q}{k}.$$

We obtain

$$\binom{q}{k} \leqslant 2^{q-1} \leqslant 2^{\alpha} = 2^{((h+1)!\log n)^{1/h} + (h-1)/2 + O((\log n)^{-1/h})}$$

$$= n^{O((\log n)^{1/h-1})}.$$

A partial result of the proof of Theorem 1 was

$$C_k(H) > n^{(h/(h+1))(h!\log n)^{1/h} + (h-3)/2 + 1/(3(h+1)) + \varepsilon_4(n)},$$

where $\lim_{n\to\infty} \varepsilon_4(n)=0$. Thus, as $n\to\infty$ for almost all hypergraphs H with n vertices we have

$$\mathrm{CM}(H) > n^{(h/(h+1))(h!\log n)^{1/h}+(h-3)/2},$$

because when n is large enough and $h\geqslant 2$ we deduce

$$\frac{1}{3(h+1)}+\varepsilon_4(n)-\mathrm{O}((\log n)^{1/h-1})>0.$$

Note that Theorems 1 and 2 are also valid for the number of independent sets, respectively for the number of maximal independent sets of the $(h+1)$-hypergraphs of order n.

Corollary 3. *For almost all $(h+1)$-hypergraphs H with n vertices, the number $\mathrm{CM}(H)$ of all maximal cliques (or the number of all cliques) verifies*

$$\lim_{n\to\infty} \log \mathrm{CM}(H) \Big/ \frac{h}{h+1}(h!)^{1/h}(\log n)^{1+1/h}=1.$$

Note that this result is valid also for the case of graphs ($h=1$) [3].

In conclusion, as has been proved above, the number of cliques (or of maximal cliques) of an $(h+1)$-hypergraph of order n is of the order of

$$n^{(h/(h+1))(h!\log n)^{1/h}}$$

for almost all such hypergraphs, as $n\to\infty$.

Thereby it follows that every algorithm solving the problem of deciding if a given uniform hypergraph has a clique of a given cardinality by listing all cliques of the hypergraph, must work at least in

$$n^{(h/(h+1))(h!\log n)^{1/h}}\text{-time},$$

i.e., it cannot work in polynomial time for almost all $(h+1)$-hypergraphs of order n, as $n\to\infty$.

References

[1] C. Berge, Graphes et Hypergraphes (Dunod, Paris, 1970).

[2] V.V. Glagolev, Some estimations of disjunctive normal forms of Boolean functions, Problems of Cybernetics, 19 (1967) 75–94 (in Russian).

[3] Phan Dinh Diêu, Asymptotical estimation of some characteristics of finite graphs, RAIRO., Inform. Théor./Theoret. Comput. Sci. 11 (2) (1977) 157–174.

P. Hansen, ed., Studies on Graphs and Discrete Programming

A SUGGESTED EXTENSION OF SPECIAL ORDERED SETS TO NON-SEPARABLE NON-CONVEX PROGRAMMING PROBLEMS*

J.A. TOMLIN

Ketron, Inc., San Bruno, CA 94066, USA

This paper suggests a branch and bound method for solving non-separable non-convex programming problems where the nonlinearities are piecewise linearly approximated using the standard simplicial subdivision of the hypercube. The method is based on the algorithm for Special Ordered Sets, used with separable problems, but involves using two different types of branches to achieve valid approximations.

1. Introduction

The first successful extension of linear programming to problems involving non-linear (other than quadratic) functions appears to be the introduction of piecewise linearization for convex functions of a single variable by Charnes and Lemke [8]. This idea was later extended by Miller [14], who developed a modification of the simplex algorithm which would allow at least local optimization of problems involving not necessarily convex piecewise linear functions. Miller's 'separable programming' method is now a routine component of any large scale Mathematical Programming System, and further refinements and extensions are described for example by Beale [2, 3].

While the immediate practical impact of Miller's work was in the treatment of separable non-linear functions; that is, functions $f(x)$ which could be expressed as a sum of functions of a single variable:

$$f(x)=\sum_{i=1}^{n} f_i(x_i),$$

Miller also indicated a way in which his treatment could be extended to non-separable functions. An n-dimensional space may be triangulated, or simplicially subdivided, and the adjacency restrictions applied for piecewise linear components in the separable (single variable) case can be generalized to adjacency restrictions on the simplices of the subdivisions. In particular, the

* This research was partially supported by the Office of Naval Research and carried out while the author was at the NASA-Ames Institute for Advanced Computation, Sunnyvale, CA.

'standard' method for simplicial subdivision of the hyper-cube [13] was recommended as a potentially practical mechanism for such an extension.

In the race to implement and refine the separable programming concept, the extension to the non-separable case seems to have been left at the post, and forgotten. In retrospect one can discern several reasons why this situation has persisted. We recite some of them:

(1) When a non-linear function is known analytically, it is nearly always possible to 'separate' it into functions of single variables by change of variable, taking logarithms, etc. (See e.g., [2]).

(2) If the non-linear functions are convex, or if a local optimum is satisfactory, then there are a number of alternative techniques, which have proved very effective. These include Wolfe's Reduced Gradient method, further refined by Murtagh and Saunders [15], the Generalized Reduced Gradient Method [1], Rosen's Projected Gradient Method [16], and the Method of Approximation Programming [11, 4].

(3) When a *global* optimum is required, then Miller's techniques are unreliable for non-convex functions. If the functions can be separated, then Special Ordered Sets, due to Beale and Tomlin [6], and further refined by Beale and Forrest [5] can be used in a branch and bound framework to find such a global optimum (within the limits of approximation).

(4) The number of 'special variables' (see below) can become very large. If there are g_i 'breakpoints' for each variable, in a non-linear function of n variables, then there are $\prod_{i=1}^{n} g_i$ special variables, and hence columns, required to represent this function. There are also more complicated programming requirements for identifying and accessing the vertices of n-dimensional simplexes, which do not arise in the one dimensional case.

The computational art of mathematical programming has progressed tremendously since the publication of Miller's paper, and it now seems appropriate to reconsider the non-separable case. Again, a number of reasons can be advanced for this:

(1) The success of Special Ordered Sets for globally optimizing separable problems invites extension.

(2) There are some non-linear, usually non-convex functions, which are not even analytic, and which arise in practice. One example involves the functions describing the behavior of various petro-chemical processes. Buzby [7] has shown that local optima can in practice be found for such problems, using the Method of Approximation Programming, but even approximate global solutions remain elusive.

(3) The success of piecewise linearization and simplicial subdivision methods [9, 17] in solving fixed point, and other difficult problems (embracing much of mathematical programming), indirectly encourages the use of similar constructs for non-convex programming.

The last point (3) should be viewed with some caution. It is a simple matter to check, by direct substition, whether an alleged fixed point of a function satisfies the definition. No such simple tests exist for global optima of non-convex functions.

In the following sections, we will present first the notation, terminology, and the basic topological result, as described by Kuhn, followed by a description of the suggested branch and bound method.

We note that a quite different algorithm, based on a simplicial subdivision process, but with stronger assumptions, has been described by Horst [12].

2. Simplicial subdivision

Let us suppose that our otherwise tractable methematical program involves a non-separable, non-convex function, which may exist only as a subroutine, whose value $v^k = f(x^k)$ can be determined at points x^k in some n-dimensional space. The most natural way of selecting points x^k is to define a grid over the space. Let there be g_i (increasing) grid, or break points for each variable:

$$x_i = h_{i1}, h_{i2}, \ldots, h_{i,g_i} \quad (i = 1, \ldots, n), \tag{1}$$

then there are $\prod_{i=1}^{n} g_i$ defined grid points x^k with associated values v^k. Let the ordering of the k index be such that the grid point $(h_{1,j_1}, h_{2,j_2}, \ldots, h_{n,j_n})$ corresponds to:

$$k = 1 + \sum_{i=1}^{n} (j_i - 1)\pi_i \tag{2}$$

where

$$\pi_i = \begin{cases} 1 & \text{for } i = 1, \\ \prod_{j=1}^{i-1} g_j & \text{for } i = 2, \ldots, n. \end{cases} \tag{3}$$

That is, the elements of the n-dimensional array are mapped onto a one-dimensional array.

The function $f(x)$ will be approximated over this set of grid points by:

$$f(x) = \sum_k v^k \lambda_k, \tag{4}$$

$$x = \sum_k x^k \lambda_k, \tag{5}$$

$$1 = \sum_k \lambda_k, \quad \lambda_k \geqslant 0, \tag{6}$$

plus the additional constraint that:

> The non-zero λ_k are restricted to those associated with the $(n+1)$ vertices of the simplex defined by the vector x and the simplicial subdivision. (7)

The simplicial subdivision to be employed is that imposed by the 'standard' subdivision of the cube [13]. It will be seen that any valid point x is associated with a hyper-rectangle defined by:

$$h_{i,j_i} \leqslant x_i \leqslant h_{i,j_i+1}. \tag{8}$$

By an appropriate change of origin and variables, we may express this hyper-rectangle as the set:

$$R = \{y \mid 0 \leqslant y_i \leqslant b_i;\ i = 1, \ldots, n\} \tag{9}$$

where

$$b_i = h_{i,j_i+1} - h_{i,j_i} \quad \text{and} \quad y_i = x_i - h_{i,j_i}.$$

Lemma 1 of [13] asserts that every point in R can be expressed in the form:

$$y = \sum_{k=0}^{n} I_k \mu_k. \tag{10}$$

Proceeding by construction, we define $\bar{y}_j = y_j/b_j$ and an ordering of these values such that $\bar{y}_{s_1} \geqslant \bar{y}_{s_2} \geqslant \cdots \geqslant \bar{y}_{s_n}$, plus two dummy variables $\bar{y}_{s_0} = 1$, $\bar{y}_{s_{n+1}} = 0$. These in turn define $\mu_k = \bar{y}_{s_k} - \bar{y}_{s_{k+1}}$.

Now letting $I_0 = 0$ and $I_k = I_{k-1} + b_k U_{s_k}$, $(k = 1, \ldots, n)$ where U_{s_k} is the s_kth unit n-vector, it is trivial to show that (10) holds.

Note that there are at most $(n+1)$ non-zero μ_k, that by construction the vectors I_k give (translated) coordinates of vertices of the n-hyper-rectangle, and that the construction carries with it a simplicial subdivision of the hyper-rectangle, defined by the $n!$ possible orderings of the $\bar{y}_j$.

If there are values $f(x^k)$ associated with the vertices I_k, then the appropriate linear approximation for $f(x)$ is:

$$f(x) = \sum_{k=0}^{n} f(x^k)\mu_k. \tag{11}$$

In order that the λ_k values obtained in a solution of the MP be admissible, we demand that only those associated with the simplex defined by (9), (10), within the hyper-rectangle defined by (8), be non-zero. We also see from (2), (3) and (8) that the set of λ_k associated with the hyper-rectangle are defined by the index set:

$$K_R(x) = \left\{k \mid k = 1 + \sum_{i=1}^{n} (j_i - \delta_i)\pi_i,\ \delta_i = 0, 1 \text{ for } i = 1, \ldots, n\right\} \tag{12}$$

while those associated with the simplex are defined by the index set:

$$K_S(x) = \left\{k \mid k = 1 + \sum_{i=1}^{n} (j_i - 1)\pi_i + \sum_{i=1}^{l} \pi_{s_i},\ l = 0, 1, \ldots, n\right\} \tag{13}$$

where the void sum is understood to be zero and the indices s_i are as in the ordering of the $\bar{y}_i$ values.

A valid piecewise linear representation then, has the characteristic that:

Only those λ_k with $k \in K_S(x) \subset K_R(x)$ may be non-zero for the current value of x. (7′)

These λ-values will correspond in value to the μ-values in (10), (11).

While the notation used may seem rather intricate, particularly compared to the one dimensional case, the actual computation required to check that (7′) is satisfied is fairly simple. Assuming (though not requiring) that (5) is explicitly represented in the problem formulation, it involves only n discrete searches to find the j_i values from (8), $2n$ subtractions and n divisions to compute the $\bar{y}_j$, a sorting operation on these values, and then a comparision between the non-zero λ_k and $K_s(x)$.

3. Branch and bound procedure

The 'non-separable' procedure discussed by Miller [14] would have required that a feasible solution to the MP be obtained intially with the λ_k restricted to $K_S(x)$, and hence to $K_R(x)$. Assuming that this can reasonably be done, his algorithm would require that one or more of these values go to zero before other λ-values may be increased in such a way as to move the point x into an 'adjacent' simplex, to give a local improvement in the objective function. Since this procedure can only guarantee an approximate local optimum, for which other methods are now available, we shall consider global optimization, within the limits imposed by the grid points and piecewise linear approximation. To do this, we adopt a branch and bound approach.

The use of a grid to partition the space, and the simplicial subdivision of the hyper-rectangles within the grid, suggests a two-stage branch and bound process, or more precisely a process with two types of branches. It is convenient to arrange matters so that the non-zero λ_k are confined to a set $K_R(x)$, and our first priority will be to branch to achieve this. We denote these as 'R-branches'. The second type of branch will be used to exclude those λ_k which do not belong to some simplex with index set $K_S(x) \subset K_R(x)$. Such branches are denoted 'S-branches'.

R-Branches

The principle used here is precisely that used with Special Ordered Sets of type 2 ($S2$ sets) [6]. Let $x = \sum x^k \lambda_k$ be the value of x at a continuous optimum of the problem, with condition (7) dropped, and define a 'current interval' for each variable x_i by:

$$h_{i,j_i} \leq x_i \leq h_{i,j_i+1}. \tag{14}$$

For (7) to hold we can state that:

Either all λ_k corresponding to $x_i^k < h_{i,j_i}$ must be zero, *or* all λ_k corresponding to $x_i^k > h_{i,j_i}$ must be zero.

A similar condition *also* holds for h_{i,j_i+1}.

Now consider some x_r such that there are λ_k values corresponding to $h_{r,j}$ outside the current interval. We then adopt the weaker of each pair of alternatives, in terms of the portion of the space excluded and create the following branch:

(i) *Either* set all λ_k corresponding to h_{rj}, $j > j_r + 1$ to zero.

(ii) *Or* set all λ_k corresponding to h_{rj}, $j < j_r$ to zero.

These alternatives admit the current interval for x_r, but in general exclude the current set of inadmissible λ_k values. The same caveat as in [6] applies; in that if the current solution is not excluded, the branch must be made about *one* of the end points h_{r,j_r}, h_{r,j_r+1}, and not the interval.

The two sets of k indices for which the λ_k must be forced to zero (flagged) in alternatives (i) and (ii) above are:

$$K_r^1(x) = \left\{ k \mid k = 1 + \sum_{i=1}^{n} (l_i - 1)\pi_i, \forall l_i = 1, \ldots, g_i \ (i \neq r), l_r = j_r + 2, \ldots, g_r \right\}, \tag{15}$$

$$K_r^2(x) = \left\{ k \mid k = 1 + \sum_{i=1}^{n} (l_i - 1)\pi_i, \forall l_i = 1, \ldots, g_i \ (i \neq r), l_r = 1, \ldots, j_r - 1 \right\}. \tag{16}$$

Since we branch on one variable x_r at a time, it is clear from (15) and (16) that some λ_k could in principle be flagged many times in branches on different variables. This could cause difficulty when unflagging λ-variables to explore an alternative branch in the course of the search. Some indicator of which branch induced the flagging, or of the x_i variable being branched on, must therefore be kept for each λ-variable, in addition to its status.

The choice of which variable x_i to branch on, and which branch to explore, is very similar to the choice faced when a problem has several conventional $S2$ sets. Since we are discussing a branching method based on [6], one could use a 'penalty' method, computing the penalties associated with flagging the λ_k associated with the sets $K_i^1(x)$, $K_i^2(x)$. Alternatively, it may be possible to give priorities to each of the non-linear variables x_i or to define pseudo-costs, as in

[10]. This treatment can be extended to the more general case in which there are several non-linear functions. Another candidate method is through the use of pseudo-shadow-prices, as in Beale and Forrest [5]. However, it seems pointless to speculate on the relative merits of these approches at this stage.

S-Branches

Once the non-zero λ_k have been confined to the vertices of a hyper-rectangle $K_R(x)$ in some subproblem they must now be further confined to the vertices of some simplex $K_S(x) \subset K_R(x)$ if this proves feasible. If the values assume this configuration naturally, of course no further branching is required for the subproblem and a feasible value has been obtained. Let us assume that this is not the case.

As we have seen each simplex in the subdivision of $K_R(x)$ corresponds by construction to an ordering

$$\bar{y}_{s_1} \geq \bar{y}_{s_2} \geq \cdots \geq \bar{y}_{s_n}$$

of the translated and transformed values of x, and a set of non-negative μ_k corresponding in value to an associated set of permitted non-zero λ_k, with $k \in K_S(x)$. If some λ_k with $k \notin K_S(x)$ is non-zero, we must branch in such a way as to eliminate this (or some other) impermissible situation. The key to such a branching scheme is in the manner in which the simplicial subdivision is constructed.

If p and q are distinct indices such that $\bar{y}_p > \bar{y}_q$, then it is clear that no vertex at which $\bar{y}_q = 1$ and $\bar{y}_p = 0$ can be assigned a positve weight in (10). Since it is the inequality relationships $\bar{y}_i \geq \bar{y}_j$ which define the simplex, this observation gives us a branching criterion. For any pair p, q of indices let us define two subsets of indices of the hyper-rectangle $K_R(x)$:

$$S_{pq}(x) = \left\{k \mid k = 1 + \sum_{i \neq p,q} (j_i - \delta_i)\pi_i + (j_p - 1)\pi_p + j_q\pi_q;\ \delta_i = 0, 1;\ \forall i \neq p, q\right\} \quad (17)$$

$$S_{qp}(x) = \left\{k \mid k = 1 + \sum_{i \neq p,q} (j_i - \delta_i)\pi_i + (j_q - 1)\pi_q + j_p\pi_p;\ \delta_i = 0, 1;\ \forall i \neq p, q\right\}. \quad (18)$$

Set $S_{pq}(x)$ represents the set of vertices (or rather their index set) which must have zero λ_k if $\bar{y}_p > \bar{y}_q$, and $S_{qp}(x)$ the set which must have zero λ_k if $\bar{y}_q > \bar{y}_p$. Since both sets must have zero weights if $\bar{y}_p = \bar{y}_q$, we will arbitrarily choose the set $S_{pq}(x)$ for such a case.

With these observations, we now state the criterion for valid simplicial representation of a point x with weights already confined to $K_R(x)$: For any p, q with $\bar{y}_p \geq \bar{y}_q$, we require $\lambda_k = 0$ for all $k \in S_{pq}(x)$.

This leads immediately to the branching mechanism:

For *any* pair of indices $p, q \in K_R(x)$ then:
Either all $\lambda_k = 0$ for $k \in S_{pq}(x)$, (19)
Or all $\lambda_k = 0$ for $k \in S_{qp}(x)$.

It now remains to choose a criterion for picking a pair of indices p, q, such that either $S_{pq}(x)$ or $S_{qp}(x)$ contains an inadmissible k. Since we have up to $\frac{1}{2}n(n-1)$ possible such choices there are several possible strategies. Note that this number is also an upper bound on the number of branches required to enforce (7′) when confined to $K_R(x)$. The obvious procedure of examining all the sets $S_{pq}(x)$, perhaps computing penalties for each of the alternatives in (19), would appear rather expensive unless n is very small; though we may have either $S_{pq}(x)$ or $S_{qp}(x)$ empty for many pairs of indices, in which cases branching need not be considered. In fact if there are m non-zero λ_k with $k \in K_R(x)$ the number of penalties (or pairs of penalties) to be computed is at most $\frac{1}{2}m(m-1)$; a potentially more reasonable number.

Simpler, cheaper (and cruder) methods can be derived by analogy with early branch and bound tactics for integer programming. Two such common tactics were:

(a) Branch on the integer variable with fractional part closest to 0.5.

(b) Branch on the variable closest to an integer value.

On the whole, the first tactic seems to be better since it tends to do the 'expensive' branches early in the search (see [10]). A rough analogy with tactic (a) is to choose the largest λ_k in $K_R(x)$ with $k \notin K_S(x)$, say λ_M, and the largest remaining λ_k, say λ_m, which may or may not have $m \in K_S(x)$, in such a way that there is at least one pair of indices p, q such that in the expressions for M and m as in (12) have $\delta_p = 1$, $\delta_q = 0$ for M and $\delta_p = 0$, $\delta_q = 1$ for m. For branching to take place, the only essential is that λ_M and λ_m are not associated with the same simplex in $K_R(x)$. This is assured by the above choice, for if the non-zero λ_k all have indices k differing pairwise in only one setting of δ_i in (12), then in (13) there will exist a permutation s_i assigning all of these k values to some simplex $K_S(\cdot)$.

A similar analogy with tactic (b) would be to choose λ_M as the smallest λ_k in $K_R(x)$ with $k \notin K_S(x)$ above some tolerance and proceed in the same way.

As in the case of R-branches, individual λ_k may be redundantly flagged to zero, since the sets $S_{pq}(x)$, $S_{qp}(x)$ overlap for different p and q values. Thus again it is important to keep track of which branch has induced the flagging. Note also that an S-branch can be defined by the 'base-index' $1+\sum_i (j_i - 1)\pi_i$ for the hyper-rectangle $K_R(x)$ and by p and q, via (17), (18).

It is possible for an S-branch to force the subsequent trial solution $\bar{x}$ out of the current hyper-rectangle $K_R(x)$, unless the preceding R-branches have completely confined the non-zero λ_k to $K_R(x)$. If the non-zero $\bar{\lambda}_k$ on this branch are confined to some new $K_R(\bar{x})$ then we may proceed with S-branches, otherwise we must revert to R-branches to establish this condition. Thus even a simple last-in first-out (LIFO) tree may contain alternating sequences of R- and S-branches.

4. Example

Of necessity an example small enough to be worked in detail and visualized must be rather trivial. Fig. 1 shows the nine grid points representing a function in two dimensions, which is piecewise linearized (triangulated), and whose maximum is sought subject to two constraints. In detached coefficient from the problem is:

x_0	x_1	x_2	λ_1	λ_2	λ_3	λ_4	λ_5	λ_6	λ_7	λ_8	λ_9	
1			−100	−10	−180	−60	−70	−200	−70	−150	−140	$=0$
	−1			1	2		1	2		1	2	$=0$
		−1				1	1	1	2	2	2	$=0$
			1	1	1	1	1	1	1	1	1	$=1$
	2	1										$\leqslant 4$
	−1	2										$\leqslant 2$

$$\text{Max } x_0,\ x_j,\ \lambda_j \geqslant 0.$$

The L.P. solution to this problem is

$$x_0 = 180.0, \quad x_1 = 1.6, \quad x_2 = 0.8, \quad \lambda_1 = 0.2, \quad \lambda_6 = 0.8.$$

The combination of non-zero λ-values is clearly inadmissible.

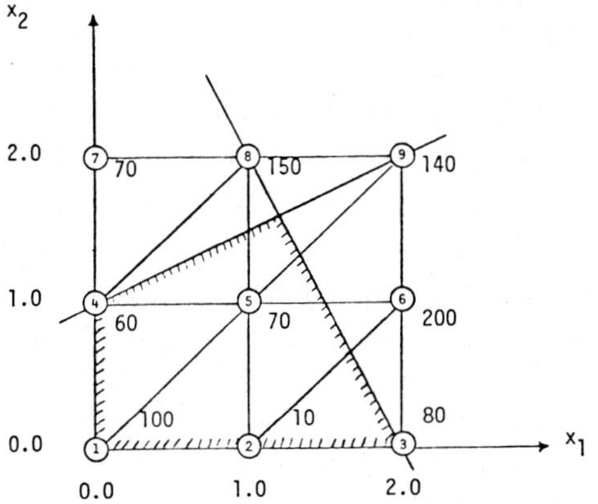

Fig. 1. Small example problem.

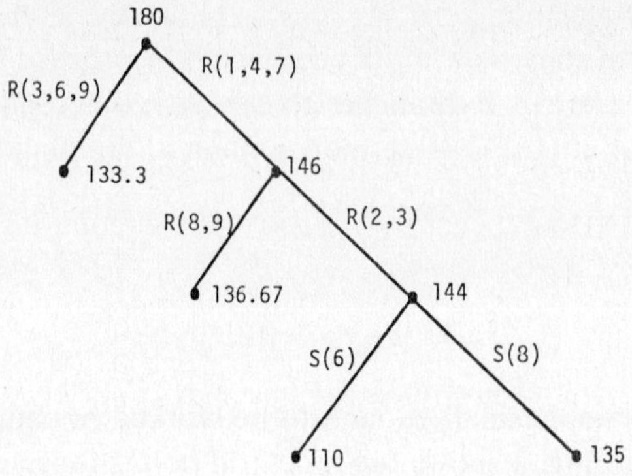

Fig. 2. LIFO tree for example.

The tree search for this example was hand-guided using the simplest possible strategy of always flagging first to drive the feasible region to the right and upwards, and using last-in first-out. Following this strategy the first branch is an R-branch which flags λ_1, λ_4 and λ_7 (represented as $R(1, 4, 7)$ in Fig. 2). The solution of this problem is:

$$x_0 = 146.0, \quad x_1 = 1.2, \quad x_2 = 1.6, \quad \lambda_2 = 0.1, \quad \lambda_6 = 0.7, \quad \lambda_8 = 0.7,$$

The non-zero λ-values are still not confined to a rectangle and another R-branch is made which flags λ_2 and λ_3. In the resulting solution:

$$x_0 = 144.0, \quad x_1 = 1.2, \quad x_2 = 1.6, \quad \lambda_5 = 0.2, \quad \lambda_6 = 0.2, \quad \lambda_8 = 0.6,$$

the non-zero λ-values are now confined to a rectangle, but the combination of non-zero λ_6 and λ_8 is inadmissible for this piecewise linearization. We now require an S-branch which flags λ_8 to zero as the first alternative and which gives the first admissable solution:

$$x_0 = 135.0, \quad x_1 = 1.5, \quad x_2 = 1.0, \quad \lambda_5 = 0.5, \quad \lambda_6 = 0.5.$$

The alternative S-branch flagging λ_6 gives the poorer solution

$$x_0 = 116.0, \quad x_1 = 1.2, \quad x_2 = 1.6, \quad \lambda_5 = 0.4, \quad \lambda_8 = 0.4, \quad \lambda_9 = 0.2.$$

Back-tracking to the previous R-branch, λ_2 and λ_3 are unflagged and λ_8 and λ_9 flagged, leading to the new L.P. solution:

$$x_0 = 136.67, \quad x_1 = 1.67, \quad x_2 = 0.67, \quad \lambda_2 = 0.33, \quad \lambda_6 = 0.67.$$

This solution is admissible and being the best so far becomes the new incumbent.

Finally, we back-track to the first node (having unflagged λ_8, λ_9) unflag

λ_1, λ_4, λ_7 and flag the alternative set λ_3, λ_6, λ_9. This leads to the L.P. solution:

$$x_0 = 133.3, \quad x_1 = 0.67, \quad x_2 = 1.33, \quad \lambda_1 = 0.33, \quad \lambda_8 = 0.67,$$

which is not only inadmissible, but worse than the incumbent. The search is therefore terminated.

5. Conclusion

The methodology proposed in this paper clearly requires larger scale implementation and computational verification. We are particularly interested in this approach in a dynamic programming context [18], where payoff functions are approximated by piecewise linearization, and the basic recursion is of the form:

$$p^t(z) = \max_{u \in U_t} p^{t+1}(F_u(z)).$$

Here $p^t(\cdot)$ is the payoff at stage t, U_t the set of permissible control vectors u, and $F_u(z)$ is the state at stage $t+1$ resulting from choice u at the state space point z at stage t. If U_t can be modelled by mathematical programming techniques, and $p^{t+1}(\cdot)$ is of the piecewise linear form we have used, the above optimization problem is of the form for which our global branch and bound method is suitable.

A program to implement the suggested approach is currently being prepared, based on an existing branch and bound code.

References

[1] J. Abadie and J. Carpentier, Generalization of the Wolfe reduced gradient method to the case of nonlinear constraints, in: R. Fletcher, ed., Optimization (Academic Press, New York, 1969).

[2] E.M.L. Beale, Numerical Methods, in: J. Abadie, ed., Nonlinear Programming (North-Holland, Amsterdam, 1967).

[3] E.M.L. Beale, Mathematical Programming in Practice (Pitman, London, 1968).

[4] E.M.L. Beale, A conjugate gradient method of approximation programming, in: R.W. Cottle, J. Krarup, eds., Optimization Methods for Resource Allocation (E.U.P., London, 1974).

[5] E.M.L. Beale and J.J.H. Forrest, Global optimization using special ordered sets, Math. Programming 10 (1976) 52–69.

[6] E.M.L. Beale and J.A. Tomlin, Special facilities in a general methematical programming system for nonconvex problems using ordered sets of variables, in: J. Lawrence, ed., Proc. 5th IFORS Conference (Tavistock, London, 1970) 447–454.

[7] B.R. Buzby, Techniques and experience solving really big nonlinear programs in: R.W. Cottle, J. Karup, eds., Optimization Methods for Resource Allocation (E.U.P., London, 1974).

[8] A Charnes and C.E. Lemke, Minimization of non-linear separable convex functionals, Naval Res. Logist. Quart. 1 (4) (1954).
[9] B.C. Eaves, Homotopies for computation of fixed points, Math. Programming 3 (1972) 1–22.
[10] J.J.H. Forrest, J.P.H. Hirst and J.A. Tomlin, Practical solution of large mixed integer programming problems with UMPIRE, Management Sci. 20 (1974) 736–773.
[11] R.E. Griffith and R.A. Stewart, A nonlinear programming technique for the optimization of continuous processing systems, Management Sci. 7 (1961) 379–392.
[12] R. Horst, An algorithm for nonconvex programming problems, Math. Programming 10 (1976) 312–321.
[13] H.W. Kuhn, Some combinatorial lemmas in topology, IBM J. of Res. and Dev. 4 (1960) 518–524.
[14] C.E. Miller, The simplex method for local separable programming, in: R.L. Graves, P. Wolfe, eds., Recent Advances in Mathematical Programming (McGraw-Hill, New York, 1963) 89–100.
[15] B.A. Murtagh and M.A. Saunders, Large-scale linearly constrained optimization, Math. Programming 14 (1978) 41–72.
[16] J.B. Rosen, The gradient projection method for nonlinear programming, Parts I and II, SIAM J. 8 (1960) 181–217 and 9 (1961) 514–532.
[17] M.J. Todd, On triangulations for computing fixed points, Math. Programming 10 (1976) 322–346.
[18] J.A. Tomlin, Piecewise linear and polynomial approximation for dynamic programming and games, TM-5613, Institute for Advanced Computation, Sunnyvale, CA (April 1978).

P. Hansen, ed., Studies on Graphs and Discrete Programming

PARAMETRIC MULTICRITERIA INTEGER PROGRAMMING*

Bernardo VILLARREAL
Instituto Tecnologico y de Estudios Superiores de Monterrey, Monterrey, Mexico

Mark H. KARWAN
State University of New York at Buffalo, Buffalo, NY 14260, USA

The parametric programming problem on the right hand side for multicriteria integer linear programming problems is treated under a (hybrid) dynamic programming approach. Computational results are reported.

1. Introduction

This paper presents an extension of the dynamic programming scheme for multicriteria discrete programming problems suggested in [7, 8], to solving the parametric problem on the right hand side. The multicriteria parametric integer linear programming problem of concern is formulated as follows.

$$\begin{array}{lll} & \text{v-max} & \sum_{n=1}^{N} C^n \cdot x_n, \\ \text{(MPILP)} & \text{s.t.} & \sum_{n=1}^{N} a^n \cdot x_n \leqslant \{b + \Theta d\}, \\ & & k_n \geqslant x_n \geqslant 0, \quad \text{integer}, \quad (n = 1, \ldots, N), \end{array}$$

where $C^n = (c_{1n}, \ldots, c_{pn})^t$; $a^n = (a_{1n}, \ldots, a_{Mn})^t$ denotes a set of integers; k_n is a positive integer constant; $d = (d_1, \ldots, d_M)^t$ denotes a direction vector; $\Theta \in [0, 1]$ is a positive parameter; and v-max (vector maximization) is used to differentiate the problem from the common case of single-objective maximization. The solution to (MPILP) corresponds to obtaining the sets of efficient or Pareto optimal solutions for the family of multicriteria integer programming problems whose right hand side vectors lie along the ray defined by $(b + \Theta d)$, $\Theta \in [0, 1]$. This problem has not yet been approached under a multicriteria framework. Previous work on parametric integer programming on the right hand side has concentrated on the single objective case [5, 6, 4, 3].

* This research was supported in part by the Consejo Nacional de Ciencia y Technologia, Mexico.

The mathematical model of concern is of practical interest when management would like to perform a sensitivity analysis of the Pareto optimal set for several values of the vector of resources. This situation may occur when precise point estimates or values of the amount of resources available are not known. Thus, obtaining the solution set for the family of problems generated by varying the resources over a range of values of interest, will allow management to determine how sensitive this set is to changes in the amount of resources. In many investment and location problems containing 0–1 decision variables, a set of alternatives must be judged under different criteria. Some essential constraints are of the budgetary type and the effects of varying budgets should be one of the main concerns of the decision makers.

The plan of this paper is as follows. The following section deals with the development of a scheme for solving (MPILP). This procedure is modified in Section 3 by the inclusion of bounds to obtain a more efficient procedure. Section 4 offers some computational results as well as conclusions and comments.

2. Development of algorithm

The parametric extension of the earlier works of Villarreal and Karwan [7, 8] can be handled by modifying the dynamic recursive procedure which they develop. This is done by taking advantage of the characteristics of the recursions. Specifically, one can obtain the set of efficient solutions for any nonnegative vector of resources, $y(\leqslant b)$. This implies that one could solve (MPILP) by solving a multicriteria integer linear programming problem with a right hand side vector with components $b_i = \max\{b_i, b_i + d_i\}$, $i = 1, \ldots, M$. Then at the final stage, instead of requiring the set of efficient points for only one vector of resources, one would compute the set of efficient solutions for each vector, $b + \Theta d$; $\Theta \in [0, 1]$. In order to clarify these ideas, let us outline the imbedded state dynamic programming procedure suggested in [7, 8] without considering the bounding feature reported there.

Step 0. Set $m = 1$ and let ϕ^0 be an empty set.

Step 1. Obtain the set of m-dimensional integer points $(x_1, \ldots, x_m)$ such that $0 \leqslant x_m \leqslant k_m$ and $(x_1, \ldots, x_{m-1}) \in \phi^{m-1}$.

Step 2. Eliminate all those m-dimensional points that are infeasible, i.e., those for which some component, i, of their resource consumption vector satisfies the following expression.

$$\sum_{n=1}^{m} a_{in} \cdot x_n > b_i + \sum_{\substack{n=m+1 \\ i \in H_n}}^{N} |a_{in} \cdot k_n|.$$

Step 3. Obtain the set of resource efficient solutions, ϕ^m, via pairwise comparisons.

Step 4. If $m = N$ go to Step 5. Otherwise, set $m = m+1$ and go to Step 2.

Step 5. For any given vector of resources, y_N ($\leq b$), obtain the set of efficient solutions by solving the following problem via pairwise comparisons.

$$\text{v-max}\left\{\sum_{n=1}^{N} C^n \cdot x_n \mid (x_1, \ldots, x_N) \in \phi^N \quad \text{and} \quad \sum_{n=1}^{N} a^n \cdot x_n \leq y_N\right\}.$$

Here, ϕ^m denotes the set of resource efficient solutions for the m-stage problem. These solutions have the property that if $x^0 \in \phi^m$ there is no other feasible point, x, such that the following expressions are satisfied with at least one strict inequality in the first expression.

$$\sum_{n=1}^{m} C^n \cdot x_n \geq \sum_{n=1}^{m} C^n \cdot x_n^0, \qquad \sum_{n=1}^{m} a^n \cdot x_n \leq \sum_{n=1}^{m} a^n \cdot x_n^0.$$

The set H_n required in the procedure is the set $\{i \mid a_{in} < 0\}$. For the case of multicriteria multidimensional knapsack problems, H_n is empty.

We can immediately notice that changing the right hand side vector of resources will only affect the feasibility test of Step 2, and Step 5 when obtaining the set of efficient solutions for the particular vectors of resources of interest. This follows from observing that the scheme computes, as an initial general step, the set of resource efficient points for the relaxed problem:

$$(\text{P}') \qquad \begin{array}{ll} \text{v-max} & \displaystyle\sum_{n=1}^{N} C^n \cdot x_n, \\ \text{s.t.} & k_n \geq x_n \geq 0, \quad \text{integer}, \quad (n = 1, \ldots, N). \end{array}$$

Then, the constraints imposed on the resources, y_N, are used in the feasibility testing step during the recursion and at the final step of the scheme. (This is one of the main differences of the dynamic programming approach to that suggested by Bitran [1] for the case in which $k_n = 1$, $n = 1, \ldots, N$. He first finds the set of efficient (not resource efficient) points for the relaxed problem (P′) before imposing the feasibility restrictions. Then, he proceeds to find other points which were not efficient for (P′) but which are efficient to the more restricted problem.) This observation leads one to further conclude that regardless of the right hand vector of resources, the procedure would remain almost the same with the exception of the feasibility test, until the final step, when a specific vector of resources must be determined. Hence, since the parametric integer linear programming problem with multiple criteria (PMILP) differs from the original multiple criteria integer linear programming problem by the structure of their right hand vectors of resources, one could use the dynamic recursions, with slight modifications, to solve (PMILP).

Let us define $\bar{B}$ and $\underline{B}$ as

$$\bar{B}=(\bar{b}_1,\ldots,\bar{b}_M)^t \quad \text{and} \quad \underline{B}=(\underline{b}_1,\ldots,\underline{b}_m)^t,$$

where

$$\bar{b}_i=\max(b_i, b_i+d_i) \quad \text{and} \quad \underline{b}_i=\min(b_i, b_i+d_i).$$

Given these definitions, one can modify the recursive scheme as follows:

(a) Replace Step 2 by

Step 2′. Eliminate all those m-dimensional points for which some component, i, of their resource consumption vector satisfies the following condition.

$$\sum_{n=1}^{m} a_{in}\cdot x_n>\bar{b}_i+\sum_{\substack{n=m+1\\ i\in H_n}}^{N}|a_{in}\cdot x_n|.$$

(b) Replace Step 5 by

Step 5′. For each vector of resources, y_N, such that $\underline{B}\leqslant y_N\leqslant\bar{B}$, obtain the set of efficient solutions by solving the following problem via pairwise comparisons.

$$\text{v-max}\left\{\sum_{n=1}^{N} C^n\cdot x_n \mid (x_1,\ldots,x_N)\in\phi^N \text{ and } \sum_{n=1}^{N} a^n\cdot x_n\leqslant y_N\right\}.$$

Even though the formulation of this problem suggests the direct variation of the vector of resources, y_N, in order to solve the family of problems associated with all the values of Θ, one actually works only with those vectors of resources which make possible new alternate feasible solutions. In this procedure, we start at $y_N=\underline{B}(\Theta=0)$, and consider only those solutions that are feasible, to determine the efficient set of solutions. The set of infeasible solutions is stored to be considered for obtaining the set of efficient solutions for subsequent vectors of resources. The process is continued considering next feasible solutions one by one until the set becomes empty. One determines if each new feasible solution is efficient by simple comparison with those currently efficient. Each time a solution is dominated, it is discarded from further consideration. The next value of Θ ($\geqslant 0$) is obtained by using the following relationship.

$$\Theta^0=\min_q\left\{\max_{i=1,\ldots,M}\{y_{iq}-b_i)/d_i\}\right\}$$

where y_{iq} is the ith component of the resource consumption vector for the qth resource efficient point which is infeasible for the current value of Θ. Obviously, the next feasible solution would be that associated with Θ^0.

As was the case for multicriteria integer linear programming problems, one could also adapt the prior scheme to allow for a bounding procedure. Developing such a procedure is the aim of the following section.

3. A hybrid approach

The hybrid dynamic scheme developed in [8] can also be extended to allow for the parametric case on the right hand side. Assume that one has partially solved the original problem by the dynamic programming recursion until the mth stage, considering as the maximum vector of resources the vector $\bar{B}$, and having at hand the set of resource efficient points for that stage. The residual problem for a particular vector of resources, $b+\Theta d$; $\Theta \in [0, 1]$, and a specific resource efficient point, say q, can be expressed as follows:

$$(\mathrm{RP}\Theta)\qquad \begin{aligned} &\text{v-max} && \sum_{n=m+1}^{N} C^n \cdot x_n \\ &\text{s.t.} && \sum_{n=m+1}^{N} a^n \cdot x_n \leqslant (b+\Theta d - y_q), \\ & && 0 \leqslant x_n \leqslant k_n, \quad (n=m+1, \ldots, N) \end{aligned}$$

where y_q denotes the resource consumption vector for the corresponding qth resource efficient solution. Let us denote by $\mathrm{UB}_{m+1}(y_q, \Theta)$ a set of upper bounds which for given Θ is a set of points which satisfies the following conditions:

(1) Each element is either efficient or dominates at least one of the efficient solutions of the problem.

(2) Each efficient solution of the problem is dominated by at least one member of the set or is contained in the set.

Let us also denote by $\mathrm{LB}(\Theta)$ a set of lower bounds for the original problem (given Θ) with the characteristic that each of its elements is either efficient or dominated by at least one efficient solution of the problem.

We shall describe now the extension in the following result.

Theorem 1. *Let a resource efficient point for the m-stage problem, say x, with resource consumption vector, y_q, and p dimensional multicriteria value H_{mx}, available. If for each value of $\Theta \in [0, 1]$, and every element*[1]

[1] $\oplus$ means that vector addition is performed with H_{mx} and each member of the set $\mathrm{UB}_{m+1}(y_n, \Theta)$. If

$$H_{mx} = \begin{pmatrix} 6 \\ 7 \end{pmatrix} \quad \text{and} \quad \mathrm{UB}_{m+1}(y_n, \Theta) = \left\{ \begin{pmatrix} 3 \\ 2 \end{pmatrix}, \begin{pmatrix} 1 \\ 4 \end{pmatrix} \right\}$$

then

$$H_{mx} \oplus \mathrm{UB}_{m+1}(y_n, \Theta) = \left\{ \begin{pmatrix} 9 \\ 9 \end{pmatrix}, \begin{pmatrix} 7 \\ 11 \end{pmatrix} \right\}.$$

$g_k \in H_{mx} \oplus UB_{m+1}(y_n, \Theta)$ there exists an element $LB_{j(k)} \in LB(\Theta)$ such that $g_k \leqslant LB_{j(k)}$ with at least one strict inequality, then no completion of x would lead to an efficient solution for any $\Theta \in [0, 1]$.

Proof. The proof follows the one given in [8] for each fixed value of Θ. Thus, it is omitted.

With this result one can modify the previous procedure to allow for the use of bounds for fathoming purposes. The changes to be made are the following:

(a) Step 0 must include the computation of the set of lower bounds for each member of the family of original problems generated by the range of vector of resources, $b+\Theta d$; $\Theta \in [0, 1]$. One must also initialize a counter of stages that will serve as an indicator to know when the bounding scheme will be used.

(b) After Step 3 one must check if the bounding scheme is to be used. If not, continue on Step 4. Otherwise, set(s) of upper bounds, $UB_{m+1}(y_q, \Theta)$; $\Theta \in [0, 1]$ for each resource efficient solution (q), are computed. Also note that the set(s) of lower bounds, $LB(\Theta)$, can be improved using heuristics (if required) to compute feasible solutions for the associated residual problems corresponding to the qth resource efficient point.

Now, it is a matter of generating the family of sets of lower and upper bounds for all the values of Θ. All the sets suggested in [8] can be extended to be used for these purposes. Unfortunately, most of them depend upon the value of the vector of resources that in turn depends on the value of Θ. This circumstance has as a result that the degree of difficulty to obtain them increases. As a consequence, it would be desirable to compute a fixed set of bounds which can be used for fathoming purposes for any member of the family of problems described by the vector of resources $b+\Theta d$, $\Theta \in [0, 1]$.

A weaker set of lower bounds for the original problem can be constructed by simply considering as the vector of resources the vector $\underline{B}$, and generating either efficient solutions by maximizing composite objective functions from the original set of objective functions or computing good feasible solutions using heuristics. Similarly, weaker sets of upper bounds can be determined by using as vector of resources the vector $\bar{B}$ which does not vary with Θ. Even though these sets are expected to be less efficient, it is also true that less computational effort will be devoted to obtain them. In order to illustrate the effect of using these type of bounds in a hybrid dynamic procedure various problems generated at random were solved using the heuristic suggested by Loulou and Michaelides [3] for zero–one problems to obtain sets of lower bounds, and the following concepts were used to obtain sets of upper bounds. The set of upper bounds is composed of only one member that is obtained from the use of duality theory and each and every objective function of the problem. Its development is as follows. Let us first recall the residual problem (RPΘ):

$$(\mathrm{RP}\Theta)\quad \begin{array}{ll} \text{v-max} & \sum_{n=m+1}^{N} C^n \cdot x_n, \\ \text{s.t.} & \sum_{n=m+1}^{N} a^n \cdot x_n \leqslant (b+\Theta d - y_q), \\ & 0 \leqslant x_n \leqslant k_n, \quad \text{integer}, \quad (n=m+1,\dots,N). \end{array}$$

An upper bound for the solution of this problem is obtained by maximizing each and every objective function of the problem, subject to the same constraint set (relaxing integrality). Then each solution is used to form a p-dimensional vector which obviously will dominate each efficient solution for (RPΘ). Let us denote by C_i the objective function to be maximized in each of these problems. The dual problem of the resulting problem is given as

$$\mathrm{UB}_{i,m+1}(y_q, \Theta) = \min \quad \{k^t w + (b+\Theta d - y_q)^t u\},$$

$$\text{s.t.} \quad T = \begin{cases} wI + uA \geqslant C_i, \\ \quad w,\ u \geqslant 0, \end{cases}$$

where w and u represent the dual variables. Let us denote by Ω_i the set of extreme points for the constraint set T. The dual solution, $\mathrm{UB}_{i,m+1}(y_q, \Theta)$, can be obtained and equivalently expressed as follows.

$$\mathrm{UB}_{i,m+1}(y_q, \Theta) = \min_{(w,u)\in\Omega_i} \{k^t w + (b+\Theta d)^t u - y_q^t u\}.$$

As can be seen, any feasible solution $(w, u) \in \Omega_i$ is also an upper bound for *any* resource efficient point q. That is

$$k^t w + (b+\Theta d)^t u - y_q u \geqslant \mathrm{UB}_{i,m+1}(y_q, \Theta).$$

Obviously, if one uses the vector $\bar{B}$ ($\geqslant (b+\Theta d)$ with at least one strict inequality), one has that since $u \geqslant 0$

$$k^t w + \bar{B}^t u - y_q^t u \geqslant k^t w + (b+\Theta d)^t u - y_q^t u \geqslant \mathrm{UB}_{i,m+1}(y_q, \Theta).$$

This is a less tight upper bound but it does not depend on the value of Θ. This result is used to solve several problems whose solution times are given in Section 4.

4. Some comments and computational experience

Table 1 illustrates a comparison of the solution times spent by three different procedures in solving a sample of 0–1 bicriterion multidimensional knapsack

Table 1
Comparison of solution times for various solution methodologies ($N=10$, DPA—Dynamic Programming Approach, HDPA—Hybrid Dynamic Programming Approach, BBA—Branch and Bound Approach)

Problem number	M	b	d	DPA	HDPA	BBA
1	10	0.25	$0.50b$	1.25	1.51	1.06
2				1.01	1.18	0.78
3				1.73	2.02	1.08
4				1.56	1.60	0.97
5				1.12	1.12	0.72
6				1.30	1.61	1.03
7				1.72	2.05	1.29
8				1.66	1.91	1.18
9				1.56	1.77	1.03
10				1.52	1.81	1.11
11	4	0.45	$0.75b$	52.72	47.23	43.48
12				111.28	83.55	45.33
13				98.06	84.88	55.44
14				67.23	63.07	51.04
15				143.45[a]	131.51	51.75

[a] Current time at stage 10.

problems. The objective and constraint matrix coefficients were randomly generated within the range [0, 99]. The constraint matrix has a 90% density and the right hand side vector, b, corresponds to 0.25 and 0.45 times the sum of the coefficients of the associated row matrix respectively. It is assumed that $d=0.5b$ and $d=0.75b$ respectively. The coding was done in Fortran on a CDC 6400 computer system. The hybrid dynamic programming approach uses the heuristic developed by Loulou and Michaelides [3] to generate sets of lower bound vectors after forming various composite objective functions. The upper bound vectors computed at each stage (after stage 3) are obtained by setting the remaining variables to their upper bound, and at stage 6, using the solutions to the dual problems associated with the linear relaxations for the remaining problems. The branch-and-bound scheme is the same procedure described without the use of dominance (i.e., the resource efficient concept).

One can observe from Table 1 that the branch-and-bound scheme appears to be a better approach to solve the type of problems described. Notice that for problems with a relatively small vector of resources ($b=0.25$ and $d=0.50b$), the dynamic programming recursion is computationally advantageous over the hybrid procedure. However, as these vectors increase, the use of a bounding scheme, in addition to the normal recursions, is justified by its ability to eliminate points not leading to efficient solutions. This is illustrated by the last

Table 2
Number of different solutions[a] for RHS less than $b+\Theta d$ ($M=10$, $N=10$, $P=2$, $b=0.25$, $d=0.5b$)

Problem no. \ Θ-value	0.0	0.1	0.2	0.3	0.4	0.5
1	1	2	4	6	9	11
2	1	2	3	3	5	5
3	1	3	3	3	3	3
4	1	1	3	4	5	5
5	1	1	2	2	2	3
6	1	1	3	4	4	4
7	1	2	4	6	8	8
8	1	1	3	3	7	7
9	1	3	3	4	5	5
10	1	1	3	4	7	8

[a] Sets of efficient solutions.

five sample problems of Table 1. Obviously, further computational experience is necessary to decide on which of these approaches is best.

Table 2 illustrates the number of different solutions obtained for the problems of Table 1, for several specific values of Θ, when we allow for the variation of the values of Θ from zero to 0.50. This sample corresponds to the family of problems generated for all possible values of $b \leqslant \hat{b} \leqslant b+\Theta d$; $0 \leqslant \Theta \leqslant 1$. One can notice that the same set of efficient solutions may be shared by several problems with different vectors of resources.

References

[1] G.R. Bitran, Linear multiple objective programs with zero–one variables, Math. Programming 13 (1977).

[2] R. Loulou and E. Michaelides, New greedy-like heuristics for the multidimensional 0–1 knapsack problem, Working paper, McGill University (1977).

[3] R.E. Marsten and T.L. Morin, Parametric integer programming: The right-hand side case, Ann. Discrete Math. 1 (1977).

[4] C.J. Piper and A.A. Zoltners, Implicit enumeration based algorithms for postoptimizing zero–one progams, Management Science Research Report No. 313, Graduate School of Industrial Administration, Carnegie Mellon University, Pittsburgh, PA (1973).

[5] G.M. Roodman, Postoptimality analysis in zero–one programming by implicit enumeration, Naval Res. Logist. Quart. 19 (1972).

[6] G.M. Roodman, Postoptimality analysis in integer programmed by implicit enumeration: The mixed integer case, The Amos Tuck School of Business Administration, Dartmouth College (1973).

[7] B. Villarreal and M.H. Karwan, Dynamic programming approaches for multicriteria integer programming, Research Report No. 78-3, State University of New York at Buffalo, NY (1978).

[8] B. Villarreal and M.H. Karwan, Multicriteria integer programming: A (hybrid) dynamic programming recursive approach, Math. Programming 21 (1981) 204–223.

P. Hansen, ed., Studies on Graphs and Discrete Programming

SCHEDULING IN SPORTS

D. de WERRA

Département de Mathématiques, Ecole Polytechnique Fédérale de Lausanne, Switzerland

The purpose of this paper is to present the problem of scheduling the games of a hockey or football league. It is shown how a graph theoretical model may be used and how some constraints related to the alternating pattern of home-games and away-games can be handled. Finally some other requirements occurring in practice are also discussed and introduced into the model.

1. Introduction

When scheduling the games of a season for say a football or a hockey league, one has to satisfy a lot of different requirements [3, 4]. In this paper we will be concerned with some of them; we will focus on the Home-and-Away pattern of the schedule. Our purpose will be to show how these requirements can be handled in a graph-theoretical model. Graph theoretical terms not defined here are to be found in Berge [2].

Let the league consist of $2n$ teams; each one has to play one game against each other. One associates with this league a complete graph K_{2n} on $2n$ nodes: edge $[i, j]$ represents the game between team i and team j. If there are no specific requirements a schedule is given by a 1-factorization of K_{2n} [5, Ch. 9] i.e. a decomposition of the edge set of K_{2n} into 1-factors $F_1, \ldots, F_{2n-1}$; each F_d consists of a collection of n non adjacent edges which represent the n games scheduled for the dth day of the season. Results on decomposition of complete graphs into subsets of non adjacent edges were given by Baranyai [1].

Since each game between i and j is played in the home city of either team i or team j, one can represent the game by an *arc* (i, j) oriented from i to j (or j to i) if it is played in the home city of j (or i); we say that it is a *home-game* for j (or for i) and an *away-game* for i (or for j). A 1-factorization of K_{2n} together with an orientation defined for each edge is an *oriented coloring* of K_{2n}; it will be denoted by $(\vec{F}_1, \ldots, \vec{F}_{2n-1})$. It defines a schedule in the sense that for each game between teams i and j, the orientation of the arc joining i and j and its color (i.e. the index p of the 1-factor $\vec{F}_p$ which contains this arc) specify where and when the game is to be played.

Usually one tries to alternate as much as possible home-games and away-games for each one of the teams. With each schedule we can associate a *Home*-

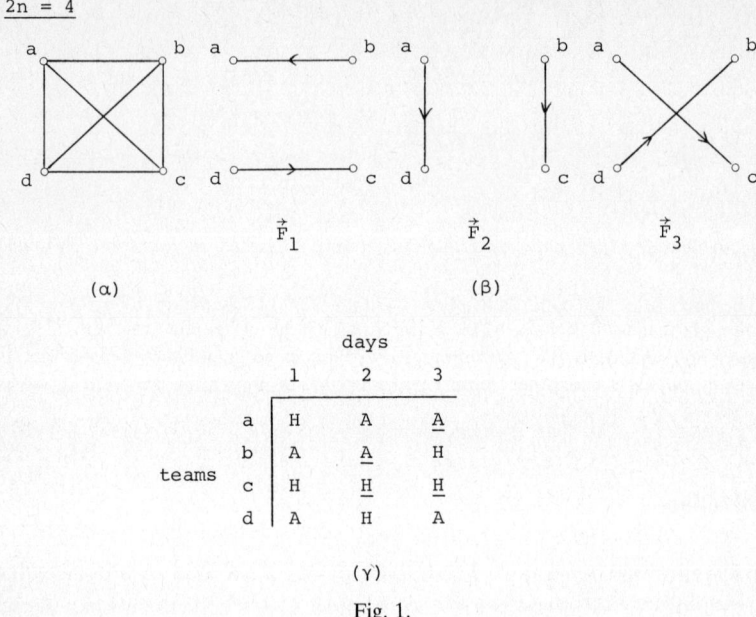

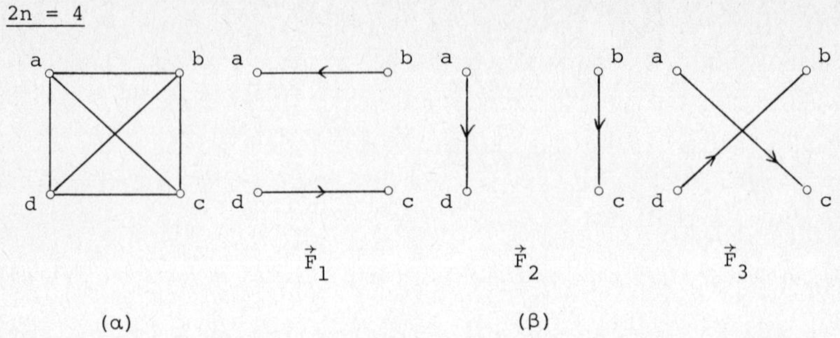

days

teams	1	2	3
a	H	A	A
b	A	A	H
c	H	H	H
d	A	H	A

(γ)

Fig. 1.

and-Away pattern (HAP): it is a $2n \times (2n-1)$ array where entry (i, d) is H if team i has a home-game on the dth day or A otherwise.

If in some row i, there are 2 consecutive entries, say $d-1$ and d with the same symbol (A or H), we say that team i has a *break* at day d (the alternating pattern is broken at day d). For instance Fig. 1(β) shows a schedule for a league of 4 teams a, b, c and d. In the associated HAP (Fig. 1(γ)) one sees that there are 4 breaks: for team a on day 3, for team b on day 2 and for team c on days 2 and 3.

In the next section we shall deal with breaks and formulate some properties using a graph-theoretical terminology. Notice that in terms of oriented colorings a break occurring on day d for team i appears as follows: the (unique) arc adjacent to node i in $\vec{F}_{d-1}$ and the (unique) arc adjacent to i in $\vec{F}_d$ are both oriented either out of i or into i. We shall similarly say that node i has a break for color d.

2. Oriented colorings

We shall first state some results related to the breaks in an oriented coloring of complete graphs K_{2n}.

Proposition 1. *In any oriented coloring of K_{2n}, there are at least $2n-2$ breaks.*

Proof. Suppose there exists an oriented coloring $(\vec{F}_1, \ldots, \vec{F}_{2n-1})$ of K_{2n} with less than $2n-2$ breaks. This means that there are $p \geqslant 3$ nodes without any break. Among these p nodes, there must be 2 nodes x and y such that in the HAP corresponding to this coloring rows x and y are identical. This means that there cannot be any arc (x, y) in K_{2n}. This contradicts the fact that the graph is complete and Proposition 1 is proven. □

Remark 1. One can as well consider oriented colorings of graphs which are not complete and define breaks in a similar way: if G is a regular bipartite multigraph then there exists an oriented coloring with no breaks (this is an immediate consequence of the fact that G has a 1-factorization [5, Ch. 9]).

If G is not regular, there is no 1-factorization, but one can define oriented colorings (each $\vec{F}_d$ is simply a subset of non adjacent arcs) and breaks: let u be the unique arc of $\vec{F}_d$ adjacent to some node x and let $\Delta \leqslant d-1$ be the largest index such that $\vec{F}_\Delta$ contains an arc, say v, adjacent to node x. There is a break at node x (for color d) if u and v are both oriented either out of x or into x.

With this definition, one can show that if G is a bipartite multigraph with even degrees, then there exists an oriented coloring with no breaks (this follows directly from results in [7]).

A 1-factorization $(F_1, \ldots, F_{2n-1})$ of K_{2n} will be called *canonical* if for $i = 1, \ldots, 2n-1$ F_i is defined by

$$F_i = \{[2n, i]\} \cup \{[i+k, i-k];\ k = 1, 2, \ldots, n-1\}$$

where the numbers $i+k$ and $i-k$ are expressed as one of the numbers $1, 2, \ldots, 2n-1 \pmod{2n-1}$.

It will be convenient to denote by a_{i0} the edge $[2n, i]$ for $i = 1, 2, \ldots, 2n-1$ and by a_{ik} the edge $[i+k, i-k]$ for $i = 1, 2, \ldots, 2n-1$ and $k = 1, 2, \ldots, n-1$. Thus

$$F_i = \{a_{i0}, a_{i1}, a_{i2}, \ldots, a_{i,n-1}\}.$$

Table 2(a) shows the array (a_{ik}) for the case $2n = 8$.

Given a canonical 1-factorization of K_{2n} and a node p we will associate with p a sequence $S(p)$ consisting of the edges in $F_1, F_2, \ldots, F_{2n-1}$ consecutively which are adjacent to node p.

For instance if $p = 2n$, $S(2n) = (a_{10}, a_{20}, \ldots, a_{2n-1,0})$ i.e. $([2n, 1], [2n, 2], \ldots, [2n, 2n-1])$. For $p = 1$, $S(1)$ starts with a_{10} (it is Part I) and continues with $n-1$ edges a_{ik} with $i-k=1$ where i takes consecutively values $2, 3, \ldots, n$ while k increases from 1 to $n-1$; this is Part II. Next Part III consists of $n-1$ edges a_{ik} with $i+k=1$ where $i = n+1, n+2, \ldots, 2n-1$ while $k = n-1, n-2, \ldots, 2$. $S(1)$ is represented in Table 1(a) and an example for $2n = 8$ is given in Table 2(b). Now if p is a node of K_{2n} with $2 \leqslant p \leqslant 2n-1$, we

Table 1

a)

		I	II	III
S(1)	=	[2n, 1]	[i+k, i-k = 1] i = 2,...,n	[i+k = 1, i-k] i = n+1,...,2n-1
	=	a_{10}	$a_{i,i-1}$ i = 2,...,n	$a_{i,2n-i}$ i = n+1,...,2n-1

b)

		I	II	III	IV
S(p) 2 ⩽ p ⩽ n	=	[i+k=p, i-k] i=1,..,p-1	[2n,p]	[i+k,i-k=p] i=p+1,..,n+p-1	[i+k=p, i-k] i=n+p,..,2n-1
	=	$a_{i,p-i}$ i=1,..,p-1	a_{p0}	$a_{i,i-p}$ i=p+1,..,n+p-1	$a_{i,p-1+2n-1}$ i=n+p,..,2n-1

c)

		I	II	III	IV
S(p) n+1 ⩽ p ⩽ 2n-1	=	[i+k,i-k=p] i=1,..,p-n	[i+k=p,i-k] i=p-n+1,..,p-1	[2n,p]	[i+k,i-k=p] i=p+1,..,2n-1
	=	$a_{i,2n-1+i-p}$ i=1,...,p-n	$a_{i,p-i}$ i=p-n+1,...,p-1	a_{p0}	$a_{i,i-p}$ i=p+1,..,2n-1

will describe the sequence $S(p)$; two cases are to be considered:

(A) $2 \leqslant p \leqslant n$: $S(p)$ starts with $p-1$ edges a_{ik} with $i+k=p$ ($i=1,2,\ldots,p-1$ and $k=p-i$); this is Part I of $S(p)$; then $a_{p0}=[2n,p]$ is Part II of $S(p)$. Next Part III consists of $n-1$ edges a_{ik} with $i-k=p$ ($i=p+1, p+2,\ldots,n+p-1$ while $k=i-p$). Finally Part IV contains $n-p$ edges a_{ik} with $i+k=p$ ($i=n+p,\ldots,2n-1$ while $k=2n-1+p-i$). Table 1(b) shows $S(p)$ and Table 2(b) gives $S(3)$ for the case $2n=8$.

(B) $n+1 \leqslant p \leqslant 2n-1$. In this case $S(p)$ also consists of 4 parts. Part I consists of $p-n$ edges a_{ik} with $i-k=p$ ($i=1,2,\ldots,p-n$ while $k=2n-1+i-p$). Part II is a sequence of $n-1$ edges a_{ik} with $i+k=p$ ($i=p-n+1,\ldots,p-1$ while $k=p-i$). Part III is a_{p0} and Part IV is formed by $2n-1-p$ edges a_{ik} with $i-k=p$ ($i=p+1,\ldots,2n-1$ while $k=i-p$). $S(p)$ is represented in Table 1(c) and $S(6)$ for the case $2n=8$ is given in Table 1(b).

Table 2

a)

i \ k	0	1	2	3
1	81	27	36	45
2	82	31	47	56
3	83	42	51	67
4	84	53	62	71
5	85	64	73	12
6	86	75	14	23
7	87	16	25	34

b) 2n = 8

$$S(1) = \Big\{ \underbrace{[8, 1]}_{\text{I}} \;\Big|\; \underbrace{[3, 1][5, 1][7, 1]}_{\text{II}} \;\Big|\; \underbrace{[1, 2][1, 4][1, 6]}_{\text{III}} \Big\}$$

$$= \Big\{ a_{10} \;\Big|\; a_{21} \quad a_{32} \quad a_{43} \;\Big|\; a_{53} \quad a_{62} \quad a_{71} \Big\}$$

$$S(3) = \Big\{ \underbrace{[3, 6][3, 1]}_{\text{I}} \;\Big|\; \underbrace{[8, 3]}_{\text{II}} \;\Big|\; \underbrace{[5, 3][7, 3][2, 3]}_{\text{III}} \;\Big|\; \underbrace{[3, 4]}_{\text{IV}} \Big\}$$

$$= \Big\{ a_{12} \quad a_{21} \;\Big|\; a_{30} \;\Big|\; a_{41} \quad a_{52} \quad a_{63} \;\Big|\; a_{73} \Big\}$$

$$S(6) = \Big\{ \underbrace{[3, 6][5, 6]}_{\text{I}} \;\Big|\; \underbrace{[6, 7][6, 2][6, 4]}_{\text{II}} \;\Big|\; \underbrace{[8, 6]}_{\text{III}} \;\Big|\; \underbrace{[1, 6]}_{\text{IV}} \Big\}$$

$$\Big\{ a_{12} \quad a_{23} \;\Big|\; a_{33} \quad a_{42} \quad a_{51} \;\Big|\; a_{60} \;\Big|\; a_{71} \Big\}$$

We will now show that the lower bound on the number of breaks for complete graphs given in Proposition 1 is best possible.

Proposition 2. *There exists an oriented coloring of K_{2n} with exactly $2n-2$ breaks.*

Proof. We shall start from a canonical 1-factorization $(F_1, F_2, \ldots, F_{2n-1})$ and orient each arc in order to obtain an oriented coloring $(\vec{F}_1, \vec{F}_2, \ldots, \vec{F}_{2n-1})$ of K_{2n}:

(a) For each i edge a_{i0} becomes arc $(i, 2n)$ if i is odd or arc $(2n, i)$ if i is even.

$\vec{F}_1$	$\overleftarrow{61}$	$\overrightarrow{25}$	$\overleftarrow{34}$
$\vec{F}_2$	$\overrightarrow{62}$	$\overrightarrow{31}$	$\overleftarrow{45}$
$\vec{F}_3$	$\overleftarrow{63}$	$\overrightarrow{42}$	$\overleftarrow{51}$
$\vec{F}_4$	$\overrightarrow{64}$	$\overrightarrow{53}$	$\overleftarrow{12}$
$\vec{F}_5$	$\overleftarrow{65}$	$\overrightarrow{14}$	$\overleftarrow{23}$

(α)

teams \ days	1	2	3	4	5
1	A	H	A	H	A
2	A	H	<u>H</u>	A	H
3	H	A	<u>A</u>	H	A
4	A	H	A	H	<u>H</u>
5	H	A	H	A	<u>A</u>
6	H	A	H	A	H

(β)

2n = 6

Fig. 2.

(b) For each i edge a_{ik} becomes arc $(i+k, i-k)$ if k is odd or arc $(i-k, i+k)$ if k is even.

An example for $2n=6$ is given in Fig. 2(α); in the corresponding HAP (Fig. 2(β)), one notices that the number of breaks is $2n-2=4$.

One immediately observes that node $2n$ has no breaks. Furthermore by examining the sequences $S(p)$ and the orientation rule (b) we notice that no break can occur within any one of the parts of the $S(p)$ because p is alternately initial endpoint and terminal endpoint of the arcs of the same part. So breaks can occur only at the boundary of 2 consecutive parts of a sequence $S(p)$.

For each node p with $1 \leqslant p \leqslant 2n-1$ there are 2 consecutive parts, say A and B, such that the last arc in A and the first arc in B both correspond to $k=n-1$. Since one of them is generated from an edge $[i+k, p]$ and the other from an edge $[p, i-k]$, the orientation rule (b) implies that one of the arcs is directed out of p while the other is directed into p. Hence there is no break at the boundary of Parts A and B. Hence there is no break at the boundary of Parts A and B. Thus node 1 cannot have any break at the boundary of Parts II and III. Since the (last) arc of Part I is $(1, 2n)$ from (a) and the first arc of Part II is $(3, 1)$ from (b), it follows that node 1 has no breaks.

We will show that each one of the nodes p with $2 \leqslant p \leqslant 2n-1$ has exactly one break: this will show that there are $2n-2$ breaks. Let us consider case (A) where $2 \leqslant p \leqslant n$. Part I of $S(p)$ ends with $(p, p-2)$, Part II is $(2n, p)$ if p is even or $(p, 2n)$ if p is odd; Part III starts with $(p+2, p)$; since no break can occur at the boundary of Parts III and IV (from the above remark about consecutive arcs in $S(p)$ corresponding both to $k=n-1$), p only has one break (it occurs for color p if p is odd and for color $p+1$ if p is even).

Case (B) can be handled similarly giving node p a break for colour p if p is odd or for color $p+1$ if p is even. This ends the proof. □

If we define breaks for graphs without 1-factorization as in Remark 1, we can state

Corollary 2.1. K_{2n+1} *has an oriented coloring without breaks.*

Proof. Starting from an oriented coloring of K_{2n+2} constructed as above, we may remove all arcs adjacent to node $2n+2$. This gives an oriented coloring of K_{2n+1}; an inspection of the sequences $S(p)$ in Table 1 shows that no node has a break.

Remark 2. It is known that in a canonical 1-factorization of K_{2n}, $F_i \cup F_{i+1}$ forms a hamiltonian cycle ($i = 1, 2, \ldots, 2n-2$). One can easily verify that the oriented coloring constructed in the above proof is such that $\vec{F}_1 \cup \vec{F}_2$, $\vec{F}_3 \cup \vec{F}_4, \ldots, \vec{F}_{2n-3} \cup \vec{F}_{2n-2}$ are hamiltonian circuits.

3. Multiperiod schedules

In the previous model, each team had to play exactly one game against every other team; in some cases, each pair of teams i, j has to meet twice during the season: once in the home city of i and once in the home city of j. In fact the season is divided into 2 periods: each pair of teams has to meet once during the first period and once during the second period. This corresponds to replacing K_{2n} by an oriented graph G_{2n} on $2n$ nodes where each pair of nodes i, j is joined by an arc (i, j) and an arc (j, i). A schedule is represented by a decomposition $(\vec{H}_1, \vec{H}_2, \ldots, \vec{H}_{4n-2})$ of the arc set of G_{2n} such that

$(\vec{H}_1, \ldots, \vec{H}_{2n-1})$ and consequently $(\vec{H}_{2n}, \ldots, \vec{H}_{4n-2})$ are oriented colorings of K_{2n}. (1)

It is easy to find a decomposition $(\vec{H}_1, \ldots, \vec{H}_{4n-2})$ of G_{2n} satisfying (1) and such that the total number of breaks is $4n-4$ and no node has 2 consecutive breaks. Such a decomposition is obtained for instance by taking $\vec{H}_i = \vec{F}_i$ and $\vec{H}_{2n-1+i} = \overleftarrow{F}_{2n-i}$ ($i = 1, 2, \ldots, 2n-1$) where $(\vec{F}_1, \ldots, \vec{F}_{2n-1})$ is an oriented coloring of K_{2n} giving $2n-2$ breaks and $\overleftarrow{F}_i$ is obtained from $\vec{F}_i$ by reversing the directions of the arcs. An illustration is given in Fig. 3 for $2n = 4$.

In practice such a schedule would not be good since the teams which meet on day $2n-1$, meet again on day $2n$.

One may thus require that the following holds

$$\text{if } \vec{H}_i = \vec{F}_i, \text{ then } \vec{H}_{2n-1+i} = \overleftarrow{F}_i \ (i = 1, 2, \ldots, 2n-1). \tag{2}$$

$\vec{F}_1 = \vec{H}_1$	$\overleftarrow{41}$	$\overrightarrow{23}$	$\vec{H}_4$	=	$\overrightarrow{43}$	$\overleftarrow{12}$
$\vec{F}_2 = \vec{H}_2$	$\overrightarrow{42}$	$\overrightarrow{31}$	$\vec{H}_5$	=	$\overleftarrow{42}$	$\overleftarrow{31}$
$\vec{F}_3 = \vec{H}_3$	$\overleftarrow{43}$	$\overrightarrow{12}$	$\vec{H}_6$	=	$\overrightarrow{41}$	$\overleftarrow{23}$

(α)

teams \ days	1	2	3	4	5	6
1	A	H	A	H	A	H
2	A	H	$\underline{\text{H}}$	A	$\underline{\text{A}}$	H
3	H	A	$\underline{\text{A}}$	H	$\underline{\text{H}}$	A
4	H	A	H	A	H	A

HAP

(β)

Fig. 3.

Notice that if $\vec{F}_1, \ldots, \vec{F}_{2n-1}$ is the oriented coloring constructed in the proof of Proposition 2, $(\vec{H}_1, \ldots, \vec{H}_{4n-2})$ would give $6n-6$ breaks and some nodes could have 2 consecutive breaks (this would mean that some teams have 3 consecutive home-games (or away-games)).

Proposition 3. *Any decomposition* $(\vec{H}_1, \vec{H}_2, \ldots, \vec{H}_{4n-2})$ *of the arc set of* G_{2n} *satisfying conditions* (1) *and* (2) *has at least* $6n-6$ *breaks.*

Proof. Since from (1) $\mathcal{H}_1 = (\vec{H}_1, \ldots, \vec{H}_{2n-1})$ and $\mathcal{H}_2 = (\vec{H}_{2n}, \ldots, \vec{H}_{4n-2})$ are oriented colorings of K_{2n}, there will be $2n-2$ breaks for $\mathcal{H}_1$ and $2n-2$ breaks for $\mathcal{H}_2$. If a node p has an even number of breaks in $\mathcal{H}_1$, there will be no break at the boundary of $\mathcal{H}_1$ and $\mathcal{H}_2$ (the first and the last arc in the sequence $S^1(p)$ associated to $\mathcal{H}_1$ are both directed into p or out of p and the first arc in the sequence $S^2(p)$ associated to $\mathcal{H}_2$ has the opposite orientation). If the number of breaks in $\mathcal{H}_1$ for some node p is odd, there will be an additional break at the boundary. Since each break introduced into $\mathcal{H}_1$ will create a similar break in $\mathcal{H}_2$ (from (2)), the minimum number of breaks is obtained when $\mathcal{H}_1$ and $\mathcal{H}_2$ each have $2n-2$ breaks. this gives $6n-6$ breaks since there are $2n-2$ nodes having exactly one break in $\mathcal{H}_1$. □

Proposition 4. *There exists a decomposition* $\mathcal{H} = (\vec{H}_1, \vec{H}_2, \ldots, \vec{H}_{4n-2})$ *of the arc set of* G_{2n} *satisfying* (1) *and* (2) *and such that the total number of breaks is* $6n-6$. *Furthermore if* $2n \neq 4$, *no node has* 2 *consecutive breaks in* $\mathcal{H}$.

For $2n = 4$, at least 2 breaks will occur in any oriented coloring $(\vec{F}_1, \vec{F}_2, \vec{F}_3)$ of K_4. If a node has 2 breaks, then these breaks will necessarily be consecutive. So assume each node has one break at most in $(\vec{F}_1, \vec{F}_2, \vec{F}_3)$.

This break can occur for color 2 or for color 3; if it is for color 2 (color 3), then there will be breaks for colors 4 and 5 (4 and 6). In all cases, there will be 2 consecutive breaks when $2n = 4$, Clearly for $2n = 2$, there are no breaks at all in $\mathcal{H}$. So we shall assume that $2n \geqslant 6$ in the proof.

$\vec{F}_1$	$\overleftarrow{61}$	$\overrightarrow{25}$	$\overleftarrow{34}$
$\vec{F}_2$	$\overrightarrow{62}$	$\overrightarrow{31}$	$\overleftarrow{45}$
$\vec{F}_3$	$\overrightarrow{63}$	$\overrightarrow{42}$	$\overleftarrow{51}$
$\vec{F}_4$	$\overleftarrow{64}$	$\overrightarrow{53}$	$\overleftarrow{12}$
$\vec{F}_5$	$\overrightarrow{65}$	$\overrightarrow{14}$	$\overleftarrow{23}$

(α)

days

teams	1	2	3	4	5	6	7	...
1	A	H	A	H	A	H	A	...
2	A	H	$\underline{H}$	A	H	H	A	
3	H	A	H	$\underline{H}$	A	A	H	
4	A	H	A	$\underline{A}$	H	H	A	
5	A	A	H	A	H	A	H	
6	H	A	$\underline{A}$	H	A	A	H	...

HAP

(β)

Fig. 4.

Proof of Proposition 4. We shall again start from a canonical 1-factorization $(F_1, F_2, \ldots, F_{2n-1})$ of K_{2n}; the orientation is defined by rule (b) in Proposition 2 and by rule (a′):

(a′) for each i edge a_{i0} becomes arc $(i, 2n)$ if $i \leqslant 2n-5$ is odd or if $i = 2n-2$; it becomes arc $(2n, i)$ otherwise.

This construction is illustrated for K_6 in Fig. 4(α); the corresponding HAP is represented in Fig. 4(β). One can see that in the decomposition of G_{2n} derived from the oriented coloring $\vec{F}_1, \ldots, \vec{F}_{2n-1}$, no node has 2 consecutive breaks.

In fact we have just reversed the orientation of the arcs corresponding to the last three edges of $S(2n)$, namely $a_{2n-3,0}$, $a_{2n-2,0}$ and $a_{2n-1,0}$. There is clearly no change for the first $2n-4$ nodes; their breaks occur for colors i with $3 \leqslant i \leqslant 2n-3$.

Node $2n$ now has one break; it occurs for color $i = 2n-3$, but node $2n-1$ no longer has a break: this can be seen in Fig. 5(α) which represents a portion of $\vec{F}_i$ for $i = 2n-3$, $2n-2$, $2n-1$ as constructed in Proposition 2. Fig. 5(β) represents the same portion after reversing the orientation of 3 arcs according to rule (a′) ($\vec{F}_i$ has become $\vec{F}_i'$).

One sees also that the break of node $2n-3$ has moved from color $2n-3$ to color $2n-2$ and similarly the break of node $2n-2$ has moved from color $2n-1$ to color $2n-2$.

⋮	(2n, 2n-4)	(2n-3, 2n-5)
F_{2n-3}	(2n-3, 2n)	(2n-2, 2n-4)
F_{2n-2}	(2n, 2n-2)	(2n-1, 2n-3)
F_{2n-1}	(2n-1, 2n)	(1, 2n-2)

(α)

⋮	(2n, 2n-4)	(2n-3,2n-5)
F'_{2n-3}	(2n, 2n-3)	(2n-2,2n-4)
F'_{2n-2}	(2n-2, 2n)	(2n-1,2n-3)
F'_{2n-1}	(2n, 2n-1)	(1 , 2n-2)

(β)

Fig. 5.

So the total number of breaks in $(\vec{F}'_1, \ldots, \vec{F}'_{2n-1})$ constructed with (a′) and (b) is still $2n-2$; since all breaks occur for colors c with $3 \leqslant c \leqslant 2n-2$, $(\vec{F}'_1, \ldots, \vec{F}'_{2n-1}, \bar{F}'_1, \ldots, \bar{F}'_{2n-1})$ will give a decomposition of the arc set of G_{2n} without consecutive breaks. The total number of breaks will be $6n-6$. □

Remark 3. If each pair of teams has to play an even number $2p$ of games during a season, then one can repeatedly use a decomposition of G_{2n}; new breaks will be introduced at the boundaries for nodes having 3 breaks. The total number of breaks will be $6p(n-1)+2(p-1)(n-1)$.

Remark 4. Instead of requiring that (2) holds one could as well ask that the decomposition of the arc set of G_{2n} satisfies: for each pair i, j of nodes if arc (i, j) is in $\vec{H}_c$ and arc (j, i) in $\vec{H}_d$, then $|c-d| \geqslant q$ where q is some parameter. We have dealt with the case $q = 2n-1$; this gave at least $6n-6$ breaks; if $q=0$, we have seen that we can have as few as $2n-2$ breaks.

4. Complementary patterns

In a league, the different teams are generally located in different cities (some of these cities may in fact have more than one team). For obvious reasons it may not be desirable to have all teams located in some small region to play home games at the same time or to be all away simultaneously. Thus, in order to get an admissible schedule, one could for instance divide the $2n$ teams and group them into n groups of 2 teams which are geographically located quite close to each other. In a good schedule one would like to have in each group 2 teams with HAP's (the HAP of a team i is simply the ith row of the HAP) as different as possible.

The HAP's of 2 teams i and j are said to be *complementary* if for each day d, teams i and j are not both away or both at home; the HAP of a team will correspond to the *pattern* of a node: H means of course that the arc is directed into the node and A that it is directed out of the node.

Proposition 5. *Let $\vec{F}_1, \ldots, \vec{F}_{2n-1}$ be any oriented coloring of F_{2n} with exactly $2n-2$ breaks; then the set of nodes of K_{2n} can be partitioned into n subsets $T_1, \ldots, T_n$ such that $|T_i| = 2$ and in each T_i the nodes have complementary patterns.*

Proof. In such a coloring, there are 2 nodes without breaks; these nodes clearly have complementary profiles (since no two nodes can have the same pattern). Each one of the remaining nodes has one break. One can never have an odd number of breaks for some colour d, because in each column of the HAP, there are exactly n H's and n A's.

So let i be one of the nodes with one break; choose i to be one of the nodes which is the first to have a break (i.e. it occurs for colour d and for any $d' < d$, no node has breaks).

Suppose that the pattern of i contains a H in column d and a H in column $d+1$. From the above remarks, there must exist a team j with a break for color d and more precisely with an A in the dth and in the $(d+1)$th column of its pattern. Since i and j have exactly one break, their patterns must be complementary (also for colors $c \neq d, d+1$); nodes i and j are placed into a subset t_i and removed. By repeating this argument, one gets the partition into subsets $T_1, \ldots, T_{2n}$. □

Remark 5. Obviously if there are more than $2n-2$ breaks, each node i may not be grouped with a node j having a complementary pattern (see Fig. 1).

In the oriented coloring constructed in the proof of Proposition 2, we have the following partition: $T_i = \{2i, 2i+1\}$ $(i = 1, 2, \ldots, n-1)$ and $T_1 = \{2n, 1\}$.

An immediate consequence of Proposition 5 is the following:

Corollary 5.1. *In any oriented coloring $\vec{F}_1, \ldots, \vec{F}_{2n-1}$ of K_{2n} with $2n-2$ breaks, for any c $(2 \leq c \leq 2n-1)$ there are 0 or 2 nodes having a break for color c.*

Proof. If there were 4 nodes having a break for the same color c, there would exist 2 nodes with the same HAP, which is impossible. □

So for K_{2n} a HAP with $2n-2$ breaks can be completely characterized (up to a permutation of its rows) by the $n-1$ indices c $(2 \leq c \leq 2n-1)$ where breaks occur (in pairs). Let $B(\mathcal{H}) = (b_1, \ldots, b_{n-1})$ where $2 \leq b_1 < b_2 < \cdots < b_{n-1} \leq 2n-1$ be the sequence thus associated with a HAP $\mathcal{H}$ having $2n-2$ breaks.

5. Feasible and canonically feasible sequences

A sequence $B = (b_1, \ldots, b_{n-1})$ is *feasible* for K_{2n} if there is an oriented coloring of K_{2n} (with $2n-2$ breaks) whose HAP $\mathcal{H}$ can be characterized by a sequence $B(\mathcal{H}) = B$. Not all sequences B with $2 < b_1 < \cdots < b_{n-1}$ are feasible: for instance for $2n = 6$, only $(2, 4)$, $(3, 4)$ and $(3, 5)$ are feasible.

Proposition 6. *Any feasible sequence $B = (b_1, \ldots, b_{n-1})$ satisfies $b_1 \leq n$ and $b_{n-1} \geq n+1$.*

In other words, in any oriented coloring of K_{2n} with $2n-2$ breaks, the first breaks must occur for some color $c \leqslant n$ and the last ones must occur for some color $c \geqslant n+1$.

Proof. Let $(\vec{F}_1, \ldots, \vec{F}_{2n-1})$ be an oriented coloring of K_{2n} with $2n-2$ breaks; suppose $b_1 > n$; then there are no breaks for colours $1, 2, \ldots, n$.

The only way of getting such a coloring is to split the $2n$ nodes of K_{2n} into 2 subsets A and B with $|A| = |B| = n$.

Each $\vec{F}_i$ (for $i = 1, 2, \ldots, n$) must be a 1-factor in the bipartite complete graph constructed on A and B.

Then the arcs in $\vec{F}_{n+1} \cup \cdots \cup \vec{F}_{2n-1}$ must form 2 disjoint complete graphs G_1, G_2 on n nodes each. If n is odd, the arc set of G_i $(i = 1, 2)$ cannot be partitioned into 1-factors; if n is even, there will be $\frac{1}{2}n$ nodes in each G_i with a break for color n. If $n \geqslant 4$, according to Corollary 5.1 $(\vec{F}_1, \ldots, \vec{F}_{2n-1})$ cannot give $2n-2$ breaks. Hence $b_1 \leqslant n$. The case $b_{n-1} \geqslant n+1$ is obtained by taking $\vec{F}_1, \ldots, \vec{F}_{2n-1}$ in reverse order. □

Remark 6. If n is even, there may exist oriented colorings $(\vec{F}_1, \vec{F}_2, \ldots, \vec{F}_{2n-1})$ of K_{2n} such that no breaks occur for colors $1, 2, \ldots, n$. But if $n \geqslant 4$, Proposition 6 says that the total number of breaks will be more than $2n-2$. For $2n = 6$, Proposition 4 implies that the sequences $B_1 = (2, 3)$ and $B_2 = (4, 5)$ are not feasible. Notice that $(2, 5)$ is also not feasible (if it were feasible, then from any oriented coloring $(\vec{F}_1, \ldots, \vec{F}_5)$ one could get another oriented coloring $(\vec{F}_2, \vec{F}_3, \vec{F}_4, \vec{F}_5, \vec{F}_1)$ corresponding to the sequence B_2.

Before stating the next result related to canonical 1-factorization, we will need the following property.

Lemma 1. *Let F_1, F_p, F_q $(1 < p < q \leqslant 2n-1)$ be 1-factors in a canonical 1-factorization of K_{2n}; then the partial graph generated by $F_1 \cup F_p \cup F_q$ has a triangle.*

Proof. Consider nodes $q+p-1$ and $q-(p-1)$ (these numbers are taken modulo $2n-1$ and satisfy $1 \leqslant q+p-1 \leqslant 2n-1$, $1 \leqslant q-(p-1) \leqslant 2n-1$). Nodes $q+p-1$ and $q-(p-1)$ are distinct: otherwise we would have $p-1 \equiv -(p-1)$ (mod $2n-1$), i.e. $2p-2 \equiv 0$ (mod $2n-1$), i.e. $2p-2 \geqslant 4n-2$ because $2p-2$ is even; but $p \leqslant 2n-2$ implies $2p-2 \leqslant 4n-6$, which is impossible.

Furthermore $q+p-1$ and $q-(p-1)$ are joined by an edge of $F_{q'}$ since the edges of F_q are of the form $(q+k, q-k)$.

Besides neither $q+p-1$ nor $q-(p-1)$ can be node $2n$. We will now show that $q+p-1$ and $q-(p-1)$ are joined by a chain of length 2 in $F_1 \cup F_p$; this will prove that the partial graph generated by $F_1 \cup F_p \cup F_q$ has a triangle.

Let e be the edge of F_p which is adjacent to node $q+p-1$; since $q+p-1$ is neither $2n$ nor p (this would imply $q \equiv 1 \bmod(2n-1)$, which is impossible since $1<q \leqslant 2n-1$), e is of the form $(p+k, p-k)$, so its second endpoint is $p-q+1$; now let f be the edge of F_1 which is adjacent to $p-q+1$. Again $p-q+1$ is neither $2n$ nor p, so f is of the form $(1+k, 1-k)$; so its second endpoint is $q-p+1$. Hence the chain consisting of edges e and f has length 2 and joins nodes $q+p-1$ and $q-p+1$. □

A sequence $B=(b_1, b_2, \ldots, b_{n-1})$ will be called *canonically feasible* for K_{2n} if there exists an oriented coloring of K_{2n} (with $2n-2$ breaks) obtained from a canonical 1-factorization of K_{2n} whose HAP is characterized by sequence B.

Proposition 7. *B is canonically feasible for K_{2n} iff for any color i ($2 \leqslant i \leqslant 2n-1$) there is a break either for color i or for color $i+1$.*

Proof. (A) Suppose that B satisfies the condition; then there exists at most one pair of consecutive colors $c, c+1$ with breaks (if there were 3 consecutive such colors or more than one such pair, there would exist 2 consecutive colors without breaks because we have $n-1$ colors with breaks in a collection of $2n-2$ colors). We shall start from the oriented coloring constructed in the proof of Proposition 2 and reorient if necessary the arcs adjacent to node $2n$. As before we refer to entry (i, j) of the corresponding HAP when we consider the element located in row i (i.e. node i) and column j (i.e. color j). The reorientation consists in modifying the values of entries (i, i) and $(2n, i)$ for $i=1, \ldots, 2n-1$. Notice that all entries $(i, i+1)$ are H's and all entries $(i+1, i)$ are A's (see Fig. 2(β)).

If there are no consecutive colors with breaks, one can fill the cells (i, i) with A's and H's alternately starting with H if $b_1=2$ or with A if $b_1=3$; in entry $(2n, i)$ we put an *H* if (i, i) has an A and conversely. One verifies that in both cases breaks are occurring for colors $b_1, b_2, \ldots, b_{n-1}$. Otherwise let $c, c+1$ be the pair of consecutive colors with breaks; we start filling the entries (i, i) as before until we reach row $c+1$; we place the same symbol in $(c+1, c+1)$ as in (c, c) and then we continue filling cells $(c+2, c+2)$, $(c+3, c+3), \ldots, (2n-1, 2n-1)$ alternately with H's and A's. Again one verifies that breaks are occurring for colors $b_1, b_2, \ldots, b_{n-1}$; here nodes c and $c-1$ have a break for color c, while nodes $c+1$ and $2n$ have a break for color $c+1$.

Fig. 6 shows an example of this construction for $2n=8$ and for $B=(3, 4, 6)$.

(B) Conversely if B is a sequence which does not satisfy the condition, there must exist 2 consecutive colors $i, i+1$ (with $i \geqslant 2$) for which no breaks occur. If $(\vec{H}_1, \ldots, \vec{H}_{2n-1})$ is an oriented coloring associated with B, then the partial graph generated by $\vec{H}_{i-1} \cup \vec{H}_i \cup \vec{H}_{i+1}$ where $\vec{H}_p$ is the set of arcs with colour p

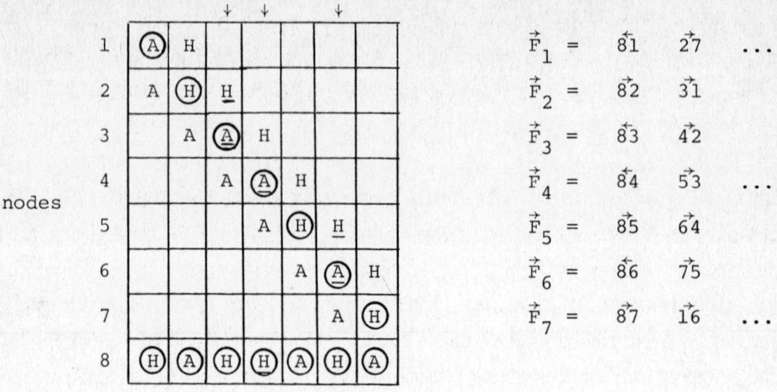

B = (3, 4, 6)

Fig. 6.

must satisfy the following: for each node j the arcs of $\vec{H}_{i-1}$ and of $\vec{H}_{i+1}$ which are adjacent to j must be both either directed out of j or directed into j, while the arc of $\vec{H}_{i-1}$ adjacent to j has the opposite orientation. This implies that $\vec{H}_{i-1} \cup \vec{H}_i \cup \vec{H}_{i+1}$ generates a bipartite graph. But this is impossible from Lemma 1, hence B is not canonically feasible. $\square$

Notice that for $2n = 8$, among the $20\,(= C_{n-1}^{2n-2})$ possible sequences $B = (b_1, b_2, b_3)$ only 4 are canonically feasible, namely (3, 5, 7), (3, 5, 6), (3, 4, 6) and (2, 4, 6). There are 4 more sequences which are feasible but not canonically feasible: (2, 4, 5), (3, 4, 7), (2, 5, 6) and (4, 5, 7). One can show that the 12 remaining sequences are not feasible.

6. Final remarks

One reason for being interested in oriented colorings derived from canonical 1-factorizations is given in the following result:

Proposition 8. *Let $\hat{F} = (\vec{F}_1, \ldots, \vec{F}_{2n-1})$ be an oriented coloring derived from a canonical 1-factorization of K_{2n}; let $h_1, h_2, \ldots, h_p$ be a sequence of positive integers such that $h_1 + \cdots + h_p = n(2n-1)$ and $h_i \leq n$ for $i = 1, \ldots, p$ (where $h_i = n$ only if $h_1 + \cdots + h_i \equiv 0 \pmod n$).*

Then there exists a partition $\hat{H} = (\vec{H}_1, \vec{H}_2, \ldots, \vec{H}_p)$ of $\vec{F}_1 \cup \cdots \cup \vec{F}_{2n-1}$ such that

(1) *each H_i is a subset of h_i nonadjacent arcs,*

(2) *the total number of breaks in $\hat{H}$ is the same as in $\hat{F}$.*

Recall that in $\hat{H}$ a node i has a break for color d if there is a color $c<d$ such that no $\vec{H}_e$ $(c<e<d)$ has an arc adjacent to node i and the arcs of $\vec{H}_c$ and of $\vec{H}_d$ adjacent to i are both oriented out of i or both oriented into i (see Remark 1).

Proof. Let $(a_{i0}, a_{i1}, \ldots, a_{in})$ be the edges corresponding to the arcs of F_i as defined previously. Clearly by taking the edges of F_1 first, then the edges of F_2 and so on in the order $a_{i0}, a_{i1}, \ldots, a_{in}$, we get a sequence S of $n(2n-1)$ edges. Any set of c (with $c<n$) consecutive edges in the sequence corresponds to nonadjacent edges: this is obvious if all c edges belong to the same F_i (in this case we can have $c=n$); otherwise $H_i=\{a_{ik}, a_{i,k+1}, \ldots, a_{in}, a_{i+1,0}, a_{i+1,1}, \ldots, a_{i+1,p}\}$ also correspond to nonadjacent edges if $p \leqslant k-2$ because the edge of F_{i+1} adjacent to node $i+q$ $(q \geqslant k)$ is $a_{i+1,q-1}$ and the edge of F_{i+1} adjacent to node $i-q$ $(q \geqslant k)$ is $a_{i+1,q+1}$ and neither $a_{i+1,q-1}$ nor $a_{i+1,q+1}$ are in H_i. Hence by splitting the sequence S into subsequences of $h_1, h_2, \ldots, h_p$ elements, one gets the required coloring. □

This situation corresponds to the case where all games cannot be played simultaneously [6]: for commercial reasons, this may happen when the different teams are concentrated on a rather small area where the public may easily attend games played in different places.

In practice there are other constraints which are to be introduced into the model such as unavailability constraints which compel some teams to have more than one break in their schedule.

These constraints generate many problems which can be formulated in graph theoretical terms; some of these will be discussed later.

References

[1] Z. Baranyai, On the factorization of the complete uniform hypergraph, in: Hajnal, Rado, Sós, eds., Infinite and Finite Sets I (North-Holland, Amsterdam, 1975) 91–108.

[2] C. Berge, Graphs and Hypergraphs (North-Holland, Amsterdam, 1973).

[3] W.O. Cain, Jr., The computer-assisted heuristic approach used to schedule the major league baseball clubs, in: S.P. Ladany, R.E. Machol, eds., Optimal Strategies in Sports (North-Holland, Amsterdam, 1977) 32–41.

[4] R.T. Campbell and D.S. Chen, A minimum distance basketball scheduling problem, in: R.E. Machol, S.P. Ladany, eds., Management Science in Sports (North-Holland, New York, 1976), 15–25.

[5] F. Harary, Graph Theory (Addison-Wesley, Reading, MA, 1969).

[6] U. Weisner, Planung von Turnieren-Kombinatorische Analyse und Algorithmen, Dissertation Nr. 5313, Swiss Federal Institute of Technology, Zürich (1974).

[7] D. de Werra, Partial compactness in chromatic scheduling, Operations Res. Verfahren 32 (1979) 207–219.